ISBN 978-0-282-58940-0
PIBN 10857075

English
Français
Deutsche
Italiano
Español
Português

www.forgottenbooks.com

Mythology Photography **Fiction**
Fishing Christianity **Art** Cooking
Essays Buddhism Freemasonry
Medicine **Biology** Music **Ancient**
Egypt Evolution Carpentry Physics
Dance Geology **Mathematics** Fitness
Shakespeare **Folklore** Yoga Marketing
Confidence Immortality Biographies
Poetry **Psychology** Witchcraft
Electronics Chemistry History **Law**
Accounting **Philosophy** Anthropology
Alchemy Drama Quantum Mechanics
Atheism Sexual Health **Ancient History**
Entrepreneurship Languages Sport
Paleontology Needlework Islam
Metaphysics Investment Archaeology
Parenting Statistics Criminology
Motivational

THE

PHYTOLOGIST.

SECOND ANNUAL PART.

WISDOM OF GOD IN CREATION.

TANTUS AMOR FLORUM.

LONDON:

JOHN VAN VOORST, PATERNOSTER ROW.

M.DCCC.XLIII.

LONDON :

PRINTED BY LUXFORD & CO., RATCLIFF HIGHWAY.

CONTENTS.

————————————————

☞ THE PHYTOLOGIST is published in monthly numbers, price 1s. each, or in
 annual parts, containing 12 monthly numbers (with a Table of Contents)
 neatly bound in cloth.

THE PHYTOLOGIST.

| No. XIII. | JUNE, MDCCCXLII. | PRICE 1s. |

ART. LXIII.—*A History of the British Lycopodia and allied Genera.*
By EDWARD NEWMAN. (Continued from page 160).

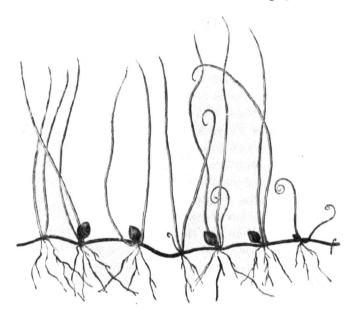

PILL-WORT OR PEPPER-GRASS.

PILULARIA GLOBULIFERA of Authors.

THE Pill-wort or Pepper-grass seems pretty generally, though not
plentifully, distributed over the United Kingdom. It is found on the
extreme margins of ponds, or on swampy ground, submerged during
the winter and more or less exposed during the summer. It is sometimes so abundant as to form a dense and almost inextricably matted
covering to the ground. In no instance that has come to my knowledge has it been found in deep water, or in a state of constant sub-
mersion.

The figures of this plant are generally characteristic, but nearly all
of them err in not representing it sufficiently slender; the best I

T

have seen are those by Bernard de Jussieu, published in the 'Mémoires de l'Académie Royale des Sciences,'* and by Mr. Valentine, in the 'Transactions of the Linnean Society.'† Both these authors have given full and interesting details of its history, and seem so completely to have preoccupied the ground as to leave little or nothing for me to add. An abstract of Mr. Valentine's paper has already appeared in 'The Phytologist' (Phytol. 55), and a second detail of his observations would not be justifiable.

The roots are generally two or three inches in length, very flexible, slender, and but slightly branched; they are hollow, and divided by several longitudinal septa; they appear to descend perpendicularly into the mud or moistened earth in which the plant is found: they spring from a creeping rhizoma, which is also hollow and longitudinally divided; it is very slender and cylindrical, and the terminal or growing portion is invariably covered with a close investment of scales or scale-like hairs; these, like a similar investment common to the creeping rhizomata of Polypodium vulgare, Davallia canariensis, and several other ferns, fall off with age, leaving the rhizoma perfectly naked and smooth. The roots spring from the rhizoma at intervals of considerable regularity, usually measuring about the third of an inch: they are generally three or four in a cluster, and immediately above them rise an equal number of erect, slender, smooth, setiform, pointed leaves: these are hollow like the roots and rhizoma; they are rather longer than the roots, and when they first make their appearance are rolled up in a manner precisely analogous to that exhibited in the circinate vernation of ferns. At many of the points of the rhizoma whence spring the leaves and roots, it also emits a small lateral branch, which bears leaves and roots at intervals like the parent rhizoma; and when this, in the course of nature, decays, these lateral branches continue vigorous, and become the nuclei whence future plants originate. The lateral branches occur with great regularity

alternately on the right and left of the parent rhizoma; in proportion to their distance from the terminal point of the rhizoma these lateral branches increase in length, and the angles at which they join it become more and more obtuse.

The capsule is placed on a short stalk in the axil of the leaves: when full grown it occasionally attains the size of an ordinary pepper-corn, and is nearly spherical, but slightly elongated at its apex; it is closely covered with

* 1739, p. 256, tab. xi.   † Trans. Linn. Soc. xviii. 483.

a dense investment of hair. When mature* it opens at the apex, dividing longitudinally into four parts, each of which continues attached by its inferior extremity to the common footstalk.

Each of these four parts is hollow, and its cavity, which retains the figure of a quarter sphere, is filled with an hermaphrodite flower composed of stamens and pistils arranged on a common placenta. This placenta is a membranous band attached to the interior spherical portion of the membrane which invests the capsule. The pistils, according to Jussieu, are ranged on the inferior part of the placenta, and consequently occupy the lower portion of the common receptacle, as exhibited in the annexed cut, which is copied, with some slight alterations, from Mr. Valentine's figure.† The shaded bodies represent the so-called pistils of Jussieu, and one of these detached is represented below, with its gelatinous covering, the result of immersion in water.

The upper part of each cell is occupied by smaller granules, which are described as stamens, but this appears a somewhat vague conjecture, their office as such never having been clearly proved. Whatever may be the correct designation of the larger granules in their earlier state, it is quite certain that they ultimately become true seeds; for both the authors quoted have succeeded in tracing

their germination, and have recorded their observations on the subject with a close correspondence which mutually proves their accuracy and precision; the improvement in the structure of microscopes and the more artistical drawing of Mr. Valentine giving however a decided superiority to his illustrations. These observations are made on the assumption that Mr. Valentine was entirely unacquainted with the contents of M. Jussieu's admirable essay, which, from an observation in his introductory remarks, I most readily believe.

According to Jussieu the seeds of Pilularia are to be found in the months of September and October, floating on the surface of the water and germinating in that situation.

EDWARD NEWMAN.

---

* This description is quoted nearly verbatim from Jussieu, l. c.

† 'Trans. Linn. Soc.' xviii. tab. 35, fig. 30.

ART. LXIV. — *Sketch of an Excursion to the Clova Mountains, in July and August,* 1840.* By WILLIAM GARDINER, Esq., Jun.

IN company with a botanical friend, I left Dundee on the morning of the 27th of July by the first railway train for Glammis, where we arrived about half past 10 o'clock. Between that place and Kirriemuir nothing of consequence attracted our attention, except abundance of Galeopsis versicolor in the cornfields, and the unusual beauty and exuberance of the more common flowers. We despatched our baggage to Clova from Kirriemuir, and walked on ourselves for the purpose of botanizing by the way. From the latter place to Cortachy there was little to cheer us save our own glad thoughts and joyful anticipations, and the beauty of the scenery through which we passed, steeped in the bright radiance of a summer sun. Beyond Cortachy, however, we began to come among the mountains, and some traces of subalpine vegetation made their appearance. Agrostis vulgaris var. γ. pumila was our first acquisition, and soon after we came upon Habenaria bifolia and Narthecium ossifragum, both very common to subalpine places, such as the Sidlaw Hills. We next met with Viola lutea in grassy places by the waysides; and this beautiful mountain violet occurred more or less profusely all the way up the glen to its very top. On little knolls beyond a miserable public-house dignified by the high-sounding appellation of the Red Lion Inn, we met with Habenaria albida : and being now at the foot of the mountains we made a short deviation from the road, and gathered from a large detached rock Alectoria jubata var. β. chalybeiformis and Andræa rupestris. By the sides of rills abundance of Saxifraga stellaris and aizoides showed themselves, and on the heaths we picked a single stalk of Erica Tetralix with white flowers, and one or two of the pink-flowered variety of E. cinerea. Farther up the Glen we found Meum athamanticum in abundance by the waysides, and by the margin of the Esk grew Carex aquatilis in great luxuriance, along with Galium boreale, and here and there a specimen of Cnicus heterophyllus. Some little distance beyond the " smithy" Pyrola media occurred, though sparingly and past its prime, as was Polygonum viviparum with which it was associated. Alchemilla alpina now showed itself, and was our constant companion wherever we went, during our stay at Clova.

Pretty well loaded for the first day, we reached the hamlet of Clova in time to enjoy the glorious spectacle of an alpine sun-set, and soon

* Read before the Botanical Society of Edinburgh, December 10, 1840.

after did complete justice to the good cheer prepared for us by Mary Findlay.

We next morning went up the mountains to Loch Brandy, for the twofold purpose of enjoying the sight of its wild and picturesque scenery, and searching out the botanical treasures of its surrounding rocks. On the marshy banks of the stream by which we ascended we found Veronica serpyllifolia var. β. alpina, Epilobium alpinum and alsinifolium, and great profusion of Saxifraga aizoides and S. stellaris. These plants we afterwards found more or less abundant by the sides of all the streams in the district which we examined. The first view we had of the Loch was from an unfavourable position, and we felt disappointed; but as we advanced up the steep hill on the left hand side, our admiration increased at every step, till at length, having gained the summit, and descended a few yards among the over-hanging cliffs, the whole grandeur of the scene burst upon our enraptured gaze. Far below lay the dark lake, slumbering in its mountain solitude, while around it rose, guardian-like, the mighty cliffs, sunning their rugged brows in the sweet light of the morning, and looking proud in their magnificence. Quietness brooded in the still air, and no sound was heard to break the awful tranquillity, save when the distant bleat of a sheep fell upon the startled ear, or the plaintive tones of my companion's flute awoke the echoes of the wild rocks. We feasted ourselves for some time on the sublimity of this alpine picture, and then turned our attention to the gems of beauty which the fair hand of Flora had profusely scattered around. We here gathered Azalea procumbens, Gnaphalium supinum, Rubus Chamæmorus, Lycopodium Selago and alpinum, Polytrichum alpinum, Conostomum boreale, Splachnum mnioides, and, at the risk of breaking our necks, Cerastium alpinum and Saxifraga hypnoides. At the head of the Loch we descended by a terrific water-course to its margin, collecting in our way Saxifraga oppositifolia, Salix herbacea, Hieracium alpinum, Rhodiola rosea, Weissia acuta, Didymodon capillaceus and rigidulus, the latter but sparingly; and among the rocks on the banks of the Loch there was an exuberance of Arbutus Uva-ursi and Vaccinium Vitis-idæa, but neither of them in flower.

We went round about the small lochs looking for Isoetes lacustris, and picked up a specimen or two, but saw no appearance of it in a growing state, and did not think of wading in search of it, otherwise we should have reaped a rich harvest, as it was found, along with Lobelia Dortmanna, in the greatest abundance, by my friends Messrs. Croall and Kerr, in September. Among the rocks, just where the southmost stream issues from the Loch, we met with the elegant Hypnum Silesianum. A little farther down we came upon some fine tufts of Aspidium Lonchitis, and before reaching the hamlet plenty of Polypodium Phegopteris.

Our next excursion was to Glen Dole, and a perilous one it was, partly owing to our inexperience of the place. On leaving the hamlet we picked a few specimens of Carex ovalis; and at Bradooney we spent a considerable time searching for Oxytropis campestris among "rocks facing the south," but found only Habenaria viridis and Marchantia hemisphærica for our trouble. Oxyria reniformis was plentiful on the banks of the Esk near Acharne, but we found it in a much finer state on a small island in the Dole, near where it is joined by Kilbo Burn. On this island there was also abundance of Valeriana officinalis, Solidago Virgaurea and Festuca ovina var. ε. vivipara: the latter plant was plentiful in many places, but nowhere finer than here. We went up Glen Phee a considerable distance, but finding nothing except Juniperus communis and Arbutus Uva-ursi, retraced our steps and ascended the face of Craig Rennet by an untrodden path, so precarious that the very recollection of it almost curdles the blood in one's veins. Our toil was rewarded, however, with fine specimens of Silene acaulis, Andræa alpina, Conostomum boreale, Weissia acuta, Jungermannia nemorosa, var. β. purpurascens, &c.; and on the summit there was abundance of Cetraria Islandica. This interesting lichen, which we had hitherto looked upon as a rarity, was plentiful on all the Clova mountains which we visited, and in fine condition, though barren. Proceeding along the ridge towards Craig Maid, we found in boggy places Vaccinium uliginosum, and several beautiful patches of Splachnum sphæricum; and on the summit of Craig Maid Luzula spicata. We had intended to have gone to the White Water and come round by the head of Glen Dole, but time forbade, as the day was wearing to a close, and we therefore determined to attempt descending into Glen Dole. The ascent of Craig Rennet had inured us to danger, and we did not hesitate to choose for our path a wild and dismal-looking water-course that led us down the very face of Craig Maid. By caution and perseverance we accomplished our task, and reaped in our way a rich botanical harvest. Among our acquisitions were the beautiful Lycopodium annotinum, Hieracium alpinum and Halleri, Cnicus heterophyllus, Aspidium Lonchitis, Rhodiola rosea in great perfection, and various other species which we had previously met with.

During our stay at Clova we made two other trips to Glen Dole.

The first of these was on the 31st of July. We ascended Craig Mellon from Acharne, collecting in our way fine specimens of Gnaphalium supinum, Saxifraga oppositifolia and Lycopodium selaginoides, and on the stones at its summit Lecanora ventosa. Proceeding along the ridge we picked specimens of Bryum alpinum and a few Jungermanniæ, and came upon the Dole at its junction with the White Water. About the falls we gathered Rhodiola rosea, Lycopodium annotinum, Cnicus heterophyllus and a variety with deeply cut leaves, Thalictrum alpinum, Hieracium pulmonarium, Carex curta, fulva, sylvatica, binervis and several others, Salix alba and arenaria, Pyrus aucuparia, Veronica alpina and Phleum alpinum. The two latter plants occurred, though sparingly, both above and below the falls. In the crevices of the rocks below we found Bryum elongatum, crudum and ventricosum, and a slender state of Trichostomum aciculare. On the banks of the White Water above the falls we met with Juncus triglumis, and proceeding round by the shieling at the head of Glen Dole we collected Melampyrum pratense, Cornus suecica, and Rubus Chamæmorus in flower and fruit.

On our third visit, August 5th, we examined some of the rocks a short distance below the falls of the White Water, and had the good fortune to discover the beautiful Hypnum Crista-castrensis in fructification, and the rare Dicranum polycarpon; and on the moist precipices almost within reach of the spray, a few specimens of Erigeron alpinus and Asplenium viride. Jungermannia Blasia occurred in crevices of rocks by the side of the Dole. At the head of Glen Dole, after an eager search, we came upon the bog where Carex rariflora grows, and found it in the greatest profusion. Before we had satisfied ourselves with the much-prized rarity, the evening mist had gathered around us, and so densely, that we could not see more than a few feet beyond us. Had we not taken our bearings well we should have been truly bewildered, and run the risk of ending our career at the bottom of the neighbouring precipices; but by the aid of our compass we managed to reach the shieling, and the shepherd carefully put us on "Jock's road," which, though not a very "redd" road, led us safely down the glen.

We spent a day on Carlowie hill gathering lichens, for few flowering plants of interest were there, except Arbutus Uva-ursi. We picked from the rocks and stones numerous specimens of Gyrophora cylindrica, proboscidea and polyphylla, Cornicularia tristis and lanata, Stereocaulon paschale, Sphærophoron coralloides and one tuft of var. β. (the Sph. fragile of Acharius) in fructification. These lichens were also abundant on the hill above the hamlet of Clova, as well as several others, particularly Lecanora tartarea and Parmelia omphalodes. On the " Greenhill," south of Loch Brandy, we found Spergula subulata and Azalea procumbens.

To the Bassies mountain we made two visits, and found several good things. Azalea procumbens was plentiful on its summit, associated with Salix herbacea and Juncus trifidus: the specimens of the latter plant gathered here, however, were stunted and insignificant in their appearance compared with the fine tufts which we culled from the fissures of the rocks about half way down. Carex rigida was frequent; Alchemilla alpina literally mantled the mountain with its silvery foliage; and we noticed Sibbaldia procumbens and Tofieldia palustris, but without flowers. On the summit there was a profusion of Cetraria nivalis, and we also gathered Solorina crocea, Cornicularia aculeata in fruit, Lecidea fusco-lutea and Splachnum sphæricum; and in a bog was Polytrichum juniperinum, var. β. gracilius (the P. strictum of Menzies). Between Bassies and Scorie there was plenty of Conostomum boreale and Polytrichum hercynicum, together with Dicranum Starkii, subulata and virens. Polytrichum alpinum was in abundance everywhere on the mountains at a good elevation. By the side of a watercouse in descending from the Bassies we met with a specimen or two of Alopecurus alpinus, Hieracium alpinum and Halleri; but the most interesting of our acquisitions on the Bassies was Jungermannia ciliaris *with calyces*, which are of exceedingly rare occurrence. The calyx-bearing plants were very small and perfectly procumbent, growing among tufts of Dicranum scoparium. We also found on this mountain beautiful specimens of Tetraphis pellucida in fruit.

On the banks of the Esk about Clova there was plenty of Carex ampullacea and Geranium sylvaticum, with Galium palustre var. β. Witheringii and Cnicus heterophyllus; and we also found Rumex aquaticus and Rubus suberectus. The latter plant occurred in many places in the valleys as well as on the mountains.

On our return down the glen we met with nothing that we had not previously seen in the district, except Triodia decumbens, which grew on small hillocks by the waysides.

This being our first visit to these mountains the ground was entirely new to us, and particular localities quite unknown, so that we were unsuccessful in obtaining many of the rarities which are there to be found. Professor Balfour, who is intimately acquainted with the district, procured, just before we left, Dryas octopetala, Veronica saxatilis, Linnæa borealis, Malaxis paludosa, Sonchus alpinus and Salix lanata. From my second visit, which I hope to be able to make this summer, better results may reasonably be anticipated.

WILLIAM GARDINER, Jun.

Dundee, March 4, 1842.

ART. LXV. — *Notes on Arenaria rubra, marina, and media.* By SAMUEL GIBSON, Esq.

Hebden Bridge, May 5, 1842.

SIR,

A difference of opinion having long existed as to whether *Arenaria rubra* and *marina* are two distinct plants, or only varieties of one species ; in order to set this question at rest in my own mind, I procured specimens of the two plants from every locality where I could possibly get them from, either by sending to my friends or by collecting them myself. After collecting what specimens I could get, I gave them a careful examination, and am now of opinion that *Arenaria rubra* and *marina*, and the *media* of Linnæus, all three possess permanent characters sufficient to keep them distinct as species. *Arenaria marina* and *media* I have found plentifully growing together, each retaining its characteristic marks; the *seeds* of *media* being bordered with a striated membrane, while those of *marina* are destitute of such a border.

What Sir W. J. Hooker's *Arenaria rubra* may be I know not. The calyx is said to be nerveless; such a character I have not been able to find in any of the three forms. It may require further investigation in order to decide whether or not Sir W. J. Hooker's plant be something very different from anything I have ever seen; but what specimens I possess will answer to the following descriptions.

1. *Arenaria rubra*, (Red Sandwort). *Stems* procumbent, smooth, except near the flowers, where they have a few glandular hairs such as cover the calyx and flower-stalks : *leaves* linear, flat, terminating in a small bristle: segments of the *calyx* ovate, acute, *three-nerved*, the two lateral nerves very short, the middle one as long as the segment: *seeds* small, somewhat pyriform, thick in proportion to their breadth, rough all over with raised points.

My specimens are from Huddersfield, Halifax, &c.

β. brevifolia, mihi. This variety differs from the more common state of *Arenaria rubra* in the internodes of the stems and the leaves being only about half the length, and in the flowers being more numerous, &c.

Common in the neighbourhood of Manchester, as at Kersal Moor, &c.

2. *Arenaria marina*, (Marine Sandwort). *Stems* procumbent, smooth : *leaves* as long as the internodes, blunt at the point, semi-cylindrical : *capsule* as long as the calyx : segments of the *calyx* ovate, three-ribbed, ribs pellucid : *seeds* ear-shaped, thickened and rough on their edges, depressed and nearly smooth in the middle, nearly twice the size of those of *Arenaria rubra*.

This appears to be somewhat rare ; some of my specimens are from the Yorkshire coast at Bridlington and Scarborough, and others from near Liverpool.

β. hirsuta, mihi. This differs from the above in the stems being wholly covered with glandular hairs.

My specimens are from Newlyn Cliff, near Penzance, Cornwall.

8. *Arenaria media*, (Smooth-seeded Sandwort). *Stems* procumbent, smooth : *leaves* as long as the internodes, blunt at the point, semicylindrical, fleshy, flaccid when dry : *capsule* twice the length of the calyx : segments of the *calyx* obovate, three-nerved, nerves pellucid : *seeds* ear-shaped, smooth, much larger than those of *Arenaria marina*, with a broad, white, striated border.

This appears to be the most common of the three plants, as I have specimens from the following localities, viz., North shore, Liverpool, Southport &c., Lancashire ; Wallazey pool, Cheshire ; &c. I also have it from two three localities in Ireland, and others from Scotland.

If you should think the above remarks worthy a place in your periodical, they are at your service.

Yours, &c.
SAML. GIBSON.

To the Editor of 'The Phytologist.'

ART. LXVI.—*On the escape of Fluid from the Apex of the Leaf of Richardia æthiopica*, Kunth, (*Calla æthiopica*, Linn.) By EDWIN J. QUEKETT, Esq., F.L.S., B.S., &c.

MY attention having been directed to this fact by my friend Mr. Ward, and having a large case constructed after his plan for the growth of plants, I placed within it, in the summer, with numerous others, a healthy specimen of Richardia æthiopica, which was kept well supplied with water. It was soon perceived that the dripping from the apex of the leaf had commenced, and it continued as long as the weather was warm.

On experimenting on this plant it was found that the greatest number of drops in a given time were to be obtained soon after the sun had ceased to shine on the plant, which was at mid-day; the number never exceeded one in a minute, and generally they were not so frequent as this. I was anxious to collect some of the fluid in order to analyze it, and suspended a small vessel near to the apex; during the day and night I could thus obtain two or three drams from one leaf. It was perfectly bright in colour and tasteless, and on applying tests which usually exhibit impurities in water no reaction could be obtained. The plant was sometimes watered with a decoction of logwood, yet no indication of its presence could be detected in the colour of the fluid, or by the salts of iron; in fact it appeared to be pure water, notwithstanding that lime and other matters must have been present in the water applied to the roots.

Reflecting on the purity of the fluid, and how the plant could so effectually separate the soluble impurities from the water absorbed, I was at a loss to conceive how such a quantity could escape from the minute apex, which it always did, and not from sundry other spots then trickling towards the point, previous to its falling from the leaf.

As the escape of this fluid does not often occur when the plant is out of doors, or if the sun shines on it when confined under glass, it was imagined that under either of these conditions the evaporation from the surface of the plant might be sufficient to carry off any excess of water sent by the roots into the interior, under the stimulus of increased temperature; and that when evaporation could not so proceed, the channels which conveyed the fluid became surcharged, and the apex, which seems as it were the confluence of numerous minute streams, gives exit to the excess collected on account of suppressed evaporation.

It became necessary to apply to the anatomy of the leaf in order to account for and prove the escape of the fluid from the apex. The leaves of this plant are arrow-shaped, and terminated by a nearly cylindrical point, varying from half an inch to three quarters in length, and about the twentieth part of an inch in diameter. The venation makes one of the exceptions to the general rule, that the leaves of all Endogens are straightly veined, for in this leaf is to be noticed an arrangement of veins somewhat analogous to that of an Exogenous plant, being in this instance especially allied to the leaf of a plant belonging to the order Myrtaceæ, on account of the presence of a distinct marginal vein. A strong midrib proceeds from the petiole for some distance up the leaf, and is gradually attenuated and expended by giving off numerous branches, which take a curved direction outwards and forwards, and which are intercepted before they reach the margin, by the vein running parallel to the circumference, which has its commencement at the petiole and its termination in the apex; consequently each vein has a tendency to carry its fluid towards the apex by the intervention of this marginal vein. And there is no doubt that the fluid is conducted by these courses, for if any one be wounded by a sharp instrument, dropping will proceed from the wound as well as from the point; besides, the veins do not form prominent opaque ribs as in many exogens, but are level with the surfaces of the leaf; and when examined by transmitted light they appear the most transparent portions, an indication of their being full of fluid.

If the attenuated apex of the leaf be examined by dissection, microscopically, it will be found to present some few stomata on its cuticle and a dense bundle of vessels in its centre, surrounded by cellular tissue containing multitudes of acicular raphides; and I could never carry the dissection fine enough to discover whether or not these vessels communicate with the surface or the caverns with which the stomata are connected.

From the anatomy of the leaf it would appear that the surcharged vessels all propel their contents towards its point, and the excess finds the means of escape as through a filter.* It cannot be from gravity, for it occurs whilst the young leaf is making its appearance, before it unrols, when the sharp-pointed apex is in the most elevated position, being quite perpendicular. If it occurred only when the leaf was pendulous, it might be imagined that the liquid was condensed in the cavernous tissue of the leaf, and escaped through the stomata from gravity; but as this is not the case, it must be considered that this peculiarly constructed apex is a beautiful provision—a kind of natural safety-valve, for permitting the superabundance of watery fluid to be readily removed, the retention of which might be connected with unhealthy action in the economy of this elegant plant.

The escape of fluid from the apex of the leaf is not peculiar to Richardia, as many other plants, when grown under glass will exhibit

---

* The apex of the spathe, in its early state, will also exhibit the phenomenon ; as will also a leaf when a fifth of its point has lost the signs of vitality.

the phenomenon, as young plants of barley and wheat; consequently I conceive that there must be some special contrivance of nature in these leaves, perhaps in all leaves, to guard against the accumulation of watery fluid.

The opportunity of witnessing these experiments can in no way be so successfully obtained, or their results so satisfactorily examined, as in Mr. Ward's cases; which gives us another instance of the useful adaptation of this plan to the pursuit of experimental inquiries respecting the functional operations performed during the growth of plants.

EDWIN J. QUEKETT.

50, Wellclose Square,
May 16th, 1842.

ART. LXVII. — *Analytical Notice of a treatise ' On the Growth of Plants in Closely Glazed Cases.'*   By N. B. WARD, F.L.S. London : John Van Voorst.   1842.   8vo.

THE lovers of Nature and of Nature's works are deeply indebted to the author of this treatise, for showing them by what means plants may be made to grow and thrive in situations where few could even exist before.   For although the fact that plants will live and grow without direct communication with the external atmosphere, may have been often observed long before the fern and grass sprang up in Mr. Ward's closed glass cylinder, yet to that gentleman is undoubtedly due the sole merit of so closely reasoning upon a simple circumstance unexpectedly brought under his notice, as to have deduced from it principles which have already led to important results, while it is not improbable that consequences still more important yet remain to be disclosed.

By means of this mode of cultivating plants, the botanist may create for himself, even within the " brick-wall bounds " of any of our large towns, such a scene of natural beauty as will in some measure compensate for his exclusion from the pleasure of studying his favorites in their own native haunts.   In many respects indeed he will be a gainer, in the facilities for study afforded by the naturalization of the denizens of the wild wood or the snow-capped mountain under his own roof, nay, even by his own fire-side : for rarely is a botanist placed in such favorable circumstances as to be able to do more than collect specimens of plants, as they are met with in his too often hurried excursions; and specimens so collected generally form the only materials with which is reared many a plausible hypothesis relating to specific distinction or identity. But now the student may have his plants living under his own eye — may watch their growth from day to day—from hour to hour. In the case of a fern, he may scatter its sporules and observe the first appearance of the slightest possible tinge of green on the surface of the soil; then after a time he will note the expansion of the first seedling fronds; and so step by step proceeds the progress of development, until he perceives with delight —

> " Each stem and leaf wrapped small,
> Coil'd up within each other
> Like a round and hairy ball."

Then again how closely may he

> ———— " Watch that ball unfolding
> Each closely nestling curl,
> And its fair and feathery leaflets
> Their spreading forms unfurl ! "

And all this pleasure and instruction he may secure, even in the most impure atmosphere, by the simple expedient of surrounding his protégés with glass, which, while it allows of the free passage of light, — a most essential condition in the culture of plants, effectually prevents the access of fuliginous matter and loss of moisture by evaporation, and at the same time ensures a calm atmosphere and a more equable temperature.

In the treatise before us Mr. Ward has published the results of his thirteen years' experience in the mode of growing plants in closely glazed cases. The work is divided into six chapters, and the contents of these we give below.

Chap. I.—*On the Natural Conditions of Plants.* Unless we possess some knowledge of the conditions which regulate the growth of plants in a state of nature, it is evident that our treatment of them in cultivation must be more or less defective. These conditions vary in an almost endless degree; plants being "influenced by the atmosphere, heat, light, moisture, varieties of soil, and periods of rest." The growth of plants is most sensibly affected by the purity or impurity of the atmosphere: the heat to which they are subjected has a range, at different seasons and in different countries, of not less than 150°: the intensity of light "varies from almost total darkness to a light double that of our brightest summer's day :" then again " the states of moisture vary as much as those of heat and light": varieties of soil also visibly affect plants: and lastly —

"All plants require rest, and obtain it in some countries by the rigour of winter; in others by the scorching and arid heat of summer. Some, " after short slumber wake to life again," while the sleep of others is unbroken for many months. This is the case with most alpine plants, and is necessary to their well-being. * * In Egypt the blue water-lily obtains rest in a curious way. Mr. Traill, the gardener of Ibrahim Pacha, informed me that this plant abounds in several of the canals at Alexandria, which at certain seasons become dry; and the beds of these canals, which quickly become burnt as hard as bricks by the action of the sun, are then used as carriage-roads. When the water is again admitted the plant resumes its growth with redoubled vigour."—p. 5.

On the power possessed by plants of adapting themselves in a certain degree to the circumstances in which they are placed, the author remarks : —

" To suit all the varied conditions to which I have thus briefly alluded, and under which plants are found to exist, they have been formed by their Almighty Creator of different structures and constitutions, to fit them for the stations they severally hold in creation; and so striking are the results, that every different region of the globe is characterized by peculiar forms of vegetation. A practised botanical eye can with certainty, in almost all cases, predict the capabilities of any hitherto unknown country, by an inspection of the plants which it produces. * * But in order to give us a clearer idea of the " strong connexions, nice dependencies," existing between climate and vegetation, let us survey plants in a state of nature. We shall find some restricted to certain situations, while others have a wide range, or greater powers of adaptation. It is not perhaps going too far to assert, that no two plants are alike in this particular, or in other words, that the constitution of every individual plant is different. Of the former, Trichomanes speciosum is an example, it not being able to exist, even for a short time, in a dry atmosphere: of the latter, familiar examples are presented to us in the London Pride and the Auricula ; these of course grow in greater or less luxuriance, as the conditions are more or less favorable."—p. 6.

Chap. II. — *On the Causes which interfere with the Growth of Plants in large towns.* Among these are more particularly mentioned " deficiency of light, the dryness of the atmosphere, the fuliginous matter with which the air of large towns is always more or less loaded, and the evolution of noxious gases from manufactories." The author expresses his belief that the generally depressed state of vegetation in large towns, is chiefly due to the quantity of soot floating in the atmosphere; and that although deficiency of light and moisture certainly exerts an influence to a certain extent, yet that neither of these, nor the evolution of noxious gases, nor the three combined, is the sole or even the chief enemy to the growth of p ts in a town atmosphere. After quoting from Mr. Ellis's paper

'Gardeners' Magazine' for September, 1839, that part which relates
to the observations and experiments of Drs. Turner and Christison
on the effects of sulphurous and muriatic gases upon vegetation, the
author observes :—

" The correctness of the above observations of Messrs. Turner and Christison, as
to the effects of sulphurous and muriatic acid gases upon plants, cannot for one moment be doubted ; and that plants suffer when exposed to a direct current of these
gases, before there is time for diffusion through surrounding space, is equally matter
of fact ; but I contend, that it yet remains to be proved that there exists generally in
the atmosphere of London, or other large cities, such a proportion of these noxious
gases as sensibly to affect vegetation.   We shall find in the windows of shops and
small houses, in numerous parts of London, hundreds of geraniums and other plants,
growing very well and without any crisping or curling of the leaves, care being taken
in these instances to keep the plants perfectly clean and free from soot ; and it is certain,
that although my cases can and do exclude the fuliginous portion of the atmosphere,
and certainly protect the plants from the effects of any direct current of hurtful airs,
they cannot exclude that portion which becomes mixed with the atmosphere."—p. 17.

The author gives various examples of the rapidity with which gases
mingle with each other and with the atmosphere, under the influence
of "a law constantly in action under all circumstances" and in all
places, by means of which the several constituents of the atmosphere
are ever preserved in their respective proportions.

Chap. III.—*On the Imitation of the Natural Conditions of Plants
in closely glazed Cases.*   At the commencement of this chapter the
author gives such a pleasant description of the frustration of his early
attempts to obtain something like country within the smoke of London, and of the occurrence which led to his subsequent success, that
we must quote the passage entire.

" The science of Botany, in consequence of the perusal of the works of the immortal Linnæus, had been my recreation from my youth up ; and the earliest object of
my ambition was to possess an old wall covered with ferns and mosses.   To obtain
this end, I built up some rock-work in the yard at the back of my house, and placed
a perforated pipe at the top, from which water trickled on the plants beneath ; these
consisted of Polypodium vulgare, Lomaria spicant, Lastræa dilatata, L. Filix-mas,
Athyrium Filix-fœmina, Asplenium Trichomanes, and a few other ferns, and several
mosses procured from the woods in the neighbourhood of London, together with primroses, wood-sorrel, &c.   Being, however, surrounded by numerous manufactories and
enveloped in their smoke, my plants soon began to decline, and ultimately perished, all
my endeavours to keep them alive proving fruitless.   When the attempt had been given
up in despair, I was led to reflect a little more deeply upon the subject in consequence
of a simple incident which occurred in the summer of 1829.   I had buried the chry-
salis of a Sphinx in some moist mould contained in a wide-mouthed glass bottle, covered with a lid.   In watching the bottle from day to day, I observed that the moisture which during the heat of the day arose from the mould, became condensed on the

internal surface of the glass, and returned whence it came; thus keeping the mould always in the same degree of humidity. About a week prior to the final change of the insect, a seedling fern and a grass made their appearance on the surface of the mould."—p. 25.

After briefly stating the reflections to which this unexpected event gave rise, and having mentioned the conclusions arrived at, the author proceeds : —

"Thus, then, all the conditions necessary for the growth of my little plant were apparently fulfilled, and it remained only to put it to the test of experiment. I placed the bottle outside the window of my study—a room facing the north, and to my great delight the plants continued to grow well. They turned out to be Lastræa Filix-mas and Poa annua. They required no attention, the same circulation of the water continuing ; and here they remained for nearly four years, the Poa once flowering and the fern producing three or four fronds annually. At the end of this time they accidentally perished, during my absence from home, in consequence of the rusting of the lid, and the admission of rain water."—p. 26.

The author next details his experiments on different plants, including Trichomanes speciosum, Hymenophyllum, Jungermanniæ, and Crocuses both with natural and artificial light ; and describes his Tintern-Abbey house, his alpine case, drawing-room case and case with spring flowers ; thus exhibiting to his readers the gradual development of his plans, until he conducts them to his " largest experimental house." In this house, erected in the scene of his early disappointments, perhaps on the very site of the rock-work on which no attentions could prolong the existence of some of our hardiest native plants, we now see not only the " wall covered with ferns and mosses," — the earliest object of the author's ambition, but a host of botanical treasures from all parts of the globe growing side by side in the greatest luxuriance and beauty : the ferns especially, both native and foreign, appear to be quite at home ; the tender and delicate Trichomanes speciosum being one of the most lovely objects in the collection, and Osmunda regalis, planted in March last, now has its noble fronds crowned with fructification.

The chapter concludes with the following remarks upon " the importance of reflecting on what we see around us."

"The simple circumstance which set me to work must have been presented to the eyes of horticulturists thousands of times, but has passed unheeded in consequence of their disused closed frames being filled with weeds, instead of cucumbers and melons; and I am quite ready to confess, that if some groundsel or chickweed had sprung up in my bottle instead of the fern, it would have made no impression upon me : and again, after my complete success with the ferns, had I possessed the inductive mind of a Davy or a Faraday, I ought, in an hour's quiet reflection, to have anticipated the results of years. I should have concluded that all plants would grow as well as the ferns, inasmuch as I possessed the power of modifying the conditions suited to the wants of each individual."—p. 42.

Chap. IV.—*On the conveyance of Plants and Seeds on Ship-board.* After some observations on the means formerly employed for the preservation of plants during long voyages, the author thus proceeds :—

" But by far the greater number of plants require to be kept growing during the voyage ; and, prior to the introduction of the glazed cases, a large majority of these plants perished from the variations of temperature to which they were subjected,—from being too much or too little watered,—from the spray of the sea,—or, when protected from this spray, from the exclusion of light."—p. 46.

The author's reflections on these causes of failure induced him, in June, 1833, to send out to Sydney two experimental cases filled with ferns and grasses, nearly the whole of which arrived there " alive and flourishing :"—

" The cases were refilled at Sydney in the month of February, 1834, the thermometer then being between 90° and 100°. In their passage to England they encountered very varying temperatures. The thermometer fell to 20° in rounding Cape Horn, and the decks were covered a foot deep with snow. At Rio Janeiro the thermometer rose to 100°, and in crossing the line to 120°. In the month of November, eight months after their departure, they arrived in the British Channel, the thermometer then being as low as 40°. These plants were placed upon the deck during the whole voyage and were not once watered, yet on their arrival at the docks they were in the most healthy and vigorous condition."—p. 46.

Subsequent experiments with plants of a higher order were equally successful; and the following extract of a letter to the author from Mr. George Loddiges, is confirmatory of the importance of the plan.

" My brother and I have, since 1835, made trial of more than 500 cases to and from various parts of the globe, with great variety of success; but have uniformly found, wherever your own directions were strictly attended to,—that is, when the cases were kept the whole voyage in full exposure to the light, upon deck, and care taken to repair the glass immediately in cases of accident,—that the plants have arrived in good condition. * * Some of the cases have been opened in fine order after voyages of upwards of eight months: in short, nothing more appears to be wanting to ensure success in the importation of plants, than to place them in these boxes properly moistened, and to allow them the full benefit of light during the voyage."—p. 86.

Full instructions are given for the construction of the cases and the preparation of the plants for the voyage.

Chap. V.—*On the application of the closed plan in improving the condition of the Poor.*

" Among the numerous useful applications of the glazed cases, there is one which

I believe to be of paramount importance, and well deserving the attention of every philanthropist: I mean its application to the relief of the physical and moral wants of densely crowded populations in large cities. Among the members of this population there are numbers, who, either from early associations, or from that love of Nature which exists to a greater or less degree in the bosom of all, are passionately fond of flowers, and endeavour to gratify their taste at no small toil."—p. 57.

After some observations on the importance of the free admission of light into human habitations, and on its influence upon the animal economy as well as upon vegetation, the author shows in what manner the innate love of Nature above alluded to may be gratified at a trifling expense. He however cautions the poorer classes "against indulging a taste for what are called fancy flowers—things which this year are rewarded with gold medals, and the next are thrown upon the dunghill"—as being opposed to the legitimate pursuits of horticulture. The benefical effects of the study of Botany are next dwelt upon; and the interesting anecdote of Parke and the moss concludes the chapter.

Chap. VI.—*On the probable future application of the preceding facts.* The Wardian cases evidently furnish great facilities for experimenting on numerous doubtful matters connected with Botany, Horticulture and Agriculture, such as the effects of different soils and manures; the power of the roots in the offices of absorption and selection; the determination of the existence and nature of excretions from the roots, whether poisonous or otherwise; "the effects of poisons upon plants;" "the influence of light in protecting plants from the effects of low temperature;" various points respecting the development and growth of Fungi, and the other lower orders of vegetation; and the investigation of "that debatable ground on the confines of the animal and vegetable kingdoms, where in our present state of ignorance it is often impossible to determine the point at which one ends and the other begins;"—these are but a few of the *quæstiones vexatæ* in the settlement of which the glazed cases may be used with great advantage.

The author next adverts to the application of the principle on a large scale as "a remedial means of the highest order" in the treatment of numerous diseases which would readily yield "to the renovating influence of pure air," although without this auxiliary the skill of the medical man may be of little avail. He more particularly mentions measles and consumption as diseases in which a supply of pure air and a properly regulated atmosphere are of the greatest importance; and after speaking of the direct mortality arising from measles in crowded districts of large towns, as well as the numbers of persons who die of diseases superinduced by neglect during the measles, he thus concludes :—

"With respect to consumption, could we have such a place of refuge as I believe one of these closed houses would prove to be, we should then be no longer under the painful necessity of sending a beloved relative to a distant land for the remote chance of recovery, or too probably to realize the painful description of Blackwood : — " Far away from home, with strangers around him,—a language he does not understand,—doctors in whom he has no confidence,—scenery he is too ill to admire,—religious comforters in whom he has no faith,—with a deep and every day more vivid recollection of domestic scenes,—heart-broken,—home-sick,—friendless and uncared for, — he dies." —p. 71.

The interest we feel in the subject and our conviction of its importance, have perhaps led us to extend our notice of Mr. Ward's useful book somewhat beyond our prescribed limits ; we however feel assured that we shall be readily excused for this by such of our readers as have used the Wardian cases, whether on the same humble scale as ourselves, or in the more superb style of Mr. Ward's own large fernery. In conclusion, we would advise all whose love of Nature is not to be suppressed by the din and smoke of large towns, immediately to set to work, and we can venture to promise that the trifling trouble and expense they may be at, will be amply repaid by the gratification and the instruction they will derive from their observations 'On the growth of plants in closely glazed cases.'

ART. LXVIII. — *Notice of a ' History of British Forest Trees, Indigenous and Introduced.'* By PRIDEAUX JOHN SELBY, F.L.S., M.W.S. &c. London : John Van Voorst. Parts 4—9.

THE illustrations increase in beauty and the descriptions in interest as the work proceeds. The details of the inflorescence are drawn, engraved and printed in masterly style : the catkins of the goat willow at p. 168, of the aspen at p. 188, and of the alder at p. 221, are extremely pretty. Some of the vignettes also merit the highest praise : the artist's "bit" at p. 193, with its cool and quiet shade, is "beautiful exceedingly," and so is the peep into a wood at p. 237. The portraits of the trees themselves still somewhat dissatisfy us : notwithstanding the skill and labour bestowed on them, the result is far from satisfactory. We are ever ready to exclaim, "What a beautiful tree!" but we always refer to the accompanying letter-press to learn its name. It would appear to be a task of infinite difficulty to portray a tree so exactly that its figure shall convey to the beholder a warrant of its identity. A familiar bird or insect is recognized at sight; but who shall select an oak, an ash, or an aspen from among the most faithfully drawn group of forestry? The fault is perhaps in the subject, not in the manner of execution. How many are there who, seeing the trees themselves tossing their sinewy branches in the breeze, would be unable to refer each to its particular species; how difficult therefore must it be for the pencil to seize on characters which the experienced eye shall often fail to detect! The work, in fine, must rest its claim to public patronage on the correctness of its details and the completeness of its descriptions, rather than on any striking likeness in the portraits of the trees themselves.

The author's remarks on the comparative value of timber are highly interesting and valuable. The Black Italian or Necklace Poplar (*Populus monilifera*) is the subject of high encomium. "The wood is of a greyish white colour, tough when seasoned, and if kept dry very durable; its great size renders it fit for the largest buildings, and as flooring for manufactories and other erections nothing can surpass it, as in addition to the property of not splitting by percussion, it possesses the peculiar advantage of not easily taking fire, and even when ignited burning without flame or violence."—p. 201. The last-named quality is an excellence of more than ordinary importance : what an amount of human life and property might be saved by the use of timber which would thus arrest or even retard the awful power of flame!

Speaking of the beech Mr. Selby tells us that when this noble tree is grown singly or in hedgerows it is, "from its dense and widely-extended shade, and the deleterious nature of its drip, more injurious to the herbage beneath than any other tree : and here we may also remark, that one of the greatest disadvantages attending Beechen woods or groves, is that no underwood or herbage, with the exception of some Orchideous and Cryptogamic plants, will thrive beneath their shade : even the hardy holly, a plant that flourishes and bears, comparatively unhurt, the drip and shade of many other trees, pines and languishes under the Beech; laurels and other evergreens, as well as deciduous shrubs, all speedily die when planted beneath its shade." —p. 312. Does not our author here lay too much stress on the drip of the beech? We should hesitate before attributing this deleterious effect to any other cause than the exclusion of light, for we have often seen young beech contending for the mastery with a vigorous undergrowth of holly and wych elms; and although the beeches have outstripped their fellows, which much consequently receive their drip, yet all are growing together in the most perfect amity and vigour. "For narrow upright hedges, to divide or enclose nursery grounds, gardens, or even small fields, the beech is superior to the hornbeam, or any other deciduous tree, as it not only bears the shears equally well, and may be trained to as great a height, but retains the leaves during winter, thus affording additional shelter and warmth, and giving a richness of appearance the others do not possess."—p. 315.

It is perhaps generally known that the Spanish chesnut, although here only occasionally eaten by the lower classes, forms in Italy and Spain a most important article of their food, serving in great measure as a substitute for bread or potatoes. The nuts are variously prepared, sometimes simply boiled or roasted, at others ground to flour; "of this flour, *la Galette,* a thickish kind of girdle cake, mixed up with a little milk and salt, and sometimes with the addition of eggs and butter, is made; *la polenta* is also another preparation made by boiling the chesnut flour in milk till it becomes quite thick; when made with water, it is eaten with milk in the same manner as oatmeal porridge in the north of England and Scotland. *Chatigna,* that is, chesnuts boiled and then mashed up as we do potatoes, is also another preparation common in France and Italy."—p. 329.

Of the Pine Mr. Selby gives a very complete and valuable history, detailing the botanical characters of the family to which it belongs, the peculiarity of its wood differing from that of dicotyledonous trees, the geographical distribution, the requisite soil, and various other particulars. With the following pleasantly written account of the fructification of the common pine we must conclude.

" The male flowers or catkins, when in bloom, are from half an inch to upwards of an inch long, and are placed in whorls at the base of the young shoots of the current year; the flowers contain two or more stamens with large yellow anthers, which discharge a sulphur-coloured pollen in great abundance. The embryo cones or female flowers appear on the summits of the shoots of the year, in number from two to as many as six, and of a green or purplish green colour. When impregnated, they become lateral and reflexed, and cease to increase in size till the following spring, when they again begin to swell, and by July attain their full size, ripening by degrees into ovate, pointed and tessellated, hard, woody cones, from one inch and a half to two inches long. These remain on the tree for a considerable time afterwards, though the seeds are discharged the following spring, and it is then that trees are frequently seen with cones in four different stages: viz., in the youngest or embryo state; in an unripe or green condition, but of full size; in a matured state, or when they have become brown; and lastly with the scales expanded, after the seed has been shed."—p. 396.

ART. LXIX.—*Varieties.*

152. *Enquiry respecting the Parasite on the Goldfish*, (Phytol. 190). In the last No. of 'The Phytologist' there is a notice of a paper read by Mr. Goodsir on the 11th of January last, at a meeting of the Botanical Society of Edinburgh, on a vegetable found on the gills and fins of a goldfish. Although Mr. Goodsir " gave a minute description of the parasite, explaining practically its form, structure, mode of fructification, &c." you have not even favoured your readers with the name of the vegetable, or given the slightest account of it. This I very much regret, as a friend of mine a few weeks since sent me a small carp, which he had kept in a pond, and which, it would appear, died from the same disease as that mentioned by Mr. Goodsir. We examined this fish, and found it covered with a minute vegetable substance, which we supposed to be a Conferva. Around the operculum it was very thick, and apparently obstructed the opening of this valve; the gills too were very much ulcerated and almost united into one mass by the parasitic covering. The internal organs were also inflamed, and near the heart we found a transparent mass, filled with a fluid, and which we conceived to be a hydatid.—*Hy. Jno. Turner; 47, Lower Stamford St., Blackfriars Road, April*, 1842.

[Our notice of Mr. Goodsir's paper was given verbatim from the report published in the Edinburgh Evening Post. We cannot learn that the parasite has yet received a name, and the only description we can meet with we have given below; it is from a report in Taylor's 'Annals' of the February meeting of the Royal Society of Edinburgh. Mr. John Quekett has obligingly furnished us with a notice of a very interesting fact observed by him in connexion with this parasite, which we have great pleasure in inserting.—*Ed.*]

153. *Description of the Vegetable parasitic on the Gold-fish.* "The concluding part of Dr. J. H. Bennet's paper on Parasitic Fungi growing on living animals was read, and as portions of it bear directly on Natural History, we shall briefly allude to these. Fungi of this description have previously been noted as occurring in the stickleback and common carp, but we are not aware that any particular description has yet been supplied of these fungi. Dr. Bennet had an opportunity of examining them upon the gold carp, Cyprinus auratus, having been persistent before death. To the eye they presented the appearance of a white cottony or flocculent matter attached to the animal. Under the microscope it presented two distinct structures, which were severally cellular and non-cellular. The former consisted of long tubes divided into elongated cells by distinct partitions. At the proximal end of several of these cells was a transparent vesicle about ·01 of a millimetre in diameter, which the author considered to be a nucleus. Some of the cells were filled with a granular matter; others however were empty, the granules having escaped through a rupture of the tube or of the cellular walls. Besides these there were long filaments about ·06 of a millimetre in diameter, which apparently sprung from the sides of the cellular tubes. They were uniform in size throughout their whole length, and were formed of an external delicate diaphanous sheath, and an internal more solid transparent matter. This vegetable structure sprung from a finely granular amorphous mass. Fungi of a similar kind were also found in the lungs of a man who died of pulmonary consumption, and from whose lungs they were also copiously discharged in the expectoration during life. The vegetable structure in this instance consisted of tubes, jointed at regular intervals, and

giving off branches generally dichotomous. They varied in diameter from ·01 to ·02 of a millimetre, and appeared to spring without any root from an amorphous, soft, finely granular mass. They gave off at their extremities numerous oval, round or oblong corpuscules, arranged in bead-like rows, which were considered reproductive sporules. The same appearances were found in the soft cheesy matter lining some of the tubercular cavities after death." — *From the 'Annals and Magazine of Natural History,' March,* 1842, *p.* 66.

154. *Note upon the Fungus parasitical on fishes.* Having seen that Mr. Goodsir of Edinburgh has described the parasite which infests the bodies of gold carp and other fishes, I imagined that the following incident concerning it might not prove uninteresting to your readers. About a month since I placed six newts in a tank of water in which there were some aquatic plants, and three small fish commonly called sticklebacks. One of these, the largest of the three, had the hinder part of its body covered with the plant. The newts had not long been in the tank, when my brother, Mr. Edwin Quekett, and myself, saw one of them in the act of nibbling away at the parasite, whilst the fish remained perfectly quiet: on disturbing them the same thing was repeated by another newt; the fish appeared much pleased, and even moved its tail frequently towards the newt, as though it were anxious to get rid of the parasitic growth. Whether the act of the newt were dictated by kindness I cannot say; probably these animals perform (as tench are said to do) the office of physicians to the diseased portion of the finny race.—*John Quekett : 50, Wellclose Square, May 23,* 1842.

155. *Note on the Oxlips from Bardfield, &c.* The oxlips kindly sent to me by Mr. H. Doubleday from Bardfield (Phytol. 204), and concerning which you enquire, appear to me to be the species intended by the figure in ' English Botany' (513), and also to be identical with Swiss and German specimens in my herbarium, which were sent to me under the name of " Primula elatior, *Jacq.*" The Bardfield specimens differ slightly from the figure in ' English Botany,' but not importantly, except in having the calyx decidedly shorter than the tube of the corolla. They are unlike any other English oxlips in my herbarium (all of which may be gradually traced either to the primrose or to the cowslip, by intermediate links), and, as appears to me, they may be safely pronounced the real representatives of Primula elatior. The dubious oxlip, gathered last year at Claygate, and mentioned in your first number (Phytol. 9), has this year flowered in my garden. It there grows in a much drier and a less shaded situation than that in which the wild root was found. In the form of the calyx, corolla, and leaf, it is now decidedly a primrose, although the umbel is elongated on a stout scape of five inches in height. In the deep colour of the corolla and the tint of the leaves, it has more nearly the cowslip hues: the pubescence is intermediate, but nearer that of the primrose. I consider the plant to be an umbelled primrose, but cannot account for the cowslip colours. Preparations for a botanical tour to the Azores (for which I expect to sail in two or three days) have prevented my giving attention to the subject of the oxlips this spring; and sundry experiments bearing upon the question of their relations to the cowslips and primroses that had been commenced will be interfered with by my absence, which I anticipate will continue for the whole summer and autumn. As far as my observations go, there is not any one point in the specific characters ordinarily given for the primrose and cowslip, which is constant in either. Each characteristic of the primrose may be seen in specimens that otherwise would be called cowslips, and *vice versâ.* The least variable, perhaps, are the very short and close pubescence of the cowslip, and the long weak hairs of the primrose. I have

never seen the primrose with obtuse sepals, however, though the cowslip has the sepals acute, obtuse, or quite rounded at the apex. The experiments on which Professor Henslow lays so much stress, are certainly of great value in relation to the distinctions of species, not merely in the genus Primula, but for systematic Botany generally. Still, they may be said to require confirmation; and one confirmation which appears to be requisite, is, that the experiment should be repeated by some botanist who would studiously avoid letting his gardener or any other party know the object of his experiment. I was once told by a gardener, that he had *helped* his master's horticultural experiments, during the absence of the latter, so as to produce the results which he supposed would gratify his master.—*Hewett C. Watson ; Thames Ditton, May* 1, 1841.

156. *On the immersion of Specimens of Plants in Boiling Water*, (Phytol. 189). Having been induced from a report in ' The Phytologist' to try the effect of boiling water in preserving the colours of botanical specimens, I was much disappointed to find that it was, with me, quite ineffectual. Lathræa squamaria, one of the plants mentioned in the report, turned completely black on remaining in boiling water ten seconds; and in one specimen which was but partially immersed, that part only turned black, whilst the remainder has preserved its colour in a slight degree. By the same treatment the colour was extracted from the flowers of Orchis mascula, leaving the petals of a dirty brown hue ; whilst specimens dried by the usual method partially retained their colour, and certainly their *form*, which the *boiled* ones did not. Specimens of other plants which I tried were all acted upon in a similar manner. It certainly appears very strange to me that others should succeed so perfectly, whilst I, using exactly the means prescribed, could not succeed in the least; and I think there must be something more than has yet appeared, either in the water or the subsequent treatment, in order to preserve the colour of such plants as Lathræa squamaria &c.—*Joseph Sidebotham ; 26, York St., Manchester, May* 5, 1842.

157. *True office of the Earth in relation to Plants.* As the article ' On the true office of the earth in relation to plants' (Phytol. 173), seems to have been penned for the purpose of exciting discussion, I am surprized that you have as yet had no communications on the subject. I take the liberty of sending you a few remarks, as I cannot subscribe to the opinion which, in that article, Mr. Newman has endeavoured to maintain. The feeding of plants in order that they may afford food for man, is an important subject at this time; and if I understand Mr. Newman rightly, he maintains that the earth or soil in which they grow has nothing to do with the supplying this food. He makes this inference from the fact that hyacinths grow in water; and to this instance he might have added those of floating water-plants, of most of the Orchideæ, and a number of mosses, lichens, &c., which evidently derive their nutriment from sources independent of those constituents of the earth or soil in which they do *not* grow. But this does not at all prove Mr. Newman's position, that those plants which grow in the earth do not derive their sustenance from the soil in which they are placed. The fact is that plants, like animals, require different kinds of food, and they are always naturally placed in those positions in which they are best supplied with their peculiar food. All plants, it may perhaps be stated, require for their growth water and carbonic acid, and they obtain these from the soil or the atmosphere, according to their structure, which is adapted to the peculiar localities in which they live. The plants of the deep sea, and most of the lower forms of Cryptogamia, obtain these agents by their whole surface; but in the higher forms of Phanerogamous vegetation, the function of absorption is exceedingly localized, and these plants seem to be almost con-

X

fined to the extremities of their roots — the spongioles, as a means of obtaining food. These spongioles are placed in the soil, and from the soil, and from no other source, do they derive their water and carbonic acid. It is this that makes carbonaceous soils so valuable when any agent is added to them that will facilitate the union of their carbon with oxygen, and thus supply to plants an abundance of carbonic acid. On the necessity of water as a food for plants I need not dwell; and that this alone enters plants from the soil is proved by the flourishing vegetation of a swamp during a drought, compared with the withered aspect of the same on hills and well-drained fields. But water and carbonic acid are by no means the only food of plants; there are other matters which they derive from the soil, and which they cannot get from any other source. The various saline and earthy constituents of plants are derived from the soil; and unless these are supplied the plant perishes. The nature and proportion of these vary very considerably in different families, but in most cases they are essential, if not to the existence, at least to the health and productiveness of the plant. Of these substances the phosphates, nitrates and carbonates of lime, potassa and soda, silica and ammonia, may be given as examples. It is a knowledge of this fact that is now giving such an impetus to the enquiry concerning the manuring of plants, and which, far from leading to the conclusion that the composition of the soil is of little importance, attaches to it the utmost value. For this purpose the vegetable physiologist has called in the aid of the chemist; and Dr. Daubeny, at a late meeting of the Agricultural Society, presented a plan for keeping a debtor and creditor account between the soil and plants that grew on it, seeing that the latter took away that which the former possessed. It is a knowledge of this fact that gives the true theory of the rotation of crops, the necessity for which does not arise from the excretions of a plant being poisonous to itself and not to another, but from the fact that plants abstract from the soil the whole of an ingredient that, as food, is necessary for their health; this is not supplied till after the next manuring. That the earth supplies ingredients necessary to the existence of plants, is also proved by their distribution on the surface of the globe, independent of height, of heat and light, which are so important; some plants grow on one stratum and some on another; and many plants are known to geologists as determining the existence of particular rocks, whose particles are mingled with the soil. I cannot therefore admit that the earth is "simply a receptacle for roots," for the very constituents of which the earth is composed, are constantly undergoing decomposition and entering into the structure of the plant; and had not the earth naturally or by artificial means a peculiar constitution, the plants which grow on it could not exist.— *Edwin Lankester ; 43, Hart Street, Bloomsbury, May 6, 1842.*

158. *Note on Sagina apetala and maritima.* In the numerous examples of Sagina apetala which I have witnessed, I have never failed to detect rudimentary petals, (see ' British Flora '). This obtains also in the maritime variety found in Anglesea, near Beaumaris. S. maritima is strictly apetalous. The latter species I have never seen growing at Warrington: the nearest habitat known to me is Runcorn Gap, on the Mersey.—*W. Wilson ; Warrington, May 6, 1842.*

159. *Carex tenella*, (Phytol. 128). A word on this subject. I have no doubt that the figure of Schkuhr has been seen by the author of ' British Flora,' and I think there is good evidence of its having been consulted at the very time when that part of. the ' British Flora ' was written. The mistake, if any has been made, may even have been caused by relying too implicitly on Schkuhr's figure of the ripe fruit. This, in Sir J. E. Smith's opinion, has been taken " from a starved specimen of *C. loliacea*," which

has "fruit *flat on one side*," ('Eng. Flor.' 83). Mr. Gibson appears to have misconceived the meaning of the question which he so freely criticises; the doubt evidently refers to the Scottish specimen, and not to the *species* called C. tenella.—*Id.*

160. *Isoetes lacustris.* Had I known of Mr. Newman's intention to describe the fructification, I would have sent my own recorded observations earlier. In some of the fertile capsules, only one central columnar bar from back to front was visible, but in other cases, and especially in those which seemed to contain male organs, there were ten such bars, ranged on each side of the thickened central line which runs down the back. No good evidence appeared that these bars were receptacles for the seeds. On the inner side of the frond, immediately above the capsule, a rounded membranous scale is observable, having a depression at its base. It has an evident communication with the capsule. In one instance two scales were seen, one placed above the other. The so-called anthers were alike furnished with scales. In the "normal form" from Llyn y Cwn, I observed that the seeds were much more numerous and smoother than in the slender variety from Ffynnon Frech, the capsules were also larger, with twelve bars or pillars from back to front. In the capsule of the slender variety the seeds were from thirty to forty in number, with winged sutures; not the least trace of any pedicel could be found. The above remarks seemed to me less likely to be accurate than those of Mr. Valentine, which, founded as they are on subsequent observation and most diligent scrutiny, must be regarded as worthy of the highest credit. One conclusion drawn by him was, that no essential difference existed between the fertile and the so-called male fructification.—*Id.*

161. *Myrica Gale with androgynous flowers.* I enclose a few specimens of Myrica Gale with *androgynous flowers*, that is, with the flowers united in the same glume, not simply monœcious. In the same catkin you will find the lower portion principally occupied with barren flowers, the upper portion with fertile flowers, and the intermediate portion with the flowers united. The whole bush, whence the specimens were obtained, had catkins of this character; it still grows on the borders of Risley Moss, near this place, and can easily be found again.—*Id.*

162. *Notes on Monotropa.* It may not be amiss to make a few more observations on Monotropa. The description of the root of M. Hypopitys in 'English Botany,' t. 68, is applicable enough to Mr. Lees' plant, but not to that which grows at Southport, and which seems also to differ from the true M. Hypopitys in its drooping flowers. With respect to the scent of the true species, it would seem, according to Smith's observation, that it becomes evident only when the plant has "arrived at maturity, and then acquiring a fragrant smell, generally compared to primrose roots, but rather resembling those flowers." My Southport specimens, therefore, though they should not prove to belong to a distinct species, are not necessarily opposed to Mr. Lees' perceptions, for they were gathered early in the season. Perhaps I have erred in not sooner replying to Mr. Lees' observations, (Phytol. 171). My silence has been entirely owing to an aversion from anything which has a tendency to defeat the object for which 'The Phytologist' is intended, as a vehicle of information. The perfect good humour with which Mr. Lees has received my former comments, assures me that I have given no offence, and surely none was intended. No one can be more sensible than myself of the great pains he must have taken in the disinterment of the roots of Monotropa; I only intended to say that closer investigation is requisite to determine the parasitism of the plant, than he has yet given to the subject. As to the scent of his specimens, I have as little doubt that I should have agreed with him, as I have of the accuracy of my

own perceptions with respect to the Southport plant. All my remarks were written purely with a wish to excite enquiry into facts not yet fully established by evidence. My friend Dr. J. B. Wood last year directed my attention to some very singular remarks on this plant by Monsieur Jaume St. Hilaire, in the 'Plantes de la France,' iii. (1809). At that period he says the plant was so little understood, that it was difficult to ascertain "whether it was a distinct species or a monstrosity of another plant," (si c'est une espèce distinct, ou une monstruosité d'une autre plante) ; and that it was always found "adhering and parasitical upon the roots of elms &c., but never [afterwards] in the same spot." So that here we have a plant at once migratory and parasitical, properties which would be very astonishing, thus united, in any but a plant of such mysterious origin.—*Id.*

163. *Note on the second British species of Monotropa.* The smooth-petalled form of Monotropa Hypopitys noticed by Mr. Gibson (Phytol. 201) will probably be found in other localities now that attention is directed to it. I have a specimen in my possession from Cholsey, Berks, gathered by Mr. Baber, and distributed by the Botanical Society of London in 1839, which differs from a Reigate specimen of the more usual form, sent to me by the same Society, and agrees with the characters given by Mr. Gibson. The plant or variety has, however, been recognized before by various foreign botanists, and appears to be the same as the Hypopitys hypophegea of G. Don, 'General System of Gardening,' iii. 866. It is the H. glabra of Decandolle (Prodr. vii. 780), who gives as its habitat the roots of beech-trees in various parts of Germany, and suggests that it may be also found in France and England. A smooth form of Monotropa Hypopitys is also mentioned by Koch, in his 'Syn. Flor. Germ. et Helv.' where it is considered as being merely a variety, between which and the hairy form this author finds many intermediate states.— *Robert J. N. Streeten; Worcester, 9th May,* 1842.

164. *Trifolium incarnatum,* (Phytol, 198). The only localities in which I have met with Trifolium incarnatum are Snelsmore and Greenham Commons, near Newbury, Berkshire ; where, in 1838, it grew on the turf, not far from the road-side, in many parts of the commons, and by an inexperienced botanist, who was not aware of the plant's being cultivated in the neighbourhood, would certainly have been supposed to be wild ; indeed, I confess, that when I first saw its deep red flowers, I hoped that I had found a prize. It grew in a scattered manner, and was always very starved and stunted in its growth.—*Anna Worsley; Brislington, May* 10, 1842.

165. *Potamogeton prælongus.* In the 'Northern Flora' by Dr. Alexander Murray, of Aberdeen, published in 1836, occur the following remarks respecting this plant.— It " is one of the most recent additions to the British Flora. It is said, however, that there are specimens of it in the herbarium of Mr. Brodie of Brodie,* 20 or 30 years old ; and it is certain that the writer of these remarks, though then unable to determine the name, gathered this species in Cromar, Aberdeenshire, several years before it was known to be a native of Britain, and showed the specimen, still in his possession, upon the same day, as something remarkable, to his friend Mr. John Anderson, now editor of a London daily paper." The readers of 'The Phytologist' are of course aware that since the time at which these words were written, various localities, both in Eng-

* Lochlee is on the Brodie property ; but few of the specimens in the Brodie herbarium were gathered within the province of Moray.

land and Scotland, have been discovered for the interesting plant to which they refer. But they are not, perhaps, aware by whose means or in what manner it was first made known that Potamogeton prælongus had been found in this country. In the summer of 1832, I found this plant first in the moss of Litie, and soon afterwards in Lochlee, both stations within the county of Nairn. After examining it over and over, and comparing it with the descriptions given in Hooker's ' British Flora,' I could not identify it with any species there described. I at last showed a specimen to my friend, Mr. William A. Stables, of Cawdor Castle, who had not previously seen it, and could not name it. Ever more ready to promote a friend's fame than his own, Mr. Stables in 1833, sent some specimens to Dr. Walker Arnott, who pronounced the plant to be Potamogeton prælongus (adding several synonymes), and " new to the British Flora." It is highly probable that it had been, in several instances, mistaken for another species before the time to which I have alluded, but certainly it was not known by any distinct name as a native of Britain until, through the communication of Mr. Stables, it was named by Dr. Arnott. I forget whether it was before or after he knew its name that Mr. S. found the plant growing in Lochindorb. With reference to what I have now written, I may just remark, that I have no wish to depreciate the botanical services of others in order to extol my own, or to take away from the merit — if merit there be— of the mere discovery of a plant previously unknown. But I think that those who do actually make discoveries, knowing or suspecting them to be so, certainly do very little for the cause of botanical science by neglecting to communicate them. And in my honest opinion my excellent friend already named has, with respect to the plant in question, more merit than any who have had a hand in adding it to the British Flora. The description of P. prælongus is now so well known, and has been so recently inserted in your pages, (Phytol. 28), that it need not be here repeated. There is, however, one peculiarity in the leaves, which I think has not been quite correctly described by our British botanists. By some they are said to be " obtuse," by others " hooded " at the point ; and by others, I think, they are not, in that particular, described at all. I would describe them as terminating in what resembles the *bow* of a boat, and I think the " foliis apice navicularibus " of the continental botanists — for which expression I am also indebted to Mr. Stables—forms the best possible description of them. When dried and flattened they are of course split at the point. The lamented author of ' The Northern Flora,' who did live to finish what would have formed the most interesting of all our local Floras, assigns, on the authority of Francis Adams, Esq., Surgeon, Banchory, the *feminine* gender to the word Potamogeton. In an appendix to Dr. Murray's work, Mr. Adams has furnished ' Notes from the Ancients on certain indigenous species ; ' and among other things remarks—" Modern botanists have fallen into strange mistakes about the gender of this word. Thus Sprengel, in his ' History of Botany,' makes it masculine ; and Hooker, in his ' Flora Scotica,' makes it neuter Now it so happens that the word is unquestionably feminine in Latin, as is proved from the following passage in the N. H. of Pliny:—"Potamogeton adversatur et crocodilis : itaque secum habent *eam* qui venantur. Castor *hanc* aliter noverat &c." Dr. Murray adds in a note, that " general principles, as well as the authority of Pliny, may be said to be in favour of Potamogeton being a feminine word."— *J. B. Brichan ; Manse of Banchory, by Abberdeen, May* 16, 1842.

166. *Enquiry respecting Pyrola media.* As it is very important that published lists of plants should be correct, and I feel especially desirous that 'The Phytologist' should become an authority on this point, allow me, through your medium, to ask Mr. Buck-

ley if he is quite sure he gathered at Lytham in July last, Pyrola media, which he includes in his list of Lytham plants, (Phytol. 165). I ask this because whilst I gathered Pyrola rotundifolia in that locality in July, 1834, and have since received it from the same place, I have never seen, nor before heard of, Pyrola media growing there. Pyrola rotundifolia is also abundant at Southport, on the opposite shore of the Ribble, in similar situations.—*Samuel Simpson ; Lancaster, May* 17, 1842.

167. *Chrysosplenium alternifolium.* I may mention that Chrysosplenium alternifolium grows along the banks of a narrow rivulet in the immediate neighbourhood of this town, in very great abundance and luxuriance, forming in some parts large patches, and entirely eclipsing its more humble, and there, less abundant sister — Chrysosplenium oppositifolium.—*Id.*

168. *Note on the Oxlips from Bardfield.* I have, by the kindness of Mr. H. Doubleday, been furnished with specimens of the oxlip (Primula elatior) from Bardfield, (Phytol. 204), which I believe to be quite distinct from the plant usually called by that name. The leaf is very differently formed, the tube of the corolla much longer, the flowers *always drooping*, and the general appearance of the plant is altogether different.—*Joseph Sidebotham ; 26, York St., Manchester, May* 20, 1842.

169. *The valuable Botanical Museum of the late Aylmer Bourke Lambert, Esq.* is advertized for sale by Mr. S. Leigh Sotheby. This collection has been in course of formation for more than half a century. It comprises about one hundred separate and distinct herbaria; the largest carpological collection perhaps ever made by a private botanist, the fruits are dry or preserved in spirits and acids; and a collection of woods and sections of barks, &c. The sale will take place at Mr. Lambert's late residence, 26, lower Grosvenor St.; it will commence on the 27th of June, and will continue for three days.

170. *Erratum.* Phytol. 194, under Bonnemaisonia asparagoides and Polysiphonia cristata, for *Mr. Carnow* read *Mr. Curnow.*

ART. LXX.—*Proceedings of Societies.*

LINNEAN SOCIETY.

April 19, 1842.— Edward Forster, Esq., V.P., in the chair. A bequest of £100. from the late Archibald Menzies, Esq., was announced. Joseph Janson, Esq., exhibited specimens of *Primula scotica,* gathered at Wick, near Caithness.

May 4.—The Bishop of Norwich, President, in the chair. The Rev. C. A. Johns exhibited a living specimen of *Jungermannia reptans,* in fruit; as well as dried specimens of many other species of the same family. The Duke of Northumberland sent for exhibition the ripe fruit, and a female plant in flower, of *Diospyros edulis,* which had grown in His Grace's conservatory at Sion. Read, the continuation of Dr. Hamilton's Commentary on the Hortus malabaricus.

BOTANICAL SOCIETY OF EDINBURGH.

Thursday, May 12, 1842.—Professor Christison in the chair. Miss Jane Farquharson was elected a life member of the Society. Donations to the library and museum were announced from the President, Rev. Mr. Hincks, Mr. Ward, Mr. Isaac Brown, Mr. Joseph Dickson, Mr. Sowerby, Mr. Stricker and Mr. Watson.

The following communications were read :—

1. *On Fumaria parviflora, as a native of England :* by Mr. C. C. Babington, M.A. F.L.S., &c., Cambridge. Mr. Babington, in reference to an opinion formerly expressed by him, that this species was a very doubtful native of England, not having then seen any specimens agreeing with the true characters of it, now states that he has obtained satisfactory proofs of its being a native, but that most botanists have been in the habit of calling *F. Vaillantii* by that name. He says, however, that the flowers of English specimens of *F. Vaillantii* are decidedly smaller than those of some which he possesses from Montpelier, and that in some white-flowered English specimens of the same plant, he perceives traces of an apiculus ;—also that in French specimens of *F. parviflora* the flowers are of the same size as those of *F. Vaillantii,* but the fruit has an apiculus. Mr. Babington then proceeds to give a minute description of the principal characters which distinguish this and other allied species of the genus, and among which there has hitherto been much confusion.

2. *On the occurrence of Gelidium rostratum,* Harv., *at Aberdeen :* by Mr. George Dickie, Lecturer on Botany, Aberdeen.* This remarkable plant, which Mr. Turner was disposed to consider, though with some hesitation, as merely a variety of *Delesseria alata,* but which Dr. Arnott and Mrs. Griffiths refer to *Gelidium,* Mr. Dickie states to be abundant at Aberdeen, though it has not hitherto been found *in situ.* It occurs on the large stems of *Laminaria digitata,* and appears to be an inhabitant of deep water — being only found cast up after storms. Mr. Dickie says, — " after comparing numerous fresh specimens of *G. rostratum* and *D. alata,* I feel convinced that there is no essential difference in the structure and outward form of the fruit in these plants. In both the ternate granules are terminal and axillary, and the capsules occupy the same position. The seeds, however, differ in form ; those of *D. alata* are mostly oval, in the other they are spherical."

3. *On some anomalies in form in Scolopendrium vulgare :* by Mr. Joseph Dickson. The fronds exhibited by Mr. Dickson presented every possible variety of shape, from lanceolate to reniform, and from entire to lobed or rather digitate. The more usual form is certainly entire and *oblongo-lanceolate,* and it is difficult to account for the freaks of form which not unfrequently occur in this species of fern.

After these papers were read Professor Graham exhibited some very beautiful and interesting specimens of exotics from his own green-house. — *The Edinburgh Evening Post and Scottish Standard, Saturday, May 21,* 1842.

———

BOTANICAL SOCIETY OF LONDON.

April 18, 1842. — Dr. W. H. Willsbire in the chair. Various donations to the library and herbarium were announced and members elected.

Mr. Edward Doubleday exhibited a *Primula* found at Bardfield, Essex ; and stated that some few years ago his brother, Mr. Henry Doubleday, observed that the oxlips growing near Bardfield, in Essex, were strikingly different from those found in the vicinity of Epping, where the oxlip is not common ; and that further observation had induced him to believe that the Bardfield plant was a distinct species, an opinion in

———

* See note by Mrs. Griffiths, Phytol. 203.

which he (Mr. E. D.) was disposed to concur. Mr. Doubleday next referred to an article in the ‘Gardeners’ Chronicle,’ since republished in ‘The Phytologist,’ (Phytol. 204), and pointed out the resemblance of the Bardfield plant to the one there alluded to. He expressed his opinion very decidedly that there were in England three distinct species of *Primula*, known by the names of primrose, cowslip or pagel, and oxlip, but that the oxlip commonly so called is nothing more than a hybrid between the primrose and cowslip. This hybrid is extensively distributed over the country, especially in localities where the primrose and cowslip abound: it constantly exhibits a tendency to revert to the primrose by throwing up single flowers of precisely the primrose character, as well as others possessing characters of its other parent the oxlip.

As a natural consequence such a hybrid would reproduce at times both the parent species, a fact Mr. Doubleday believes to be fully proved.

The Bardfield plant, which Mr. Doubleday considers the true oxlip, differs from the hybrid in the form of the calyx, in its drooping umbel, and in its leaves dying off in autumn: he has examined thousands of plants at and near Bardfield, and never observed a single instance of a solitary flower being thrown up as in the hybrid. The primrose does not occur for some miles round Bardfield, though the cowslip is abundant; therefore hybridization cannot well take place in that locality. The plant under cultivation does not change its character. Should it prove a distinct species Mr. Doubleday claimed for his brother the credit of first detecting the distinction.

May 6.—J. E. Gray, Esq., F.R.S. &c. President, in the chair. The following specimens were exhibited:—*Dicranum spurium*, Hedw., collected in Stockton forest, near York, in March last, by Mr. Spruce, presented by him. *Leskea pulvinata*, Wahl., collected on willows by the Ouse near York, by the same gentleman, and presented by him. *Desmidium Swartzii* and *D. mucosum*, collected near Penzance in December last, by Mr. Ralfs, and presented by him. The following were presented by Mr. Wm. Gourlie, jun.—*Jungermannia stellulifera*, Taylor, collected at Crich, Derbyshire, by Mr. W. Wilson; *Gymnostomum Hornschuchianum*, Arnott, collected at Cromaglown, in July, 1840, and first discovered by Dr. Taylor; *Jungermannia voluta*, Taylor, found at Gortagonee in March, 1841, by Dr. Taylor. Mr. J. G. Lyon presented specimens of *Jungermannia Lyoni*, Taylor, collected at Dunoon, Argyleshire. Mr. T. Sansom exhibited specimens of the following mosses collected by the Rev. C. A. Johns, F.L.S.:—*Bryum Tozeri*, Grev., collected at Swanscombe, Kent;* *Hypnum catenulatum*, Schwæg. from Betsham, Kent; *Tetraphis pellucida*, Hedw., Abbey Wood, Erith, Kent. British plants had been received from Dr. Francis Douglas, Dr. Spencer Thomson, the Rev. W. S. Hore, Mr. W. Wilson, Mr. M. Moggridge, and Mr. Fordham; and donations to the library were announced from Dr. Willshire, Mr. H. O. Stephens and Mr. Adam White.

Dr. Spencer Thomson communicated a paper “ On the Anatomy and Physiology of the Seed of *Phaseolus vulgaris*.” The paper was accompanied with drawings.— G. E. D.

---

* See note by the Rev. Mr. Johns, Phytol. 200.

THE PHYTOLOGIST.

No. XIV. | JULY, MDCCCXLII. | Price 1s.

ART. LXXI.—*Analytical Notice of the ' Transactions of the Linnean Society of London,' vol.* xviii. *pt.* 4. *August,* 1841.

(Concluded from p. 72).

ART. XXXIII.—*A Monograph of the Genus* Disporum. By DAVID DON, Esq., Libr. L.S., Prof. Bot. King's Coll. Lond.

THE name of this genus, unaccompanied by a description, first appeared in Mr. Salisbury's list of Petaloid Monocotyledons, published in the first volume of the Horticultural Society's Transactions. The chief characters of the genus were pointed out by Mr. Brown, "and among others its binary ovula, which doubtless suggested to Salisbury the name of *Disporum.*" No description of the genus appeared until the publication of Mr. Don's 'Prodromus Floræ Nepalensis,' in 1824, where, to Salisbury's single species—Disp. pullum (misprinted *fulvum* in the Prodromus, as well as in the report of the present paper in the 'Proceedings of the Linnean Society,' p. 45), two others are added, namely, Disp. Pitsutum * and Disp. parviflorum.†

" The characters of the genus consist in its campanulate perianthium, with the sepals produced into a short pouch or spur at the base, in the cells of its ovarium bearing two [ascending] ovula, in its baccate pericarpium, and in its umbellate inflorescence. These distinctions will be found to be common to all the Asiatic species hitherto improperly referred by most botanists to Uvularia. * * This genus terminates the series of the Melanthaceæ, forming the transition from that family to the Smilaceæ, the chain of connexion between them being rendered complete by the intervention of a new genus, of which Streptopus lanuginosus is the type."—p. 513.

The normal Melanthaceæ, principally North American plants, have the floral organs " persistent, and the partial decomposition of the trimerous pericarpium is almost universal." From the author's remarks on the three groups into which " the Melanthacea appear naturally to divide themselves," we extract the following characters.

* The Uvularia Pitsutu of Buchanan Hamilton, MSS.; Uv. umbellata, Wallich, 'Asiatic Researches,' xiii. 379, Wal. Catalogue,' No. 5090; Streptopus peduncularis, Smith, Rees' 'Cyclopœdia,' under Uvularia.

† Uvularia parviflora, Wallich, 'Asiatic Researches,' xiii. 379.

1. *Melantheæ* or *Veratreæ*. Carpels partially concrete; pericarp capsular; dehiscence generally septicidal; flowers frequently unisexual; perianth less coloured, persistent; stamens persistent; rhizoma fibrous. British genus, *Tofieldia*.

2. *Colchiceæ*. Perianth more highly developed; sepals with long claws, often combined into a tube; styles long; carpels concrete; pericarp capsular, dehiscence septicidal; rhizoma bulbous; floral axis naked, hypogæous. British genus, *Colchicum*.

3. *Anguillarieæ*. Floral organs frequently deciduous; styles short, as in the first group; carpels completely concrete; pericarp capsular or baccate, dehiscence loculicidal; rhizoma bulbous or fibrous; axis leafy.

" The genus Colchicum establishes an evident relationship through Sternbergia and Crocus between Melanthaceæ, Amaryllideæ, and Irideæ. The present genus connects the family with Smilaceæ, and Tofieldia as clearly with Junceæ, whilst a comparison of the structure of Uvularia and Erythronium fully makes out their affinity with Liliaceæ or Tulipaceæ. * * The class of Monocotyledonous plants offers a beautiful confirmation of the truth of the doctrine of the continuity of the series of organized beings; and however much the universal existence of transition or osculant genera in this class may perplex the botanist who looks to the technical definition of his groups as the highest object of the science, we are not to exclude such genera from our researches merely because their presence renders the circumscription of our pretended natural orders more difficult, for they certainly form the most interesting part of the study of natural affinities."—p. 514.

Eight species, natives of Asia, are fully described in this Monograph; and the paper is concluded by a description of Reichenbach's closely allied genus Kreysigia. The species—Kreys. multiflora, is a native of New Holland, and was discovered by Mr. Allan Cunningham, who introduced it in 1823 to the Royal Botanic Garden, Kew, where it annually flowers and matures its fruit. It was at first supposed to be a species of Schelhammera, but on examination was found to differ essentially both from that genus and Disporum.

" This genus is essentially distinguished from Schelhammera by its sessile biappendiculate sepals; by the stamens proceeding free from the torus, unconnected with the sepals; by the cells of its ovarium bearing only two ovula; by its somewhat baccate pericarpium; and, lastly, by its axillary peduncles, which are furnished with three small verticillate bractes. The presence of appendages, the spreading sepals, free stamens, strophiolate seeds, minute embryo, axillary inflorescence, and valvular fruit remove it equally from Disporum."—p. 523.

The attention of the author was drawn to the appendages at the base of the sepals, resembling those of Parnassia, by Mr. John Smith of Kew; Mr. Don, as well as Endlicher, at first supposed them to be imperfectly developed stamens, but Mr. Brown having pointed out " the intimate connexion of these curious appendages with the sepals, and the entire absence from them of vascularity," the author here corrects the error into which he had previously fallen.

ART. XXXIV.—*A Monograph of* Streptopus, *with the Description of a new Genus now first separated from it.* By DAVID DON, Esq., Libr. L.S., Prof. Bot. King's Coll. Lond.

" The genus Streptopus was first proposed by the elder Richard in Michaux's ' Flora Boreali-Americana,' and was intended to include not only the Uvularia amplexifolia of Linnæus, but two other plants therein described for the first time, namely, S. roseus and lanuginosus. The two last are exclusively confined to North America, while the first is common to Europe and America."—p. 525.

In Streptopus amplexifolius, which Mr. Don considers as the type of the genus, are united the following characters :—a six-leaved campanulate perianth with deciduous sepals, which have " a nectariferous furrow at their base; erect sagittate anthers, with short dilated filaments; three separate stigmata;" and a berry-like pericarp with polyspermous cells. From the genus thus characterized the author has found it necessary to remove Strept. lanuginosus, on which he founds a new genus. The genus Streptopus still comprises three species,— Str. amplexifolius, roseus, and simplex: the last species is from the Himalayas, and was first described by the author in his ' Prodromus Floræ Nepalensis.'

" These plants have all a peculiar habit, cylindrical leafy stems, broad amplexicaul leaves, glaucous beneath, and axillary, solitary, mostly single-flowered peduncles, which in amplexifolius are curiously twisted at their middle. The genus undoubtedly belongs to the Smilaceæ, and is nearly allied to Convallaria and Smilacina, but is essentially distinguished from both by its distinct sepals, each furnished with a nectariferous furrow, separate stigmas, and polyspermous berry. With Uvularia it accords in habit, and in its solitary, axillary, campanulate flowers; but its innate anthers, furnished with short filaments, baccate pericarpium, and noncarunculate seeds, remove it widely from that genus."—p. 526.

Of the new genus, Prosartes, the author observes :—

" This very natural genus, as I have already stated, forms the transition from the Smilaceæ to the Melanthaceæ, and possesses several characters in common with Streptopus and Disporum. From the former genus it is essentially distinguished by its much more lengthened filaments, binary pendulous ovula, and terminal umbellate inflorescence; [and it differs from Disporum] in its innate anthers, nearly concrete styles, and pendulous seeds."—p. 531.

The genus Prosartes includes two species,—Pros. lanuginosa, from North America, the Streptopus lanuginosus of Michaux; and Pros. Menziesii, also from North America, named after the late Mr. Menzies, and now first described.

" In the Smithian Herbarium there is a single specimen of this highly interesting plant gathered by my venerable friend Mr. Menzies on the north-west coast of America in the voyage of discovery under Vancouver, to which he was attached in the capacity of naturalist. It bears a close resemblance to some species of Disporum; and it moreover agrees with that genus in its sepals being produced into a pouch at their base. The flowers, which are also terminal and in pairs, are twice the size of those of the preceding, and the style is copiously hairy."—p. 534.

From their position in the natural system, and the author's remarks on their affinities, it will be evident that the plants described in the papers above noticed bear a close relationship to many interesting British genera, including Trichonema, Convallaria, Ruscus, Paris, Tofieldia, Narthecium, &c.

Art. XXXV.—*On some new Brazilian Plants allied to the Natural Order* Burmanniaceæ. By John Miers, Esq., F.L.S.

Von Martius, in his ' Nova Genera et Species Plantarum Brasiliensium,' gives the characters of the genus Burmannia, and fully describes five species, discovered by him in the interior provinces of Brazil. Michaux gave the generic name of Tripterella to two North American species, which Mr. Miers appears to consider as not distinct from Burmannia. Seven other species have also been found in Africa, India, and New Holland. Previously to his departure from Brazil the author discovered five new plants, closely allied to Burmannia, but differing from that genus in many important particulars. These five species, together with another discovered by Mr. Schomburgk in British Guinea, and Nuttall's Apteria setacea, are divided into three genera, namely, Dictyostega, containing four species, Cymbocarpa with one, and Apteria, *Nuttall,* (formerly Stemoptera, *Miers*) with two species. All these genera and species are fully described in the present paper, and their characters minutely illustrated by figures.

The author observes that the Burmanniaceæ may be divided into two groups. The first will contain the genera Burmannia and Gonyanthes, having a trilocular ovarium and central placentation; the second, possessing a unilocular ovarium and parietal placentation, will include Dictyostega, Cymbocarpa, Apteria and Gymnosiphon. If the principle be adopted " on which Apostasieæ have been separated from Orchideæ and Xyrideæ from Restiaceæ," the two sections must be kept distinct, and the author suggests that the second would, in that case, form a separate family, under the name of Apteriaceæ; if however the difference in the structure of their ovaria be not thought sufficient to warrant their separation, they must remain associated as Burmanniaaceæ, the first section being named Burmannieæ, the second Apterieæ. The author mentions Gentianeæ as an order presenting many similar instances of transition from a unilocular capsule and parietal placentation to a bilocular fruit and central placentation.

The author next details some very striking points of resemblance between these plants and many Orchideæ; particularly in their seeds and the structure and texture of the pericarp, as well as in their stem and imperfectly developed leaves: and he observes that but for the differences in the stamens and stigmata, "it would be difficult to draw a line of distinction between the structure of these plants and that of Orchideæ."

" Another analogous fact is deserving of notice: on examining the stigma of Dictyostega after flowering, it will be found to be crowded with bundles of white cottony filaments, which may be seen even with a common lens to consist of pollen-tubes issuing in a body from the cells of the anthers and penetrating the stigma, leaving their ends exserted, and clavately terminated by their respective grains, thus displaying in a very beautiful manner the singular mode of fecundation so ably illustrated by Mr. Brown in his admirable paper on that subject, published in the 16th volume of the Transactions of this Society. The pollen also in its texture presents great resemblance to that of the Orchideæ, its component granules cohering in like manner into a solid waxy mass previous to the dehiscence of the anthers."—p. 551.

ART. XXXVI. — *Some Account of the* Curata, *a Grass of the Tribe of* Bambuseæ, *of the Culm of which the Indians of Guiana prepare their Sarbacans or Blow-pipes.* By ROBERT H. SCHOMBURGK, Esq. *Communicated by the Secretary.*

·DURING his first expedition in Guiana M. Schomburgk discovered the plant from which the Indians prepare their deadly arrow-poison, (Phytol. 47). This discovery rendered our enterprizing traveller the more anxious to identify the plant from which are obtained the reeds used in the manufacture of the Indian blowpipes. Nearly forty years had elapsed since Baron Humboldt saw a canoe nearly filled with them, and was led to ask the question—" What is the Monocotyledonous plant that furnishes these admirable reeds?" and during that period botanists had received no further information relative to the plant or its place of growth.

" No wonder, therefore," says M. Schomburgk, " that next to the plant which furnishes the active principle of the famous Urari or Wurali poison, the discovery of the reed by means of which the Indian is enabled to send his poisoned arrow with so much precision into his intended victim, should have been a point of the greatest interest to me.

" But in answer to all my questions to the Indians as to the locality from whence they procured the reeds that play such an important part in the construction of the blowpipe, they merely pointed to the west, and gave me to understand that it was far away. The value which the Indians of Guiana set upon these reeds, and the uncertainty from whence they came, increased their interest; and one of my first questions on arriving at a settlement of Indians which I had not previously visited, was, whether they knew from whence were obtained these reeds, so different in structure from all known Bambuseæ. I ascertained at last that the Macusis received them from the Arecunas, but that they did not grow in the country of that tribe; on the contrary, the Arecunas undertook journeys of several months duration to procure them from another tribe, who lived still further westward."—p. 557.

M. Schomburgk, in his third expedition, visited the Arecunas, and from them he ascertained "that the plant which produced the reeds grew in the country of the Guinau and Maiongcong Indians, near the head-waters of the Oronoco."

"We saw among the Arecunas a large number of these reeds, which they were manufacturing into blowpipes. The reed being so valuable, and so liable to destruction if carried openly through the woods, the Indian puts it for protection into the slender trunk of a palm (a species of Kunthia?), which he simply hollows out for the purpose. Being aware that the tube thus manufactured is in constant demand by the other tribes, he does not leave the regions which he inhabits to offer his ware for sale, but patiently awaits the visits of the Macusi, skilled in manufacturing the Urari poison, who brings him that deadly preparation, and exchanges it against these reeds or the ready-finished blowpipe. By this mutual exchange, they are each rendered masters of life and death over the feathered game; for, armed with his blowpipe, the wily huntsman gradually steals nearer and nearer to his victim, and launches his weapon of death, which seldom fails of its deadly aim, before the unconscious bird is even aware of the approaching danger.

"The great object of my last expedition led me to that far west. We camped on the 26th of January near the river Emakuni, at a settlement inhabited by Maiongcong Indians; and the first object which struck me on entering the miserable hut which served as a dwelling to the Indians, was a large bundle of these reeds, some of which were sixteen feet long; a circumstance which naturally induced the inquiry, from whence they came. The houses being built on elevated ground, we had an extensive view before us: at the distance of twenty miles we observed a large chain of mountains, which trended N.N.E. and S.S.W.; and among this chain a high mountain was pointed out to us, which they called Mashiatti, and where we were told that these reeds were growing; but as we were given to understand that we should find them likewise at Marawacca, and as Mashiatta was entirely out of our road, we did not visit it. It was consequently only in the middle of February, and after we had crossed the river Parima, that my wish of becoming acquainted with that curious plant was accomplished.

"The Maiongcong and Guinau Indians, whom the Spaniards call Maquiritares, conducted us to that part of Marawacca (a high mountain which terminates in an almost perpendicular wall of sandstone) where the plant grows. It is a day's journey from a Maiongcong settlement on the river Cuyaca, from whence the hospitable and good-natured savages showed us the beaten track. After having ascended Mount Marawacca, to about 3500 feet above the Indian village, the traveller follows a small mountain-stream, on the banks of which the *Curas* or *Curatas*, as the Indians call these reeds, grow in dense tufts. They form generally clusters of from fifty to one hundred, which are pushed forth, as in many other species of that tribe, by a strong, jointed, subterranean rootstock. The stem rises straight from the rhizoma, without a knot, and of equal thickness, frequently to a height of sixteen feet, where the first dis-sepiment stretches across the inside, and the first branchlets are formed. The articulations then continue at regular intervals of about fifteen or eighteen inches to a further height of from forty to fifty feet. The full-grown stem is at the base an inch and a half in diameter, or nearly five inches in circumference. It is of a bright green, perfectly smooth, and hollow inside."—p. 557.

The long jointless stem Mr. Schomburgk considers to be the growth of a very short period; it is surmounted by a head of numerous verticillate, slender, leafy, jointed branches, three or four feet long, which spring from the nodes of the articulated portion of the stem; the whole being terminated by the inflorescence.

"The whole stem is from fifty to sixty feet high; but the weight of the numerous branchlets forces the slender stem to droop, and the upper part describes an arch, which adds greatly to its graceful appearance."—p. 560.

The Curata (its native name) grows in a rich soil and shady situation, about 6000 feet above the level of the sea. It appears to be restricted in its range to the chain of sandstone mountains extending between the second and fourth parallel, and forming the separation of waters between the rivers Parima, Merewari, Ventuari, Orinoco and Negro: M. Schomburgk determined only three localities, — Mounts Mashiatti, Marawacca and Wanaya.

"It is a remarkable circumstance, that the plant which furnishes the chief ingredient for the preparation of the Urari poison is likewise peculiar to a few mountainous tracts; consequently the tribes who inhabit the regions where these plants grow, and who are acquainted with the mode of their preparation, acquire a general importance."—p. 560.

The Indians who inhabit the district where these reeds grow, are called Curata-people by the other tribes, a circumstance indicative of the rarity of the plant. The Indians of the Rio Negro and the Amazon, who have no intercourse with the Curata-people, manufacture their blow-pipes out of a slender palm, the stem of which is hollowed out, either by being steeped in water for some days, when the internal substance is pushed out with a stick, or else the stem is split along its length, and the interior is removed by burning; when the inside has been polished, the two parts are accurately joined together by an indigenous glue, and a wooden mouth-piece is added.

The constant demand for a plant having so limited a range, would be likely soon to exhaust the stock, were it not for the numerous shoots from a single rootstock and their rapid growth, combined with the great care taken of his blowpipe by the Indian. Carrying it erect he winds his way through thickets which would be almost impenetrable to the unincumbered European, without injuring his weapon. It is said by Humboldt that "a hunter preserves the same sarbacan during his whole life," and boasts of its precision and lightness as we do of the good qualities of our fire-arms.

The young reeds only are used in the manufacture of the blowpipe; these are cut into the proper lengths, turned slowly over a moderate coal fire to prevent their warping, exposed to the sun until they have acquired a deep yellow colour, and are then encased for protection in the trunk of a slender palm. "This case is called by the Macusi Indians Yúrúa-Cura-pong."

Mr. J. J. Bennett has determined this reed to be a distinct species of Arundinaria, near to the Arund. verticillata of Nees von Esenbeck and Kunth; he has named it *Schomburgkii*, and gives the following characters.

Arundinaria Schomburgkii. Leaves linear, acuminate, smooth; mouth of the sheaths bristly on each side: spike simple, few-flowered; spikelets sessile; hypogynous scales lanceolate, acute.

ART. XXXVII. — *On* Cuscuta epilinum *and* halophyta. By CHARLES C. BABINGTON, Esq., M.A., F.L.S., F.G.S., &c.

IN a paper 'On the Structure of Cuscuta europæa,' (Linn. Trans. xviii. 213), Mr. Babington confirms the accuracy of Mr. Brown's observations, on the existence of scales in the tube of the corolla of that species, in opposition to Sir J. E. Smith's opinion, that its flowers are, "in all the British specimens, as well as in Ehrhart's German ones, destitute of scales in the throat of the tube," ('English Flora,' ii. 25). Mr. Babington, on examining fresh specimens of Cuscuta europæa from Sompting, in Sussex, gathered in company with Mr. Borrer, found the scales lying quite close to the corolla, being perfectly transparent and very minute; and these circumstances would seem to account for their having been overlooked by Smith and Hooker, as well as by some of the continental botanists. Mr. Babington remarks :—

"They are, indeed, so difficult of detection as not to have been at first noticed by Mr. Borrer and myself, even when examining fresh specimens, and it is scarcely possible to discover them in flowers that have been dried."—p. 213.

Reichenbach, in his 'Icones Plantarum,' pl. v. fig. 690, represents "each scale exactly under its corresponding stamen,* yet at p. 62 of the same volume he calls the corolla a calyx, and appears to have

---

* See our fig. 2, which, with the other figures are copied from the Linn. Trans.

looked upon the scales as constituting the true corolla, and as alternating with the stamens : —

" This view is manifestly incorrect, for the scales constitute a complete internal whorl, each of them being connected with its neighbour so as to form a short tube, the upper edge of which is always free and distinct from the corolla (calyx of Reich.), and the lower parts of the filaments of the stamens may be traced under the cuticle of the corolla, descending exactly behind the centre of each scale. It is perfectly clear, therefore, that the scales cannot represent petals, since the whorl of stamens is invariably found within that of petals, but in this plant the stamens are situated further from the axis of the flower than the so-called corolla.

" I do not attempt to form any theory concerning these minute organs, but hope that some fortunate botanist will soon discover them in such a state of monstrous development as to show what is their real nature.

" That the number of scales is equal to that of the segments of the corolla is proved by their structure in C. epithymum, in which plant they are not even divided into two lobes, [fig. 3]. There is not, indeed, the slightest trace of a division to be discovered with a very high power of the microscope. I ought to add, that Reichenbach does not continue the above theory in his ' Flora Excursoria;' but reverts to the old nomenclature."—214.

Referring to Reichenbach's figure of the opened flower of Cuscuta europæa (Ic. Pl. v. f. 690, B., our fig. 2) Mr. Babington makes the following observations, which we particularly recommend to the notice of our readers.

" It will be seen that this last differs materially from my fig. 1. May not his plant be a distinct species characterized by its constantly 4-cleft corolla and palmate sexfid scales? This genus is well deserving of attention from British botanists, for several other species are known in Germany, Sweden, and France, which most probably exist in these islands."—p. 215.

Since the publication of the paper from which the above extracts are made, Mr. Babington has examined Cuscuta epilinum and halophyta, and in both these species he has detected the presence of scales. His observations are contained in the paper the title of which is given above.

" In the first of these plants, Cuscuta epilinum, *Weihe*, we find a ventricose tube furnished with a whorl of adpressed bifid scales, each branch of which is usually divided in a rather irregular manner into two or three fingerlike points, as I have endeavoured roughly to represent in fig. 1, [our fig. 4] ; the divisions of the corolla terminate in acute points, and the stamens have very short filaments and are inserted much higher up than the extremity of the scales.

" In Reichenbach's figure of this plant in his ' Icones Plant.' tab. 693, the scales are very incorrectly given, each of them being there represented as two minute, separate, roundish bodies, pointing downwards. Specimens received from him (No. 19 of his Fl. Germ. exsic.), gathered near Borna, in the neighbourhood of Chemnitz, by M.

Weicker, have however these parts of exactly the form described above, and agree in all points with the English plant, with the exception of the want of a bractea under each bunch of flowers. It is however possible, from the manner in which this bractea is hidden by the flowers in the English plant, that it may also exist in that found in Germany, although the employment of its absence as a part of the specific character, is strongly opposed to this supposition."—p. 563.

The next species, Cuscuta halophyta, has not, we believe, been detected in this country. It was recently discovered on the southern coast of Norway, growing on succulent saline plants; and was first described by Fries, in his 'Novitiarum Fl. Suec. Mantissa prima,' p. 8. Fries does not mention the scales; but Mr. Babington has found them in a specimen gathered by Dr. Blytt "on the coast of the Fiörd, near Christiana," who gave it to Mr. Bowman, from whom Mr. Babington received it.

The following illustrations exhibit the corolla of each species laid open, in order to show the form of the scales, and their position with respect to the stamina; they are copied from Mr. Babington's figures in the two papers under notice: from which source are also derived the characters of the four species, and the description of Cuscuta epilinum.

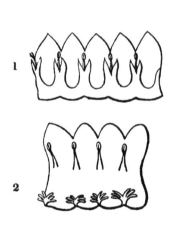
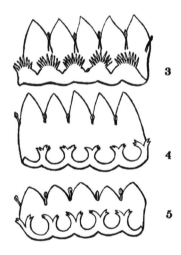

1. *Cuscuta europæa*, Linn. (Sp. Pl. 180). "Clusters of flowers bracteated," sessile: scales bifid, erect, adpressed to the tube of the corolla; tube cylindrical when in flower, ventricose in fruit: calyx much shorter than the corolla. (Fig. 1).

Reichenbach's figure of the opened corolla is shown at fig. 2.

2. *Cuscuta epithymum*, Sm. (Eng. Bot. p. 378). "Clusters of flowers bracteated," sessile; scales palmately cut, connivent; tube of the corolla cylindrical, limb *cam*panulate: calyx much shorter than the corolla. (Fig. 3.)

3. *Cuscuta epilinum*, Weihe, (in Boenningh. Prod. Fl. Monast. 75). Clusters of flowers bracteated, sessile ; scales palmately somewhat six-cleft, adpressed to the tube of the corolla ; tube always ventricose ; segments of the calyx fleshy, del-toid at the base, scarcely shorter than the corolla. *Segments of the corolla acute.* (Fig. 4).

   " Segments of the calyx 5, ovate, attenuated above into an acute point, very fleshy, with peculiarly large cells ; some of them often so much thickened as to become del-toid. Tube of the corolla 1½ times as long as the limb, slightly inflated, the lobes tri-angular, acute ; stamens inserted very near to the summit of the tube ; filaments short ; anthers cordate ; the limb of the corolla is often very fleshy. Scales bifid, each lobe either entire or 2- or 3-fid, short. Styles 2, short, bent round each other. Bractes not always present, broadly ovate, obtuse, with a minute point, often purplish. Flow-ers whitish yellow, sometimes tinged with pink. Anthers bright yellow."—p. 565.

Found on flax in many parts of Britain, but most probably not in-digenous. Mr. Babington observes that flax raised from American, and, he believes, Riga seed, are free from this parasite ; but that it is " introduced with flax-seed from Odessa, and other ports of Southern Russia."

4. *Cuscuta halophyta*, Fries, (Nov. Fl. Suec. Mantis. p. 8). " Clusters of flowers somewhat bracteated," sessile ; scales bifid, the segments also bifid, adpressed to the ventricose tube of the corolla : calyx much shorter than the tube of the co-rolla.—*Segments of the corolla ovate, obtuse. Segments of the calyx obtuse. Styles* 2. *" I have seen most of the clusters of flowers bracteated,"* Fries. (Fig. 5).

A Norwegian species, not yet detected in Britain.

ART. XXXVIII.—*On the Reproductive Organs of* Equisetum. *By* Mr. Jo-seph Henderson. *Communicated by the* Rev. M. J. Berkeley, M.A. F.L.S.

The results of Mr. Henderson's observations on the reproductive organs of this genus, differ, in some respects, from those arrived at by Treviranus, Meyen, Bischoff and Mohl, by whom, without his being aware of the fact at the time he wrote, he has been in part anticipated. Mr. Henderson well remarks that : —

   " There is no part of the structure of Equisetum more curious or more anomalous than the organs of reproduction ; and although the position of the order in the natu-ral system depends on the nature of these organs, yet this is so far matter of doubt, that very eminent botanists do not seem decided as to whether Equisetaceæ are to rank among Phænogamic or Cryptogamic plants."—p. 567.

Without however entering into this question here, we must content ourselves with giving, by way of introduction, a brief description of the fructification of the genus, which will render the subsequent ana-lysis more clear and satisfactory.

The reproductive organs of Equisetum are borne in terminal cone-like catkins, which are composed of a number of angular peltate scales, spirally arranged round a central common stalk or axis, to which each is attached by a short pedicel. Seated round the margin of each of these scales, on its inner or under surface, are from four to eight oblong membranous cells or thecæ, opening inwardly and longitudinally, and discharging numerous somewhat globose sporules, to each of which are attached, at one point, four spiral filaments; the filaments are dilated at the extremity, and they are generally more or less studded with small granules. Hedwig and other observers have described these filaments as stamens, looking upon their dilated extremities as the anthers, and supposing the minute granules to be the pollen. The filaments are at first wound spirally round the sporules, but when discharged from the theca suddenly unroll themselves, and cause the sporules to leap about as if alive. The curious motions of the sporules are owing to the structure of the filaments, which are excellent hygrometers, being twisted like the awns of Avena fatua or animated oat, and many other grasses, and are influenced in the same way by the varying degrees of moisture in the atmosphere.

Mr. Henderson's observations were made on Equisetum hyemale; and he describes the changes which take place in the various parts of the fructification, from the first appearance of the catkin above the terminal sheath of the stem up to the discharge of the mature sporule from the theca.

" When the spike or fructification of Equisetum hyemale begins to swell beyond the terminal sheath, the spores may be observed in a rudimentary state on carefully dissecting the theca, the interior of which is at this time divided into cells of extreme tenuity, in which the spores originate. These cells are filled with a viscid, greenish-coloured fluid, which, when mixed with a small portion of water and highly magnified, will be found to contain innumerable minute granules, possessing spontaneous motion, and moving apparently on their axes with considerable rapidity: they are of various sizes and of various shapes, the larger generally oblong, the lesser spherical: they are all equally active, and being transparent, they communicate a whitish colour to the water when viewed with the naked eye.   *   *   It is extremely difficult at this time to detach any of these cells entire, owing to the filmy condition of the walls and the viscid nature of their contained fluid: a better opportunity is afforded of viewing their form and arrangement, by macerating the theca in dilute nitric acid, when they appear somewhat shrunk and collapsed, and the minute granules are therefore easily discernible in the cells and also in the spores: the whole mass is easily forced asunder even to the theca, which separates into parts corresponding with the sides of the outer cells."—p. 568.

The next stage in the progress of the sporules, namely, their transition from the granules with which the cells are at first filled, to the aggregation of these granules in the centre of the cell, is partially supplied in a passage from Mohl, given as a foot-note at p. 567, from which we make the following extract.

" The young capsules (of Equisetum variegatum) are filled with a very delicate, polyhedral, cellular tissue. These cells are connected together in greater or lesser masses, without, however, being surrounded by mother-cells (if they are not rather themselves to be so regarded), and are filled with a granular mass. In older capsules these cells are larger and distinct from each other, and the green granular contents form for the most part a disc lying in the middle of the cell. In still more advanced capsules this green disc is changed into an oval grain, wrapped round with the two elaters. *   * — Flora, 1833, pp. 45, 46."

The spores are at first oval or ovate in shape, but they soon become globular, which form they afterwards retain: on the contrary the cells, which previously had an angular form from mutual pressure, " gradually acquire substance, separate from each other, and, changing their form, become first globular, and afterwards oval integuments of the spores; the spaces caused by their separation being filled up with a dark green viscid fluid containing abundance of minute granules."

" The next change which the integument undergoes is in the development of the spiral sutures, by which it is divided into two narrow bands with broad and rounded ends: at first the dividing lines are indistinctly seen traversing the integument; after a time they become more distinct, and their spiral direction becomes evident. Two lines of separation run in a spiral direction round the integument, and meet in a sinuous transverse suture at each end: these lines cut the integument into two equal parts, the ends of which are dilated and uniform; and these are the clavate ends of the filaments which have been considered by Hedwig and others as forming part of a sexual apparatus. The separation of the integument into parts takes place immediately after the edges of the sutures have arrived at their proper thickness; it is therefore very difficult after this to find the integument entire.

" The spore at this time contains a greenish-coloured fluid mixed with some minute granules; soon after it changes to a deeper green colour, its contents become thicker, less soluble in water, and filled with a greater number of granules; the fluid which had previously filled the integument and the rest of the theca is gradually absorbed, leaving the granules which it contained sticking in masses to the spores and to the separated portions of the integument. It is these masses of granules, when found adhering to the filaments in the ripened state of the spore, that have been mistaken for pollen-grains: when removed by means of water, they are found to consist exclusively of the lesser granules, the larger ones having now altogether disappeared. As the spore swells, the divisions of the integument are forced asunder; a portion at each end however generally adheres longer; and although further separated, these divisions are still held in their spiral position until the ripening of the spore, when, being ejected from the theca, they recoil with a jerk, and immediately twist into narrow clavate filaments, the state in which they have been most frequently observed."—p. 569.

When the spore is ripe it has a wrinkled appearance; on the addition of water its size is considerably enlarged and the wrinkles disappear. A curious effect is produced on tincture of iodine being added to the water; the nucleus of the spore "is contracted to a much smaller size, leaving the outer membrane occupying the space to which it had been distended by the water, and appearing under transmitted light like a transparent limb to the opake spore." The minute granules in the spore are said to be "exactly similar to those contained in the pollen-grains of flowering plants;" and the author is of opinion that they are of the same nature as the smaller granules on the integument and in the spaces between the cells of the theca; he also finds that the larger granules are soluble in boiling water though not in alcohol; water does not dissolve the smaller ones, and alcohol produces no other effect on them than to suspend their motion. Iodine imparts to the larger granules a bluish colour, but produces little or no effect on the small ones.

Both kinds of granules have been found by the author in the unripe thecæ of ferns, of Lycopodium and of Ophioglossum; he has also observed active granules in the unripe thecæ of mosses and several species of Jungermannia, in the apothecia of lichens, in the lamellæ of Agarics and the perithecia of some other Fungi.

" On comparing these granules with those contained in the unopened anthers of flowering plants, they appear to me to be in every respect identical; in both cases, where the larger ones occur, they are similarly acted upon by iodine, and are therefore probably of the same nature; in the theca they appear to occupy a similar place with those in the cells of the anthers, and they decrease in like manner during the progress to maturity of the pollen-grain and of the spore. In the granular contents of the spore also there is the most perfect resemblance to those of the pollen-grain. Perhaps the most obvious difference is in the entire absence of green colour from the fluid of the latter."—p. 571.

The paper concludes with a detail of the changes which take place in the organization of the theca during the progress of the spores to maturity; and the descriptions are illustrated by figures.

ART. XLIII.—*Account of two new Genera allied to* Olacineæ. By GEORGE BENTHAM, Esq., F.L.S.

THE two species on which Mr. Bentham has founded one new genus, Pogopetalum, are among the plants collected by M. Schomburgk in British Guiana: and fine specimens of another genus, named Apodytes dimidiata by Ernst Meyer in Drége's plants, but first described in the present paper, are in a collection from Port Natal, in South

Africa, communicated to the author by Mr. Harvey. The determination of these genera having led the author into an inquiry as to the affinities of others belonging to the same group, the results of his inquiries are given in the paper here published.

Great difference of opinion seems to have been entertained among botanists, respecting the true position and affinities of the Olacineæ : after giving a synoptical view of the whole order, in which the eleven genera belonging to it are briefly characterized and divided into three tribes, the author comes to the conclusion that Mr. Brown's view of the close connexion of this order with the Santalaceæ is the correct one. The paper concludes with detailed descriptions of the two new genera, which are also figured.

Art. XLIV.—*Extracts from the Minute-Book of the Linnean Society of London.*

1837. *Nov.* 21. Read the following " Notice of the discovery of *Cucubalus baccifer*, Linn., in the Isle of Dogs." By Mr. George Luxford, A.L.S.

" The accompanying specimen of Cucubalus baccifer was, with many others, collected by me in the Isle of Dogs, in the early part of last August. This plant was originally introduced into the British Flora by Dillenius, in the third edition of Ray's Synopsis, [267, under the name of Cucubalus Plinii]. He there speaks of it as having been gathered in hedges in Anglesea (Mona) by Mr. Foulkes of Llanbeder, and sent by him to Dr. Richardson ; but in a letter from Mr. Foulkes to the latter gentleman, published in the Linnean Correspondence, vol. ii. p. 171, he states that he only had ' an account of it from one who pretended to know plants very well,' but that he himself ' could find no such plant.' In a note to this letter, in the work just mentioned, Sir J. E. Smith says, ' Nobody, as far as I could learn, has ever met with the plant since, except in curious botanic gardens, in any part of the British isles ; and accordingly I was obliged to be content with a garden specimen for the figure in ' English Botany,' tab. 1577. I am, therefore, under the necessity, however unwillingly, of excluding the Cucubalus baccifer from our British Flora.' It was accordingly omitted when Sir James published his ' English Flora.'

" The locality in the Isle of Dogs is on the banks of the ditch on the left hand of the road from Blackwall to the Ferry-House ; and there, if not truly indigenous, it is at least perfectly naturalized. I also feel convinced that I have met with it in similar situations in other parts of England ; but the plant not being in flower, I have passed it, as I did the first time I saw it in the Isle of Dogs, thinking it to be merely Cerastium aquaticum, which in that state it much resembles. It is probable that, like Polygonum dumetorum, this plant only requires to have the attention of botanists directed to it, to lead to its discovery in other localities ; and I shall be happy if my meeting with it so near London may be the means of getting it restored to the British Flora, where it is certainly as much entitled to a place as Centranthus ruber, Petroselinum sativum, and other avowedly naturalized plants."—p. 687.

1838. *June* 19. Read a description of Cattleya superba, *Schomburgk;* the description and figure have since been published in Lindley's 'Sertum Orchidaceum,' t. 22.

December 18. Read a "Notice of Cereus tetragonus." By Edward Rudge, Esq., F.L.S. The plant is a single stem between nine and ten feet high. For about a foot from the roots it is four-angled; at between three and four feet it is five-angled, with the angles lobed, and within about eighteen inches of the top it has six angles. The flowers are produced from the angles, and near the top of the stem.

1839. *December* 3. Read "Descriptions of some Vegetable Monstrosities." By the Rev. W. Hincks, F.L.S. The case first described occurred in a flower of Iris versicolor, which had "5 outer reflexed segments, 4 inner upright segments, 5 stamens, 5 distinct stigmas and a 5-celled ovarium. This variation appears to have been the result of the union of two flowers; and the author has witnessed a similar case in some Œnotheræ. The second case occurred in Iris sambucina, in which "3 segments of the inner series only remain, while there are 5 parts in all the other circles: the line of junction is much less evident than in the former, but may be observed in the ovarium and tube of the perianthium."

After mentioning similar cases of union of parts in other plants, Mr. Hincks observes : —

"But the most remarkable instance of this kind of union with which he has met, occurs in a specimen of Scrophularia nodosa, found at Water Fulford, near York, in which four flowers are united into one. In this case several monstrous flowers occur on the same branch, but are generally unions of only two flowers, and the terminal flowers are invariably of the ordinary structure. This Mr. Hincks regards as what might be expected in a plant with centrifugal inflorescence, where the monstrosity consists of a union of flowers; whereas in the same kind of inflorescence, when the monstrosity consists in a more full and equal development, the central flower might be expected to be the first affected; and this actually occurs in a specimen which he possesses, of a species of Linaria with all the terminal flowers (and those alone) *peloriated.*

"In the stalk of the flower of Scrophularia nodosa referred to, Mr. Hincks thinks he can recognize the junction of 4 peduncles; the number of sepals is 15, one of them being narrow and somewhat displaced; that of the petals, which all cohere together, 16. Of these 7 are the lower or more developed petals, which are upright in the limb and are united in pairs. * * Of the upper or reflexed petals only 9 remain; and as there are 3 of these in each ordinary flower, if one be supposed to have perished at each juncture, according to the analogy of the Irides and Œnotheræ, the whole number will be accounted for. The number of stamens is 20, or 5 to each flower; one of these has its anther abortive and changed into a scale, and there are several instances of two being united together, but all may be distinctly traced. There are 3 distinct ovaria," [two of which are each 2-celled, the third is 8-celled].—p. 692.

ART. LXXII. — *List of Jungermanniæ &c. observed in the neigh-bourhood of Dumfries.* By JAMES CRUICKSHANK, Esq.

THE neighbourhood of Dumfries appears particularly favourable to the growth of Cryptogamic plants, especially mosses and Hepaticæ, as may at once be seen both from the number of species and the size and beauty of the specimens.

The extensive tracts of uncultivated mossy ground on the Dumfries side of the Nith, and the long range of green pasture hills on the opposite side of the river in Kirkcudbrightshire, intersected as they are by deep rocky glens, and in some parts covered with extensive woods, are all favourable to the growth of these orders.

I have for the last two years paid particular attention to the mosses and Hepaticæ, and have succeeded in collecting a good number of species of each. I subjoin a list of the Hepaticæ, with their localities, and, so far as I can judge from my opportunities for observation, the frequency of their occurrence throughout the district.

RICCIA, *Linn.*
R. *crystallina,* Linn. Moist shaded ground, not uncommon: abundant in the orchard at Brownhall, near Dumfries.

ANTHOCEROS, *Linn.*
A. *punctatus,* Linn. Sides of ditches and moist ground, not uncommon: abundant on the back of the embankment below the New Quay.

MARCHANTIA, *Mich.*
M. *polymorpha,* Linn. Shaded banks and moist ground not uncommon.
M. *conica,* Linn. Shaded bank near the old College: by the side of a stream in Dalskairth woods.

JUNGERMANNIA, *Linn.*
J. *asplenioides,* Linn. Moist woods, damp rocks &c., common. Rare in fruit, though I have found it in that state in several localities, particularly at Dalskairth, in Kirkcudbrightshire.
J. *spinulosa,* Dicks. Dry rocks at the Craigs, near Dumfries. Near Moffat with perianths; not common.
J. *pumila,* With. On rocks and stones in a stream in Dalskairth woods; rare.
J. *cordifolia,* Hook. Criffel, Kirkcudbrightshire; rare.
J. *Sphagni,* Dicks. Terregles and Criffel, Kirkcudbrightshire: abundant in Locharmoss, Dumfriesshire; always barren.
J. *crenulata,* Sm. Near Glen-mills and Goldilee, Kirkcudbrightshire; rare.
J. *hyalina,* Lyell. Ruttin-bridge, Kirkcudbrightshire: near Closburn, and in great abundance by the side of the stream a little above Moffat-well, Dumfriesshire.
J. *emarginata,* Ehrh. Banks of the Nith, near Friar's Carse: abundant among the Kirkcudbrightshire and Moffat hills.

J. *inflata*, Huds. Plentiful in Lochar-moss; always barren, though the perianths are abundant.

J. *excisa*, Dicks. Abundant in various parts of Lochar-moss, Dumfriesshire: by the side of the Glasgow road, a little beyond Maxwilton, in a small Marsh.

J. *ventricosa*, Dicks. Kilton Coves, Dumfriesshire, in fruit; various parts of Lochar-moss, barren.

J. *bicuspidata*, Linn. Lochar-moss, &c., common in damp places.

J. *byssacea*, Roth. Marsh a little above the Ruttin-bridge, Kirkcudbrightshire: on the top of a stone dike, opposite the farm of Akerhead, near Dumfries: and in various parts of Lochar-moss.

J. *connivens*, Dicks. Wet rocks above the Ruttin-bridge, Kirkcudbrightshire: Creech-hope Linn, Dumfriesshire. It bears fruit abundantly in May.

J. *incisa*, Schrad. Summit of Criffel: abundant in various parts of Lochar-moss; rare in fruit.

J. *pusilla*, Linn. Moist fields, ditch-banks, &c., common.

J. *nemorosa*, Linn. Friar's Carse, Lochar-moss, &c.: very abundant on various parts of Criffel. Var. β. *purpurascens*, wet rocks near the summit of Criffel.

J. *undulata*, Linn. Small stream above Ingleston, Kirkcudbrightshire, growing among Hypnum cordifolium.

J. *resupinata*, Linn. By the side of the English road, Lochar-moss.

J. *albicans*, Linn. Lochar-moss: the Craigs: Glen Mills: Criffel: Dalskairth: and various other places in Kirkcudbrightshire, rather common.

J. *complanata*, Linn. Trunks of trees, roots of hedges and rocks, common.

J. *anomala*, Hook. Lochar-moss, near the side of the Lochmaben road, in considerable abundance; always barren.

J. *scalaris*, Schrad. Road-side near the Craigs: various parts of Lochar-moss, &c. not uncommon.

J. *polyanthos*, Linn. Abundant in a ditch at Terregles: small stream between Criffel and Knockindock.

J. *viticulosa*, Linn. Lochar-moss, by the side of the English road; barren.

J. *Trichomanis*, Dicks. Dalskairth woods, in fruit: various parts of Lochar-moss &c. not uncommon.

J. *bidentata*, Linn. Woods, hedge-banks and moist ground; the most common species we have.

J. *heterophylla*, Schrad. On decaying stumps of trees in the belt of wood at the Powder-Magazine.

J. *Francisci*, Hook. Road-side between Rosehall and Brownhall Farms, near Dumfries. I have not as yet found this rare species in fruit, but *perianths* are abundant.

J. *barbata*, Schreb. Craigs near Dumfries: various parts of the Moffat hills: not com.

J. *Lyoni*, Taylor. Dalskairth, with perianths; side of the stream near Moffat-well. The want of stipules at once distinguishes this from J. barbata. Described and figured in 'Transactions of the Botanical Society of Edinburgh,' i. 116, pl. 7.

J. *reptans*, Linn. Abundant above the Ruttin-bridge, growing among tufts of Dicranum glaucum: Glenmills and Dalskairth, among mosses on shaded rocks.

J. *setacea*, Web. Abundant in various parts of Lochar-moss, growing among tufts of Sphagnum: small marsh by the side of the Glasgow road, a little beyond Maxwilton: abundant on Criffel.

J. *trichophylla*, Linn. Rocks above the Ruttin-bridge: side of a stream in Dalskairth woods: sides of streams among the Moffat hills: by no means common.

J. *platyphylla*, Linn. Craigs near Dumfries, &c. common on rocks and trees, near the ground: rare in fruit.

J. *lævigata*, Schrad. Abundant at the Craigs, near Dumfries; barren.

J. *tomentella*, Ehrh. Wet places in Dalskairth woods. This and the following may be considered two of the rarest, as they are certainly the most beautiful, species of the genus found in this neighbourhood.

J. *ciliaris*, Linn. On a rock above Dalskairth; rare.

J. *serpyllifolia*, Dicks. Dalskairth woods: Craigs near Dumfries: Creech-hope Linn, Dumfriesshire: and near Moffat; always barren.

J. *dilatata*, Linn. Trunks of trees, roots of hedges, rocks and stones; common.

J. *Tamarisci*, Linn. Rocks and trees, near the ground, common; rather scarce in fr.

J. *pinguis*, Linn. Damp ground near the Powder-Magazine, &c., not uncommon in bogs and watery places. Var. β. Lochar-moss.

J. *multifida*, Linn. Marshes and wet ground, pretty common.

J. *Blasia*, Hook. Field opposite Clouden-mills, Kirkcudbrightshire; rare.

J. *epiphylla*, Linn. Moist ground, ditches &c., common. Var. β. less common though by no means rare.

J. *furcata*, Linn. Trees, hedge-roots and rocks, common.

J. *Lyellii*, Hook. Lochar-moss, near the side of the English road, very rare. So very sparing, indeed, is this rare species, that the whole patch might be covered by a man's hat; and though I have made the most careful and anxious search in the same neighbourhood, I could only find this single patch.

JAS. CRUICKSHANK.

Dumfries, April 18, 1842.

ART. LXXIII. — *Notes on the Genus Utricularia.*
By the Rev. J. B. BRICHAN.

THE second species of this genus — Utricularia intermedia — has long had assigned to it, as one of its habitats, the Loch of Spynie in Morayshire. Having never examined that locality I cannot positively affirm that the species is not to be found there; but I think that the following remarks, if they do not prove the contrary, at least render it very doubtful whether it has ever been gathered in Morayshire.

About ten years ago I received from a botanical friend a single specimen of an Utricularia labelled *intermedia*, and taken from the locality in question. Without particularly comparing it with the description, and not doubting the authority by which it was named, I added it to my collection. Not long after I had the pleasure of finding in the Moss of Inshoch, Nairnshire, a plant flowering abundantly, which on examination I found to be indubitably Utricularia minor; it then struck me that the plant I had found and examined was the same as that which had been sent me from the Loch of Spynie, and on comparing them they appeared identical. Not over confident, however, in my own botanical discrimination, I communicated my opinion to my disinterested friend Mr. Stables, who sent some of my Nairnshire specimens to Sir W. J. Hooker. That distinguished botanist pronounced them to be specimens of U. minor, and this, of course, I received as a final confirmation of my own opinion.

It may not at once appear from what I have written, that Utricularia intermedia has not been found in the Loch of Spynie. From circumstances, however, which have come under my cognizance, I have not the least doubt that all the specimens gathered there as U. intermedia have belonged to the other species — Utricularia minor. To confirm what I say, let me present the reader with the following extract from a ' Collectanea for a Flora of Moray,' published in 1839.— " U. intermedia. Observed to flower annually since 1830, in some holes whence turf seems to have been cut, under the north bank of the Loch of Spynie, about half a mile west from Ardivol. *If there be a specific difference between this and U. minor, the Spynie plant upon closer inspection will probably be found to belong to the latter species.*"

I request the reader's attention to that part of the extract which I have marked for Italics; it implies a doubt concerning the existence of any specific distinction between U. intermedia and U. minor. This at once convinces me that the writer of it had never gathered U. intermedia, and that the specimens in his possession were specimens of U. minor. The former I have never seen, except as figured in Sir J. E. Smith's ' English Botany ; ' but presuming from the correctness with which U. vulgaris and minor are there given, that U. intermedia also is correctly figured, I have not a doubt as to the truly specific distinctness of all three; and I think that U. intermedia would more readily be confounded with vulgaris than with minor. It seems rather surprising that the Spynie plant should ever have been mistaken, as it appears to have been.

Perhaps the following description of the three species of Utricularia, from ' The Northern Flora,' may be acceptable to the reader.

1. *Utricularia vulgaris.*

" Plant floating, of considerable size, sometimes fully a foot long. Leaves green, composed of numerous capillary or bristle-like segments, fringed at the margin, and carrying small, beautiful, reticulated bladders. Flowers yellow, placed upon a leafless stem, which elevates them several inches above the water: lower lip longer than the upper, and having a projecting palate, which is about the same length as the upper lip. Spur conical."

2. *Utricularia intermedia.*

"Leaves with a similarity to those of the former, usually dividing into three parts, each of which splits into two flat acute segments, fringed with bristle-like teeth. Bladders placed upon separate stalks. Flowers yellowish : the upper lip twice as long as the palate. Spur conical."

3. *Utricularia minor.*

"Leaves smooth at the edges and bearing bladders. Flowers few and small, pale yellow, with a very short blunt keeled spur : upper lip about the length of the palate."

The author of 'The Northern Flora' further observes, that "independently of the flowers it will be possible to determine the plants, as follows. The *first* species may be always recognized by the large size, and by the leaves being minutely fringed and supporting the little bladders. In the *second* the bladders are not mixed with the leaves, but placed upon distinct stalks, and the plant is more leafy, with the segments broader, and, so far as I have observed, of a paler green. The *last* species is known by its small size and smooth leaves, which support the bladder-like bodies."

I may add, that as far as my experience goes, U. vulgaris flowers in *deep* water, while U. minor flowers only in *shallow.*

J. B. BRICHAN.

Manse of Banchory,
June 11, 1842.

ART. LXXIV. — *Varieties.*

171. *List of some of the rarer Plants observed in the neighbourhood of Birmingham.*

Veronica montana and *Anagallis.* Sutton Park.

Pinguicula vulgaris. Sutton Park.

Valeriana dioica. Moseley Bog.

Rhynchospora alba. Ditto.

Scirpus cæspitosus. Sutton and Coleshill.

Eleocharis acicularis. Sutton.

Eriophorum vaginatum. Ditto.

Alopecurus fulvus. Edgbaston Park.

Agrostis Spica-venti. Canal-side near Strawberry-vale.

Aira præcox. Moseley Common & Sutton.

—— *caryophyllea.* Common about Edgbaston.

Danthonia decumbens. Moseley & Sutton.

Festuca loliacea. Pershore road and other places in the neighbourhood.

Montia fontana. Sutton and Moseley.

Dipsacus pilosus. Several places in the neighbourhood.

Alchemilla vulgaris. Fields on both sides the Pershore road.

Sanguisorba officinalis. Common.

Radiola Millegrana. Coleshill Pool.

Echium vulgare. Dudley Castle.

Anchusa sempervirens. In the hedge of a field in Strawberry-vale.

Myosotis cæspitosa. Moseley.

Anagallis tenella. Moseley, Sutton, and Coleshill.

Menyanthes trifoliata. Ditto.

Villarsia nymphæoides. Packington Park, abundantly.

Atropa Belladonna. Dudley Castle.

Verbascum nigrum. Lane connecting the two Walsall roads at Perry Barr.

Jasione montana. Plentiful near Sutton.

Campanula patula. On the Coventry road 5½ miles from Birmingham.

———— *latifolia.* Yardley Bridge.

———— *Trachelium.* Perry Barr.

Rhamnus catharticus. Yardley Bridge.

———— *Frangula.* Coleshill Pool.

Viola palustris. Common in boggy places.

Ribes alpinum. Side of Edgbaston Pond.

—— *nigrum.* Yardley Bridge.

Helosciadium inundatum. Coleshill Pool.

Œnanthe fistulosa. Yardley.

Chenopodium polyspermum. Bristol Road.

Ulmus carpinifolia. Edgbaston.

Parnassia palustris. Sutton and Moseley.

Linum usitatissimum. Castle-Bromwich road.

Drosera rotundifolia. Sutton and Moseley.

Myosurus minimus. Found in 1839, 40, and 41, at Mr. Dickinson's nursery, Bristol Road; but never more than two specimens in one season.

Peplis Portula. Sutton.

Narcissus Pseudo-narcissus. Abundant at Yardley.

Narthecium ossifragum. Moseley & Coleshill.

Luzula campestris, var. β. *congesta.* Moseley Common.

Rumex palustris. Sutton.

Triglochin palustre. Sutton and Moseley.

Erica Tetralix, var. *alba.* Ditto.

Vaccinium Myrtillus & *Vitis-Idæa.* Sutton

———— *Oxycoccos.* Moseley.

Epilobium roseum. Bristol-Road nursery in 1839, but I have not seen it there since.

———— *palustre.* Moseley.

Polygonum Bistorta. Perry Barr.

Sedum reflexum. On houses by the roadside at Solihull.

Spergula nodosa. Sutton.

Rubus Idæus. Very common.

Potentilla Comarum. Moseley and Sutton.

Nymphæa alba. Sutton.

Ranunculus Lingua. Ditto.

Galeopsis versicolor. Perry Barr, &c.

Scutellaria minor. Moseley.

Pedicularis palustris. Ditto & Coleshill.

Linaria Cymbalaria. Dudley Castle.

Thlaspi arvense. Field by roadside leading from Mereden to Stonebridge.

Cardamine amara. Edgbaston Pond, abundant.

Turritis glabra. Lane leading from the Castle-Bromwich Road to Yardley.

Nasturtium sylvestre. Sutton.

———— *amphibium.* Common in watery places.

Geranium pyrenaicum. Harbourn.

Corydalis claviculata. Hedge by roadside, Strawberry Vale.

Genista anglica. Sutton.

Hypericum maculatum. Canal-side near Strawberry Vale, and at Packington.

———— *pulchrum.* Moseley.

———— *elodes.* Sutton and Coleshill.

Carduus eriophorus. Dudley Castle.

———— *pratensis.* Moseley and Coleshill.

Bidens cernua and *tripartita.* Worcester Canal and elsewhere.

Chrysanthemum segetum. Packington.

Achillæa Ptarmica. Very common.

Typha angustifolia. Sutton.

Carex vesicaria and *ampullacea.* Edgbaston Pond and Moseley.

Littorella lacustris. Coleshill.

Myriophyllum spicatum. Sutton.

Salix pentandra. Edgbaston Pond.

Empetrum nigrum. Sutton.

Lastræa Oreopteris, Lomaria Spicant and *Osmunda regalis.* Moseley Common.

Lycopodium clavatum. Sutton.

The above list is no doubt very imperfect, but it includes all, except the commoner plants, that I have had an opportunity of observing.— *Samuel Freeman;* 11, *Sun St. West, Birmingham, October,* 1841.

172. *Note on Phascum alternifolium.* I enclose a specimen of the Phascum alterni-folium of Bruch and Schimper's 'Bryologia Europæa.' It is a very different plant from the Phascum alternifolium of British botanists, which is the Archidium phascoides of continental authors.—*Jas. Cruickshank; Crichton Institution, Dumfries, Apr.* 18, 1842.

[The specimens are in fine condition, and are labelled — "Near Dumfries, April, 1842. I discovered it in March, 1840. *It is very rare.*"—*Ed.*]

173. *Arenaria verna.* I likewise enclose a specimen of Arenaria verna, from Col-veind, Kirkcudbrightshire, where it was discovered by Mr. John Brown, Dumfries, in 1838, thus proving that it is found on the west coast of Scotland, where Sir W. J. Hooker says it does not occur.—*Id.*

174. *Economical use of the Brake (Pteris aquilina,* Linn.) *in the Forest of Dean.* Passing a few days a fortnight since at a friend's house in the Forest of Dean, Glou-cestershire, I was surprized by some girls bringing a quantity of recently cut Pteris aquilina, or "*Farn,*" which they retailed about at twopence per bushel. On enquiring the use to which it was put, I was informed that it was extensively employed in the forest for feeding pigs, which are very fond of it. For this purpose, however, it must be cut while the fronds are still uncurled, and a quantity of them boiled in a furnace. The *slushy* or mucilaginous mass thus produced is then consigned to the wash-tub, or any other receptacle, and in this state it will keep as pig-food for a considerable length of time. I was informed that it was found very serviceable, especially to cottagers, as coming in at an early period of the summer, when the produce of the garden is but scanty. Perhaps a boiled dish of the Pteris in its circinate state, might not be ve-ry unpalateable or unacceptable with a broiled rasher of bacon — at least to a hungry man. If so, we might thus have a variation in our spring vegetable condiments; but I am not aware whether the experiment has been made.—*Edwin Lees; South Cottage, Malvern Wells, June* 4, 1842.

175. *Enquiry respecting Carex axillaris and remota.* Allow me, through the me-dium of 'The Phytologist,' to enquire if any of the readers of that work can tell me how to distinguish Carex axillaris from Carex remota. I have spent much time in trying to make out the difference, and all to no purpose. Carex axillaris is said to be a very rare plant; this may account for my not being able to make it out. Dr. Good-enough tells us in the second volume of the Linnean Transactions, that the capsules in C. remota are entire, and in C. axillaris cloven; if I rely on that character C. re-mota *I have never seen.* But on turning to the third volume of that work, I find Dr. Goodenough withdrawing the statement, and telling us that he believes that all Cari-ces dispose of their seeds by the opening of the points of their capsules. It would ap-pear by this that Dr. Goodenough was not a very close observer of the Carices, or he would have known that they, at least our British species, do not dispose of their seeds in other way than by their capsules (if so I may call them) and seeds falling together. Smith seems to rely very much on the spikes being simple or compound; he tells us that the spikes of C. remota are simple, and those of C. axillaris compound: if this be the character to distinguish them by, I have often seen both species on the same root. He also says that the lower bracteas of C. axillaris are very long, and that the beak of the fruit is more deeply cloven than that of C. remota, though he says the dif-ference is not very striking, It appears that Sir W. J. Hooker relies very much on the length of the lower bracteas in distinguishing the two species; directly opposite to what Smith has said, he tells us that the lower ones in C. axillaris are scarcely so long as the spike. He also tells us that C. axillaris is a stouter and taller plant than C.

remota, the latter character needs no comment; suffice it to say that I have now before me good specimens with roots, varying in height from three inches to three feet; and as to the bracteas, a little attention to the Carices will serve to show that the length of these parts is not at all to be depended upon. I see there is something said in the 'British Flora' on the scales of the calyx of the two species; these I will now examine and give you the result. The three *scales* I now enclose are from *the same spike;* they *all* have *two close green nerves;* in one you will observe that the nerves are discontinued below the point, in the others they are extended to the point: you will also observe that one is only about one half the proportionate breadth of the others. As the scales of most of our Carices are much broader about the middle or in the lower part of their spikes than at or near the summit, we may rely too much on such characters. Now, after gaining what information I possibly can from books, I turn to my specimens, and on examining them, among the rest I find two fine specimens of C. divulsa, sent to me a few years ago (from what was then considered good authority) for C. axillaris; another I find, of recent date (from what is now considered good authority), that was sent for C. axillaris, which turns out to be C. paniculata; so that I gain no information by this, and as a last resource I avail myself of 'The Phytologist.' Perhaps some of the readers of that work will be able and kind enough (if there be no secret in the matter), to inform me how to distinguish the two species. In addition to the above enquiry, I beg to say that good specimens or information on *any* of the Carices is always most gratefully received by me; as it is my intention, at some time or other, to give a paper at some length on the British species and varieties of this genus, when I shall have to notice some strange forms, and one in particular, which is now considered to be a new species, and will be called *Carex Leylandi.* The plant was discovered by me about a mile from Hebden Bridge in May 1840. It has been seen by Mr. Leighton and many others of our eminent botanists, and they all consider it quite new; and Mr. Babington, in his letters also expresses his opinion that it is distinct. —*Samuel Gibson; Hebden Bridge, June* 8, 1842.

176. *Enquiry respecting Bromus commutatus.* While on the subject of enquiries, allow me to ask what are the "more technical distinctions" of Bromus commutatus? (Phytol. 136). The writer of the notice referred to appears to doubt whether Bromus arvensis be a British plant; a few days ago I had the pleasure of giving to Mr. Borrer specimens of the true B. arvensis. I first found the plant in 1840, growing about half a mile from Hebden Bridge; and in 1841 I again found it in abundance in this neighbourhood. The three spikelets I enclose will serve to show that no reliance can be placed on the size of the flowers: the three are from the same plant.—*Id.*

177. *Stellaria holostea with laciniated petals.* In my rambles about this neighbourhood I have met with a singular laciniated variety of Stitchwort (Stellaria holostea ?). The petals have at the base of the incisure a very minute tooth, slender and sharp-pointed, about half a line in length; the tip of each segment of the petals is also divided into two unequal acute teeth, the outermost being about half the size of the inner one. The leaves on the stems are the shape and size of those of S. holostea, and have also the peculiarly fine serrature of the edges. The flowers are about one third smaller than the regular size of those of S. holostea. The plants formed several large bushes in the hedge on one side of the road near the summit of one of our mountains, they occupied a portion of the hedge about a dozen yards in length, forming four or five rather large bushes at each end of the above distance. Bushes of plants with flowers of the regular form were closely adjoining them, the branches in one case in-

tertwining. The laciniation of the petals gave the bushes of the plant a singular appearance, that attracted the attention when at some distance. I intend to try to procure some of the seeds when mature, and sow them against next season, in order to observe whether the dentation of the petals will be trasmitted by seeds or not.—*James Bladon; Pont-y-Pool, June* 13, 1842.

178. *Lavatera Olbia in Epping Forest.* A few years since a new piece of road was made through Epping Forest to Woodford. At a spot called Fair-mead Bottom a large quantity of earth was dug from the forest and thrown up to raise the road, for the distance of about half a mile. The following summer the sides of this piece of road were covered with various plants, such as Senecio Jacobæa, thistles, &c., and among them a great number of plants of Lavatera Olbia, a species not known, I believe, as a native of Britain. There is not the slightest doubt that the seeds had been buried for a vast number of years, and vegetated when brought to the surface, as it seems impossible for the plants to have got there in any other way. For three or four years they seemed to flourish, and flowered abundantly; but now the banks having become covered with grass &c., they seem to be disappearing, and last year I could find only three or four plants: when I first noticed it there were hundreds scattered along the whole length of the raised portion of the road.—*Henry Doubleday; Epping, June* 14, 1842.

ART. LXXV.—*Proceedings of Societies.*

LINNEAN SOCIETY.

Anniversary, May 24, 1842.—The Bishop of Norwich in the chair. The bye-laws relating to the election of officers having been read, the ballot was commenced and the following were declared duly elected.

President. — Edward, Lord Bishop of Norwich.

Edward Forster, Esq., *Treasurer.* J. J. Bennett, Esq., *Secretary.* Richard Taylor, Esq., *Under Secretary.* New members of Council :— Lord Beverley, J. A. Hankey, Esq., John Miers, Esq., R. I. Murchison, Esq., and Alfred White, Esq., in the place of A. B. Lambert, Esq., G. Mantell, Esq., LL.D., The Marquis of Northampton, R. H. Solly, Esq., and W. Yarrell, Esq. It appeared from the Treasurer's accounts that the receipts during the past year had amounted to £785. 10s. 10d., the expenditure to £761. 17s. 5d., leaving a balance of £23. 13s. 5d. in the Treasurer's hands.

The following Statement was read. — "The Council, having had under their serious consideration the financial affairs of the Society, submit the following statement to the Fellows at large. The cost of the Collections and Library of Linnæus, together with those of the first President, Sir James Edward Smith, purchased of the executors of the latter in 1828, amounted to £3000. Of this sum about £1500 were then raised by subscription; and to meet the remainder a debt, on bonds, was incurred, which now amounts to £1300, paying interest at 5 per cent. In consequence partly of this amount of interest, and partly of a diminution in the annual receipts, there has been accumulated, within the last few years, a further debt of about £500. By recent arrangements a saving of some amount has been effected in the expenditure ; but the Council are convinced that no farther material reduction can be made, without greatly impairing the efficiency of the Society, and they desire to avoid, as far as possible, the necessity of calling upon the Fellows to agree to a small charge being placed upon the Society's publications, that appearing to be the most obvious means of supplying the deficiency in the annual receipts. With this view they propose a general subscription, which they trust may reach such an amount as to meet the present liabilities, and to relieve the funds of the Society from the burthen of debt and interest. They, therefore, earnestly recommend the subscription to the members of the Society." The names of subscribers to the amount of £600 were appended to this statement, which was circulated among the members present at the meeting.

The following members have died during the past year :—J. Ansley, Sir W. Beatty, Sir Charles Bell, Rev. I. Bell, J. E. Bowman, Rev. T. Butt, W. Cattley, Dr. E. J. Clarke, Geo. Coles, R. Goolden, W. Harrison, Robert Higgin, Philip Hurd, Dr. J. R. Johnson, A. B. Lambert, C. Lane, Richard Leigh, Robt. Maughan, Archibald Menzies, and David Pennant; Fellows. Of Foreign Members, Auguste Pyrame De Candolle and Janus Wilken Hornemann. Of Associates, the Rev. R. F. Bree, Professor Don, and Mr. C. E. Sowerby.

BOTANICAL SOCIETY OF EDINBURGH.

June 9, 1842. — Dr. Neill in the chair. Donations to the library and herbarium were announced from the Worcestershire Natural-History Society, Dr. Miguel, M. Parlatore and Mr. Sowerby.

Professor Graham communicated the agreeable intelligence, that the late Mr. Archibald Menzies had bequeathed to the Botanic Garden his interesting and valuable herbarium, which was chiefly formed in the course of his voyages round the world with Vancouver and other circumnavigators. Mr. Menzies was the last survivor of Vancouver's companions, having lived to the age of eighty-eight. He was a native of Perthshire, and studied at this University, towards which he continued throughout life to entertain the warmest feelings of attachment.

The chairman adverted, with deep regret, to the loss which the Society, in common with the botanical world, had sustained by the death of Mr. Falconar of Carlowrie, who was a most zealous and successful cultivator of the science, and who enjoyed, in a high degree, the esteem and respect of his friends.

The following papers were read : —

1. *Notice of the discovery of Phascum alternifolium* (Bruch &c.) *in Dumfriesshire, and of Arenaria verna, on the west coast of Scotland :* communicated by Mr. James Cruickshank. This *Phascum* is not the plant of Hooker, which is the *Archidium phascoides* of continental botanists. The present plant was formerly discovered in Britain, but long ago and in very small quantity. It is, in the opinion of Mr. Wilson, a good species. *Arenaria verna* was found at Drumlanrig, by Mr. Cruickshank. It is very rare, if existing at all, on the west coast of Scotland.*

2. *Notice on the occurrence of Avena alpina and Saxifraga umbrosa in Yorkshire :* by Mr. J. Tatham, jun.† Mr. Tatham says,—" *Avena alpina* grows here (Settle) at an

---

.

* There is a discrepancy between the Report and Mr. Cruickshank's note, Phytol. 263.

† See also a note by Mr. Simpson, Phytol. 75.

elevation of between 600 and 800 feet above the sea. When growing in our elevated open pastures the plants are generally single, also on our limestone cliffs; but when in our natural woods, which are mostly hazle, it is found in large tufts, where you may get perhaps fifty specimens in the space of a few inches. I believe I could send from the same tuft specimens of *alpina* with the panicle quite as simple as any *pratensis*. I consider *Saxifraga umbrosa* as really wild here. It is met with in Heseltine Gill, which is a deep ravine at the foot of Pen-y-ghent, and Fountains Fell. There are only two houses in about three miles, and these not near the place. *Actæa spicata*, *Ribes petræum*, &c., grow along with it. The valley runs from west to east, and the *Saxifraga* is found only on the south side, which receives no sunshine except in summer. Some of the plants are inaccessible, the cliffs are so steep."

The impression of the meeting was that no specific distinction existed between the plant now shown as *Avena alpina* and *A. pratensis*, and the same remark may be applied to all other specimens of the former hitherto exhibited from British stations.

3. *On three new Species of Grasses of the Genus Poa* : by Richard Parnell, M.D., F.R.S.E. The author stated that these grasses were so unlike, in general appearance, to any of the other *Poæ*, and possessed such strong marks of specific distinction, that he considered them entitled to rank as distinct species. 1. *Poa sectipalea*, Parnell.— This plant differs from *Poa pratensis*, the only species it can well be confounded with, in the branches of the panicle being stouter, more erect and rigid; the spikelets larger. Outer palea seven or nine ribbed, seven of the ribs being very distinctly marked; inner palea one-third shorter than the outer, and invariably divided to the very base: whereas in *Poa pratensis* the inner [? outer] palea has never more nor less than five ribs, and the inner very little shorter than the outer, and always entire. Found growing in sandy situations between Cramond and Queensferry. 2. *Poa polynoda*, Parnell. This species differs from *Poa compressa*, in the florets not being ribbed at the base; outer palea five-ribbed. Joints eight or ten in number, the uppermost joint situated but a short distance from the panicle; whereas in *Poa compressa* the florets are very distinctly ribbed, suspending the carix [?] by their silky fibres. Outer palea three-ribbed. Joints seldom exceeding four in number. 3. *Poa nemoralis*, *montana* of Koch. Inflorescence simple, panicled, occasionally racemed. Panicle erect, narrow and slender; the branches erect, long and slender, bearing few spikelets. Spikelets lanceolate-ovate, of two or three awnless florets. Calyx of two unequal, acute glumes, three-ribbed. Florets not in the slightest degree webbed. Outer palea five-ribbed, the dorsal and marginal ribs slightly hairy. The whole plant is of a glaucous hue. This grass was first obtained by Dr. Greville, who, in the year 1833, gathered several specimens on Ben Lawers; since then it has been found in many parts of the Highlands, but has hitherto been considered as a glaucous variety of *P. nemoralis*.

These grasses are figured in Dr. Parnell's work on the Scottish grasses, now in the press, in which he has given 130 figures, with minute descriptions.

Dr. Graham exhibited some beautiful exotics from the greenhouses, chiefly natives of Asia—and also a specimen of the interesting *Megaclinum falcatum* in flower, which had been raised by Mr. Gray, of Greenock. A handsome plant of *Armeria fasciculata* from the south of Europe, which had been raised in the Horticultural Garden, was likewise exhibited. — *The Edinburgh Evening Post and Scottish Standard, Saturday, June* 18, 1842.

May 20, 1842. — J. E. Gray, Esq. F.R.S., &c., President, in the chair. The following donations were announced. A specimen of sugar-cane from Madeira, presented by Mr. James Halley: a specimen of *Bupleurum tenuissimum*, found at Highgate and presented by Mr. W. Mitten (Phytol. 203): British plants from Mr. G. H. K. Thwaites, Mr. John Ellis and Mr. Edwin Lees: Books from the President. A paper was read from Edwin Lees, Esq., F.L.S., "On the Flora of the Malvern Hills. Part 3: being a Sketch of the Cryptogamic Vegetation indigenous to the chain."

Notwithstanding the limited extent of this narrow chain of hills, scarcely exceeding nine miles in length, and rising to only 1500 feet in altitude, yet they offer almost every variety of aspect and condition favourable to the development of Cryptogamic vegetation. In fact the Malvern Hills, when considered only as a ridge, without reference to the country around them, are far more remarkable for their *Acotyledonous* than their *Vascular* productions.

Commencing with the Northern termination of the hills in Cowleigh Park, several miniature syenitic spurs here appear, abrupt and rocky yet prettily shaded with wood, amidst deep glens and shaggy defiles overtopped by lateral steeps of limestone, amidst whose gullies streamlets are ever gushing with musical intonation. From this "happy valley" a verdant park-like glacis leads the wanderer up among the exposed treeless turf, and rugged, jutting-out, lichened rocks of the End and North Hills, those of the latter being more precipitous and remarkable than those of any other hill of the chain, and boasting a great number of lapideous lichens. Between this hill and the Worcestershire Beacon, a deep and winding valley extends, watered by bubbling streamlets, and abutted by moist dripping rocks on the southern side, where several species of *Jungermannia* shelter; but it must be observed that excepting in this place, and in "The Gullet" (as it is termed) of the Holly-bush Hill, almost all the other Malvern rocks are without exception dry and bleached by the wind and sun. At the Western base of the Worcestershire Beacon occurs one of the few bogs that yet remain about the hills, *Aspidium Oreopteris* marking this and the other boggy places by the profusion in which it covers the margin of the black soil. A mile farther south, at "the Wych," the syenite and limestone are in contact, and the latter having been extensively quarried, numerous abandoned excavations occur, in many instances embowered in wood, and offering favourite habitats for many mosses unable to fruit on the sunburnt sides of the hills. These limestone rocks also offer an instructive example of the lichens more particularly affecting limestone, when compared with the loftier and more exposed syenite.

From the Worcestershire Beacon undulating green knolls, many cultivated to near the summit, stretch past the Wells for a distance of four miles to the Herefordshire Beacon, without any intervening valley; but diversified in some places by rocky dry ravines, strewed with broken fragments, and in others by plantations, or natural thickets of stunted whitethorn along their sides. Round tufts of *Ulex nanus* dot the hills in every direction. The portion of the chain just mentioned terminates rather abruptly at "the Wind's Point," where a deep valley commences, plunging down to the eastern base of the Herefordshire Beacon, and extending to the romantic wooded vicinity of Little Malvern Priory, where some thick alder-holts shroud a purling stream.

The Herefordshire Beacon, with several detached eastern buttresses, lifts up its bare, shivering and indented fortified sides in sullen grandeur, turf and mosses alone occupying its windy ramparts, except a number of scathed, old, scattered elder-trees.

Mr. L. considered that nearly one half of the plants occupying the Malvern Hills are Cryptogamic; and the following synopsis will show this to be not an unreasonable supposition, especially as the census the author had taken is not to be considered a perfect one, embracing however all the species Mr. L. had been enabled to identify after an attentive examination of five years and upwards.

ENUMERATION.	Species.
Ferns and Equisetaceæ,	25
Mosses,	121
Jungermanniæ,	23
Other Hepaticæ, Characeæ, &c.	15
Lichens,	223
Fungi,	305
Total,	712

Mr. Lees had been unable to give any attention to the *Algæ Confervoideæ*, and had not taken much note of the minuter species of *Sphæria* among the Fungi, so that were these carefully noted by the algologist and practised mycologist, doubtless the list might be easily extended to a thousand or more. But in an enumeration of the Flora of any locality, especially when considered with reference to a comparison with other lists, it is hardly fair to drag into the account every black spot or stain upon leaves or dead sticks; and therefore Mr. L. had left untouched nearly all but the really tangible and decided species, independent of minute microscopical examination.

Ferns.—The rarer ferns of the hills are *Polypodium Dryopteris* and *Allosorus crispus;* the latter only occurs upon the Herefordshire Beacon, thus offering a good illustration of the geographical distribution of plants, the fern thus occurring so sparingly on these hills being plentiful upon the Welch mountains. *Asplenium viride*, also another fern of common occurrence about the waterfalls of Wales, though not found on the hills themselves, occurs on an old stone bridge over the Teme called Ham Bridge, a few miles northward of Great Malvern. *Aspidium Oreopteris* occupies in profusion the margin of various boggy spots on both the eastern and western sides of the hills. *Aspidium dilatatum* is of common occurrence among the stones of the hills; and *Asplenium Filix-fœmina* almost fills the little watery glens running among them. *Polypodium vulgare*, with stunted growth and multilobed varieties, is not unfrequently met with in and on the sides of deep hollow lanes. *Grammitis Ceterach*, though scarcely truly belonging to the district, and not found on any of the rocks, yet flourishes on a massive stone wall by the side of the road at Great Malvern; and Mr. L. had also noticed this fern growing in the interstices of an old brick wall at Forthampton. On Rosebery Rock, north of Great Malvern, Mr. L. had gathered very singular specimens of *Scolopendrium vulgare*, eighteen inches in length, with the extremities of the frond lobed in a most remarkable multifid manner; the specimens are in the Society's herbarium, having been sent some time since. *Pteris aquilina* robes the bases and three parts up most of the hills of the chain, and would be a great pest were it not mowed down every autumn, and stacked in ricks for litter by the humble economical farmer. It is perhaps remarkable that Mr. L. had never yet been able to detect a single species of *Lycopodium* at any point or in any spot throughout the entire length of the chain or about its base.

Mosses.—The mosses have exercised a considerable agency in the creation of the soil now upon the Malvern Hills; doubtless indeed they were the primary originators of vegetation upon the bare rocks, whose hollows they have filled up in the lapse of ages with a soft spongy carpet, and so encompassed and obscured them that numerous masses of grey rock, almost immersed in the verdant mossy inundation, now scarcely exhibit their points above it. The lichens have been generally considered as the first pioneers of vegetation, but their efforts to create a *humus* for the nourishment of other plants are but trifling when compared with the economical powers of the mosses. To test this by experiment, observing the tiled roof of an outbuilding at Malvern Wells, evidently erected but a few years, studded with tufts of *Bryum capillare*, (Linn.), Mr. Lees gathered one of them in March last, with the black earth collected around its base. The mass altogether weighed six ounces; but when, after repeated and careful washings, Mr. L. had extracted all or nearly all the black mould that enveloped the roots, the actual residuum of frondescence when weighed amounted to only half an ounce; thus satisfactorily showing (for no soil could have been collected on the sloping roof, independently of the agency of the *Bryum*) that the moss, through atmospherical and imbral agency, had formed a soil exceeding its own weight at the very least above ten times! By operations on a more extensive scale, it is easily conceivable how a bare mass of rock may, in the course of a few years, be covered with a thick coating of soil, sufficient for the nourishment of any of the phanerogamous species adapted to the climate and elevation where they may stand. Mr. L. regretted that in the experiment adduced he could not certainly determine the exact space of time that the *Bryum* had occupied the roof; but as it is found expedient, from the excessive growth of mosses, to cleanse roofs about Malvern almost every year, Mr. L. felt certain that, at the utmost, the plants in question had been located on the tiles between two and three years. *Bryum hornum* has been noticed to be a great accumulator of soil in marshy spots; while the excessive growth alone of the stems and foliage of such mosses as *Sphagnum palustre*, *Dicranum glaucum*, *Bryum palustre*, *Hypnum molluscum*, *Hypnum cuspidatum*, &c., in the course of time entirely fills up bogs, drinks up their water, and conduces to their ultimate establishment as component parts of *terra firma* fit for useful cultivation. In this manner Mr. Lees considered the Malvern Hills to have received originally that rich *humus* which covers their sides, and which, combined with the disintegrating touch of time's mouldering fingers, renders their soil, in the present day, capable of immediate cultivation, even in the steepest places, producing crops that well repay the toil of the industrious cultivator, and tend to give an impulse to fresh inclosures of the verdant turf every year. On a first cursory glance at the turf of the hills, there seems a great sameness in the mosses that luxuriate there. *Dicranum scoparium*, *Hypnum triquetrum*, *splendens*, *purum* and *molluscum*, seeming as if they had united to exclude the rest, *Hypnum triquetrum* especially everywhere predominating. However, a little attention will show a considerable variety, especially upon or in the immediate vicinity of the rocks or on the margin of the numerous tinkling rills that show a cincture of the tenderest green wherever they trickle down. Inclusive of the woods about the bases of the hills, Mr. Lees had numbered 121 species, without by any means exhausting the interminable *Hypna*, so that it is probable a few more may yet be detected. Specimens of nearly all that Mr. L. had met with accompanied the paper, and many of them were exhibited. Here followed a list of those observed.

June 3.—J. E. Gray, Esq., F.R.S., &c., President, in the chair. Mr. J. A. Brewer exhibited living specimens of *Ophrys muscifera*, *Aceras anthropophora*, *Orchis ustulata*, *Paris quadrifolia*, *Mespilus germanica*, and other interesting plants from Reigate, Surrey. Mr. T. Twining, jun., exhibited a large collection of cultivated specimens from Twickenham. Mr. M. J. F. Sidney presented a specimen of *Lycopodium lepidophyllum* from Valparaiso. Mr. Adam Gerard presented a collection of plants from Sierra Leone. Dr. J. B. Wood presented specimens of *Carex elongata*, collected at Chorlton, near Manchester, (Phytol. 198). Donations to the library were announced from the Natural-History Society of Northumberland and Mr. M. J. F. Sidney: and British plants had been received from Mr. Sidney. Read, the continuation of Mr. Edwin Lees' paper (commenced at the last meeting), — " On the Flora of the Malvern Hills. Part 3 : being a Sketch of the Cryptogamic Vegetation indigenous to the Chain."

Hepaticæ.—Among the Hepaticæ occur *Anthoceros punctatus* and *Targionia hypophylla*, the latter at the foot of the Worcestershire Beacon; and there are three species of *Marchantia, polymorpha, conica* and *hemisphærica*. The *Jungermanniæ* are not in such variety as might have been expected, though some of them, especially *Tamarisci, dilatata* and *ciliaris*, are abundant upon the turf of the hills as well as upon the rocks. *Jungermannia pinguis* (Linn.) is excessively common, though very rarely fruiting : *J. tomentella* is a beautiful species that has only occurred in one place, on a dripping rock in " the Gullet." The following are all the species Mr. L. had met with : —

Jungermannia asplenioides	Jungermannia complanata	Jungermannia Mackaii
ventricosa	scalaris	serpyllifolia
bicuspidata	viticulosa	dilatata
connivens	Trichomanis	Tamarisci
pusilla	bidentata	pinguis
resupinata	platyphylla	epiphylla
albicans	ciliaris	furcata
obtusifolia ?	tomentella	

Lichens.— The Malvern Hills are particularly remarkable for the various lichens they produce, most of which grow in a very luxuriant and beautiful manner; and in the moist autumnal and wintry months many of the rocks present an appearance from them truly gratifying to the lover of nature. Some of the harder granitic rocks are entirely covered with *Umbilicaria pustulata*, which at this time is of an olive-green colour, and as flabby as a piece of moist leather, though in the summer months it appears as black and sooty as if subjected to the action of fire. On other rocks the deep purple *Parmelia omphalodes* extends itself, contrasted with wide patches of the grey *Par. physodes*, the darker *P. saxatilis*, the dingy *P. olivacea*, or the conspicuous glaucous pitted thalli of *Sticta scrobiculata*. On the higher rocks the curled *Cetraria glauca* grows in abundance; while a remarkable hoary aspect is imparted to the protruding masses by the silvery *Toidium coralloides*, and the still more coralline appearance of *Sphærophoron compressum*. The rein-deer lichen, *Cladonia rangiferina*, called by Crabbe

" The *wiry moss* that whitens all the hill," —

is plentiful on the turf with its allied species, and the sadder and darker *Cornicularia;* while in every part the brown and scarlet apothecia of the *Scyphophori*, in all their multiform varieties, contribute to decorate the scene. In the whole, including the cortical lichens, about 220 species are met with.

It may be specially remarked that the Parmeliaceæ and Collematæ, and the Peltigerous tribe of lichens, are particularly abundant on the Malvern Hills, as well on the moss and turf as on the rocks. Here followed a list of the lichens, most of which accompanied the paper.

Fungi.— The extensive dispersion of the majority of the Fungi causes them to be met with in most parts of the country at all favourable to fungoid development. The moist grassy declivities of the hills are, however, in autumn peculiarly adapted to the growth of the Agaric tribe; and at that season great quantities appear, and of every conceivable colour, from the vivid scarlet of *Agaricus muscarius,* or the brilliant green and yellow of *A. psittacinus,* to the dullest brown assumed by the common fairy-ring *A. oreades.* Mushrooms are generally very plentiful, though the common kind, *A. campestris,* is much exceeded in numbers and luxuriance by *A. Georgii : A. procerus* is also very abundant, and frequently of enormous size. *A. granulosus, pratensis, conicus* and *coccineus* commonly occur. The short turf, too, is often adorned by different coloured *Clavariæ,* intermixed with the dark *Geoglossum glabrum* and *Spathularia flavida.* The brilliant little blood-red *Peziza humosa* is very common amongst dark masses of *Polytricha ;* while in moist weather a characteristic feature is presented to view in the great number of " Jews' ears " (*Exidia Auricula-Judæ*), hanging upon the countless old elders that cover the eastern declivity of the Herefordshire Beacon.

The little beech-wood behind the Wells Hotel, harbours several rare or curious species; and here Mr. L. had gathered the following species in great abundance : —

Agaricus hypothejus, *Fries.*	Hydnum repandum	Helvella crispa
torminosus	Thelephora coralloides	Peziza onotica, *Pers.*
violaceus	Clavaria abietina	Phallus impudicus
glaucopus	rugosa	

In the same little wood *Bryum roseum* may always be found, and Mr. L. had gathered it in fruit there.

Many other fungoid productions may be met with : and here followed a list of those observed.

On the whole the Flora of Malvern may be considered as most remarkable and abundant rather in its Cryptogamous than Phanerogamous productions. After an attentive examination of the hills and the district around them for some years, Mr. Lees had been enabled to determine : —

> Dicotyledonous Plants, ... 553
> Monocotyledonous Plants, ... 173
> _____
> Total Phanerogamic, ... 726

Mr. L.'s Cryptogamic census amounts altogether to 712, and with a little more industry and research among *mycological* productions, the author doubted not it might considerably exceed this amount, while few, he thought, could be added to the Phanerogamous list. The flowering plants observed by Mr. Lees are extended to the limits (given in the first paper, Phytol. 152) to the banks of the Severn, while the Cryptogamia are more exclusively the product of the hills and the woods at their bases. Altogether the entire number of plants which Mr. Lees had determined and appropriated as belonging to the Flora of the Malvern Hills, amounts to 1438.

Thanks having been voted to Mr. Lees for his interesting communication, the President announced that the Council had appointed Mr. Arthur Henfrey, Curator; and that the herbarium might be inspected every Monday, Wednesday and Friday, from 10 to 4, and on Friday evenings from 7 to 10. The President also announced that the first excursion of the members would take place on the 7th instant.—*G. E. D.*

THE PHYTOLOGIST.

| No. XV. | AUGUST, MDCCCXLII. | PRICE 1s. |

ART. LXXVI.—*A History of the British Equiseta.*
By EDWARD NEWMAN.

ROUGH SHAVE-GRASS, OR DUTCH RUSH.
EQUISETUM HYEMALE of Linneus.

β. *Equisetum Mackaii*, Newman.   *Equis. elongatum*, Hooker, 'London Journal of Botany,' 42; but neither of Willdenow nor Reichenbach.

γ. *Equisetum variegatum*, Schleicher.

ALTHOUGH our more eminent modern botanists have considered as distinct species the plants which I have here given as varieties; and although I most willingly admit that each, in its extreme state, is without difficulty to be distinguished from the others; yet I am totally unable to find *constant* characters by which to distinguish either: I shall therefore treat of them as constituting but a single species, figuring nevertheless each peculiar form of its natural size, and giving on one block magnified representations of those characters which have been pointed out as *distinctive*.

The figures usually quoted as representing this plant are so incorrect that they can only be referred to with doubt. Tragus[*] and Dalechamp,[†] whose figures are usually quoted as representations of Equisetum hyemale, have adorned the upper part of each stem with whorls of short branches, which give the plant a very extraordinary appearance, and suggest the idea of their being intended for Equisetum limosum or Hippuris vulgaris, the upper part of the stems much resembling the latter plant; and it should be added that Tragus assigns to his figure the name of Hippuris. It should however be observed that one of these figures is a servile copy from the other, the outline having been traced and transferred to another block, by which the figure has been reversed. The figure in Matthiolus,[‡] also supposed to refer to this species, may possibly be intended for the *variegatum* of Schleicher; but I can only venture this as a suggestion, for it is by no means characteristic of any plant with which I am acquainted. Gerarde's figure§ evidently represents variegatum: the specimen selected for the purpose being more than usually divided: the learned author however ascribes to his plant the property usually assigned exclusively to the normal form, speaking of it as the " small and naked shave-grasse wherewith Fletchers and Combemakers doe rub and polish their work."‖ Camerarius¶ also represents variegatum. The figure in 'English Botany'[**] appears to me to be spoiled by an attempt to represent the roughness of the stem, which of course cannot be accomplished. That in the 'Flora Londinensis'[††] is the most characteristic I have met with.

* Tragus. Histor. 822.   † Dalechamp. Histor. 1073.   ‡ Matt. Valgr. i. 875.
   § Ger. Em. 1115.   ‖ Ger. Em. 1. c.
¶ Camer. Epit. 886.   ** Eng. Bot. 075.   †† Flora Londinensis. 161.

The medicinal and other properties of this Equisetum have been amply recorded by the earlier herbalists, but most of them appear to quote Galen as their authority.   In consequence of the confused nomenclature and very indifferent figures of these authors, the properties in question become very doubtful as intended for the present species; and even were they so intended, all faith in them, as far as regards medicine, has long been exploded. Tragus* tells us that it is most useful as a medicine, taken internally or externally; internally its properties appear to be somewhat astringent, but it has long been out of use: held to the nostrils and applied at the same time on the neck, it stops bleeding at the nose, and when bruised and laid on a recent wound it staunches the blood.

We are told by Lightfoot† that " some entertain an opinion that if cows chance to feed upon" this Equisetum "their teeth will drop out: how far this may be true," he adds, " I know not, but I am persuaded that the pasture must be very bad where they are compelled to eat such food." Sir William Hooker appears to give the plant credit for this injurious quality : he says "that though while to sheep it proves injurious, and that the teeth of cows who eat it soon fall out, it is wholesome for horses."‡

The stems of this plant have for centuries been used by artificers in smoothing the surface of wood, bone, and even metal, previously to giving these substances their final polish.   This employment of the

---

* Vires ac temperamentum.

Hippuris ea quæ capillamento potius quam folio articulatur, magni in medicina et maxime probandi est usus, propter miram sanandi vim quam obtiret spissandi facultate prædita et utrisque corporis partibus utillissima.

Intrinsecus.

Hippuris cujus jam meminimus, sistendi sanguinem mira facultate pollet.   Succus namque ejus bibitus profluvia sanguinis * * * * cohibet. * * Eadem potest aqua stillatitia per diem bis aut ter mensura trium cochlearium sumpta. Succus in vino potus tormina ventris sedat, tussim, orthopnæum, ruptaque adjuvat necnon dissectiones vesicæ et intestinorum enterocelasque.

Ad eadem herba vino decocta et calida pota facit.   Potest vero vel herba vel radix, vino aquave ad omnes istas affectiones decoqui, pro ratione morbi.

Extrinsecus.

Succus herbe expressus naribusque inditus et cervici simul impositus sanguinem e naribus erumpentem sistit.   Idem aqua stillatitia efficit.

Herba tusa ac cum succo suo imposita cruorem e vulneribus manantem supprimit ipsaque intra paucos dies glutinat teste Galeno qui Hippurim ad sanguinis rejectionem * * ad dysenterias et ad alia ventris profluvia generosum esse medicamentum asserit.

† Lightfoot, ' Flora Scotica,' 650.   ‡ Flora Londinensis, 156.

stem was noticed by Gerarde in the passage already quoted. Haller*
also mentions it as being in his day used in common with other species of Equisetum to polish metal vessels, but speaks of this species as
being the roughest and best. We are further told by Lightfoot† that
"in Northumberland the dairymaids scour their milk-pails with it."
The value of this plant for the purpose of smoothing or polishing is
not, however, merely traditionary, or, like its medicinal virtues, ima-
ginary: it is still used for polishing wood, bone, ivory, and various
metals, particularly brass; for this purpose it is imported, under the
name of "Dutch Rush," in large quantities, from Holland, where it is
grown on the banks of canals and on the sea ramparts, which are often bound together and consolidated by its strong and matted roots.
Bundles of this imported Dutch Rush are exposed for sale by many
London shopkeepers. They may be seen at Mr. Woodward's, Old
Compton St., corner of Frith St., Soho. I find however that a doubt
exists with some excellent botanists, whether the Dutch Rush as cultivated in Holland is identical with either of the plants which I have
enumerated. Mr. Shepherd, the curator of the Liverpool Botanic
Garden, having this plant in cultivation, has most kindly supplied me
with specimens in a recent state. These are of much larger size than
any British specimens of hyemale I have yet seen, and present structural characters more strikingly different from those of either of the
British plants, than those by which these are distinguished from each
other. The most obvious differences are the much greater number of
striæ, amounting in some instances to thirty-two, and the colour of
the sheath, which, at the base, is identical with that of the stem, and
towards the apex paler or grey-green, the extreme margin being
tipped with black. The differences, however, do not appear so
great on examining a bundle of these rushes as exposed for sale; the
stems being generally of much smaller size, and the sheaths variegated with black and white, as in our British specimens.

a. Equisetum hyemale.

This plant appears almost unknown in the midland and southern
English counties. Indeed, throughout the kingdom it is but sparingly distributed, and may be considered a strictly local plant. —
In Turner and Dillwyn's 'Botanists' Guide,' the following English
habitats are recorded: — in Northumberland, Scott's wood, wood be-

---

* Omnia Equiseta ad polienda vasa metallica adhibentur. Hoc tamen ut omnium
asperrimum ita aurifabris et scriniariis suos ad usus optissimum est.
† Lightfoot, 'Flora Scotica,' 650.

low Mill Green, and Heaton wood; in Cumberland, Lowgelt-bridge; in Durham, woods about Derwent bridge and Castle Eden Dean; in Yorkshire, woods at Castle Howard and Kirkham, Rigby woods near Pontefract, near Ingleton and Halifax, about Leeds, Hackfall, near Huby, Laver Banks and Mackershaw wood, on the banks of the Skell by Ripon; in Cheshire, near Arden Hall, in a wood close to the river; in Norfolk, on St. Faith's bog and Arming Hall wood, near Norwich; in Nottinghamshire, about the middle of Nettleworth Green, two miles from Mansfield, plentifully among the rushes; in Warwickshire, in a moist ditch near Middleton; in Cambridgeshire, Stretham Ferry and Gamlingay Bogs; in Bedfordshire, Potton Marshes and Ampthill Bogs; in Wiltshire, in a rivulet near Broadstitch Abbey, plentifully. The same authors have recorded two Welch habitats:—in Denbighshire, on the west side of the brook that runs from Henllan Mill into the river Elwy, about 300 yards from Trap Bridge, less than a mile from Garm, and in Flintshire, near Maesmynnan. Through the kindness of Dr. Greville, Dr. Balfour and Dr. Campbell, I have received specimens from several Scotch habitats, more particularly in the vicinity of Edinburgh. Mr. Kippist informs me that he has seen it in abundance in the woods about Corra Lyn, Lanarkshire. In Ireland, Mr. Mackay and Mr. Moore have found it in the counties Dublin and Wicklow.

The roots are strong, black and frequently divided; the rhizoma or underground stem is creeping, jointed, branched, and with age extends to a great length: at the joints it is solid, but between them it is partially hollow, the interior being occasionally more or less divided by longitudinal septa. The stems are generally erect, and, when divided, the branch is lateral, and issues from the main stem immediately below the base of one of the sheaths; a stem has rarely more than a single branch. The annexed figure represents a branched specimen, for which I am indebted to Dr. Greville: it is from Roslin woods, near Edinburgh.

The engraving at the head of this article represents a stem of fine but not extraordinary growth: the stem has been divided into six portions in order to exhibit the whole at a single view; its diameter and length, together with the distances between the sheaths, have been faithfully copied. The sheaths in the specimen figured are fourteen in number, the internodes decreasing in length towards either end. Both the internodes and sheaths are striated, the former more strongly so : the striæ are usually about twenty in number in luxuriant stems, but this number is liable to great variation, and appears to depend solely on the size of the stem, always decreasing towards its attenuated apex. The stems are hollow, and jointed or divided by a strong transverse septum at each of the sheaths : the striæ of the sheaths correspond in number with those of the internodes, and they terminate in an equal number of acute and elongate, but membranous and often deciduous teeth. Under certain but unascertained conditions these teeth become setiform and persistent, but in general all trace of them is early lost, the upper margin of the sheath exhibiting a regular series of rounded divisions, uniform in number with the striæ of the stem. The basal portion of each sheath is black, the central part whitish, and the upper part again black, the deciduous teeth excepted, the sides of which are membranous and transparent.

Sir Humphrey Davy detected in the stem of this plant an extraordinary quantity of silex ; it is this that communicates the rough and file-like character to its exterior, from which it derives its value as an article of commerce. The silex appears in the form of minute crystals, and is arranged with beautiful and perfect regularity. Under the microscope we find on the stem numerous longitudinal series of elevations, each bearing a cup-shaped depression in its centre, at the bottom of which is placed a stoma. In the volume on Optics in Lardner's ' Cabinet Cyclopedia,' Dr. Brewster has recorded that he found each particle of silex to possess an axis of double refraction. We are told by botanists that the quantity of silex is so great and the particles are so closely set, that the whole of the vegetable matter may be removed by maceration without destroying the form of the plant.

The catkin is small, dark coloured, apiculate and terminal ; its scales are from forty to fifty in number, and each is impressed with two or three vertical striæ. Before the scales have separated in their approach to maturity, these striæ are continuous throughout the catkin, even entering its terminal apiculus, which, in consequence, assumes a polyhedral figure : they generally correspond in number with the striæ of the last internode, thus leading to the conclusion that the catkin is a metamorphosed portion of the stem.

EDWARD NEWMAN.

(To be continued.)

ART. LXXVII. — *List of Plants observed in the neighbourhood of Manchester.* By J. B. WOOD, Esq. M.D.

Broughton, near Manchester,
March 10, 1842.

SIR,

I have noticed and read with much interest and pleasure the several local lists which have appeared in 'The Phytologist.' Of the utility of such catalogues I think there can be but one opinion, as it is only by such means that the Botany of the Kingdom can be satisfactorily known, the geographical range and distribution of the various species ascertained with any degree of accuracy, and the discovery of new habitats for some, hitherto considered unique or very rare, made public. Should you deem the accompanying list of the principal plants of the Flora of this neighbourhood (and I have mentioned only such as I conceived were not generally-diffused, from the fear of trespassing too much on your pages) worthy of having a place in your valuable columns, I shall feel myself obliged by its insertion.

I am, Sir,
Yours most respectfully,
J. B. WOOD, M.D.

To the Editor of ' The Phytologist '

———

Acorus Calamus. Frequent in the neighbourhood, and flowering more or less freely every season.

Agrimonia Eupatoria. In meadows, hedge-banks &c. but not frequent.

Agrostis vulgaris, var. γ. *pumila.* On Baguley Moor.

Alchemilla vulgaris. In meadows, pastures, &c. very common.

Alisma ranunculoides. In ponds on Baguley Moor, plentiful.

Anagallis tenella. Common in boggy situations, Hale Moss, &c.

Andromeda polifolia. Abundant on all the peat-mosses.

Anthemis arvensis. Clover fields, occasionally.

Aquilegia vulgaris. Baguley Moor, Cotterill Wood, but sparingly.

Arctostaphylos Uva-ursi. Upon the hills near Mottram and Glossop.

Arundo Calamagrostis. Rostherne Mere and near Staly-bridge, in plenty.

Avena strigosa. In cornfields &c. not unfrequent.

Barbarea præcox. In cultivated fields, plentiful.

Bidens tripartita and *cernua.* Frequent; var. β. *radiata,* abundant on Wilmslow Com.

Bromus racemosus. Very common in meadows and cultivated fields.

——— *secalinus.* In fields occasionally.

Calamintha Acinos and *Clinopodium.* Rather scarce ; the former is rather plentiful in the neighbourhood of Bowdon.

Callitriche autumnalis. Reservoir near Mere Clough.

Calluna vulgaris, var. β. *hirsuta.* Baguley Moor, Boghart-hole Clough, &c.

Campanula latifolia. Frequent in many places, Broughton, &c.

Cardamine amara. In moist woods, not uncommon.

Carduus nutans. Near Disley, Cheshire.

———— *heterophyllus.* In a meadow above Mere Clough.

Carex acuta. Rostherne Mere, plentiful.

—— *ampullacea.* Abundant in many places.

—— *axillaris.* Near Leigh, very rare.

—— *binervis.* In various places near Prestwich and Pilkington.

—— *curta.* Upon peat-mosses and swampy places, very common.

—— *dioica.* Hale Moss, Baguley Moor and Knutsford Moor.

—— *elongata.* Abundant in three localities.

—— *fulva.* Baguley Moor.

—— *intermedia.* Rostherne Mere and near Moston, sparingly.

—— *lævigata.* In swampy fields and woods, frequent.

—— *muricata.* In the vicinity of Tildsley, plentiful.

—— *pendula.* Common in woods.

—— *Pseudo-cyperus.* On the margins of ponds, very common.

—— *pulicaris.* Rostherne Mere, Hale Moss, &c. plentiful.

—— *riparia.* Near Tildsley and at Rostherne Mere, in abundance.

—— *stricta.* In the greatest profusion on the borders of Rostherne Mere.

—— *strigosa.* Cotterrill Wood, Reddish, Marple, &c., very plentiful.

—— *vesicaria.* Near Tildsley and at Mere Mere, in immense quantities.

Carum Carui. Occasionally in cultivated fields.

Catabrosa aquatica. About ponds and ditches, common.

Centunculus minimus. Wilmslow Common and Hale Moss.

Cerastium semidecandrum. Kersall-moor race-ground.

Cerasus Padus. Near Agecroft.

———— *avium.* Near Rostherne.

Chrysosplenium alternifolium. In boggy woods, not uncommon.

Cicuta virosa. In the vicinity of Rostherne.

Cladium Mariscus. On the borders of Rostherne Mere.

Convallaria multiflora. Near Mottram.

Coriandrum sativum. Found occasionally.

Cornus sanguinea. Woody banks of Rostherne Mere.

Corydalis claviculata. Frequent in hedges.

Crepis paludosa. In all our swampy woods, very common.

Crocus nudiflorus. Very common in meadows.

—— *vernus.* Near Hulme and Prestwich.

Danthonia decumbens. In dry heathy situations, plentiful.

Daphne Laureola. In Cotterrill Wood.

Dianthus Armeria. Near Moston, (Miss Potts, Chester).

Drosera anglica, longifolia and *rotundifolia.* All very common on the Mosses.

Elatine hexandra. Mere Mere, Cheshire.

Elymus europæus. Cotterrill Wood.

Empetrum nigrum. On all the mountainous districts, abundant.

Epilobium angustifolium. On hills near Bury.

Epipactis latifolia. In woods, but not common.

———— *palustris.* Knutsford Moor and near Moston.

Eriophorum angustifolium, polystachion and *vaginatum.* On all the peat-mosses in profusion.

Festuca bromoides. In sandy ground, very common.

———— *loliacea.* In meadows, frequent.

———— *elatior.* On the banks of the Irwell, near Agecroft.

Fumaria capreolata. Very common.

Galeopsis versicolor. In potato-fields &c., frequent.

Genista anglica. On heathy moors, not uncommon.

Gentiana Amarella and *campestris.* Hills in Saddleworth.

———— *Pneumonanthe.* Baguley Moor and near Tildsley, frequent.

Geranium Columbinum. Not unfrequent.

Geum rivale. Mere Clough &c. in profusion.

Gnaphalium dioicum. Greenfield, and common on the neighbouring hills.

———————— *rectum.* Near Agecroft, Boghart-hole Clough &c. plentiful.

Gymnadenia albida. In Saddleworth and near Pilkington.

————————— *conopsea.* On hills behind Ashton-under-Lyne.

Habenaria viridis. Greenfield.

Helosciadium inundatum. Frequent.

Hieracium sabaudum, sylvaticum and *umbellatum.* Common.

Hypericum Androsæmum. In two or three localities.

———————— *dubium.* Not uncommon.

———————— *elodes.* Greenfield, very rare.

Hypochæris glabra. In cornfields about Bowdon, abundant.

Impatiens Noli-me-tangere. Bamford Wood, near Heywood, in great abundance, and *undoubtedly wild.*

Isolepis fluitans. Ponds on Baguley Moor, plentiful.

Lactuca muralis. Moist shady woods, not uncommon.

Lamium amplexicaule and *incisum.* The former about Bowdon, the latter in many places.

Lathræa squamaria. Near Northen, Eccles and Broughton.

Lemna trisulca and *polyrhiza.* Ponds in various places.

Lepidium Smithii. Altrincham and other places, but local.

Limosella aquatica. Mere Mere, Cheshire.

Listera cordata. Hills near Bury.

Littorella lacustris. Baguley Moor, Mere Mere, &c.

Lolium multiflorum. In cultivated fields and waste ground, *abundant.*

———— *temulentum,* var. β. *arvense.* Cornfields, not unfrequent.

Lotus corniculatus, var. δ. *tenuis.* Near Withington, and many other places.

Luzula campestris, var. β. *congesta.* Common on all our mosses.

Lysimachia Nummularia. Rostherne Mere.

Malva moschata: Frequent about Rostherne and Broughton.

Matricaria Chamomilla. Near Old Trafford in abundance.

Mentha Pulegium. Near the Railway-station at Godley.

———— *piperita.* Clayton vale, plentiful.

———— *rotundifolia.* Greenfield.

Milium effusum. In various woods, but not common.

Myosotis versicolor. Very common.

———— *sylvatica.* Cotterrill Clough and other places, abundant.

Myrica Gale. Rostherne Mere, in profusion.

Myriophyllum spicatum and *verticillatum.* Both very common.

Narcissus biflorus. Near Pilkington, plentiful.

Nasturtium amphibium. Near Stretford.

———— *terrestre.* In wet places, common.

Œnanthe crocata and *Phellandrium.* In two or three localities.

Papaver dubium. In corn-fields, very common.

———— *Argemone.* Not unfrequent.

———— *Rhœas. Very rare.*

Paris quadrifolia. Cotterrill Wood, plentiful.

Parnassia palustris. Hale Moss, abundantly.

Pedicularis palustris. In the same locality and at Rostherne Mere.

Petasites vulgaris, var. β. *hybrida.* The banks of the Irwell and Mersey are covered with the blossoms of this beautiful plant.

Pinguicula vulgaris. In swampy situations amongst the hills near Bury, &c.

Polygonum Bistorta. In meadows, very common.

———— *minus.* On Hale Moss, Baguley Moor and Wilmslow Common, very plentiful.

Potamogeton lucens. In Rostherne Mere, in abundance,

———— *pectinatus* and *perfoliatus.* In the Worsley Canal, near Tildsley.

———— *pusillus,* and var. β. *compressus.* Both equally common.

———— *rufescens.* Abundant in many places.

Primula veris. Near Altrincham, but comparatively rare.

Radiola Millegrana. Baguley Moor and Bowdon Moss.

Ranunculus Lingua. Seaman's moss-pits, Altrincham.

Rhynchospora alba. On all the peat-mosses abundantly.

Ribes Grossularia. In Cotterrill Wood, unquestionably wild.

——— *rubrum.* Red-brow Clough.

• *Rosa villosa.* Not uncommon.

——— *tomentosa.* Near Agecroft, rare.

Rubus Chamæmorus. Greenfield.

——— *saxatilis.* Ashworth wood.

——— *idæus.* In all our woods and copses, common.

——— *cæsius* and *suberectus.* Not unfrequent.

Sanguisorba officinalis. Near the banks of the river, very abundant.

Saponaria officinalis. On the banks of the Medlock, plentiful.

Saxifraga granulata. Near Agecroft and in Chorlton, but sparingly.

———— *Hirculus.* This still exists on Knutsford Moor, but is almost destroyed by the rapacity of some individuals who have dug it up *for sale,* in the most remorseless manner.

Scirpus sylvaticus. Very common.

Scutellaria minor. Near Bowdon, rare.

Sedum Telephium. In several localities.

Senecio sylvaticus. Very common.

——— *erucæfolius.* Near Bowdon and Mosley, very common.

Silene inflata. Very rare.

Sium angustifolium. About Altrincham and Bowdon, frequent.

Sparganium ramosum, simplex and *natans*.   All very common; the latter is seldom seen in flower here.

Stachys ambigua, Sm.   Mere Clough, abundantly.

Stellaria aquatica.   Near Chorlton, but rare.

——— *nemorum*.   Moist shady situations on the banks of rivers, *extremely abundant*.

Stratiotes aloides.   In ponds, frequent.

Teesdalia nudicaulis.   About Bowdon, in several places.

Thrincia hirta.   Near Tildsley, abundant.

Trifolium medium.   Woods and hedges, plentiful.

——— *filiforme*, Sm.   Near Bucklow Hill.

Triticum caninum.   About Chorlton &c. plentiful.

Trollius europæus.   Near Pilsworth in abundance.

Typha angustifolia.   Common near Stretford &c.

Utricularia minor.   Hale Moss and Baguley Moor, plentiful.

Vaccinium Myrtillus, Oxycoccos and *Vitis-idæa*.   Common on heaths and moors.

Valeriana dioica.   Near Reddish and in Broughton, abundantly.

Valerianella dentata.   In cornfields &c. not unfrequent.

Veronica montana and *scutellata*.   Abundant in many places.

Vicia sylvatica.   Cotterrill Wood.

——— *tetraspermum*.   In cornfields, occasionally.

Villarsia nymphæoides.   Near Chorlton, but originally introduced.

Vinca minor.   In woods near Marple, and near Agecroft.

Viola odorata.   Hough-end, but rare.

——— *palustris*.   In swampy situations, very frequent.

——— *lutea*.   In mountainous pastures behind Mottram.

Wahlenbergia hederacea.   Near Mottram, but rare.

FERNS AND THEIR ALLIES.

Allosorus crispus.   Fo-edge, near Bury, in great profusion.

Asplenium Ruta-muraria.   Hough-end, but not common.

——— *Trichomanes*.   Arden-Hall, near Reddish.

Botrychium Lunaria.   Rough upland pastures in several localities.

Cystopteris fragilis.   Rostherne Church.

——— *dentata*.   Salebark and Greenfield, in both places very rare.

Equisetum sylvaticum.   In woods and hedges, very abundant.

Hymenophyllum Tunbridgense and *Wilsoni*.   Both found on rocks at Greenfield, but very sparingly.

Lastræa Oreopteris.   In various woods, common.

——— *spinulosa*.   Baguley Moor, rare.

——— *Thelypteris*.   On the borders of Rostherne Mere in great plenty.

Lycopodium alpinum, Selago and *clavatum*.   Frequent on the high mountainous moors near Bury, and at Greenfield.

——— *inundatum*.   Baguley Moor, plentiful.

Osmunda regalis.   In the same place as the last, and on Chat Moss.

Pilularia globulifera.   In ponds &c. on Baguley Moor.

Polypodium Phegopteris.   In Mere Clough, Boghart-hole Clough &c. in abundance.

——— *Dryopteris*.   In the same localities, but more sparingly.

Mr. Babington meets these objections by observing that he cannot but think he has done well in keeping distinct all the species here described; for, however anomalous may be their forms, he can always refer the living plants to their respective species: and that if they are not to be separated by the characters employed in this paper, they must be reduced to fewer species than the author conceives would be recommended by " the most energetic 'lumper of species.'"

In Meyer's 'Flora Altaica' the European species of Chenopodiaceæ are divided into four tribes, which may be thus characterized.

1. *Salsoleæ.* Seeds with little or no albumen; embryo spiral. British genera, *Salsola* and *Schoberia*.
2. *Chenopodiaceæ.* Seeds albuminous; embryo forming a ring round the albumen; (in these points agreeing with the two following tribes): flowers hermaphrodite: stems not jointed. British genera, *Chenopodium* and *Blitum*.
3. *Salicorneæ.* Flowers hermaphrodite: stem jointed. British genus, *Salicornia*.
4. *Atripliceæ.* Flowers diœcious or monœcious, sometimes with a few hermaphrodite flowers intermixed: stems not jointed. British genera, *Atriplex* and *Halimus*.

Mr. Babington remarks — " These tribes are well marked by their very different habit, although in description their characters may not appear to be peculiarly strong:" and then proceeds to describe the genera and species to which the paper refers.

I. ATRIPLEX, *Linn.* Flowers polygamous: female perigone compressed, formed of two distinct or more or less connate leaves: stigmas two: pericarp membranaceous, free: seed vertical, either attached near the base by a lateral hilum, or towards the centre by means of an elongated funiculus; testa crustaceous; radicle inferior, ascending. " *Wallr. Sched. Crit.* 114. *Nees ab Esenbeck, Gen. Pl. Germ. Icon. (Monochlam.)* 63."

The genus Atriplex, as thus restricted, may be again divided into the two following sections.

1. *Euatriplex*, Meyer, (the true Atriplices). Leaves of the female perigone two, distinct to the very base: seed vertical; horizontal in the hermaphrodite flowers, which are rarely produced.
2. *Schizotheca*, Meyer. Flowers monœcious, (true hermaphrodite flowers never appearing): leaves of the female perigone more or less connected below, the attachment not extending above the lower half.

"In both of these sections, only the latter of which has as yet been detected amongst the native plants of Britain, the pericarp is quite detached from the perigone, the testa is crustaceous, and the radicle, although always ascending, is never terminal." — 3.

1. *Atriplex littoralis*, Linn. Root fibrous. Stem erect, two or three feet high, smooth, angular; branches numerous, alternate, erect. Leaves linear-lanceolate, en-

tire, or sometimes slightly toothed, stalked. Spikes long, erect, clusters distinct, of from 4—12 or more flowers. Enlarged calyx-valves rhomboidal-ovate, acute, toothed, with spreading tubercles on the back; terminal point rather prominent. Seeds black, smooth and shining. "*Linn. Sp. Pl.* 1494! *Bluff et Fingerh. Comp. Fl. Germ.* (ed. 2), i. pt. 1. 446; *Reich. Fl. Excurs.* 577; *Sm. Eng. Fl.* iv. 260; *Eng. Bot.* 708; *Huds. Fl. Ang.* (ed. 2). 444; *Wahl. Suec.* 661; *Fl. Alt.* iv. 311; *Detharding, Cons. Megap.* 24; *Bab. Prim. Fl. Sarn.* 81."

On the sea-coast in many places. Annual. July to November.

2. *Atriplex marina*, Linn. Root fibrous. Stem erect, two or three feet high, smooth, angular; branches numerous, alternate, erect. Leaves ovate-lanceolate, irregularly toothed, sometimes nearly entire; in luxuriant specimens the lower leaves are nearly ovate, inciso-serrate: stalked. Spikes long, erect, clusters distant, of from 3—6 or 8 flowers. Enlarged calyx valves somewhat cordate-triangular, obtuse, toothed, tuberculated on the back. Seeds smooth and shining, black, with a slight reddish tinge, fuscous-red when immature. "*Linn. Mantissa,* 300; *Bluff et Fingerh.* 446; *Reich.* 577; *Dethard.* 24; *Bab. Prim.* 81. A. serrata, *Huds.* (ed. 2) 444. A. littoralis, β. *Wahl.* 661; *Sm. Engl. Fl.* iv. 260."

On the sea-coast in many places. Annual. July to September.

Mr. Babington remarks—" These two plants (*littoralis* and *marina*) differ from all the following, by the total want of larger lobes at the base of their leaves. The latter species has long been considered as only a variety of the former, although originally distinguished by Linnæus, and also adopted as a species by Hudson. Within the last few years, they have been again distinguished upon the authority of the observations of Detharding, as published in his Conspectus quoted above; and I have much pleasure in recording, in confirmation of their distinctness, the valuable and independent observations of Mr. Power."—p. 6.

It is probable that in a living state the plants may differ more than they do in description; the chief points of distinction appear to be that in A. littoralis the enlarged calyx-valves have " the apex in general considerably projecting and acute, each in an advanced state *remarkably diverging* from its fellow, and giving a peculiarly rough aspect to the plant," the tubercles on the margins of the valves are not coloured, and " the whole plant is more or less covered with a *greenish-hoary* mealiness: " while in *A. marina* the calyx-valves have " an obtuse contour at the apex, which scarcely at all projects from the outline of the valve, in many cases appearing like a mere continuation of the marginal denticulations : they are *closed* when mature: " the points of the tubercles on the margins of the valves are generally reddish, " and the whole fruit and the fruit-stalks have a peculiar *yellowish* mealiness."

3. *Atriplex angustifolia*, Sm.! Stem erect or prostrate, nearly round, striated, usually with long, simple, opposite branches; each branch and the main stem termi-

nating in a wand-like, interrupted, subsimple spike of distant few-flowered clusters. Leaves lanceolate, entire; lower ones hastate, the lobes ascending from a wedge-shaped base, all shortly stalked. Enlarged calyx-valves rhomboidal, acute, entire; lateral angles acute, prominent, ascending, without tubercles on the back, reticulated, rather longer than the smooth, shining, black seeds. "*Sm. Eng. Fl.* iv. 258 ; *Wallr. Sched. Crit.* 116; *Eng. Bot.* 1774; *DC. Fl. Fr.* v. 371. A. patula, *Huds.* 443 ; *With. Arr.* ii. 275 ; *Wahl.* 660."

Common on waste and cultivated land. Annual. July to October.

"In its normal state the whole of this plant is slender and delicate, the leaves thin, and the calyx of the fruit small, but sometimes the stems and leaves become thickened and very fleshy, and the calyces exceedingly enlarged (I have seen them an inch in length, and broad in proportion) and even tubercled. In this monstrous form the fruit is usually transformed into leaves, and no seed is produced. Various intermediate states occur, and often only a few of the calyces become monstrous, whilst the rest of the plant retains the normal appearance."—p. 7.

4. *Atriplex erecta*, Huds. Stem most frequently erect, sometimes weak and prostrate, quadrangular, striated, often reddish, 12 or 18 inches high ; branches mostly opposite, simple, ascending. Leaves mostly opposite ; lower leaves ovate-oblong, wedge-shaped at the base, with ascending lobes, irregularly dentate, sinuato-dentate or inciso-dentate ; upper leaves lanceolate or nearly linear, entire; all pale green above, mealy beneath. Spikes many-flowered, terminal and axillary, shortly branched ; flowers in small, round, dense clusters, usually so close as to appear continuous on the spike, in which respect this plant differs from Atr. angustifolia. Enlarged calyx-valves rhomboidal, acute, toothed above the lateral angles, which are acute, sometimes prominent ; valves more or less muricated on the back, scarcely longer than the seed, densely clothed with minute, pellucid, crystalline glands, which dry into a mealy coat. Seed black, smooth, shining, half the size of that of Atr. patula. "*Huds. Fl. Ang.* (ed. 1) 376 ; *Sm. Eng. Bot.* 2223! *Eng. Fl.* iv. 259 ; *DC. Fl. Fr.* v. 371 ; *Bab. Prim.* 82 ; *Fl. Bath. Suppl.* 88. A. angustifolia, *Drej. Fl. Hafn.* 106."

Common on cultivated land throughout England. Annual. July to October.

It is probable that this plant is frequently confounded with Atr. angustifolia, which Mr. Babington says it sometimes resembles in the lower leaves being without teeth and the spikes being interrupted, but from that species "it is still clearly distinguishable by its compound spike, calyx-valves and leaves." He also observes : —

"Our plant is certainly the Atr. erecta of Hudson and Smith, of which the only authentic specimen, preserved in Sir J. E. Smith's herbarium, is apparently only the upper part of a very luxuriant plant; in it the calyx of the fruit is much more spinous than is usually the case, and the panicle larger and more dense."—p. 8.

5. *Atriplex prostrata*, "Bouch." Stem prostrate, quadrangular, somewhat striated, much branched, branches prostrate. Leaves nearly opposite, fleshy, hastate-triangular, entire or with few teeth, lateral lobes horizontal or slightly descending, the base

truncate, margins entire or rarely sinuato-dentate; in the intermediate leaves the lateral lobes are ascending, the base wedge-shaped, and one or two teeth are generally to
be found above the lobes; the uppermost leaves are lanceolate and entire, gradually
decreasing until they become very small nearly linear bracteas: all the leaves, as well
as the other parts of the plant more or less clothed with mealiness. Spikes terminal
and axillary, numerous, slightly branched, clusters small and distinct, as in Atr. patula and angustifolia. Valves of the calyx of the fruit cordate-triangular, often scarcely
longer than broad, and but just covering the seed, sometimes more nearly cordate,
rarely much elongated and with two prominent angles between the lateral ones and the
apex. The valves are often unsymmetrical in form, usually slightly toothed. Seeds
black, smooth, shining, small. " ' *Bouch. Fl. Alb.* 76 ;' *DC. Fl. Fr.* iii. 387 ; *Bot.
Gall.* 398 ; *Lois. Fl. Gall.* (ed. 2.) i. 218.   A. triangularis, *Willd. Sp. Pl.* iv. 965 ?
Reich. 578 ? A. latifolia, *Wahl.* 660 ? *Drej. Fl. Hafn.* 107 ? "

Common on the sea-coast.   Annual.   August to October.

There appears to be much uncertainty respecting this plant. None
of the descriptions of the authors referred to exactly agree with it, although those of Wahlenberg and Drejer's Atr. latifolia come nearer
than any others.   Mr. Babington, considering our plant to be identical with DeCandolle's Atr. prostrata, notwithstanding the imperfect
description, has preferred adopting that name rather than to "intro-
duce another species into this difficult genus."

6. *Atriplex patula*, Linn.   Stem erect, quadrangular, striated; branches ascending.   Lower leaves ovate-hastate, toothed, lobes horizontal; upper leaves lanceolate,
usually entire.   Spikes terminal and axillary, long, wand-like; clusters of flowers numerous, distinct. Valves of the calyx of the fruit rhomboidal triangular, often entire,
usually slightly muricated on the back, lateral angles obtuse; much longer than the
fruit.   Seeds opaque, rough, often tinged with red.   "*Linn.* 1494 ! *Sm. Eng. Bot.*
936 ! *Eng. Fl.* iv. 257 ; *Reich.* 577 ; *Bluff et Fing.* 445 ; *Wallr. Sched.* 115 ; *Fl. Alt.*
310 ; *DC. Fl. Fr.* v. 370 ; *Bab. Prim.* 83.   A. hastata, *Huds.* 443 ; *With.* 274.   A.
latifolia, β. elatior, *Wahl.* 660."

Common throughout the country.   Annual.   June to October.

This plant is very variable in luxuriance; in a rich soil the calyx-
valves often take the character of leaves, and then "several sets of
apparent calyces are found within each other, with one or more enlarged fruits inclosed in them."   Attention must also be paid to the
characters of Atr. angustifolia, prostrata, deltoidea, microsperma and
rosea, in order to avoid confounding them with the present species.

7. *Atriplex microsperma*, Walds. et Kit.   Stem erect or ascending, striated, angular; branches ascending.   Leaves opposite; lower ones ovate-hastate, toothed, lobes
prominent, horizontal; upper leaves small, linear-lanceolate, almost awl-shaped, very
acute, entire, with a prominent, acute, horizontal lobe on each side of the truncate
base.   Flowers in small, close clusters, forming on the stem a compound, branched,
terminal panicle; on the branches they take the form of branched spikes.   Valves of

the calyx of the fruit ovate, triangular, acute, entire, rugose but very rarely tuberculated on the back, rather longer than the fruit. Seed smooth, shining, black, about half the size of that of A. patula. " *W. et K. Plant. Hung. Rar.* t. 250; *Reich.* 578; *Bab. Fl. Bath. Suppl.* 88; *Sadl. Fl. Pest.* 475. A. ruderalis, *Wallr. Sched.* 115; *Bluff et Fingerh.* 445."

"On waste ground near Bath. Wouldham in Kent, *Mr. C. A. Stevens.* Annual. July to September."

8. *Atriplex deltoidea*, Bab. Root fibrous. Stem erect, quadrangular, striated, branched; branches ascending, one or two feet high, often tinged with red. Leaves mostly opposite, all triangular-hastate, truncate at the base, lobes descending, irregularly dentate or sinuate-dentate, sometimes nearly entire, apex acute-angled, dull green above, mealy beneath; uppermost leaves or bracteas usually of the same form as the lower ones, only longer in proportion to their breadth, rarely entire, with a tendency to a wedge-shaped base. Spikes numerous, branched, densely flowered, forming a large terminal panicle, each lateral branch also terminates in a branched spike. Clusters of flowers small, round, close together. Calyx of the fruit small, ovate-triangular, sometimes almost cordate below, acute, truncate, strongly muricated, slightly stalked, but little longer than the fruit, thickly covered with a fine mealy coat. Seed black, smooth, shining, about half as large as that of A. patula, reddish when immature. Upper part of the plant covered with a minute crystalline afterwards mealy coat. — "*Bab. Prim. Fl. Sarn.* 83; *Leight. Fl. Shrop.*! 501."

"On cultivated and waste land near London, in Kent, Leicestershire, and near Maidenhead. Annual. July to October."

" Having now studied this plant during three successive autumns, I am confirmed in my opinion, that it is a distinct and unnoticed species. It is now found to be rather a common native of England."—p. 13.

9. *Atriplex rosea*, Linn. Stems diffuse, procumbent or ascending, usually slender, square, striated, sometimes much thickened and fleshy, clothed with whitish meal in common with all parts of the plant, often beautifully tinged with red or purple, with spreading branches. Leaves ovate-triangular, with two large, prominent, horizontal lobes at the base, irregularly sinuato-dentate, very white and mealy beneath; upper leaves similar in general character, but with the lobes smaller in proportion and the leaf lanceolate: in the more fleshy plants the leaves are more triangular and less lobed, the upper ones being more lanceolate and nearly entire; in a straggling much branched form occurring on muddy shores, all the leaves are lanceolate and nearly entire, a few only having small basal lobes. Clusters few-flowered, small, distinct, either collected towards the end of the stem and branches into a somewhat spicate form, each being subtended by a small lanceolate bractea; or a few of the uppermost clusters only are bracteated, the rest being axillary; or else the clusters are all axillary, except the two or three last, and all so much scattered that the spicate appearance is quite lost. Calyx of the fruit large, rhomboidal, acute, toothed in the upper part, with a double series of tubercles on the back, sometimes nearly smooth, varying in outline even on the same plant. Seed large, tubercularly rugose, opaque, tinged with red. " *Linn. Sp. Pl.* 1493; *Koch,* 611; *Bluff et Fingerh.* 443; *DC. Bot. Gall.* 398; *Fl. Alt.* 314; *Bab. Prim.* 84; *Sadl. Pest.* 476. A. alba, *Reich.*! A. patula, β. *Smith, Fl. Br.* iii. 1092!

2 B

Common on the sea-coast in rocky, gravelly, or muddy situations. Annual. July to September.

Apparently the most variable of the genus, but when once known easily distinguishable from all the other species. It often approaches Atr. patula and angustifolia in the form of the leaves and the scattered clusters of flowers, and Atr. laciniata in the form of its fruit; from all these the very large lobes of the leaves, the form and dentate margins of the calyx, and the scattered flowers will serve to distinguish it.

10. *Atriplex laciniata*, Linn. ! Whole plant covered with whitish meal. Stems diffuse, prostrate, branches spreading. Leaves irregular in outline, between triangular and rhomboidal, irregularly toothed and lobed, hoary beneath. Clusters of flowers small, mostly collected into leafless terminal spikes, a few only being seated in the axils of the upper leaves. Calyx of the fruit rhomboidal, with each lateral angle broadly truncate; the valves vary in breadth, but retain the general form, three-ribbed on the back, lateral ribs sometimes tuberculated towards the extremity. Seeds rough and opaque. " *Linn.* 1494; *Sm. Eng. Bot.* 165; *Eng. Fl.* iv. 257; *Wahl.* 661; *Koch*, 611; *Bluff et Fingerh.* 414; *Fl. Alt.* 313; *Bab. Prim.* 84; *Sadl. Fl. Pesth.* 476."

Common on the sea-coast. Annual. July to September.

II. HALIMUS, *Wallr.** Flowers monœcious: female perigone compressed, leaves two, tridentate, connate to the apex: stigmas two: pericarp very slender, when mature adhering to the tube of the perigone: seed vertical, pendulous by an elongated funiculus, ascending to the apex ; testa membranaceous; radicle terminal, porrected. " *Wallr. Sched. Crit.* 117; ' *Wahl. Act. Upsal.* viii. p. 228, 254, t. 5, f. 2;' *Fl. Suec.* 662; *Nees ab Esenbeck Gen. Pl. Germ. Icon.* (*Monochlam.*) 64."

" In these plants the perigone is contracted below into a peduncle, which in H. pedunculatus is elongate, and although short, is still present in H. portulacoides."—p. 4.

1. *Halimus pedunculatus*, Wallr. Stem herbaceous, erect, flexuose, shortly branched. Leaves obovate-oblong, obtuse, entire, contracted at the base into a short petiole, upper ones of the same form but narrower. Flowers scattered, in a lax terminal spike, sessile ; as the fruit ripens the base of the calyx becomes lengthened into a long slender peduncle, the upper part taking an inversely wedge-shaped form, with two obtuse lobes and an acute intermediate point. " *Wallr. Sched. Crit.* 117; *Reich.* 576; *Bluff et Fingerh.* 442; *Koch*, 609. Atriplex pedunculata, *Linn.* 1675! *Sm. Eng. Bot.* 232; *Eng. Fl.* iv. 261."

On the sea-coast, very rare. Annual. August and September.

Mr. Babington mentions the variations in length to which the peduncle appears subject; it being, in his English specimen from Yarmouth, nearly an inch long, while in a German specimen it scarcely

---

* This genus was founded by Wallroth in his ' Schedulæ Criticæ,' 117, for the reception of Atriplex pedunculatus, *Linn.*; and he observes — " Perhaps Atr. portulacoides may also belong to this genus; I have not seen the fruit."

exceeds a line: in the latter, too, the expanded part of the calyx differs in form, the lateral lobes being rounded, and not longer than the intermediate one. The author remarks, — "Can there be two species confounded under this? Unfortunately the extreme rarity of the plant puts great difficulty in the way of the determination of this point."

2. *Halimus portulacoides*, Wallr. Stem erect or ascending, woody, branched. Leaves oblong-lanceolate, obtuse, entire, contracted below into a rather long petiole. Flowers in small clusters, forming a small, branched, terminal raceme. Peduncle very short; calyx of the fruit rounded below, widening upwards, three-lobed at the top, intermediate lobe usually longest; lower part of the back of the valves muricated.— " *Wallr.* 117; *Reich.* 576; *Fl. Germ. Exsic.* 870! *Bluff et Fingerh.* 442; *Koch.* 609. Atriplex portulacoides, *Linn.* 1493! *Sm. Eng. Bot.* 261; *Eng. Fl.* iv. 256; *Fl. Dan. tab.* 1889."

Common on the sea-shore. Perennial. August to October.

The existence of two species under the name of H. portulacoides, is indicated by Nees ab Esenbeck,—"the one, *borealis*, described by Roth in the 'Flora Germanica' and figured by Nolte in 'Flora Danica,' the other, *australis*, occurring in the South of France, and differing from the first in habit, in the narrower and more scaly leaves, and in the pointless calyces of the fruit."* Mr. Babington says —

"Specimens which I possess from the south of Europe are unfortunately in much too young a state for the characters drawn from the fruit-bearing calyces (the only ones of value) to be determined. I suspect that specimens which I gathered upon Exmouth Warren, in Devonshire, in September, 1829, will prove to be this latter plant; for although the fruit, from being young, will not allow a certain conclusion to be drawn, yet, as far as I can ascertain, the calyx is totally without tubercles upon its back, and it appears to be much less rounded below than is usual in the true H. portulacoides. * * It is much to be wished that those botanists who may visit the southern coasts of Britain, would endeavour to determine the existence of the latter plant (H. australis) upon our shores, and also that they would turn their attention to the value of the character drawn from the muricated calyces in the genus Halimus."—p. 16.

Each species is illustrated by very neatly engraved figures in outline, of the lower leaves, the leaves near the top of the stems, and the enlarged calyces of the fruit.

(To be continued).

ART. LXXIX.— *Varieties.*

179. *Additions to the List of Wharfedale Mosses.* I returned a short time since from a two days' excursion in Wharfedale. This excursion was partly a geological

---

* Genera Plantarum Floræ Germanicæ Iconibus Illustrata, (Monochlam.) No. 64.

one, and partly made in order to renew my acquaintance with a few of the mosses and lichens which grow in the neighbourhood of Bolton-bridge, Barden, &c. With the idea that local lists of plants often prove interesting to botanists, I now send you the following additions to the list given by Mr. Spruce, (Phytol. 197). I hope my friend Mr. S. will pardon the use of the more practical names of Leighton's Catalogue, in preference to those of Hedwig, Bridel, &c.; for would it not have been just as well understood if he had called his Grimmia rivularis G. apocarpa, var. *a*. nigro-viridis, Hooker, and for Weissia fugax said W. striata, var. *a*. minor, Hooker? For the Weissia striata of Hooker is Bryum crispatum of Dickson, Grimmia striata of Schrader, Weissia Schisti of Schwægrichen, and the Grimmia Schisti of the 'Flora Britannica.' Mr. Spruce may be right, but if we carry out this plan where are we to stop? for so great is the multiplication of synonymes in this department of Botany, that it has become a matter of difficulty in some cases to decide which name ought to be adopted.

Mosses.

Andræa Rothii
Phascum cuspidatum
    alternifolium
Gymnostomum rupestre
Anictangium ciliatum
Encalypta streptocarpa
Weissia nuda
    lanceolata
    acuta
Grimmia apocarpa β. stricta
Trichostomum fasciculare
    heterostichum
Dicranum adiantoides
    flexuosum
    β. nigro-viride
    squarrosum
    pellucidum
    scoparium
    β. majus
    γ. fuscescens
Polytrichum piliferum
    alpinum
Orthotrichum cupulatum
    anomalum
    affine
    diaphanum
    striatum
    crispum
    pulchellum
Bryum androgynum
    turbinatum

Bryum nutans
    ventricosum
    roseum
    ligulatum
    punctatum
    rostratum
Leucodon sciuroides
Fontinalis squarrosa
Hypnum complanatum
    serpens
    polyanthos
    sericeum
    curvatum
    splendens
    flagellare
    ruscifolium
    striatum
    triquetrum
    molluscum
    multiflorum

Lichens.

Verrucaria rupestris
Pertusaria communis
    fallax
Thelotrema lepadinum
Variolaria discoidea
Urceolaria scruposa
Lecanora atra
    Parella
    tartarea
Squamaria elegans
    candicans

Parmelia caperata.
    olivacea
    pulverulenta
Sticta crocata
    sylvatica
Peltidea horizontalis
    aphthosa
    canina
    rufescens
    polydactyla
Borrera tenella
    furfuracea
Evernia prunastri
Ramalina fraxinea
    fastigiata
    farinacea
Sphærophoron coralloides
Scyphophorus radiatus
    gracilis
    filiformis
    deformis
    digitatus
Opegrapha scripta of *Lin*.
    sixteen varieties
    Confervoideæ.
Conferva purpurascens
    vesicata
Draparnaldia glomerata
Chætospora endiviæfolia
    elegans
Gomphonema ampullaceum

—*Samuel Gibson ; Hebden Bridge, June 5,* 1842.

180. *Note on Carex tenella &c.* Perhaps you will allow me a little room to make a few remarks on Carex tenella, (Phytol. 234). Allowing that I have made some mis-

take in relying too implicitly on Schkuhr's figures, I must again say, without reference to those figures, that I am much surprized to find that after Sir W. J. Hooker has given us a description of C. tenella (which is verbatim from Smith), differing in every point from C. remota, that he should ever say " may it not be, &c." But I will just see what Sir J. E. Smith says on the subject. He first tells us that his plant is the C. tenella of Schk. Car. 23. t. P, p. f. 104, exclusive of *i. k. l :* he then tells us Willdenow and Wahlenberg refer the same plant of Schkuhr to C. loliacea, which he says differs in having a ribbed fruit, flat on one side, &c., this he also tells us is Schkuhr's C. gracilis, 48. t. E. f. 24. Then he goes on to say that it appears to him that Schkuhr has drawn the ripe fruit of his C. tenella, fig. i. k. l. from a starved specimen of C. loliacea, (Eng. Fl. i. [iv.] 83). I shall now refer to the three figures of Schkuhr, viz. i. k. l. and see how far they represent a *ribbed* fruit, which is *flat on one side*. The figures i. k. l. I find, as I expected, to represent a *smooth* fruit, *equally convex on each side :* here again I am under the painful necessity of either becoming a critic, or, what appears to me to be much worse, of believing that the starving of C. loliacea would change its ribbed plano-convex fruit into fruit which is smooth, and equally convex on both sides. For my own part I can no more believe that the starving of a plant will cause all these changes in the character, than that the starving of the soil in which they grow would make it produce new plants altogether.—*Id.*

181. *Note on Sagina maritima.* Perhaps it may not be out of place here to say where my Sagina maritima was gathered, so far as Warrington is concerned, (Phytol. 179 and 234). My first specimens are from the late Mr. E Hobson, gathered at Curedly Marsh, in July, 1824. I have the plant from the same locality, gathered in 1840 and 1841 by two different persons. Again, I have Sagina maritima gathered by Mr. G. Crozeir, three and a half miles from Runcorn Gap, nearer to Warrington. Curedly Marsh, if I am rightly informed, is three and a half miles from Warrington; Runcorn Gap is eight miles from Warrington.—*Id.*

182. *Lotus angustissimus.* I take the liberty of sending you some (I am afraid rather poor) specimens of Lotus angustissimus. I found them near this place in rather a rocky soil a short time ago; they were rather scarce, I found only three.— *Robert Jordan ; West Teignmouth, Devon, June* 21, 1842.

183. *Monograph of the British Roses.* I wish some one of your more able correspondents would kindly publish in your admirable work ' The Phytologist,' a monograph of the genus Rosa, with the varieties ; I think, to a beginner, it is one of the most puzzling genera that can be.—*Id.*

184. *Trifolium filiforme.* Having for several years past narrowly watched this plant, I send for insertion in your periodical, the result of my enquiries, which will not be thought superfluous, when, even in so late a work as Leighton's ' Flora of Shropshire,' this truly distinct species is passed by as a mere variety of T. minus, and the figure in ' English Botany,' t.1257, which is sufficiently expressive, is not at all alluded to. I can confirm the account of this plant given in the ' English Flora ' by Sir J. E. Smith, as very faithful. The *racemose* inflorescence is an unfailing character, and there are not wanting other peculiarities by which it may be known from T. minus, Eng. Bot. t. 1256. The two plants grow together near Warrington in several places, and I am quite satisfied that the view taken of T. filiforme in Hooker's ' British Flora,' and especially in Leighton's work, is erroneous. It may easily be recognized by its more truly procumbent or prostrate habit, its deep yellow almost fulvous flowers and its dark green foliage. The common stalk of the leaves is always very short, about half as long

as the stipules. The corolla does not "become tawny as the seed ripens," but turns very pale, and owing to the narrowness of the petals, and especially of the standard (which is deeply emarginate), the legume or fruit as it ripens becomes quite conspicuous; whereas in T. minus it is entirely covered and concealed by the faded, deflexed and scariose standard, twice as broad as in T. filiforme, and furrowed. The diligent observer will find other points of difference, which I forbear to enumerate. I have sought in vain for intermediate states, and fully believe that the two species may be identified, if only a single flower of each be produced for that purpose. Your useful periodical will, I trust, raise up a host of *field-botanists*, who will put to the test all our dubious species, and point out the diagnostic characters of such as are genuine but imperfectly described. These will counteract the modern epidemic termed *hair-splitting*, which sometimes intrudes itself into your pages, and must be tolerated until persons infected with the disease have learned better. If any of your readers suspect me to have now fallen into this error, let him investigate the subject for himself. Nature will "deceive no student," if he diligently explore the volume with —

> " A mind well strung and tuned
> To contemplation, and within his reach,
> A scene so friendly to his fav'rite task."

— *W. Wilson ; Warrington, June 23, 1842.*

185. *Poa maritima and P. distans.* These are unquestionably distinct species, but unless studied at large in their native haunts, apt to be confounded. The creeping root of P. maritima is not always a very obvious character. The leaves in both species seem to be equally convolute; but those of P. maritima are destitute of the seven prominent rough ribs found on the leaves of P. distans. P. maritima also has the branches of the panicle smooth. In favourable situations the stems are quite prostrate, yielding an abundant and heavy crop of herbage. The produce of a single root will sometimes cover a space of more than three feet in diameter.—*Id.*

186. *Potamogeton setaceum of Hudson.* It has long been my opinion that this plant is no other than a narrow-leaved variety of the species now called P. oblongus. In peat-ditches on the borders of Risley Moss, near Warrington, this plant may be seen in various states, from the normal form to one with extremely narrow (linear-lanceolate) floating leaves; but when in that state the plant is sterile. P. oblongus is most prolific when growing in shallow water, and its broadest leaves are produced when the plant is almost left dry. In deep water it becomes P. setaceum of Hudson. Sir J. E. Smith, in 'English Botany,' t. 1985, remarks that no one knows this plant. —*Id.*

187. *Circæa alpina and C. lutetiana.* The first of these has each flower-stalk subtended by a bractea; while in the other species bracteas seem to be always absent.— Those who have opportunity for extended observation are requested to try the validity of this discriminative mark.—*Id.*

188. *Scleranthus perennis.* Never having seen this plant in a growing state, I offer with some hesitation the following remarks, which I trust some one will put to the test. In habit the plant seems very different from S. annuus; for instead of being repeatedly forked, with wide-spreading divisions and flowers in the forks of the stem, my specimens are mostly unbranched, never dichotomous, and the flowers are terminal, the leaves nearly erect and directed to one side, the whole plant having much of the habit of Spergula nodosa; so that if my specimens truly represent the species, they might be thus characterized : —

S. annuus. Stems dichotomous, leaves widely spreading, flowers in the forks of the stem.

S. perennis. Stems irregularly branched, leaves erect, unilateral, flowers terminal. —*Id.*

189. *Equisetum fluviatile.* There is a large patch of Equisetum fluviatile at Norwood. A road leads down the hill from the neighbourhood of the Woodman Inn towards Dulwich; and a little way down the hill, on the right hand, a quantity of the soil has been dug out, so as to leave a precipitous bank; on the acclivity of this bank, and about a small pond close by, the above-named plant grows in such luxuriance as to make a very handsome appearance. The whorls of dark green leaves, rising one above the other, can only be compared to a miniature grove of pines, growing up the side of some steep mountain in Germany.—*Wm. Ilott; Bromley, Kent, July 2, 1842.*

190. *Cucubalus baccifer*, (Phytol. 255). In the margin of my copy of Ray's Synopsis, against Cucubalus Plinii (C. baccifer) a former possessor of the book has written as a habitat *Springfield, Essex.* From the colour of the ink and style of writing it is evident this entry was made soon after the Dillenian edition of the Synopsis appeared. Perhaps some of your readers, on seeing this note, will search in the neighbourhood of Springfield for this plant, which may very readily have been overlooked. —*H. O. Stephens; 78, Old Market St., Bristol, July 5, 1842.*

191. *An Hour's Botanizing among the Falls of Lawers.* I should like to call the attention of your readers for a few minutes to this spot, not so much for its botanical treasures, as for the exquisite beauty of its scenery, which is, I fear, not much known to tourists, in consequence perhaps of being overlooked in the guide-books. It is, however, well worthy of being visited by every lover of picturesque scenery who passes this way, and to the botanist it is a little Garden of Eden. The principal Falls are situated in a rocky dell embosomed in a fir-wood, a few minutes walk up from the toll-bar of Lawers. Before reaching the wood a number of beautiful little cascades occur, several of them fourteen or fifteen feet high, some of them overshadowed by graceful ash-trees, and all margined by verdure-mantled rocks glowing with bright blossoms, among which Vicia sylvatica, Geranium sylvaticum and the "foxglove's purple bells" are conspicuous. Polypodium Phegopteris and Polygonum viviparum are also here in abundance. On entering the dark wood the rocks assume a wilder aspect, the roar of water greets the ear, and we soon come upon the principal falls, which, although by no means so imposing as those of Acharn, Moness or Bruar, possess a charm peculiar to themselves. In the upper one the water takes four distinct leaps, and in the lower is precipitated through a narrow channel over a perpendicular rock about fifty or sixty feet high, into a dark and dismal-looking pool beneath. The harmonious and soothing murmur of the smaller cascades is now exchanged for the roar of the cataract, the light of the sun is almost shut out by the overhanging woods, and the rocks that rise rugged and lofty are garnished profusely by tufts of ferns, and wild flowers of varied hue. The botanist will here find in great exuberance among the moist rocks the beautiful little Asplenium viride and the delicate Hymenophyllum Wilsoni. Hieracium paludosum is plentiful, and Circæa alpina, Alchemilla alpina, Oxyria reniformis, Melica uniflora, Festuca ovina, var. *e.* vivipara, Melampyrum sylvaticum and pratense, occur more or less abundantly. In the crevices of the rocks may be gathered Hypnum commutatum in fine fructification, with plenty of Bryum turbinatum and ventricosum, Marchantia hemisphærica and the elegant Hypnum rufescens. Bryum julaceum grows by the sides of the stream, and many others might be detected by a

little closer examination. The dell all the way up is a perfect picture-gallery of sweet little water-falls, and an hour spent in exploring their varied beauties cannot fail to be pleasantly remembered in after years. To the lover of Flora it will furnish an excellent preface to the mighty volume of Ben Lawers, of which I will say something to your readers by and bye; and the first thing he should do on reaching this snug little inn, is to enquire for the *Falls of Lawers.—William Gardiner, jun.; Ben Lawers Inn, July 6,* 1842.

192. *Note on the British Pyrolæ.* The "Enquiry respecting Pyrola media" by Mr. Simpson (Phytol. 237), is one of which I should like to see many more examples. I am persuaded the two species P. media and rotundifolia are often confounded: with regard to the latter Hooker observes — "Yorkshire and many places in Scotland are assigned as stations for this plant; but it is so often confounded with the two following species (media and minor), that I cannot quote them with equal certainty." From what I have seen of Pyrola minor in its living state, I should think that no one could confound it with P. media. The rose-coloured, pink or nearly white, and much smaller flowers, and its short style and broad stigma *without* erect points, serve at once to distinguish it. Pyrola media is a much stouter plant, with larger flowers, which are generally whitish, sometimes tinged with pink or rose, and a protruding style, which is almost invariably deflexed and slightly curved, seldom quite straight: the stigma is rightly described as having "five erect points." I have never seen living specimens in flower of Pyrola rotundifolia, and therefore cannot speak so confidently with regard to its being distinct from P. media: from the dried specimens in my possession I can however say that the remarkable curvature of the style, which is considerably longer than that of P. media, seems to me to render the difference between the two very visible, even at first sight. With respect to the leaves I think it is next to impossible for the most acute observer to distinguish between the three species above named, until they are in flower. Pyrola secunda is a very distinct species; its ovate and serrated leaves and "greenish white" secund flowers, which do not spread themselves open like those of the other Pyrolas, render the plant at once distinguishable. Pyrola uniflora, now separated from the genus and named Moneses grandiflora, I have never seen except in a dried state: for many years it has not appeared in the two stations near Brodie House; the other station in Moray I have not visited. With respect to the distribution of the Pyrolæ in the only localities with which I am particularly acquainted, I may mention that in my opinion P. minor is the most extensively distributed within the county of Moray, and next to it P. media. P. secunda is rather local. In *this* district, or rather in the small part of it which I have examined, the two species media and minor are equally and rather abundantly distributed; while I have not been able to detect more than a single patch of P. secunda, containing about twelve or twenty specimens, and not in flower this season. The only station in which I have had an opportunity of seeing P. rotundifolia is that mentioned in the 'Collectanea for a Flora of Moray.' This I visited in July, 1834; the plant was not then in flower. By the way I would observe that the Pyrolæ in general seem to flower during the months of June and July, not of July and August, as is generally stated. They grow best in shady places, but Pyrola media is often found in heathy ground, and there it flowers earlier, and the flowers assume a whiter hue, tinged with more of pink or rose-colour. — *James B. Brichan ; Manse of Banchory, July 7,* 1842.

193. *Rhinanthus major and Crista-galli.* Some years ago having heard doubts expressed respecting the existence of a real specific distinction between the two species

of Rhinanthus found in Britain, although I had no doubt of their being really distinct, I examined rather minutely a number of living specimens, and the following was the result of that examination.

Rhinanthus major.	*Rhinanthus Crista-galli.*
Whole plant often immensely larger than Crista-galli, generally more branched; sometimes smaller and not branched, but in every state preserving its characteristics. *Branches* often numerous, nearly erect.	*Branches* few, nearly horizontal.
Calyx equal to the tube of corolla.	*Calyx* longer than the tube of corolla.
Corolla twice as long as calyx; bright yellow, segments of upper lip purple, and a single spot of purple on each of the side lobes of the lower lip: closed, with segments of upper lip connivent.	*Corolla* not much longer than calyx; yellow, variously spotted with purple: open, with segments of upper lip divaricate and somewhat revolute.
Style almost always exserted, sometimes the sixth of an inch.	*Style* included.
Capsule smaller, seeds fewer, with one side thicker, and a narrow margin; when fully formed bursting the skin at the thicker side.	*Capsule* larger, seeds more numerous, of uniform thickness, with a broad margin.

To the above observations I may add that the *tout ensemble* of R. major is very different from that of R. Crista-galli, I mean in so far as appearance is concerned. I have seen both plants growing together, and think no one who could see them growing would consider them to belong to the same species: when dried they become quite black. Rhinanthus major may be said to be rather common in the counties of Moray and Nairn, at least on the coast. In the former county I have seen it growing on an eminence at the height of about 200 feet above the sea, and apparently aspiring still higher, in defiance of Ulex europæus and other well armed and stubborn shrubs. Its locality however is in corn and grass fields, where its abundance, its size, and the profusion of its bright yellow flowers, render it very conspicuous. There seems to be little reason to doubt that in Scotland at least it is not indigenous.—*Id.*

194. *Schistostega pennata.* I enclose a few specimens of Schistostega pennata, which I gathered last month in Nottingham forest, where it almost completely covers the roof of a dry sandstone cave. I found specimens in all stages of fructification, and think, from the situation in which it grew that there must be a constant succession in fruit. I possess specimens gathered by Dr. Howitt in the month of November.—*Joseph Sidebotham; 26, York St., Manchester, July 7, 1842.*

195. *Notes on the supposed parasitism of Monotropa Hypopitys.* It is under feelings of most unaffected diffidence that I venture to publish my notes on this subject, more especially as I find them at variance with the recorded observations of botanists so much more able to grapple with the question than myself. Still, having most unexpectedly become possessed of some luxuriant specimens of this interesting plant, I could not forego the opportunity of investigating for myself this *quæstio vexata*, and I now offer the results to the readers of 'The Phytologist.' The plants were four in number; each had two or more stems, about seven inches in height, bent over at the top and in full flower, a few young stems were ascending in a perfectly erect position.

The flowers had a scarcely perceptible scent, but the stem when broken smelled exactly like a raw potato. Each plant had been dug up with care, and was accompanied by a large ball of chalky earth, the surface-mould being abundantly intermixed with fragments of the cones of spruce and larch firs, and the leaves of spruce fir, birch and whitethorn, all in a state of decay. Each mass was permeated by roots, the majority of which were also decayed, but some few of them were still living, and I made them out to be those of a fir, a Hieracium and a scabious. I separated the earth from each mass with great care; the roots of the plants above mentioned falling from the mass as soon as the removal of the earth permitted of their doing so; and although, from the multiplicity of fibres, many of them had obviously been in immediate contact with the Monotropa, there appeared to be nothing like adhesion; and in no instance that I could detect was any portion of either of the roots contorted, swollen, shrunk or altered in appearance by the proximity of the Monotropa. In order to satisfy myself fully on this head, I subjected each detached root to a lengthened and tedious examination under a lens. Having cleared the Monotropa from extraneous substances, I next subjected the plant itself to a rigid scrutiny. It appeared to me to consist of three parts, somewhat analogous to the frond, rhizoma and roots of ferns. The rhizoma or underground stem was fleshy, brittle, succulent and branched: the branches were thickly clustered, the termination or growing extremity of each being always obtuse: scattered at intervals over the surface of the rhizomata were the gemmæ or buds destined to become fronds or ascending flower-stalks; on making a longitudinal section of one of these, the scales of the future frond were observable, neatly packed one within the other. Those portions of the rhizoma on which these gemmæ were most observable, were usually more detached than the rest; and it appears that when a frond begins to ascend, an active formation of rhizomata commences from the same point, shortly forming a dense cluster somewhat difficult of examination. Closely investing every part of these rhizomata except the growing extremities, which, for a short space, are invariably naked, I found the byssoid substance which Mr. Wilson has suggested may be the "woolly matted extremities of grasses," (Phytol. 149); this substance I believe to be an intrinsic and most essential portion of the Monotropa, and is the part to which I have applied the term *root*. My reasons for supposing it a part of the Monotropa are these:—1st. Its *constant* presence; a "byssoid fungus" (Phytol. 43), "the woolly matted extremities of grasses" (Phytol. 149), or any other extraneous matter, would of necessity be irregular in its appearance. 2ndly. The uniformity of its growth; the larger end of each fibre being invariably attached to the rhizoma; the branching, which is frequent, taking place at angles which are uniform among themselves, and follow the normal mode in other roots; and the distal extremities being extremely minute. 3rdly. Because on viewing a thin transverse slice of the rhizoma under a high power, the substance of the rhizoma and that of the roots appeared perfectly continuous and identical. 4thly. That when these fibres were forcibly detached at their origin, a manifest rupture of the cuticle of the rhizoma took place. These roots, for such I must consider them, spread freely over every substance within their reach. In many instances I found them forming a beautifully reticulated covering to the fragments of decaying fir-cones and leaves, and also in the fissures occasionally occurring in nodules of chalk. Sometimes they appeared closely investing the extraneous roots, but it is worthy of notice that I generally found these completely decayed: not simply shrivelled as by the exhausting power of a parasite, but in that state of decay in which a slight touch of the forceps would cause immediate separation. In many instances, when the Monotropa roots were purposely separated from the decayed roots, leaves, or portions of cone, their extremities remained attached to these extraneous substances. The proportion of *living* roots to which the fibrous extremities of the Monotropa roots had found their way was small as compared with that of those in a state of decay, yet such instances did occur; and although I used every endeavour to make out a decided continuity between the roots of the Monotropa and those of its supposed supporter, I relinquished the search without any proof that this occasional contact between the living roots was a matter of choice, or by any means essential to the vitality or well-being of the Monotropa. In candour it should further be remarked that in some instances the connexion between these byssoid fibres and the Monotropa was not fully made out; the more beautiful examples occurring on fragments detached from the mass before the superincumbent network had been observed; yet between attached and unattached fibres I detected no difference. The conclusions at which I have arrived as to the true nature of the byssoid covering of the rhizoma having been drawn from observations made with what might be considered, in the present day, an imperfect instrument, I obtained the kind assistance of Mr. E. J. Quekett, and by means of his superior microscope and able manipulation, the opinion which I had previously formed of the fibres became fully established.

It is, I believe, generally admitted that many species of ferns derive part of their food through the decaying portions of the bark and wood of trees to which their rhizomata are appressed: if this be strictly parasitism, then I think it will not be difficult to prove a like parasitism in the plant now under consideration. If, on the other hand, we are to understand by the word parasitism that one plant extracts the living juices of another *by immediate contact and positive adhesion*, as in the case of the dodder and mistletoe, or, as suggested by Mr. Lees in the case of Monotropa, by means of " hairy vesicular knobs seated on and of necessity nourished by the radical fibres of " another plant, (Phytol. 99), then I must confess that I met with nothing to induce such a conclusion.—*Edward Newman; Peckham, July* 8, 1842.

196. *New locality for Carex clandestina.* I am permitted by my brother-in-law, the Rev. Thomas Butler, of Langar, Notts., to state that on visiting Brean Down, Weston-super-Mare, in May last, he discovered Carex clandestina growing abundantly over the hill, in the same places with Helianthemum polifolium and Iris fœtidissima. I say *discovered*, because I have never seen any locality quoted for C. clandestina but that of St. Vincent's Rocks. The addition of a second station for so rare a plant, seems a fact of much interest.—*A. Worsley; Brislington, July* 12, 1842.

197. *Carex axillaris and C. remota.* These two species are readily distinguished from each other, as C. vesicaria is from C. ampullacea, by the structure of the culm or stem and leaves. In C. axillaris (and in vesicaria) the stem has three acute angles and the leaves are flattened: in the other two the stem is nearly round, and the leaves are bent at the sides so as to be almost semicylindrical. The bracteas of C. axillaris are by no means constant in their length; in one of my Cheshire specimens the lowest bractea scarcely overtops the spikelet to which it is attached; while in one gathered by John Martin in this county, it is twice as long as the spike. In this species, however, the second bractea is always very small compared with the lower one, having a membranous base much resembling the glumes in size and shape, the upper part rough, very narrow and awn-like. All the bracteas are auricled at the base; while those of C. remota, instead of auricles, have generally a pale very obscure ligule, passing completely round the rachis or common stalk of the spike, where a striking feature

exists : — *It is zigzag*, with only two rough edges, taking a fresh direction at each joint, as if pushed aside by the spikelet and its bractea ; in C. axillaris the rachis is perfectly straight, and has three rough edges. The glumes in C. axillaris are roundish ovate (not acuminate), tipped with a very short rough point or continuation of the midrib ; they are larger and whiter than those of C. remota, which are *ovate-acuminate* and narrower than the fruit. The lowest spikelet is generally compound in C. axillaris ; but I have never seen it so in C. remota. With the ripe fruit of C. axillaris I am not yet acquainted ; but from what I can judge of it in an immature stage it must be narrower than the glumes, and the ribs on the outer side must be essentially different from those of C. remota. I fear your correspondent (Phytol. 263) has another object in view, besides asking for information. He seems to have a particular fancy for severe criticism, and I am really surprized that he should advance the strange opinion that Dr. Goodenough was " not a very close observer of the Carices." The passage alluded to only shows that he had not at that time fully investigated the subject. It is surely sufficient for us to rectify the casual mistakes of our predecessors, without robbing them of their due meed of praise. Again, Mr. G. should be careful to quote accurately ; and it is hardly fair to quote at all the first edition of Hooker's ' British Flora,' when in the second and subsequent editions the mistake has been corrected. In the second edition Carex axillaris is thus described : — " lower bractea long, the rest scarcely so long as the spike." The criticism unfairly represents that author as standing alone in the statement that C. axillaris is a taller plant than C. remota. Smith, in different language, says the same, namely, that C. axillaris is " larger" than remota. I quite agree that comment on this point is needless ; and, as a comment, the superadded remark about the size of C. remota is inconclusive and misplaced. If Sir J. E. Smith were living, he would much disapprove of the use made of the other passage *misquoted* by Mr. G. — Smith, no doubt, had in mind what Goodenough had said of the " entire capsule " of C. remota, and was desirous of correcting the mistake, in language and in a spirit well worthy of imitation by all critics. He therefore mildly says of C. axillaris :—" beak more deeply cloven PERHAPS than that of C. remota, though this difference is not very striking." Mr. G. omits the important word " perhaps," and thus reduces the passage to sheer nonsense. This is not the way to deal with an author, nor the way to derive (much less to communicate) instruction. Mr. G.'s real difficulty in distinguishing C. axillaris from C. remota arises from his never having seen it. As for the descriptions in the two works quoted, they are not so defective that specimens actually in my possession might not, if taken singly, justify either of them. I cannot see the utility of alluding to the " good authority " of the source whence Mr. G. has received C. paniculata under the name of C. axillaris. If any *competent* botanist has thus sent it, he must have done so through mere inadvertence ; and I do not think the *kind intention* of the donor is well requited by the public and somewhat sarcastic announcement of his error.—*W. Wilson ; Warrington, July* 15, 1842.

198. *Description of Carex axillaris*, (Phytol). 263). Root creeping, *not cæspitose* (which in C. remota it certainly is), growing in a more scattered and isolated manner than C. remota, quite as much so as C. teretiuscula compared with C. paniculata.— Stem from eighteen inches to two feet or more in height, rigid, comparatively robust and acutely triangular (its angles rough), strongly striated, nearly erect and straight. Leaves arising from the lower part of the stem, which they enclose in their sheathing bases, linear, plane, though channelled on their upper surface, striated, of a bright *light green colour,* more than twice the breadth of those of C. remota, slightly keeled

on the back, the edges of their lower half smooth, of the superior part rough, about equalling the stem in height, gradually narrowing into slender, rough points. Lower bractea foliaceous, rigid like the leaves, *erect*, forming as it were a continuation of the stem as regards its direction, generally taller than the spike, though very variable in this respect; the *second and superior* bracteas *remarkably short and diminutive*, nearly wanting in the upper spikelets, their bases expanded, then suddenly contracting and assuming an awn-like or capillary aspect; all the bracteas have evident auricles.— Spike two or three inches long, of from six to twelve ovate-lanceolate spikelets; the upper spikelets simple and remarkably crowded, those in the lower part more remote, and the lowest of all almost always *compound;* sometimes there are two or more compound spikelets on the same spike, especially in robust plants, their common rachis straight and triangular. Scales membranaceous, of a brownish white colour, broadly ovate, equalling the fruit in breadth but not in length, bluntish, with a strong prominent green central rib or keel, which extends beyond their apex, forming a distinct and very evident mucro. Fruit ovate, ribbed, with a rather broad, straight, cloven beak. Those who doubt the specific difference of this plant from Carex remota, do so, I feel persuaded, from not having had a sufficient opportunity of contrasting the two plants in a living state. I feel satisfied that no one who has ever witnessed, as I have done, these two species growing within a few inches of each other, and preserving unaltered their peculiar characteristic features, could be longer sceptical on this point. Their *habit* is strikingly different, so much so as to impress upon the mind at once (without the necessity of having recourse to minute anatomical differences) the conviction that they must be essentially different plants. The distinct and separate mode of growth of C. axillaris, its robust, rigid, and nearly erect triangular stems, its broader, plane and channelled foliage, the remarkable disposition and comparative length of its bracteas, its more numerous and larger spikelets, and their aggregation at the summit of the spike, are differences, I should think, amply sufficient to enable any one to distinguish it, *when seen*, from C. remota, and to satisfy any mind that is open to conviction, and willing to acknowledge the truth of facts so plainly manifest to the most careless observer of the beautiful works of creation.—*J. B. Wood, M.D.; Broughton, Manchester, July* 19, 1842.

199. *Love of Nature.* Ah! it is the love of nature that burns within our bosoms; the instinctive admiration of those woods, dark in shadow or hallowed by the coloured Iris; those cliffs now lit up in gold, or gray in twilight; those ravines whose depths are hidden in foliage, and into which the river plunges with sullen roar; those landscapes with all their waters and all their inhabitants, that, solemnly robed in the mists of morning, or splendidly revealed before the setting sun of evening, with all their associations, and all the thoughts and reflections they create and absorb, that charm, enchant, and enchain us. Whatever our avocations may be, whatever may be the object or the pretence with which we set out, when once under the open canopy of heaven, we are *free;* that machinery spreads before us in its simplicity and complexity, which requires no sighs, groans or anguish to keep up its movements; and that pure brisk air which the country only knows, is in motion to fan our foreheads, fill our lungs and excite us to hope, thought and inspiration!—*Edwin Lees' 'Botanical Looker-out among the Wild Flowers of the Fields, Woods and Mountains of England and Wales.'* *

---

* Being obliged to defer a regular notice of this pleasant work, we have given a few extracts from the month of August.

200. *The Heaths.* Now it is that the different species of heaths (Erica) appear in their perfection of beauty, making glad the wilderness wherever they present themselves. Sandstone cliffs are empurpled with the flowers of the Erica cinerea, which often, too, covers the sides of mountains to a considerable height; while, wherever a weeping spring oozes upon the waste, the pale wax-like bells of the Erica Tetralix droop in clusters to the ground. Sir Walter Scott has finely depicted in Marmion, a sun-rise in a mountainous country, when the heath was in flower, and the first golden rays fell upon the mountains —

> "And as each heathy top they kiss'd
> It gleam'd a purple amethyst."

* * But the mountain heather of the Scotch poets, which gives such a black aspect to the bleak hills of Scotland, is the ling, or common heath (Calluna vulgaris), whose calyx, as well as corolla, is coloured; and whose elegant attire, generally diffused as it is in Europe, deserves every encomium it has received. When in full flower, nothing can exceed the beauty presented by a near prospect of hills of blooming heather, while they offer to the way-worn wanderer a fragrant couch, on which he may recline in luscious idleness, and obtain "divine oblivion of low-thoughted care." From the extent of moorland in Scotland, that country has been generally distinguished as the "land of brown heath," and the clans of McDonald and McAlister bear two of the species as their device: hence clouds, storms, and impending dreary rocks, are images that unconsciously arise in our minds, when referring to the heather bells; and a modern writer, when descanting upon the "moral of flowers," has exclaimed —

> "Since I've view'd thee afar in thine own Highland dwelling,
> There are spells clinging round thee I knew not before;
> For to fancy's rapt ear dost thou ever seem telling
> Of the pine-crested rock and the cataract's roar." — *Id.*

201. *The White Water-lily.* As the rose is the queen of the bower, so undoubtedly is the lily the empress of the lake, and I have only done my duty in thus testifying my admiration, as far as she is concerned; but I have merely sketched her figure as she reclines upon her liquid throne, realizing her poetical Indian name "Cumada," or "Delight of the Waters;" but there seems something so emblematical of *purity* about this lovely plant, that the warning of Shakspeare not to paint it is singularly appropriate, and I shall not soil the fair petals of the flower by touching farther upon it.—*Id.*

202. *Wild Flowers of August.* Summer! ah, where has summer been this year? is often a common exclamation at its close; for in ungenial years scarcely have we been able to obtain a glimpse of it, before it is already perceived waning away. Fine or wet, the flowers spring and fade, and the profusion of composite or syngenesious ones now perceptible, gives serious warning that the summer is declining and the days shortening. On the river side the tansy (Tanacetum vulgare) spreads its golden disk, gilding the bank; the hawkweeds muster numerous on the walls; the bristly-leaved Picris echioides, and grove hawkweed (Hieracium sabaudum), in the woods; other species appear throwing a golden hue upon the aftermath of meadows, or limestone banks; and the fleabane (Inula dysenterica) opens its specious disk upon the last days of August. Other signs are, alas! not wanting — the berries of the mountain-ash are flushed; those of the water Guelder-rose (Viburnum Opulus), and the Rhamnus Frangula, show their crimson beauties impending above the deep-flowing streams; the willow-herbs (Epilobium) empurple the beds of rivulets and wet ditches, and the mints

are beginning to blossom. Now the great mullein or hag-taper (Verbascum Thapsus) shows its " flannel leaves " and lofty spike of yellow flowers in perfection, like a huge torch in the dusk of evening ; and others of the same species flash gloriously by waysides or gardens. In certain spots the tall dyer's weed (Reseda luteola) is very conspicuous, and the starry scabious (Scabiosa arvensis) lifts its flowers of regal purple high in air. The little centaury (Chironia Centaurea) named from Chiron the centaur, about this time adorns many a bank with its bright pink flowers ; and the hedges are over-run with the ramping fumitory, the brilliant violet clusters of the tufted vetch (Vicia Cracca), the pink flowers of the everlasting pea (Lathyrus sylvestris), and the conspicuous white bells of the great convolvulus (Convolvulus sepium).—*Id.*

ART. LXXX.—*Proceedings of Societies.*

BOTANICAL SOCIETY OF EDINBURGH.

July 14, 1842. — This Society held its last meeting for the season in the Botanic Garden, Professor Graham in the chair. Various donations were presented to the herbarium and library : — from Mr. Loudon, plants collected in the South Sea Islands by the late Mr. Corson, surgeon ; from Miss Ferguson, seeds collected by her father in Sierra Leone ; a box of plants from the Mediterranean, collected by Edward Forbes, Esq., and volumes of important works from Dr. Muller of Emmerwick, on the Rhine, David Steuart, Esq., Edinburgh, and the Leopoldine Academy of Breslau. Thanks were ordered to be returned to the several donors, and Dr. Muller was unanimously elected a foreign member of the Society.

The corresponding Secretary read a statement of the plants which had been contributed during the season, amounting, on a rough estimate, to 5,000 species, and 25,000 specimens, many of which are rare and otherwise interesting, and he expressed his gratification in being enabled to say that they were generally much better preserved than the contributions of former years.

Professor Graham next exhibited to the Society a number of rare and beautiful plants which had recently come into flower in the green-houses ; most of the gentlemen present thereafter accompanied the professor in a walk through the garden and greenhouses, and particularly to inspect a magnificent specimen of *Caryota urens*, which was a few days ago removed from the large green-house to the open border at the north end of the range. This magnificent palm was raised in the garden from seed brought from Calcutta about twenty-seven years ago, and has now attained a height of above forty-five feet, which rendered its removal necessary, as the house could no longer contain it. It is still comparatively uninjured, but will doubtless soon fall a victim to the cold and rough weather of our climate. The public should therefore not miss this rare opportunity of seeing a tropical palm growing luxuriantly in a Scottish garden — especially as it is much the finest specimen of the kind in Britain, if not in Europe. The heaths and many other green-house plants are at present in a beautiful state, and well worthy of a visit, — nor could the garden generally exhibit a more agreeable aspect than it now does.

The various papers on the list were deferred till next session. — *From ' The Edinburgh Evening Post and Scottish Standard,'* Saturday, July 16, 1842.

[The following list of papers deferred may be interesting to some of our readers. *Ed.*]

1. Remarks on the Assam Tea-plant, with specimens. *By Professor Christison.*

2. Some remarks on the state of Vegetation in Jersey, in March, 1842. *By Professor Graham.*

3. Report on Vegetables parasitic on living animals. *By Mr. John Goodsir, Conservator of the Museum of the Royal College of Surgeons.*

4. On the characters of the British Violets. *By Mr. C. C. Babington.*

5. On the nomenclature of British Plants, and the authority upon which several species have been introduced into the Society's Catalogue. *By Mr. C. C. Babington.*

6. Remarks on the British species of Cerastium. *By Mr. Edmonston.*

BOTANICAL SOCIETY OF LONDON.

July 1st, 1842.—J. E. Gray, Esq. F.R.S. &c., President, in the chair. The following donations were announced:— British plants from Lady Sophia Windham and Mr. F. Robins, and British mosses from Mr. I. T. Hollings and Mr. H. Ibbotson.—Donations to the library were announced from the Imperial Academy of Sciences, St. Petersburgh, the American Philosophical Society, the American Academy of Sciences Philadelphia, and the Rev. A. Bloxam. Various specimens of plants, sections of wood &c. purchased at the sale of the Botanical Museum of the late A. B. Lambert, Esq., were presented by some of the members.

Mr. Arthur Henfrey (Curator) exhibited a monstrous specimen of *Scrophularia aquatica* (which is now in the Society's Museum), found by him on the 30th of June last, on an island in the Thames above Teddington. The plant was about three feet high, having a flat ribband-like stem rather more than half an inch broad, and scarcely an eighth thick. The flower-stalks grew chiefly out of the flat surfaces, nearly perpendicular to them, a very few only being at the edges, and not in any regular order. These flowering stalks extended over about eighteen inches of the stem, being about forty in number, exclusive of a very dense cluster at the summit of the plant. The flowers all appeared perfect, and the peculiarity of growth seemed to have resulted from a natural grafting of two plants. Mr. George Dickie presented specimens of *Gelidium rostratum* (Harvey), collected by him at Aberdeen. Specimens of *Lastræa cristata* (Presl), collected at Holt, Norfolk, were presented by the Botanical Society of Holt; and Mr. R. Phillips presented some seeds from the Cape of Good Hope. Mr. Thomas Sansom exhibited a monstrous specimen of *Cynoglossum Omphalodes* (Linn.) in which three peduncles were united longitudinally from the base to the extremity, and terminated by *two* calyces, the first being 6-partite, bearing a corolla of six segments, *five* stamens, one pistil and four seeds. The second was 9-partite, formed by the union of *two* calyces, respectively 4 and 5-partite, bearing two distinct petals placed side by side, each 5-lobed, each with five stamens, and each containing a pistil and a set of four seeds. Mr. S. also exhibited a specimen of *Galeobdolon luteum* (Sm.) in which the terminal petal was salver-shaped and 5-lobed; stamens four.

A paper was read by Mr. T. Sansom, being Notes of the first Excursion of the Members of the Society into Kent in June last; containing the habitats of the rarer species of flowering plants, and also notes on the most interesting specimens collected. —*G. E. D.*

THE PHYTOLOGIST.

| No. XVI. | SEPTEMBER, MDCCCXLII. | Price 1s. |

Art. LXXXI.—*A History of the British Equiseta.* By Edward Newman. (Continued from p. 278).

MACKAY'S SHAVE-GRASS.

Equisetum hyemale, β. Mackaii,* *Newman.*

* Named after the original discoverer, Mr. J. T. Mackay, of Dublin.

2 c

THIS plant occurs in the North of Ireland, more particularly in the counties Derry and Antrim. It was originally discovered in the year 1833, by Mr. Mackay, the well-known author of the 'Flora Hibernica,' when in company with Mr. Whitla, in Colin Glen, near Belfast. Subsequently to this date it has been repeatedly observed by different botanists: Mr. Moore, the talented and energetic conservator of the Dublin Glasnevin Garden, to whom I am indebted for a supply of both recent and dried specimens, has found it in many of the glens in the northern counties, particularly in Ballekavregan Glen (Derry), and in the wild deep ravines emphatically called "The Glens" (Antrim). In Scotland it was first discovered in 1841, on the banks and "in what is usually called the bed of the river" Dee, in Aberdeenshire, by the Rev. Mr. Brichan, to whose kindness I am also indebted for specimens.

Its discovery caused a multiplicity of correspondence among botanists, some maintaining that it was merely an elongate and exuberant form of Equisetum variegatum; others that it was a good species, perfectly distinct from any which had been previously recorded as British. The matter rested thus until the question was referred to Sir W. J. Hooker, and that illustrious botanist decided not only in favor of its distinctness as a species, but pronounced it to be the Equisetum elongatum of Willdenow, (see 'London Journal of Botany,' 42, and Phytol. 174). Feeling as I do the difficulty under which I shall labour in venturing to differ from so high an authority as Sir William Hooker, I must still record my opinion that the plant before me is not identical with the Equis. elongatum of Willdenow* and Reichenbach,† the essential characters of which appear to be that it has *verticillate* and 6-*angled* branches, and that the sheaths are *hoary green*, conco-

---

* Equisetum elongatum, W. E. caulibus subduplicato-ramosis, ramis subternis scabriusculis sexsulcatis, dentibus vaginarum membranaceis, W.

E. (*ramosissimum*), caule striato ramosissimo, ramis virgatis striatis erectis verticillatis, apice floriferis. Desf. Atl. ii. p. 3982.

Caulis tripedalis et altior quasi scandens, subduplicato ramosus, profundè striatus, scabriusculus. Rami terni, superiores simplices semipedales usque ferè pedales, inferiores iterum ramosi, ramulis suboppositis sexsulcatis. Vaginæ concolores dentatæ, dentibus albis, diaphanis, aristatis, aristis caducis. Ab omnibus mihi cognitis abundè diversum. Equisetum ramosissimum Clariss. Desfontaines non differre videtur. Willdenow, Sp. Pl. v. 8.

† E. elongatum, W. caule ramulisque sexangularibus longissimis superioribus spiciferis, spica mucronata, vaginis concoloribus (cano-viridibus), dentibus persistentibus albis vix puncto sphacelatis cartilagineis, in acumen quasi fimbriam longissimam hyalinam flaccidam deciduam productis.

Hyemale, β. procerum Pollin et omnino vix recedere video.

Quadripedale et altius quasi scandens ramuli ultra pedales. Flor. Germ. Exc. 155.

lorous with the internodes: I have faithfully transcribed both of the original descriptions. Moreover, on consulting Professor Vaucher's 'Monographie des Prêles,' published in the first volume of the 'Mémoires de la Société de Physique, etc. de Genève,' I find that learned author has not only described at length but figured (Pl. VI.) Willdenow's plant under Desfontaine's prior name of Equisetum ramosissimum: the subduplicato-ramose and verticillate characters of the stem as well as the concolorous sheaths are well represented. Vaucher gives the appropriate name of Equisetum multiforme to a species which appears to include the variegatum of Schleicher's Catalogue and various other forms, among these the present plant is not distinctly characterized: his β. Equisetum multiforme ramosum, which, in other respects, comes the nearest, having the sheaths differently coloured. — His description of this variety is quoted below.*

This author, in his introductory remarks on the genus, aptly cites this species as an instance of characteristic liability to variation. After mentioning that the Equiseta generally occur on the banks of streams and in damp places, as well as in the water, he says they are occasionally "even met with in sandy places that are not watery, such is the case with Equisetum multiforme: this species appears strongly influenced by the properties of the soil in which it grows, for sometimes it throws out but a small number of slender and short branches, at other times, on the contrary, and especially when in a more fertile soil, we find issuing from the principal stem branches not only much longer but much more divided, so much so indeed that it has been thought it could not then be referred to the same species."†

* β. *Equisetum multiforme ramosum.* Prêle multiforme rameuse. Cette seconde variété est peu connue des Botanistes quoiqu'elle soit assez répandue; elle émet de sa racine plusieurs tiges courtes semblables à celles de la variété *a.* dont les gaînes sont plus ou moins noirâtres; mais on y observe encore une ou plusieurs tiges principales, qui peuvent s'élever jusqu'à trois pieds et qui sont terminées par un épi plus grand que celui de la variété *a.* Les gaînes sont amples, assez lâches, blanches ou brunes, mais rarement noires; les rameaux sont assez nombreux, plus ou moins réguliers, et quelquefois prolifères. Cette variété β. se trouve souvent réunie à la première, et l'on peut facilement observer des échantillons qui présentent toutes les nuances intermédiaires. Ordinairement la Prêle rameuse se rencontre dans des terrains plus riches et plus favorables à la vegétation. —'Mémoires de la Société de Physique et d'Histoire Naturelle de Genève.' i. 379.

† On en rencontre même dans les terrains sablonneux et non humectés, comme par exemple la Prêle multiforme; mais cette dernière espèce paroît être fortement influencée par la nature du terrain dans lequelle elle croît, car tantôt elle ne développe qu'un petit nombre de tiges grèles et fort courtes, tantôt au contraire, et surtout lorsqu'elle sort d'un terrein plus riche, on voit sortir de la touffe principale des tiges beau-

The roots and rhizoma present no characters by which I can distinguish this plant from that previously described as the normal form of the species : they are both black, the roots being tortuous, much divided, and often clothed with minute and matted fibrillæ : the stems are very long, generally erect, nearly straight, and jointed as in the former species; the figure shows a perfect stem divided into six portions, its size and the relative length of the internodes having been copied with scrupulous accuracy. Both the internodes and sheaths are striated; the striæ vary in number from eight to twelve, or even fourteen. The stem is hollow, with the exception of the transverse septa occurring at the sheaths. Instead of being uniformly simple, as represented in the figure at the head of this article, it is often sparingly branched, as shown in the left hand figure; the branches rise singly from below one of the sheaths, and a stem often bears two or three such branches, the branches themselves also occasionally emit other branches in the same way, the plant, in that case, being very luxuriant, and attaining a height of three to four feet : the right hand figure is a diagram showing the mode of branching. Under the microscope the structure of the stem appears precisely identical with that of E. hyemale; the double row of elevations on each of the ridges, with their cup-shaped depressions in the centre, are exactly as I have already described them, (Phytol. 278). The sheaths are generally black, the central part sometimes white, but scarcely ever so distinctly banded as in hyemale : the teeth are very long, flexuous and setiform; their edges at the base are dilated, membranous, somewhat whitish, and nearly transparent; they are partially but not so decidedly deciduous as in the normal form of hyemale. The catkin is small, nearly black, apiculate, terminal, and striated as in hyemale : the scales are about thirty in number.

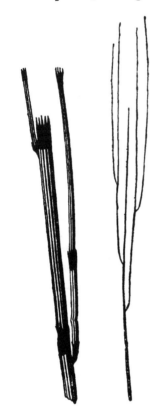

EDWARD NEWMAN.

(To be continued).

coup plus grosses et plus ramifiées, qu'on croiroit ne point appartenir à la même espèce. —Id. 333.

ART. LXXXII. — *On the authority upon which several Plants have been introduced into the 'Catalogue of British Plants' published by the Botanical Society at Edinburgh.* By CHARLES C. BABINGTON, Esq., M.A., F.L.S., &c.

1. *Alchimilla conjuncta*, Bab. MSS.   A. argentea, *G. Don, MSS. in Borr. Herb., Trevelyan in Edinb. New Philos. Journ.* 1835, *not Lam. Enc. i.* 77.

Gathered by the late Mr. G. Don on the Clova Mountains, as I learn from an original wild specimen in Mr. Borrer's herbarium.   As the name originally applied to this plant by Mr. Don (but not published) had been previously employed by Lamarck for another plant, it has become necessary, in order to avoid confusion, to give a new name to this species.

Closely allied to A. alpina, but usually much larger in all its parts, and distinguished by its leaflets not being separated down to their base, broader, more silky on the under side, and so placed that in the radical leaves the two external leaflets almost, if not quite, touch each other.

2. *Alyssum calycinum* has maintained itself for several years on uncultivated spots in Scotland, and is therefore possibly a true native of that country, although an introduced plant in England.   See Eng. Bot. Suppl.

3. *Betula glutinosa.*   This is the more common species in the northern parts of Scotland.

4. *Bunium Bulbocastanum.* Plentiful in the chalky fields of Hertfordshire and Cambridgeshire.   See Eng. Bot. Suppl.

5. *Carex irrigua.*   I have gathered this plant upon Muckle Moss, Northumberland.   It is distinguished from C. limosa by having flat leaves, with their margins scabrous only near the end, usually three female spikes, fruit without striæ, and the scales of the catkins without the longitudinal green band which is seen in that species.   This plant is well figured in Hoppe's 'Caricologia Germanica.'   It was first noticed in the above station by Mr. John Thompson, of Crow Hall Mill, near Ridley Hall, Northumberland.

6. *Centranthus Calcitrapa.* This has but slender claims to be considered as a British plant, as it has only been found in a naturalized state at Eltham (?) in Kent.

7. *Echinospermum Lappula.*   In small quantity at Southwold in Suffolk, where it was discovered by the Rev. E. A. Holmes.

8. *Epilobium lanceolatum*, (Seb.) This is the E. montanum, var. γ. *lanceolatum*, of my Prim. Flor. Sarn., which was there considered as a variety owing to my supposing that the plant of Koch was the same as that of Sebastiani. Bertoloni has since shown that Koch's plant is, as he supposed, a variety of E. montanum, but that the true E. lanceolatum, with which ours agrees, is a distinct species.

9. *Eranthis hyemalis*. Only a naturalized species.

10. *Erysimum virgatum*. In the neighbourhood of Bath the place of E. cheiranthoides is supplied by this plant.

11. *Galium insubricum*. I look upon this as rather a doubtful species; it being too closely allied to G. Mollugo. Found near Winnander Mere several years since by the Rev. C. A. Stevens.

12. *Gentiana germanica*. Stated by Dr. Grisebach to be common in Britain; I am inclined to consider it only a variety of G. Amarella.

13. *Hieracium lævigatum*. I am now convinced that this plant is not the species intended by Willdenow, but that it is the H. rigidum of Hartmann and Fries. It is found in many places.

14. *Linaria purpurea*. Only a naturalized plant.

15. *Malcolmia maritima*. In my opinion this plant has no just claims to be included in the list.

16. *Melissa officinalis*. Naturalized in many places.

17. *Nasturtium anceps*. A common and distinct species.

18. *Oxalis stricta*. Naturalized at Penzance, in Cornwall.

19. *Pinguicula longicornis*, (Gay ?). An apparently distinct species, found by Mr. Jos. Woods in a valley near Helvellyn, and called by this name, of which there is no trace in any of the works to which I have access.

20. *Rumex scutatus*. Near Edinburgh: a very doubtful native.

21. *Scirpus parvulus*. Found in Hampshire, and probably overlooked in other places owing to its minuteness.

22. *Scrophularia Ehrharti*. An account of this plant will be found in the 'Annals of Nat. Hist.' v. 1. It has been observed near Edinburgh, Berwick-upon-Tweed, Preston and London.

23. *Teucrium regium*. It is much to be feared that some mistake has occurred in stating that this plant grows on the Blorenge, near Abergavenny, as the exact spot on which it is believed to have been gathered, has been examined carefully by a distinguished botanist, but without success.

24. *Trifolium Bocconi*. Found by Mr. Borrer and myself near the Lizard Point in Cornwall.

25. *Urtica Dodartii*. Found in Cambridgeshire, Norfolk and Essex.

26. *Vicia gracilis.* A native of Somersetshire and the Isle of Wight, first recorded as British in the Supplement to the 'Flora Bathoniensis. C. C. Babington.

Art. LXXXIII. — *Analytical Notice of the 'Transactions of the Botanical Society.'* Vol. i. pt. i. Edinburgh: Machlachlan, Stewart & Co.; H. Bailliére, London; Smith & Son, Glasgow; W. Curry, jun. & Co. Dublin; J. B. Bailliére, Paris; J. A. G. Weigel, Leipzig. 1841.

(Continued from p. 291).

II. *Account of Botanical Excursions from Edinburgh in Autumn* 1839. *By* Robert Graham, M.D., F.R.S.E., F.L.S. & B.S., *Professor of Botany in the University of Edinburgh.*

One of these excursions was commenced on the 2nd of August: the first plant mentioned is the rare moss Diphyscium foliosum, which Dr. Greville observed in great quantity and fine fruit—"by the roadside, in many places both to the north and south of Loch-Earn-Head." Carex vesicaria was also found in large quantities at the head of Loch Lubnaig, "towards the village called Nineveh or Strath-Eyre." Dr. Graham doubts the accuracy of the statement that the latter plant, with Lysimachia vulgaris and Lythrum Salicaria occur sparingly near Edinburgh. Of the vegetation on the mountains beyond Milgurdy the author remarks : —

"In general the alpine vegetation was found to be very scanty, except Gnaphalium supinum, which here, as on almost every mountain of considerable elevation in Scotland, was abundant. Azalea procumbens was in considerable quantity, though much less abundant on the Breadalbane mountains than in many other stations in Scotland. Erigeron alpinus, Salix reticulata and Saussurea alpina were found sparingly on the most distant mountains which we visited. Hieracium alpinum was also scarce, but more diffused. Carex saxatilis was abundant, especially on the north side of the mountains below the summits, and at the base of a steep wet cliff there, Mr. M'Nab found Epilobium alsinifolium(?) In bogs, on the north side of the mountains, below the summits, we also found Juncus biglumis."—p. 20.

In their route to Catjaghiamman the party observed nothing "but a sparing quantity of the ordinary alpine plants, and a few fine specimens of Woodsia hyperborea, Myosotis alpestris and Veronica saxatilis," Juncus castaneus being abundant in one place. Tempestuous weather prevented the party from thoroughly examining the summit of Catjaghiamman, of which the author observes : —

" I feel quite certain that this is untrodden ground, for no collector could have left such specimens as we gathered of Draba rupestris ; specimens many times larger than I ever saw before, except some which I found at no great distance on the same ridge, during the last visit which I paid to that district."—p. 21.

On the south side of Ben Lawers Dr. Greville and Mr. M'Nab found the only specimens of Azalea procumbens seen on the whole range, except those found to the north-west of Milgurdy. Sedum villosum occurred at a much greater elevation than the author had before observed it in. Gentiana nivalis was found in considerable abundance at the base of the rocks ; and —

" Among shelving rocks, near the bottom of the hill, and to the eastward of the wood above Finlarig, I gathered Eriophorum gracile. This I think certainly the plant indicated by Don, and figured by Smith in ' English Botany,' and it was even at a distance easily distinguished from Eriophorum angustifolium, among which it grew ; but whether it is entitled to rank as a species is quite another question. I do not think it a scarce plant, at least in the north of Scotland. I am sure I have seen it in large quantities in Sutherlandshire."—p. 22.

At Inverarnan Drs. Graham and Greville collected specimens of the oaks which grow there, in the expectation of finding among them the true Quercus sessiliflora, which the former gentleman thought he had seen in that locality on a former visit.—

" In spite of our receiving a shower-bath with every twig we pulled, we persevered as long as we saw the least chance of clearing up a doubt about the species of this most important genus ; but we were obliged to desist, with only a strong suspicion that my conclusion regarding the sessile-flowered oak was hasty. Whatever character may be got for this supposed species, I fear at present that the form of the leaf will yield none, for I certainly saw trees with leaves which I should have considered characteristic of Quercus sessiliflora, which nevertheless had peduncles several inches long. It was too dark to judge of habit, in which also, in my former visit to this district, I conceived I had seen a character."—p. 23.

In a second excursion, during a walk by the bottom of Loch Eck to Kilmun, the author observed " nothing worth mentioning except a profusion of Carum verticillatum in almost every damp pasture," and a great quantity of Polygonum amphibium, var. β. terrestre, in flower. Ben More, the highest mountain in Cowal, the author found to be most unproductive of alpine plants; this could hardly have been expected, the whole country being micaceous and wet, but an inspection of the district explained the cause.

" The rocks are not crumbling, but present the same forbidding sharp angles as those we had before seen in Glen Ogle. We got absolutely nothing worth naming in *our* ascent. We descended by a crumbling ravine in a cliff where eagles build, and got there the only things worth notice. Among these were Aira alpina, Poa glauca, Alchemilla alpina, Galium boreale, Saxifraga oppositifolia, S. stellaris, S. nivalis (three or four specimens only), S. aizoides, Cochlearia officinalis, Carex rigida (very little in fruit), C. pallescens, Luzula spicata, Salix herbacea, Saussurea alpina (on one spot only), Aspidium spinulosum in various forms, Asplenium viride, Hymenophyllum Wilsoni in abundance, and in moist pasture, some way up the mountain, Carum verticillatum."—p. 24.

The last excursion mentioned is that by Messrs. Brand and Campbell in September, from Edinburgh by Perth and Dunkeld to Blair Athol, thence to Cawdor Castle in Nairnshire. On Craig-y-Barns, a hill north of the park at Dunkeld, Saxifraga umbrosa was observed covering acres of ground, and in some places forming the entire turf to the exclusion of everything else, with every appearance of being native; but the presence of Hypericum calycinum and other certainly introduced plants, induced our travellers to believe that the Saxifraga had been planted there, though at a remote period. From the Sow of Athol Menziesia cærulea has been nearly eradicated, only two plants having been found after a long search: at the foot of the mountain Carex pauciflora was abundant. In one place by the roadside between Dalnacardoch and Blair occurred what was considered to be the true Arctium Bardana, a plant so distinct from A. Lappa, that the propriety of sinking it as a species is questioned. But while an old specific distinction might thus be restored, the travellers seem disposed to discard a new one: for Rumex aquaticus, found frequently near Cawdor Castle, on being examined in different situations, was thought to "shade away *towards*, if not *into*, Rumex crispus." Among other good plants abounding in the Cawdor woods were Rubus saxatilis, Goodyera repens and Trientalis europæa; the latter is sometimes of "remarkable size; Mr. Stables having one plant whose branches, spread out, measured about one foot square." Lapsana pusilla and Lolium temulentum were observed near Elgin, usually among wheat.

" Mr. Gorrie has been the first to discover Orobus niger in the Pass of Killiecrankie, scattered over a piece of ground in the coppice wood, at least twenty yards across, and far removed from any cultivated ground."—p. 26.

III. *Observations on a Metamorphosed Variety of Antirrhinum majus. By* HERBERT GIRAUD, M.D., *Member of the Council of the Botanical Society.*

" The general growth and habit of this plant corresponded with its normal condition, and the stems, leaves and sepals answered to their specific characters. But the petals, with the other parts of fructification, existed in a state of very singular transformation, — affording a striking illustration of the great doctrine of morphology, —

that the organs of reproduction, with their appendages, are but metamorphosed leaves. * * The points in which this plant departed from its normal condition, appear to be these: — The corolla, instead of being petaloid, irregular, five-cleft and brightly coloured, was leafy, regular, in five deep segments, of a green colour. The stamens, which should have been didynamous, were here absent. The ovarium, from being two-celled, became five-celled; the stigma five-cleft, instead of bifid; and the style hollow, instead of being solid.

"The quinary condition of the ovarium, and the alternation of its carpels with the divisions of the corolla, is a state which would be expected to follow any alteration in the number of its cells, with an accompanying tendency to assume regularity in its development; as the quinary arrangement is that which prevails in dicotyledons. The supernumerary carpellary leaves, of which the ovarium is composed, are obviously those which, under the normal condition of the plant, would have gone to the formation of stamens. That they are developed from the next whorl of leaves to the corolla is plain, from their alternating with its divisions. Now, in the normal condition, the stamens would be so situated; one leaf, however, in the whorl is altogether abortive, —forming neither a stamen nor a carpel; for the carpellary development of this would produce a six-celled ovarium. The abortion of stamens in the natural family Scrophularineæ (to which this plant belongs) is by no means unfrequent."—p. 27.

This case of metamorphosis is remarkable as presenting an instance of changes proceeding in opposite directions in the same flower: — 1. from centre to circumference, the bright-coloured irregular corolla having been converted into five equal, green, leafy segments, clothed with glandular hairs, like the calyx: 2. from the circumference to the centre, the stamina taking the form of carpellary leaves. The former of these changes is considered by the author to disprove an opinion expressed by Dr. Lindley, that transformations of the reproductive organs "*always* take place in the order of development, or from circumference to centre; that is to say, that the calyx is transformed into petals, the petals into stamens, &c., but that *the converse does not occur.*"

IV. *On the British Species of Fumaria.* By CHARLES C. BABINGTON, M.A. F.L.S., F.G.S., &c.

Mr. Babington, in a paper read before the Society on the 12th of May last, has entered more fully into the specific distinctions of Fumaria parviflora and other allied species of the genus, (Phytol. 239); we therefore think it better to defer our analysis of the present paper until the publication of the second in the Transactions.

V. *On Systematic Arrangement in the Formation of Natural-History Collections.* By WILLIAM BRAND, C.S., F.B.S.

BELIEVING that nothing has been created without an object, the author also imagines " that all things exist according to a suitable arrangement, the laws or conditions of which may, in some measure, be ascertained by a well-directed investigation," the first requisite for which is such an acquaintance with the various objects as will enable the enquirer " to recognise and identify them wherever they appear." This knowledge can be gained only "by studying the features of each individual," and can be applied only by the several individuals being characterised, so that each may be distinguished from all others of its class.   This may be termed " the process of *separation* or *disjunction ;* and an object thus discriminated is termed a *species.*"

The author then briefly treats of " the process of *combination*," by which species are formed into genera, these again into other groups, and so on, until all the objects in any kingdom of nature are classed under a few grand divisions.   These groups, being but interrupted links, "where nature probably exhibits a continuous chain," must be to a certain extent artificial, a separation into *species* or *individuals* being the only real classification indicated in nature.   The author meets the objection that naturalists "waste their time and energies upon the external character and appearance of objects, instead of applying themselves to ascertain their relationship, properties and uses," —by remarking that the former is an indispensable preliminary to the successful prosecution of a higher course of research, to the promotion of which a still farther advance must be made, "namely, to ascertain the range or distribution of species, with their relative condition, and other attendant circumstances in any different region of the globe."

" Now, in prosecuting such an investigation, I consider it necessary to commence by mapping out, or dividing the surface of the globe into appropriate sections ; then to ascertain the various productions which occur in each, with the circumstances attending them ; and lastly, to register the information thus acquired, in such a way as to afford a ready and comprehensive view of these productions, both as they exist in any one section, and as they stand relatively to those of other sections."—p. 41.

In forming these sections, although no abrupt or decided change in their productions may be occasioned by topographical influences, yet may certain lines be traced whereby such changes are indicated ; such lines would form appropriate boundaries for the sections.   Thus the ridge of a mountain chain, a table-land, or the channel of a great river, may sometimes be properly chosen as a sectional line, especially when the former regulate the water-sheds or sources of streams flowing in opposite directions.   The sections should be so limited in extent that all the important changes occurring in the productions of a district may be exhibited, while they should "not be so numerous as to render definite limits unattainable, or otherwise obstruct the scheme in its practical working." The author observes that "any mode of division adopted must always be liable to controversy," but that he himself believes "about 110 sections would suffice for the proposed object."

" This object being accomplished, the next requisite for ascertaining with accuracy the productions of the earth, is to procure specimens of these from *every* section wherein they occur, along with full particulars relative to their condition ; and to record those particulars in such methodical array as to make them directly available for the purpose intended. To carry on this process efficiently, however, I consider it essential that the information should be registered immediately on being obtained, or at least as soon thereafter as possible, both that the rapid accumulation of objects in a general collection may not interrupt the regularity of procedure; and because I conceive that a material saving of labour will always be effected by disposing of these objects at once, and singly as they occur."—p. 42.

Mr. Brand considers that the best mode of rendering Natural-History collections easily available, is to record the species in a tabular form under an alphabetical arrangement, employing certain simple signs or marks to indicate all the information obtained respecting them. The author is of opinion that the adoption of this mode of arranging and registering all known facts and circumstances connected with objects of Natural History, must eventually lead to the highest practical results.

VI. *Description of Pothocites Grantonii, a new Fossil Vegetable from the Coal Formation.* By ROBERT PATERSON, M.D., &c., *Extraordinary Member of the Royal Medical Society, and one of the Council of the Wernerian Natural History Society of Edinburgh, &c. &c. Communicated by the President.*

THE fossil which is the subject of the present paper " was found in a mass of bituminous shale, from the coal strata which are exposed along the coast at Granton." The specimen here described and figured is the only one met with during a long-continued series of researches in the locality wherein it occurred ; and it appears to constitute an entirely new fossil genus.

At first sight the figure (which is of the natural size) appears to represent about three inches of the lower part of a catkin of Typha angustifolia, attached to a portion of the stem. On a closer inspection, however, it is seen to differ materially from the fertile catkin of a Typha, for in that genus there is no other floral envelope than the hairs surrounding the pedicel of the pericarp ; whereas on the exposed surface of the fossil are four longitudinal series of bodies, each of which, in the accompanying magnified views, is seen closely to resemble a four-leaved calyx. Another point in which the fossil differs from Typha is that on the stem, about an inch below the catkin, we observe a scar apparently indicating the spot whence a leaf or spatha had fallen, or where a branch had been broken off; now there is nothing of this kind in Typha. It therefore became necessary to seek in some other genus for the living analogues of this fossil, and these are apparently furnished by several species of Pothos, a genus belonging to the natural order Aroideæ, and characterized by having a one-leaved spatha, and a simple cylindrical spadix covered with flowers, which have a four-leaved calyx, no corolla, four stamina, and bear a 4-seeded berry. These characters will be found to apply nearly to the fossil, making allowance for "the compression it has undergone, and the change of appearance produced by its mineralization."

"The greatest number of the species of the genus Pothos are parasitic, and inhabit the vast forests of tropical countries. In some of the species, also, there are truncated fleshy scales on each side of the germen, and which, in the young state, completely cover the male organs of the plant; these are especially conspicuous in P. acaulis. — The similarity of the habitats also favours the idea of its belonging to this class of plants."—p. 51.

The author enters at some length on the mode in which the different carboniferous strata have been deposited, and explains the process of formation of fossil fuel in the present era, as observed by travellers in America.

VII. *Extracts from the Minute-Book of the Botanical Society, from November* 1839 *to July,* 1840.

1839, *December* 12. Read, Extracts from a Letter addressed to Dr. Greville, from R. J. Shuttleworth, Esq., Berne, dated 11th September, 1839, containing "Observations on Diatomaceæ." After some observations on Gomphonema, Meridion, Diatoma, &c., the author says —

"I have no doubt as to the non-animal nature of these and analogous genera; for though I have examined, I may say, tens of thousands of individuals of most of them, I have *never* observed the slightest trace of *spontaneous* motion or action. The motion of Oscillatoria is perfectly mechanical, caused by the rapid development of each successive cell or joint; and an analogous motion, though merely caused by separation of each joint, is common in Diatoma."—p. 53.

Mr. Shuttleworth passed "eight days on the Grimsel, the hospice of which is 6400 feet, or thereabouts, above the level of the sea;" he wished to ascertain if Diatomeæ exist in the alpine waters, which are scarcely above the freezing point.

"I found that every pool or stream was full of them and Infusoria, and even the Scytonema that covered the rocks several hundred feet above the hospice, and over which the snow-water trickled, served as the habitation or matrix of Synedra lunaris, *Ehrenb.* (Exilaria, *Auct.*) a true Diatomea."—p. 54.

The author enumerates many of the species found here, several of which were quite new: he also details the results of an examination of "the Red snow *quite fresh*," which, to his astonishment, he found composed of Protococcus nivalis (*Ag.* not *Grev.*), P. nebulosus, *Kütz.* three or four deep red and two uncoloured Infusoria; "the Infusoria were endowed with the most astonishing activity of motion."

"The coloured snow was collected in a bucket, and then allowed to melt, when the colouring matter was deposited, and immediately examined; but hardly had the water acquired the temperature of the air, than *all life ceased*, and the Alga was only to be distinguished from the Infusoria, by its greater transparency and lighter colour. The proportion of the Alga (Protococcus nivalis) to the Infusoria was about 5 or 6 to 1000 of the latter. That this is always the case I do not believe, as Hæmatococcus Noltii is always mixed with a brown Stentor, which latter varies in its proportion much. The Protococcus nivalis of the Grimsel is the same plant as that figured by Agardh, in the 'Icones Alg. Europ.' tab. 21, f. a. The sporules, or rather the contents of the plant, are merely grumose, and the granules are so small, as to be even inconspicuous with a power of 300 diameters, whereas in Hæmatococcus Noltii (to which your species of red snow is closely allied) the granules are large, and few in number. The Protococcus nebulosus, *Kütz.*, is a simple, very small, grey or uncoloured globule, which abounds everywhere, and as, from mechanical causes, it is attracted round the globules of red snow (Alga and Infusoria), it is certainly the cause of the gelatinous appearance of the substratum, described and figured by many authors, which however does not exist in our Swiss plant. The colouring matter of the red snow extends to the depth of eight inches at least below the surface; it is only found on *old* snow of some years standing probably, and only developes itself during warm weather and a southerly wind. This snow is also always more or less covered with minute particles of humus; and I regret that the state of the weather did not permit me to examine the humus immediately below the snow."—p. 55.

There is also a reference to a paper by Mr. Shuttleworth published in the 'Bibliothèque Universelle de Genève,' Feb. 1840.

———

Exhibited, by Mr. W. C. Trevelyan, a series of varieties of Scolopendrium vulgare, gathered at Auchmithie in Forfarshire. The most frequent form is that with a simply bifid termination; one specimen is three-cleft; in others the midrib being divided near the base "forms a frond with two branches, nearly at right angles with each other;" in another specimen the frond is much branched at the extremity; then again some have a rounded termination, the midrib not reaching to the end, and other fronds have a circular form; the edges of some fronds

are fimbriated, of others crenated; and in one remarkable variety the
frond is much thickened, and the veins are more distant than usual,
" there being about one-third only of the usual number in a given
space."

1840, *February* 13. Read, a Communication from the Rev. C. A.
Stevens, Kent, on Scrophularia aquatica of Linnæus and of Ehrhart.
The author, considering the plants of the above authors to be distinct
species, has named the latter S. Ehrharti, and thus characterizes them.

" *S. aquatica*, Linn. Leaves cordato-ovate, very obtuse, crenato-serrate, lower ones
auriculate; lateral cymes corymbose, many-(8—15)-flowered; sterile filament reniform, nearly round, entire; capsule ovate, rather acute.

" Cymes mostly opposite; peduncles and pedicels glandulose; bracts generally linear, obtuse, entire.

" *S. Ehrharti*, C. A. Stev. Leaves ovate or ovato-lanceolate, somewhat cordate at
the base, acute, serrate; lateral cymes divaricating, few-(5—6)-flowered; sterile filament bifid, its lobes divaricating; capsule globose, very obtuse.

" Cymes mostly alternate; peduncles and pedicels hardly at all glandulose; bracts
foliaceous, lanceolate, acute, serrate."—p. 57.

There is also a reference to Mr. Stevens's paper on these plants in
' Ann. Nat. Hist.' v. 1.

April 9.   Dr. W. B. Clarke read a communication on certain properties existing in Tilia europæa, *Linn*. The author observes that
" this plant is remarkable for the immense quantity of mucilage contained in the bark, and for the manner in which the cellular tissue in-
volving it is arranged within the meshes of the ligneous tissue." The
mucilage exists in a nearly solid form, but is soluble in water; the
spaces in the woody fibre of the inner bark containing the cellular tissue appear to be formed " by the cohesion of the woody tissue in certain parts, and its separation in others, to such an extent, that by the
gradual and lateral distension of the earlier layers of the liber in the
exogenous growth of the tree, certain spaces are made to intervene
between the reticulations of the woody fibre, which are gradually filled up by the development of cellular tissue." A transverse section of
a branch exhibits these spaces so disposed in the bark as to present
" a series of little conical figures, with their bases towards the circumference of the bark, and their apices in communication with the medullary rays, which traverse the wood of the stem."

The author describes the mode of preparing this vegetable mucus
from the bark.

May 14. Read, Observations on Gentiana Amarella, *Linn.*, and G. germanica, *Willd.*; and on Pyrola media, *Sm.*; by P. J. Brown, Esq., Thun. The author states his belief that the much-controverted question of the specific identity or distinctness of the two Gentians is yet far from being settled; and "that it never can be set at rest in the closet, or in any other way than by the continued examination of both plants in a living state, and in their native abodes."

" Smith, in his ' English Flora' (under G. Amarella), says that G. germanica, W. ' may be a good species, but has not yet been observed in England.' Grisebach, the latest and best authority for the Gentianeæ, has England or Scotland as the country of every variety of Amarella, and near Ripon as a habitat of germanica. His distinctions and descriptions, although apparently available upon paper, will not, however, stand the test of the field, as they fluctuate within the limits of probable variation, as may be judged to be likely when he has four states of each species. Minor distinctions being set aside, there seems to be little more left than *G. Amarella*, ovarium and capsule sessile, calyx-lobes lanceolate, obtuse, *though variable; G. germanica*, ovarium and capsule stipitate, calyx lobes ovato-lanceolate, acuminate. It is true that in his *characters*, he calls the tube of the corolla of Amarella cylindrical, and of germanica ' sensim ampliata;' but in his *descriptions* he calls them both *obconical*. Every distinction he makes depending on size, form, or proportion of parts, may be found here in the space of a few square yards equally applicable to the one as to the other, our plants varying from two inches to two feet, from quite simple, with one flower, to very much branched, with from 150 to 200 flowers; leaves broad and narrow, blunt and pointed, much more distant from each other than their own length, or quite the reverse; the calyx-lobes, which are avowedly inconstant, vary as much as the other parts, and the question is nearly reduced to the sessile or stipitate ovarium, and capsule. Now, although our plants are, at least in my opinion, decidedly Grisebach's G. germanica, I do not find the stipitate ovarium, and am inclined to suspect it may be a delusory appearance, arising in the following manner. * * * The lower third of the ovarium being destitute of seeds, and at the same time fleshy and succulent, I suspect that when gathered in an early stage of flowering, or later if much pressed and dried very quickly, the ovarium shrinks uniformly in drying, and has the sessile form attributed to Amarella; but that when in a more advanced stage, particularly if dried slowly under slight pressure, the substance of the swelling seeds keeps the upper portion of the ovarium expanded, while the lower empty succulent part shrinks and produces the stipitate form of germanica."—p. 59.

The author observes that his attention was drawn to the subject too late in the season to pursue the inquiry satisfactorily; he suggests that botanists should study these plants " from the commencement to the end of their flowering season, drying specimens at stated intervals of about a fortnight, and noticing the form of the ovaria in the living and dried states." In this recommendation we most heartily concur, and should be glad to learn the results. Mr. Babington's opinion on G. *germanica* is recorded in the present number (Phytol. 310).

The author's idea relative to the ovarium of Gentiana germanica, was suggested by his observations on Pyrola media, which appears to be a rare plant in Switzerland. In June, 1838, the author chanced to find a tolerable supply about five miles from Thun. The plants agreed pretty well with the characters given, except that the style instead of being club-shaped was decidedly cylindrical. In the progress of drying however this difficulty was explained "by the conversion of the cylinders into clubs," whereby the character most insisted on was shown to be false.

"The styles, which are considerably elongated, and although declined, are scarcely curved, have a diameter at least twice that of P. rotundifolia, and are very succulent, being furnished with a stout ring at the apex, as in the last-named species. In the individuals gathered while the flowers were yet young, the rigidity of this annular protuberance prevented the summit of the style from shrinking, and it consequently became club-shaped when dry; but in those which were further advanced, the ring having become more or less flaccid, it ceased to afford the same resistance, and the styles remained nearly cylindrical, although shrunk to half their original thickness. This view is now confirmed by British specimens in my herbarium. One from England, gathered when the plant was coming into flower, has a style nearly triangular; the other six, from different Scottish stations, are all in fruit, and have all the styles, excepting one, cylindrical, and that one not far from being so."—p. 61.

June 11, 1840. Read, Notice of the Occurrence of several Rare Cryptogamic Plants on the Sidlaw Hills; by Mr. William Gardiner, jun., Dundee. The following is a list of the plants, with their localities.

1. *Buxbaumia aphylla*, Linn. Northern declivity of one of the hills; May. "It grew very sparingly on several small spots of bare soil that occurred among the heath."—p. 62.

2. *Parmelia physodes*, Ach. Deerhill Wood; end of March. "The tree state of this lichen, as far as I have observed it, differs from that found on walls and stones, in being of smaller size, more deeply divided, of less dense growth, and of a clearer colour above, with its under surface darker. Only one specimen was found with the apothecia fully developed."—Id.

3. *Dicranum squarrosum*, Schrad. "Marshy banks of a small lake at the west side of the White Hill of Auchterhouse;" associated with Weissia acuta, Hypnum aduncum and H. revolvens, and "near it was abundance of Polytrichum commune and yuccæfolium. The P. commune β. attenuatum, though said to attain only the height of three or four inches, is frequently as tall as the other variety, sometimes above a foot high."—Id.

4. *Hypnum fluitans*, Linn. "Near the source of Dryburn rivulet, which flows from the Sidlaws into Glen Ogilvy. There were a good number of capsules on it but immature."—Id.

5. *Ramalina farinacea*, Ach. "Found in Deerhill Wood, with apothecia."—Id.

6. *Cetraria islandica*, Ach. "On the top of an old wall, at the foot of the White Hill of Auchterhouse."—Id.

July 30. Read, Notice of Recent Excursions in the Neighbourhood of Edinburgh; by Dr. Graham.

"Dr. Graham stated that Epilobium alsinifolium, said to have been found on the Ochil Hills, appeared to him to be only a variety of E. tetragonum; that Salix herbacea and Gnaphalium supinum had been found sparingly on Ben Clach, a mountain of that range; and that Galium pusillum, Rubus Chamæmorus, Polygonum viviparum, Lycopodium Selago and selaginoides, had been gathered in the same locality.

"Dr. Graham also mentioned that he had found several fine specimens of Orobanche rubra on the cliffs below St. David's, Fife, and one specimen of Anthemis tinctoria on ballast-heaps near the same place. Epipactis ensifolia he stated to have been found near Dunfermline by a lady."—p. 63.

———

Dr. Herbert Giraud read a paper " On the Existence of Nitrogen in Plants, considered with reference to their development, and to their serving as food for animals." The author first pointed out the relations of nitrogen in the constitution of organized beings, and that it "appears to be the most essential element of organization." That all the tissues of plants contain nitrogen, is shown by the experiments of the author, as well as by those of Boussignault, Payen and Rigg; who have also proved "that vegetables have the power of deriving that element from the atmosphere." The proportion of nitrogen varies from about three to five per cent. Speaking of M. Boussignault's researches the author remarks : —

"The results of his experiments have shown, that the proportion of nitrogen in any vegetable tissue or organ, bears an intimate relation to the activity of the vital functions with which the tissue or organ is endowed. Thus, in that part of the seed in which germination commences, nitrogen predominates. Seeds also, which germinate most readily and most rapidly, contain the largest quantity of nitrogen. With regard to woody structures, it appears that the alburnum, which greatly exceeds the heart-wood in the activity of its functions, contains by far the largest proportion of nitrogen. Those timbers also, which grow most rapidly, contain the largest proportion of nitrogen."—p. 64.

(To be continued).

———

Art. LXXXIV. — *Notice of the 'Transactions of the Microscopical Society of London.' Vol. i. pt.* i. London : Van Voorst. 1842.

This part, containing 86 pages Royal 8vo. and illustrated with 8 plates, has just been laid before the public, and affords ample evidence of the active and industrious spirit which animates this young but prosperous Society. The papers are twelve in number; seven of these are zoological, four botanical and one geological. We give a *short* analysis of the botanical papers in the order in which they occur.

I. — *On the Development of the Vascular Tissue of Plants.* By Edwin J. Quekett, F.L.S., B.S., &c.

Mr. Quekett commences his paper by commenting on the difficulties attendant on the enquiry ; and observes that he believes the idea to be in great measure correct, that " in structure vessels differ but little from cellular tissue, and that the elements of which the latter is formed are only altered and converted to fulfil a different function in the former, and that the development of the one will more or less correspond to that of the other." He then proceeds to give the views of Schleiden, Raspail, &c., and alludes to the opinion of Mirbel, Treviranus and Slack, that the membranous tube of a vessel is formed from a number of cells ranged end to end, their connexions being ultimately absorbed, thus causing the production of a continuous cylinder instead of several separate cavities. Pursuing this idea the author thus proceeds :—

" I must state that in many instances I have met with arrangements of cells in such a way as would lead one to suspect that this was the true origin of a vessel ; and it is a curious fact that cells adhere end to end much more strongly than they do side by side ; therefore, when disturbed, they appear more frequently in strings than in other forms : but I believe no one has ever seen the fact farther than here described, or followed out the complete development of a vessel from this condition of cells ; and there is one fact presently to be mentioned, connected with the development of a vessel, which entirely disproves this theory."—p. 3.

In order to watch the development of the membranous tube of a vessel, Mr. Quekett recommends that some part of a plant in a nascent state, as a bud or bulb, should be selected for examination ; he also lays great stress on the necessity there is for caution in making dissections of recent parts in order to witness the appearances which he subsequently describes. The author then details observations for the most part confirmatory of Schleiden's views, as explained by that writer in his memoir on Phytogenesis, a translation of which appeared in the 'Annales des Sciences Naturelles,' (tom. xi. Botanique) : he continues : —

" When the young vessels are recognized (which by experience becomes an easy matter, even in parts considerably developed, though not so at first), they appear as pellucid glassy tubes, with a cytoblast in some part of their interior ; earlier than this they are not to be recognized readily from cells. As they grow older the cytoblast diminishes, and the contents, which at first were clear and gelatinous, become less transparent from containing thousands of granules, which are too small to allow of the passage of light, and consequently appear as dark points ; these atoms are about the $\frac{1}{8000}$ of an inch in diameter, and have the motion known as "active molecules." If the vessel be wounded at this period the gelatinous contents pour slowly out, and then the singular movements of these molecules are still more clearly seen. These atoms, from their freedom of motion, are arranged indiscriminately in the interior of the vessel, but in a short time some of them enlarge, and then transmit a little light, which, on account of their minute dimensions, is not suffered to pass as a white pencil, but is decomposed in its course, the granule thereby becoming of a greenish hue. The granules exhibiting this greenish hue are now in a fit state to enter into the composition of the fibre that is to exist in the interior of the membranous tube, and in a spiral vessel this is the manner in which this act is accomplished.

" The granules which are in active motion in the viscid fluid near one of the ends become severally attracted to the inner wall of the vessel, beginning at the very point; those granules first attracted appear as if cemented to the spot, by the viscid fluid in that direction losing some of its watery character, for there appears a string of a whitish colour, besides granules, in the line which the fibre is to occupy. As the other granules are attracted to those already fixed in an inclined direction, the spiral course is soon to be seen, and the same action progressively goes on from the end it began towards the other, around the interior of the tube in the form of a spiral; the fibre being produced, like a root, by having the new matter added continually to the growing point, thereby causing its gradual elongation.

" This action is not throughout the vessel at the same instant, for I have witnessed a vessel having one half laid down with fibre, and in the other part the operation had not been commenced. When the granules have arranged themselves throughout the whole length of the tube, those which were first deposited, and had then some slightly visible space between them, have by this time been reinforced by others or nourished by the contents of the vessel, so that that space is obliterated, the fibre beginning to assume a thread-like shape with defined borders, and sufficiently large to allow of the transmission of white light. When this same action has progressed throughout the entire vessel, the transparency begins to be restored, and what is singular, the entire mass of granules has completely disappeared, appearing as if the exact number and no more had been generated to form the fibre. After the vessel has reached maturity, the liquid contents themselves become absorbed, as happens in the cells of the pith, and the vessel is then empty ; and probably from being seen at different periods of its existence in these different states, sometimes full and at others empty, may account for the discrepancies existing among botanists as to the functions these vessels perform."— p. 5.

It appears scarcely necessary to follow the author through the remarks which succeed the passage above quoted; a summary of his views on the direction of vegetable fibre, will, however, we trust, be highly acceptable to physiological botanists.

" It appears to me that the only theory capable of explaining the direction of the fibre, is one that will apply to some peculiar laws existing between the granules themselves and between the granules and the vital force residing in the vessel or cell in which they are contained. There can be no doubt that at first the granules are in the jelly, consequently as they become sufficiently developed they acquire freedom of motion, and attraction commences between the wall of the cell and the granules, and it can be easily imagined how these numerous atoms may be induced to approach to the circumference of the vessel, but the difficulty of the proposition is to account for their doing so in a spiral or other determinate form, and always of the same figure in the same situation in the same plant.

" Some part of the law, I believe, is made tolerably clear, viz., that fibre is composed of granules arranging themselves like beads on a string, which become nourished by the contents of the vessel until a perfect thread is the result, and the direction this takes seems to me to be the result of some special power residing in the vessel under the control of the whole plant, probably electrical; and which is modified in the several vessels I have enumerated: farther than this I believe we cannot go, though nature occasionally alters forms, she seldom varies much in her laws, but what these may be it is forbidden the eye of man at present to detect, and they appear to me, though operating in such minute spaces, to be stamped with as much permanency of power in the formation of these curious and elegant organs, as those laws on a grander scale are in the fashioning of our own frame, or in the maintaining of the stability of the universe."—p. 11.

IV.—*On certain Phenomena observed in the genus Nitella, as illustrative of the peculiar structure recently discovered by Mr. Bowerbank, in a Fossil Wood from the London Clay.* By ARTHUR FARRE, M.D., F.R.S., &c.

Dr. Farre's paper on Nitella is highly interesting, as illustrative, in some degree, of appearances previously detected in fossil wood by Mr. Bowerbank. The author had procured some specimens of Nitella flexilis for the purpose of observing the circulation of the sap : up to the 4th of April this was going on vigorously, but two days afterwards it had entirely ceased, and certain green particles, previously lining the interior of the stem, had shrunk from the parietes, and, together with the green circulatory matter, was collected in irregular masses within the tube. Five days afterwards Dr. Farre found that in many of the joints these irregular masses had resolved themselves into globular bodies of a brown colour, the tubes being left as transparent as glass, and the brown globules appearing as an irregular row of beads in the interior. In almost all the globules was a cup-shaped depression, generally so situated as to face the surface of the tube in the centre of the depression : a small collection of brown granules, of about uniform size with the globules of circulation, was always present. The brown bodies, on being torn open, were found to consist of a very thin investing capsule, filled with the green granules of the plant mixed with mucous fluid.

" It appears then that this remarkable change had taken place within a week after the circulation had been observed to be going on vigorously in the plant. And the nature of the change appears to be this. The green granules which line the internal surface of the living joints desert the parietes, and, together with the green circulating granules of the interior collect together in irregular masses in the centre of the tube, which then resolve themselves into irregular spheres, still retaining the granular outline indicative of their formation by aggregation, but which they afterwards lose on assuming a more perfect spherical form, and become bounded by an investing capsule, which turns of a rich brown colour, while the contained granules retain their original green.

"It is difficult to imagine what purpose is intended to be answered by such a change taking place after all circulation and other evidences of life in the plant have ceased. Yet the idea of such a change being the result of a merely fortuitous arrangement of the component particles of the plant attendant on decomposition, is negatived by the circumstance of the remarkable uniformity and symmetry of the resulting globules, which appear to possess the most definite characters, differing from each other only in size. Nor is the change by any means uniform for the whole plant; for in some parts two of the joints were observed to be in the green state, while the joint situated between them was free from green matter and contained the brown bodies: but this might have resulted from the circulation ceasing earlier in the central joint, and consequently allowing more time for the changes to take place."—p. 23.

V. — *On the Structure of some Tissues possessing Hygrometric Properties.* By E. J. Quekett, F.L.S., B.S., &c.

Mr. Quekett cites many interesting examples of hygroscopic properties as exemplified in Mosses, Lycopodium lepidophyllum, the seed-vessels of Epilobium, Mesembryanthemum, Rhododendron, Geranium, Banksia, Hura, Avena fatua &c. He then details the structure of the capsule of Cerastium, which we give in his own words.

"Let us take, for example, the capsule of some plant of the order Caryophylleæ, as Cerastium, and it will be found that when that organ approaches maturity, the apex, which was pointed and entire, will open by splitting into five equal valves, which curl outwards, making one complete coil, and always in the same direction; by the application of moisture the valves will resume their original position, and when dry take on the curled form again.

"If one of these valves be examined it will be found to be thin and diaphanous where it forms part of the body of the capsule, but where it is hygroscopic it is horny and opaque. When a section taken from the edge of the valve or curled part longitudinally is examined by the microscope, it will be found that the tissues will be different on the exterior and internal surfaces, both however cellular; but the inner layer, or the cells of the convex border, are of different dimensions from the outer, being neither so large nor having so thick walls, (though they are thicker than ordinary cells); whilst those of the outer layer are almost solid, and the only cavity they have is indicated by thin spaces between a series of lines, the chief one of which is in the direction of the longer axis of the cell, the others connected to it at right angles. These cells are of most curious structure, and form an interesting object for the vegetable anatomist.

"In this arrangement of parts it must be evident that by the shrinking of the tissues of the seed-vessel by loss of moisture, there will occur unequal contraction, and that side will be curved which contracts the most forcibly, and by this curvature one valve must necessarily be removed from the next, evidently showing that the tissues act as unequal antagonists to each other."—p. 26.

Mr. Quekett gives numerous instances in which the great end of disseminating the seeds of plants is accomplished by the hygroscopic properties of their tissues. The subject is one of much interest, and well worthy the attention of the student of nature. In pursuing the enquiry he will almost invariably find the means so admirably adapted to the required end, that not only would it be impossible for human ingenuity to devise an improved plan, but the plans adopted by nature often offer us models and exhibit combinations which have suggested some of our most apt and useful applications of mechanical power.

VII. — *The process of charring Vegetable Tissue as applied to the examination of the Stomata in the Epidermis of Garden Rhubarb.* By the Rev. J. B. Reade, M.A., F.R.S.

The Rev. Mr. Reade's paper on charring vegetable tissue appears to be of great importance, as tending to settle the mooted question of the existence or otherwise of a membranous covering to the stomata of vegetables. The author's conclusions are these : —

" That in the simple uncharred state of the semitransparent tissue there is much room for difference of opinion, so that the eye, fortified by a little previous theory, might most pardonably see the stomata either open or closed.

" That the application of the process of charring proves, beyond a doubt, that the stomata in this tissue of the rhubarb are distinct openings into the hollow chambers of the parenchyma of the leaf.

" That the perforation is the rule and not the exception in the structure.

" And that the exception, where it exists, *i. e.* where the stomata are closed, proves the existence of the overlying membrane discovered and described by Dr. Brown." — p. 41.

ART. LXXXV.— *Varieties.*

203. *Additions to Mr. Flower's List of Plants in the vicinity of Bristol,* (Phytol. 68).

Bromus erectus. Between Horfield and Filton, abundantly.

Danthonia decumbens. Clifton Rocks, sparingly.

Anchusa sempervirens. Between Frenchay and Downend.

Anagallis cærulea. Cornfield at Horfield.

Samolus Valerandi. Stapleton Quarries, near the river.

Rhamnus catharticus. Hedges at Stoke Gifford.

Viola hirta. Stapleton, Leigh Woods and St. Vincent's Rocks.

Scilla autumnalis. Near bridge, St. Vincent's Rocks.

Fritillaria Meleagris has been gathered in flower two or three different times in a field of Mr. Maule's at Stoke Gifford.

Juncus maritimus. Portishead.

Rumex Hydrolapathum. River-side near bridge at Stapleton.

Triglochin maritimum. Clevedon.

Colchicum autumnale. Fields near Dundry

Paris quadrifolia. Near Shirehampton.

Silene maritima. Clevedon, abundantly.

Alsine marina. Under St. Vincent's rocks.

Cotyledon Umbilicus. Stapleton, Hanham and Easton.

Sedum album. Wick Cliffs & Frenchay.

Spiræa Filipendula. Durdham Down.

Tilia europæa. Leigh Woods.

Mentha rotundifolia. Road-side between Westbury and Horfield.

Teucrium Chamædrys. Blaize Castle.

Linaria spuria and Elatine. Cornfield at Horfield and at Bishpool.

——— *minor.* Crew's Hole.

——— *repens.* Roadside at Nailsea.

Thlaspi arvense. Cornfield at Horfield.

Cochlearia anglica. Rownham, abundant.

Koniga maritima. About Baptist Mills.

Geranium pratense. Week.

——— *lucidum.* Easton.

Althæa officinalis. Portishead.

Vicia sylvatica. Leigh woods.

Hypericum calycinum. Leigh woods.

Carduus eriophorus. Near Bedminster coal-pits.

——— *pratensis.* Filton mead.

Conyza squarrosa. St. Vincent's rocks.

Aster Tripolium. Rownham.

Gymnadenia conopsea. Stoke Gifford.

Habenaria bifolia. Woods at Stapleton and Stoke.

Spiranthes autumnalis. Purdown.

Carex pendula. Near St. Ann's Wood.

——— *Pseudo-cyperus.* Winterbourn.

Mercurialis annua. Crew's hole.

Polypodium vulgare, γ. cambricum. Near Downend.

Ophioglossum vulgatum. Stoke Gifford.

Lycopodium clavatum. Clevedon, abndnt.

Diplotaxis tenuifolia is abundant in St. Phillip's and at Lower Eaton.

Sedum rupestre is very abundant on St. Vincent's Rocks, nor should I have thought it had been introduced, for it is by no means confined to any particular spot, and as it does not seed there, I think it would be difficult for it to spread over such an extent as it does by its roots alone.

Pyrus torminalis grows in Leigh Woods, but I have never seen it on St. Vincent's Rocks.—*Samuel Freeman ; Birmingham, October* 16, 1841.

204. *Rarer Plants near Southampton.* Appended is a list of some of the rarer plants observed at Southampton, in May and June, 1842, with their localities. Those marked with an asterisk were furnished by Mr. T. S. Guyer, of Southampton.

Utricularia vulgaris. Old canal, Milbrook shore.

Circæa Lutetiana. Near Wood Mills.

*Rhynchospora alba. Botany Bay.

Isolepis fluitans. S. end of Miller's pond.

*Verbascum Blattaria. Entrance of the avenue.

Drosera longifolia. Botany Bay, near the Fareham road.

Atropa Belladonna. Wood mills ; shore near Itchen and Netley.

Œnanthe Phellandrium. Milbrook shore

Allium oleraceum. Ditto.

Convallaria multiflora. Wood by footpath to Netley, near the abbey.

Daphne Laureola. Ditto.

Sedum anglicum. Netley shore.

*——— Telephium. Near a pond on going to Shirley across the fields.

Silene maritima. Shore between Itchen and Netley, in the utmost profusion

Alsine marina. Milbrook shore.

Nymphæa alba. Miller's pond.

Orobanche major. Heath near Miller's pond.

*Linaria repens. Milbrook shore, beyond the church; also near Shirley church

Diplotaxis tenuifolia. Southampton walls

Lathyrus palustris. Botany Bay.

Carduus pratensis. Southampton Common ; heath by Botany Bay.

Hypericum Androsæmum. Roadside between Wood mills and Northam bridge.

Epipactis latifolia. Southampton Com.

Myriophyllum verticillatum. Canal by Milbrook shore.

Tamus communis. Southampton Comn.

Myrica Gale. Botany Bay, near Southampton, plentiful.

Lomaria Spicant. Botany Bay.

Osmunda regalis. Ditto.

Lastrea dilatata. Ditto. — *W. L. Notcott, Fareham, July 6, 1842.*

Athyrium Filix-fœmina. Ditto.

Asplenium Adiantum-nigrum. Fareham Road.

Scolopendrium vulgare. Ditto.

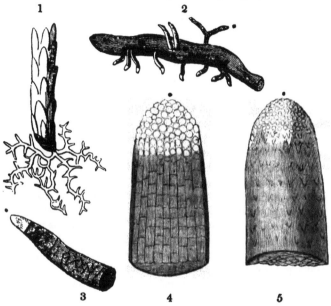

Fig. 1. Lower part of stem of Monotropa, with a portion of the branched root. 2. Part of the root with its attached fibrils, magnified. 3. A fibril detached from the root, as seen under a higher power. 4 & 5. Sections of the end of one of the fibrils, showing the loose cellular tissue of which the tip is composed, highly magnified. The * attached to the various figures indicates the transparent cellular tip of the fibrils, as seen under different powers.

205. *On the Mode of Growth of Monotropa Hypopitys.* The discussion between Messrs. Lees and Wilson on the parasitic nature of Monotropa excited so much interest in my mind, that not only did I regret its sudden termination, but determined on the first opportunity to examine the matter thoroughly for myself. Today, through the kindness of Mr. Wilson, I obtained specimens from Southport, and after some labour and trouble have arrived at the following conclusions, which, as they differ in some measure from those of both the combatants above mentioned, I beg to hand you for a place, if thought worthy of it, in your invaluable Journal. Having carefully removed the outer covering of sand, I came to the matted roots of the willow, which, from their density, and the quantity of sand amongst them, threatened to prove an effectual barrier to my further progress, for an hour was spent in unsuccessful picking and poking at them with a penknife. In this stage of my proceedings a large bowl of water was procured, and by slowly moving the mass, held carefully between my hands, through the water, the stem of the plant being supported by my thumbs, I had soon the gratification of seeing, not only all the sand, but the loose willow-roots and much other foreign matter, deposited in the bowl. The plant was then allowed to drain for a minute, and it afterwards required very little labour to extricate it, with roots, as seen at fig. 1. The plant is not parasitic; it has no organic connexion with

the "nidi" of roots among which its own are developed; but on the other hand it is provided with an independent root of its own, and that, too, as extensive as in most plants of its size: nor are these dense nodules of fibre produced only in its immediate vicinity; several such were examined, and no trace of Monotropa could be found;— they appear to be attached equally to other substances. After examining the willow-roots which were in contact with those of the Monotropa, and on which I could find, in no instance, the slightest evidence of attachment, I proceeded to note the structure of the root itself. It consists of very brittle, tortuous, fibres, which are thickly studded with more or less claviform branches or fibrils (as represented magnified in fig. 2). Many of these fibrils were necessarily broken off during their separation from the mass, but in such cases a scar distinctly marks their former position. One of them was then placed under the microscope; with a low power it appeared as at fig. 3, and the transparency of the tip suggested the idea that they were spongioles formed as in other plants. Various sections were then made, which, on the application of higher powers, proved this to be the case; figs. 4 and 5 are selected from these, and show at once all the peculiarities of the ordinary spongiole. From this it will appear that although the root of the Monotropa be remarkable in structure, it is still a *true root*, and present in sufficient quantity to supply the necessities of the plant; and with regard to its constant proximity to the root of some tree, may it not — the difficulty of the subject and the notion of parasitism having called undue attention to it — be more signally remarkable in our books than in nature? Or is the excrementitious matter of these trees particularly nutritious to it, or it's to them? This I leave to those who have better opportunity to decide the matter; but I have no hesitation in saying, that the Monotropa has as independent an existence as any other plant. The "minute hair-like fibres" mentioned by Mr. Lees were also carefully examined; they have certainly not the appearance of suckers, and though they infest the plant to a considerable extent, are not universally present, indeed I could not assert that they are part of its structure.*—*Thos. G. Rylands; Bewsey House, Warrington, July 23, 1842.*

206. *Cistopteris fragilis* grows abundantly at Castleton; I gathered some beautiful fronds that were growing on the rocks at the entrance of and rather sheltered by the justly celebrated Peak Cavern. — *John Heppenstall; Upperthorpe, near Sheffield, August 3, 1842.*

207. *Lathyrus Aphaca*. I take the liberty of enclosing a specimen of Lathyrus Aphaca with its true leaves, which I believe are rare. It was discovered in 1841 by Mr. W. Newnham, jun., in a corn-field near Farnham, where it was then very plentiful; this year it is by some accident *very* rare. The enclosed was gathered by myself. —*Christopher A. Newnham; Farnham, Surrey, August 4, 1841.*

[We beg our correspondent to accept our thanks for the interesting little specimen sent. The occasional appearance of leaves in this species, especially on young plants, is mentioned by Linnæus, Smith, Hooker, and other authors, but we do not remember having seen any notice of a variation in the form of the stipules when the leaves are present. The young specimen kindly forwarded by Mr. Newnham has two leaves, each consisting of a pair of elliptic-lanceolate leaflets; the stipules accompanying the leaves, instead of being of the usual broad arrow-shaped form, are merely half arrow-shaped. Three of the simple tendrils in a flowering specimen in our own herbarium,

---

* Since writing the above Mr. Rylands has more fully investigated these fibres; the result of his researches will appear in our October number.

bear each a single lanceolate leaflet; but in this case the stipules are all of the usual form.—*Ed.*]

208. *Inland Localities for Maritime Plants.* I beg to hand you the following notice of plants which have been recently discovered by Mr. George Reece of this city, in localities not hitherto noticed by botanists, so far as I am aware. Arenaria marina is mentioned in Purton's 'Midland Flora' as occurring on Defford Common, where salt springs were discovered many years since, and where salt is now, I believe, manufactured an a small scale. Erodium maritimum is also recorded by Purton as having been found at Stourbridge, Kinver and Bewdley; and to Triglochin palustre is assigned a habitat at Feckenham; but neither Glaux maritima nor Plantago maritima is noticed in the 'Midland Flora.' Arenaria marina, Glaux maritima and Triglochin palustre are found on the banks of the Droitwich Canal; Erodium maritimum and Plantago maritima on Hartlebury Common. Beautiful specimens of these plants have been collected by Mr. Reece, and are now deposited in the herbarium of the Worcestershire Natural-History Society.—*J. Evans; Grove House, Worcester, August* 8, 1842.

209. *Note on Woodsia Ilvensis.* On the 17th of 8th month [August] 1798, my father gathered a single frond of a fern from Crosby-Ravensworth Church, Westmoreland. Being unable to name it he showed it to several botanists in London, who could not decide what it was; Lewis Dillwyn at length sent it to Sir J. E. Smith, who returned the specimen labelled as follows:—

" Polypodium arvonicum,
       With. et Fl. Brit. *J. E. Smith.*
P. ilvense, *With. &*
Acrost. ilvense, *Huds.*
(not Linn.)
Acr. alpinum, *Bolt.*"

The original specimen is now in my possession, with Sir J. E. Smith's autograph; the frond is 3¼ inches in length from the bottom of the rachis to the apex, and about 2 inches from the lowest pinnæ to the apex. The church has been pulled down and rebuilt within the last few years.— *Silvanus Thompson; Friends' School, York, 8th Month* 9, 1842.

210. *Enquiry respecting the British Oaks.* Can any of your correspondents inform me what are the true specific distinctions between Quercus Robur, intermedia, and sessiliflora? I have found what I consider to be the three species growing in various localities round Manchester: I have also gathered a number of specimens with the form of the leaves, length of the petioles and situation of the fruit so varied and intermediate, that I cannot with certainty refer them to either species. Several trees near Mottram, from which I procured specimens last month, have the fruit quite sessile and the petioles of the leaves so short as to be scarcely perceptible; whilst specimens of the true sessiliflora from Leigh Woods, near Bristol, given to me by Mr. Grindon, and also specimens gathered by myself in Bredbury Wood, Cheshire, have petioles near an inch in length. I have also specimens from near Mottram, which appear to be hybrids between Q. Robur and intermedia; these have their fruit upon peduncles less than an inch long, but never properly sessile, whilst the leaves are perfectly so. The form which I consider to be the real intermedia of authors, has long petioles to the leaves, and peduncles nearly half an inch in length, but one of the most prominent characters is in the form of the foliage; for whilst in Robur the leaf is much dilated towards the extremity, in intermedia the greatest width exists near the centre, the end

being much narrower in proportion, and the margin being altogether more bluntly cut or divided. I am inclined to believe from the consideration of these several forms that either the number of our native species of oak is at present *undetermined*, or that in reality we possess but *one.—Joseph Sidebotham; 26, York St., Manchester, August* 12, 1842.

211. *On Monotropa Hypopitys.* In deference to such high authority as that by which Mr. Newman's views are supported, I have re-examined the subject with much care, and specimens kindly supplied by Mr. Newman in a recent state have been studied in connexion with others obtained from Southport. Before I state the conclusion at which I have arrived, let me acknowledge to my much-esteemed opponents, that although I consider them to have been mistaken, I regard their error as scarcely to be avoided under the circumstances. I have seen a portion of a specimen corresponding with that which they examined, and was myself strongly inclined at the first to coincide in their view; nor was it until after very minute dissection that I perceived any fallacy. The "rhizoma" had apparently an uniform fibrous covering; but in the specimen since received from Mr. Newman, and in those from Southport (examined at the same time) the fibrous matter occurs only in very irregular and detached masses, leaving considerable tracts of the rhizoma quite naked, and this fibrous matter appears to be different on different specimens; but as an able and enthusiastic friend is now engaged in the investigation of this portion of the subject, I shall briefly state that my inference is that the supposed "cuticle of the rhizoma" is really a foreign body. Thin transverse sections of the stem exhibit no trace whatever of a cortical layer; except that the cellular tissue is surrounded by a single indurated stratum of smaller cells of a brown colour, perfectly continuous with the rest of the cellular tissue. No membranous cuticle could be detected, by scraping or otherwise, upon the stem. On the rhizoma something like a cuticle was at times found in parts not coated with fibrous matter; but this, besides being immeasurably thinner, is different in structure from the cuticle referred to in Mr. Newman's paper* (Phytol. 298), and can only be detected by scraping the surface of the rhizoma. In some cases it seems to be quite absent, and may therefore not belong to the plant. Mr. Newman's first and second reasons do not appear to be well founded. Wherever the fibrous matter occurs I find it matted together, and attached to the rhizoma in such a manner as to forbid the idea of organic connexion; and if there be any difference in the thickness of the ends of each fibre, which is very doubtful, the position of the parts is at variance with Mr. Newman's idea. The fibres indeed are often united together into fasciculi, presenting an appearance very apt to mislead. After careful inspection of the colourless extremities of the "rhizoma," I retain the opinion expressed in my former remarks. In opposition to the opinion that they are only growing extremities of the divisions of a rhizoma, I would remark that I know of no instance where the growing parts of a rhizoma have a downward or backward direction; also that the supposed growing extremities never develope themselves into scaly buds:—these are always laterally inserted, and, if I mistake not, are at first immersed in the substance of the root, like the buds in the tuber of a potato. On this point my former remarks require some correction.— *W. Wilson; Warrington, August* 12, 1842.

---

* Considerably more than $\frac{1}{2000}$ of an inch in thickness, and therefore easily seen when a section of the rhizoma is under the microscope.

212. *On Carex tenella.* Far be it from me to condemn that liberal criticism, the object of which is the attainment or communication of truth. Next to the relation of interesting facts, I know of nothing more suited to the pages of 'The Phytologist' than judicious and candid strictures on standard works relating to Botany; I trust we shall see many such, and that Mr. Gibson will frequently direct his abilities into that channel. In my former remarks I did not intend more than to show that the author of the 'British Flora' is not chargeable with the neglect or inadvertence imputed to him. The account given by Sir J. E. Smith was referred to, proving that an experienced botanist, with superior advantages for forming a correct judgement, and with peculiar inducements to be deliberate, may see or think he sees, in Schkuhr's figure of Carex tenella, something which is not apparent to Mr. Gibson. The question indeed is properly one between him and Sir J. E. Smith only; and if it should prove, as I believe it will, that the question of doubt in 'British Flora' was confined to the Scottish specimen called C. tenella by Smith, Mr. G. must be sensible that some of his remarks were needless, as bearing upon a point not under dispute. Admitting Mr. G. to be right in his view of Schkuhr's figure, it follows that Smith must have been wrong, and if so, he may have been also in error as to the Scottish specimen. I could point out parallel cases where Sir J. E. Smith has too hastily assumed the identity of essentially different plants; and it is probable that if he possessed only a solitary specimen, his examination would, in this case, be rather superficial. To ascertain the existence of only two stamens, dissection would be requisite, and to dissect would be to mutilate what he might wish to preserve uninjured. The fruit also might be so immature as not to exhibit the character of the species. The author of 'British Flora' was not bound to admit that specimen, which he had never seen, on the mere authority of one who reported that Schkuhr's figure could not be implicitly relied on; and it would not be unreasonable in Mr. G. to have some misgivings on the same subject. Let it be remembered also that Wahlenberg and Willdenow refer the whole figure to C. loliacea, a species which, notwithstanding what is said of it by Smith, may possibly have fruit corresponding with the figure. To determine this question recourse must be had to the original description, and perhaps to an authentic specimen of C. loliacea.—*Id.*

213. *Note on a Criticism in Taylor's Annals.** Allow me to say in reference to a note in the July number of the 'Annals and Magazine of Natural History,' (p. 421),

---

* The following is the criticism referred to by Mr. Gibson. "Notes on *Arenaria rubra, marina* and *media;* by S. Gibson, Esq. [The two former appear to us to be distinct species, but we cannot agree with Mr. Gibson in separating the latter from *marina*, as our own observations would lead us to believe that the characters drawn from the seeds and length of the capsule are not constant. We trust that we shall not be considered presumptuous if we hint to this very accurate observer, that a more frequent reference to the writings of continental botanists would be desirable. We say this without the least wish to detract from the value of Mr. Gibson's papers, but merely to avoid the introduction of additional synonyms into our already encumbered science, of which an instance occurred in a late number of the 'Phytologist,' where a *supposed* new species of *Monotropa* is named and described, which had long since received several denominations in botanical works.]"—From a notice of 'The Phytologist,' in the 'Annals and Magazine of Natural History,' ix. 421.

that I have this season, with a determination to avoid any preconceived opinion on Arenaria marina and media, made all the enquiry that I possibly could; the result of this enquiry is that I have not been able to find anything opposed to what I have before said on the subject, (Phytol. 217). After assiduously examining these plants, I may be excused for observing that I cannot participate in the writer's opinion; for notwithstanding the possibility of there being some little variation in the seeds (but such I have never seen), I have not the least doubt of their *specific* distinction. I might just as confidently state the rubra to be identical, if I were to find some little difference in the seeds of that plant, and indeed it would appear that the seeds of rubra do vary most strangely, as they are described by Leighton and others as being angular, whereas all that I have seen are somewhat pyriform; perhaps the Arenaria rubra described by authors may be a plant which I have never seen, as it would appear that the rubra of authors is something very like marina, since Mr. Babington tells us that he is inclined to believe them to be two species, but that he has been unable satisfactorily to distinguish them, (' Primitiæ Floræ Sarnicæ,' p. 16). So far as regards the Monotropa, I am quite willing to give up for a better one the name I have used to distinguish the two forms of that plant, and shall leave to botanists the decision of the question how far they may be distinct as species. I am well aware of the evil attending the introduction of additional synonymes, and am much surprised to find that botanists who ought to know better are carrying out the thing to the greatest length; as I see that we have in Baines's ' Flora of Yorkshire,' published in 1840, a description of Aira cæspitosa, var. β. *rigida ;* in Leighton's ' Flora of Shropshire,' published in 1841, we have an Aira cæspitosa, var. β. *major ;* in the Edinburgh Botanical Society's Catalogue, second edition, 1841, we have an Aira cæspitosa, var. β. *vivipara ;* and if we look into a few works on our British plants, we shall find that in many instances the same plant has a different name in each. In Lightfoot's ' Flora Scotica' we have a plant under the name of Arundo arenaria; in the 4th edition of Withering's Botany we have the same plant called Calamagrostis arenaria, and in the 5th edition of the same work Arundo arenaria; in Gray's Botany it is called Psamma arenaria, in Smith's ' English Flora.' Arundo arenaria ; by Sir W. J. Hooker it is called Ammophila arundinacea; in the Edinburgh Society's new Catalogue we find it to be AMMOPHILA ARENARIA. Should we wish to trace this plant into the works of continental writers, we shall find it under such names as Spartum anglicanum, &c. But as I have no wish to go beyond the bounds of my own country, I will trace the Ammophila a little further in the works of my countrymen; the above name I find used in 1802 for *a genus of Hymenopterous insects* by Kirby, the same name has been retained by Turton, Sowerby, Leach, Samouelle, Stephens, &c. The above name AMMOPHILA ARENARIA may be convenient to the Edinburgh Botanical Society for the *Sand-marram,* but for me as an entomologist it is not convenient, for if I consult Donovan on that subject, I find the very same name—AMMOPHILA ARENARIA applied to one of our *Sand Wasps.* And perhaps botanists will soon lose their antiquated generic name Ficus, as I see Mr. Cumberland of Bristol is now using it for a genus of *fossil Crinoidea.* The editors of the 'Annals and Magazine ' say that they trust they will not be thought presumptuous in hinting to me that a more frequent reference to the writings of continental botanists is desirable. In reply to this I would tell them (as it appears they have forgotten it) that all botanists are not equally fortunate in being so placed as to have an opportunity of consulting such books; and if I may be allowed to speak on the introduction *of synonymes,* we have a fine specimen of this in the Shropshire Flora. The four species of Arenaria found in that county are there placed under four different generic names; the Arenaria rubra of Smith being Lepigonum rubrum in Leighton's Flora and the Alsine rubra of the Edinburgh Society's new Catalogue. — *Samuel Gibson; Hebden Bridge, August* 15, 1842.

214. *Note on Carex axillaris and remota.* Allow me to return my best thanks for the information given by Mr. Wilson on the two Carices, (Phytol. 299). I have carefully looked over the descriptions and separated the characters assigned to each; and leaving out all the superfluous matter they will stand thus: —

1. *Carex axillaris. Leaves* flattened. *Bracteas* variable in length, second very small, with a membranous base, upper part rough, very narrow and awn-like, all of them auricled at the base. *Rachis* perfectly straight, with three rough angles. *Glumes* roundish-ovate, tipped with a very short rough point.

2. *Carex remota. Leaves* bent in at the sides, so as to be almost semicylindrical. *Bracteas* generally with a pale very obscure ligula, passing completely round the common stalk. [*Rachis*] zigzag, in the upper part with two rough angles. *Glumes* ovate-acuminate, narrower than the fruit.

Mr. Wilson tells us that the rachis in *remota* is zigzag, and in *axillaris* perfectly straight. Dr. Wood tells us that the lower bractea in axillaris forms as it were a continuation of the stem as regards its direction. Now if the Dr. be correct in this point the stem must be pushed out to give room for the spikelet, and therefore would be at variance with Mr. W.'s straight rachis. Mr. W. tells us that the leaves in axillaris are *flat*, the Dr. tells us that they are *channelled* — but stop, perhaps I am leaving out something of importance, he says "*plane* though *channelled*," &c.; here I am a loss to know how a plane leaf can be channelled. Whatever praise might be due to Dr. Goodenough from his first paper on our British Carices, I do think that he robbed himself of such praise by writing the following passage.—" I have stated axillaris as having the capsule divided at the summit, and remota as having it entire; but this is not constant. I believe all Carices dispose of their seeds by the opening of the point of their capsule, this opening is observable in some very early, in others not till quite old. In the former the capsule is described as opening; in the latter, because it is not seen but in very advanced age, it is mentioned as closed."—(Trans. Linn. Soc. iii. 76). Whatever Mr. W. may think of the above passage, it was written when the subject had, or at least ought to have been, fully investigated. However Mr. W. may be surprised at my opinion on what Dr. Goodenough has said on the capsules of our Carices, I now say that if ever he had investigated the fruit of one Carex, he would have known that when the fruit is first formed, that the opening, if there be any, is very conspicuous, and may be seen surrounding the style, with the stigmas protruding therefrom; and as soon as the capsule is discernible, it will be seen whether its beak be entire or cloven, as it never changes in that respect after it is first formed. So far as regards my quotation from the 'English [British] Flora,' I would ask Mr. W. if Sir W. J. Hooker had not accurately described C. axillaris in the first edition of that work, or has he corrected himself in order that his description may better agree with that of Sir James Edward Smith? Mr. W. tells us that the bracteas in C. axillaris are by no means constant in their length; if he be correct in this, Sir W. Hooker had made no mistake, and therefore we could have no need of any such correction. Mr. W. tells us that the passage I have quoted from the 'English Flora' without the word "perhaps," is sheer nonsense; I would say that the passage as it stands in the 'English Flora' is something very like nonsense, for it amounts to nothing more than to say it *may or may not be so.* What Sir J. E. Smith might have had in his mind when he wrote the passage I know not; but I do disapprove of the use of such passages in descriptions of plants. But as I have said before, in my enquiry on the two species of Carex, my real difficulty in making out the two may perhaps arise from my never having seen them both; but here I must say (omitting Dr. Goodenough's description of the capsule of his remota), that I have specimens in my possession that will answer to every description that I have seen, and many other forms that are not described at all in any work. Our Carex axillaris, remota, angustifolia, cæspitosa, aquatilis, stricta, acuta, and others, require a strict investigation, in order that the species may be correctly made out, and the mere varieties arranged as such. To describe the extremes of any variable plant is quite easy, but when we have got a regular gradation from the two extremes, then the difficulty occurs.—*Id.*

[We trust that the discussion on this subject will now be allowed to terminate: had Mr. Gibson's only object been to obtain information on a doubtful point, he would, we think, have been satisfied with acknowledging that already afforded, which appears quite sufficient to enable any one to distinguish Carex axillaris from every state of C. remota. We have thought it unnecessary to extend Mr. Gibson's paper beyond its present limits, by printing some remarks therein contained on two specimens of Carex remota which accompanied it: the only difference we can perceive between these specimens is, that one is rather more slender than the other.—*Ed.*]

215. *Note on the Stomata of Equisetum hyemale.* I would remark on what is said of the stomata of Equisetum hyemale (Phytol. 278), that they are immersed in *cavities*, not in elevated disks. There are two series of stomata in each furrow, one immediately on each side of the ridge, and somewhat concealed under it when in a dry state. The structure of E. variegatum is similar, but less conspicuous.—*W. Wilson; Warrington, Aug.* 15, 1842.*

Art. LXXXVI.—*Proceedings of Societies.*

BOTANICAL SOCIETY OF LONDON.

August 5, 1842.—John Reynolds, Esq., Treasurer, in the chair. The following donations were announced:—British plants from the Rev. T. Butler, Mr. John Pearson, Mr. Arthur Henfrey, Miss Anna Worsley and Mr. Samuel Freeman. American plants from Mr. O. Rich. British Mosses from Mr. I. F. Hollings. Specimens of *Schistostega pennata*, collected in Nottingham forest by Mr. Joseph Sidebotham, were exhibited and presented by him.

A note was read from Mr. Adam White, stating that he had found specimens of *Dentaria bulbifera* in Chesham Bois Wood, Buckinghamshire. Mr. Thomas Sansom, Librarian, exhibited a monstrosity of *Rosa centifolia*, Linn., in which a second flower was developed from the centre of the first. Mr. W. H. White communicated a paper, being a Report of the botanical state of the Mauritius; translated from the Eighth Annual Report of the Natural-History Society of that island.—*G. E. D.*

---

* In a letter to E. Newman; who begs to thank Mr. Wilson for this very necessary and important correction, and requests the reader will also correct a repetition of the error at p. 308 of the present number.

THE PHYTOLOGIST.

| No. XVII. | OCTOBER, MDCCCXLII. | PRICE 1s. |

ART. LXXXVII.—*A History of the British Equiseta.* By EDWARD
NEWMAN. (Continued from p. 308).

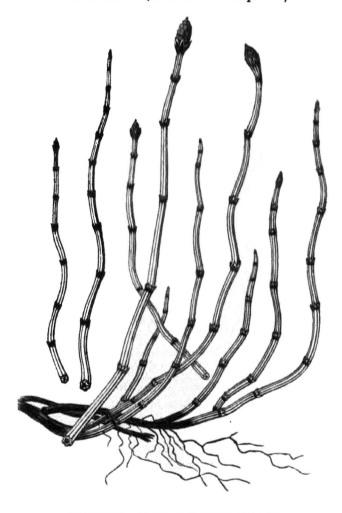

VARIEGATED SHAVE-GRASS.

EQUISETUM HYEMALE, γ. VARIEGATUM.

2 ε

Equisetum hyemale, Linneus, (who gives Bauhin's E. nudum minus variegatum basiliense, as a synonyme), Sp. Plantarum, 1517.

Equisetum variegatum (Schleicher's Catalogue) Willdenow, Smith, Hooker, &c.

Equisetum reptans, Wahlenberg.

Equisetum ramosum var., Decandolle (on the authority of Vaucher).

Equisetum multiforme, Vaucher.

THIS, like the preceding form of the species, is local, occurring generally on dry moveable sand in the immediate vicinity of the sea: when in moist situations it assumes a somewhat altered appearance, but the example figured is, I think, a faithful representative of the plant when growing in dry sand.

In England it occurs at the mouth of the Mersey; I have repeatedly observed it on New Brighton sands on the Cheshire side, and the Bootle sands on the Lancashire side: it is tolerably abundant also at Southport, in the latter county. In Teesdale it has been found in several spots, more particularly on Widdy Bank, and about Winch Bridge: the specimens from this locality are much less rough to the touch than those from the Mersey.

In Scotland it occurs on the sands of Barry, near Dundee; also in Rosshire, and in Kincardineshire on the banks and in the bed* of the Dee, intermixed with the two forms of the plant already described; the specimens are generally much more elongate, and like those from Teesdale are also smoother than the Mersey specimens. In Ireland it is abundant in several sea-coast localities, more particularly at Port Marnock in the vicinity of Dublin, and also in the Dublin Canal.

For specimens from these localities, and for much valuable information respecting them, I have to acknowledge my best thanks to Dr. Greville, Dr. Balfour, Mr. Wilson, Mr. Babington, Mr. Gibson, the Rev. Mr. Brichan, and Mr. Moore of the Dublin Glasnevin Garden.

The roots and rhizoma, like those of E. Mackaii, present no characters by which I can distinguish this from the normal form of hyemale. Like the roots of many other plants when growing in loose sand, those of the present variety are frequently clothed with a quan-

---

* An error occurs in a quotation from the Rev. Mr. Brichan's letter, (Phytol. 306); instead of "what is usually *called* the bed of the river," it should have been "what is usually the bed of the river." The meaning is this, "that the Dee has this season, owing to the excessive heat, sunk so far within its usual bounds, as not to cover what *is usually* its bed." That part of the Dee in which the Equiseta occur is in the county *of Kincardine*, not of Aberdeen.

tity of matted fibrillæ : the stems are short, often semiprostrate, and the internodes are short and frequently somewhat arcuate, giving the entire stem a sinuous appearance. The figure represents the plant of the natural size ; the four detached stems being portions of rather more elongate specimens : both the sheaths and internodes are striated; the striæ are few in number, six or eight may be taken as the average : under a microscope the cuticle precisely resembles that of the varieties already described : the ridges are grooved, the grooves being margined on each side with a longitudinal series of minute flinty tubercles : in the furrows are two longitudinal series of stomata placed very near the ridges, indeed so near, that in stems dried in an immature state, they are frequently partially obscured by the ridges. Mr. Wilson, whose surpassing accuracy deserves the thanks of all botanical students, has pointed out a very decided error in my prior descriptions, in which I have stated the stomata to be placed *on the ridges* instead of *in the furrows*. On receiving Mr. Wilson's note, correcting this error, I carefully re-examined the specimens, and found that my published description was erroneous, and Mr. Wilson perfectly correct. The lower portion of the sheaths is perfectly concolorous with that of the internode, the upper portion only is black, thus causing the sheath to appear much shorter than in the varieties previously described : the teeth are short, wedge-shaped, and commonly without the setiform apex which distinguishes the variety last described: their edges are membranous, occasionally black, as represented in the detached stems to the left of the figure, but usually white, giving the plant that variegated appearance from which its name has probably been derived. The catkin is small, apiculate, terminal and striated, as in the preceding varieties; its scales are few—eighteen to twenty-five in number.

The stem of this variety is much less liable to become branched than either of the preceding, still this branching occasionally occurs; in the margin I have represented a New Brighton specimen bearing a branch ; the black colouring of the sheath in this specimen extends much lower than is usual, yet the teeth remain wedge-shaped. Specimens occasionally occur in the same habitat repeatedly branched, and much more luxuriant than the one figured. Of one of these in his own herbarium Mr. Wilson has kindly furnished the following description.

"I have some specimens of E. variegatum from New Brighton, opposite Liverpool, which are very much branched, and very tall ; but even in these I can find only two branches at any one joint. I will describe one of the best marked of these. Height 22 inches, of which 2½ inches is black, having been buried in the ground to that depth, and rather more. At 2¼ inches two *opposite* branches (A A) shoot out, with *rooted bases.* At the next joint are two other *opposite* branches

(B B) placed at right angles to the first pair A ⁎ A with respect to the axis. At the next joint are two other opposite branches, as in A A, and at the next joint is another opposite pair, as in B B. The right branch, A, at the seventh joint, has two branches, which are contiguous or nearly so (not opposite), and two more at the ninth in the same condition; two opposite ones at the tenth joint, one branch having the same direction as the two lower branches. In this specimen the main stem is broken off at the height of seven inches; the second primary pair of branches has also one branch ramified, but less remarkably so than the one which I have already given in detail."

Mr. Moore's specimens of the Dublin plant were accompanied by the following note. "There is a third plant⁎ which claims attention. it occurs near Dublin, growing in water, upright, and to a much greater size than E. variegatum in its usual state, and it possesses the other characters noticed in Hooker's 'British Flora,' 4th edition, p. 394,† as distinguishing the plant found by Mr. Wilson at Muckruss. This species or variety, whichever you please to call it, appears to adhere to its natural habit when subjected to cultivation, under which state I have had it about six months, and the new fronds are getting much stronger than those which the plant produced naturally. Although entirely removed from the water it grows quite erect, and continues smoother, with the same number of striæ." I do not agree with my much-esteemed correspondent in thinking the Dublin and Muckruss plants identical, still the Dublin plant is very different from the usual variegatum form, and nearly resembles the specimens from Kincardineshire already alluded to. EDWARD NEWMAN.

(To be continued).

⁎ The usual forms of Mackaii and variegatum are the other two plants noticed in Mr. Moore's letter.

† "At Mucruss Mr. Wilson finds this plant growing in water and upright to thrice that size [6—8 inches long], with a *stem* smoother, about 10-furrowed and more polished in the furrows, and the *sheaths* not so conspicuously nor so constantly furnished with acuminated *teeth* or summits as is usual in the ordinary state of the plant."—Br. Fl. ed. 4. 394.

ART. LXXXVIII.—*On the nature of the Byssoid Substance found investing the Roots of Monotropa Hypopitys.* By THOS. G. RYLANDS, Esq.

ALTHOUGH I feel, and with much more cause, the diffidence experienced by Mr. Newman in expressing an opinion on a subject so difficult as the present one, yet, when after an investigation conducted with the utmost care and deliberation, and extended over many hours, I have arrived at results the most constant and apparently decided, I feel that it is not only desirable, but my duty, to publish them: and moreover, it seems but right, since the matter has proceeded thus far, that it should not be abandoned till some decision be recorded. It is on these grounds that the following remarks are produced to the public.

The opinions already expressed in 'The Phytologist,' relative to the nature of the byssoid substance found on the roots of Monotropa, from all stations that have supplied specimens for examination, are as follows.

Mr. Luxford refers them to a "byssoid fungus," (Phytol. 43); Mr. Lees regards them as "suckers," (Id. 171); Mr. Wilson suggests that they may be "the woolly matted extremities of grasses,"* (Id. 149); and Mr. Newman believes them to be essentially the root, (Id. 297). To these, in the first place, I will add a detail of my own investigations, and then proceed to compare the results of all. Having arrived at the conclusion that the "claviform branches" of the "rhizoma" were the true roots of Monotropa (Id. 329), the minute fibres now under consideration were only examined in connexion with Mr. Lees' idea that they were spongioles or suckers, and then thrown aside: but on the publication of Mr. Newman's observations, to whose kindness I am indebted for specimens of the plant from Shoreham in Kent, and Hurstperpoint in Sussex, without which I should have been ill prepared for my present essay, attention was again directed to them, and the following is the result.

Viewed with a magnifying power of about 30 linear, the byssoid substance presents an appearance not unlike that of unsized paper seen under similar circumstances. It consists of an irregularly matted mass of flocculent matter, more or less depressed, investing the roots of

* It is but just to my friend Mr. Wilson to state here that after a more careful investigation his opinion is, in the main, in strict accordance with the results of my own observations. Nor would I omit to acknowledge how much confidence the accession of one whose authority is so deservedly high, has imparted to my mind.

Monotropa, and occasionally other substances in its vicinity. Usually it is present in such quantities as to render the examination of it an operation exceedingly difficult; on the Southport specimens it is sometimes much less abundant, but even in these cases it was not till after repeated examination that the real nature of its connexion became satisfactorily apparent. The fibres of which it is composed are nearly equal in thickness throughout, irregularly branched, and united *by lateral adhesion, at intervals,* to the substance on which they are found : this union, in several instances, was so complete, that when the fibres were forcibly detached, a very small portion of the body with which they were connected was separated with them; in no case, however, was this operation performed but a continuation of the fibre became evident, as seen at *n, o,* fig. 2. Evidence of a *terminal* connexion or thickened extremities was seen in no instance.*

In order more fully to test the correctness of these observations, the roots of other plants were appealed to : several from the greenhouse which were matted owing to the small size of the pots in which they grew, were examined; but in these a marked distinction was evident, so much so, that at first sight I should have had no doubt of their origin, they possessed in every particular the appearance of spongioles : while the *really* fungoid matter found on the roots of groundsel, Epilobium, Plantago &c., had so much resemblance to the substance in question, that it would be difficult by words to render the difference appreciable.

Thus far had I proceeded when by the application of a magnifying power of about 350 linear, I became aware that the fibres on the Southport plant were of two very different kinds; and, on examination, the specimens received from Mr. Newman afforded, at least, a third. This I regard as a somewhat forcible argument in favour of the opinion that they are altogether extraneous; for all the species are similarly attached, and I apprehend that no such differences as those about to be described would be likely to occur in the roots of one plant.

The species observed are as follows : —

No. I. — Filaments tufted, fasciculate, more or less adnate, membranaceous, slightly tubular, jointed, irregularly branched, colourless.† (Fig. 1, c, d, e).

Found only on Mr. Newman's specimens from Shoreham.

* These results were obtained with powers of about 100—200 linear.

† The Sussex specimens from Mr. Newman supply a filament differing from these; but as the one had been some time dry before examination, and the others were viewed *in a fresh state,* I have preferred leaving it undescribed.

Were further proof of a lateral connexion necessary, the fact that the filaments, when not in contact with any other body, are found thus attached to each other, might supply it: in the species most thoroughly adnate they frequently adhere throughout their whole length; (see *d, e,* fig. 1).

No. II.—Filaments spreading irregularly, adhering at intervals, horny, smooth, distinctly tubular, and divided by septa, from which are produced the hemispherical buds of the branches? (*g*): branches nearly at a right angle. Of a rich brown colour. (Fig. 2, *f, g, m, n, o; Fig. 3, x, y*).

Found in moderate quantities on most of the portions examined, exterior to the other species.

No. III.—Filaments agglutinated or interwoven together, pellucid, with slight traces of a cellular structure, rarely dichotomously branched; accompanied by numerous subspherical tuberculated or granulated bodies (sporidia ?), which have on their depressed surfaces a circular lucid disk. (Fig. 2, *h, i, k, l*).

Found only on a portion of the Southport specimens.

To Mr. Wilson I am indebted for calling my attention to the spheroidal bodies; they were not observed during my examination of Southport plants, gathered three weeks after his, and as they are objects far from likely to be overlooked, I suppose they were not present: the fibre to which they belong was there in abundance. Specimens from Southport gathered in 1841, kindly supplied by Mr. Wilson, having the filaments, are likewise without the sporidia.

All my attempts to discover actual proofs of an organic connexion of these globules with the fibres among which they are found, have been unsuccessful: though from several circumstances observed, such as that the filaments are frequently flattened, and have occasionally in the depressions the appearance of scars (fig. 2, *i*), I have little doubt that such does exist, the transparent disks being the point of union. Fragments of one or two other kinds have been met with, but not in quantities to supply the characters for a description. The three given will suffice.

Such are the data on which I would ground the opinion that the "byssoid substance" is really fungoid, and performs no essential function in the economy of the Monotropa. Each of the operations has been often repeated, with all the care and diligence I could command. Sections in all variety have been viewed, wet, moist and dry, with reflected and transmitted light; and though the results were in all cases the same, yet it should be stated that those were the most satisfactory in which the portion had been saturated with, and was examined

FIG. 1.

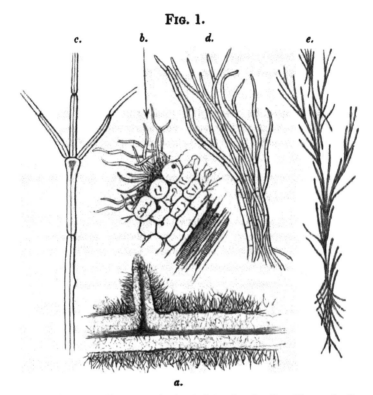

a, A section of the root of Monotropa, showing the internal continuation of the vascular tissue of the spongiole or fibril.   *b*. A portion of the same highly magnified, to show the connexion of the flocci.

c. *Epiphagos Luxfordii*, highly magnified.   *d*, *e*. The same, showing its adnate and fasciculate habit.

whilst immersed in, water. From the nature of the objects this might have been expected; for let it be remembered that six thousand of the filaments in juxta-position would but occupy an inch, and moreover, that so delicate and tender is their texture, that the air of a room, or even the dryness ordinary to the atmosphere at this season, causes them speedily to shrivel and contract. The advantages of this mode of examination, and its efficiency, are admitted in regard to other members of the class.

In reply to Mr. Newman's reasons for believing them a portion of the Monotropa, I would say : —

Firstly. — Its presence — if he refers to *a* byssoid substance generally — is, as far as we can judge, scarcely more "constant" than other species of fungus, when the circumstances are suited to their growth ; — putrefying paste has *constantly* its Mucor. On some portions of the Southport plant it was not present at all; if the species found on the plant from Shoreham alone is intended, and such

Fig. 2.

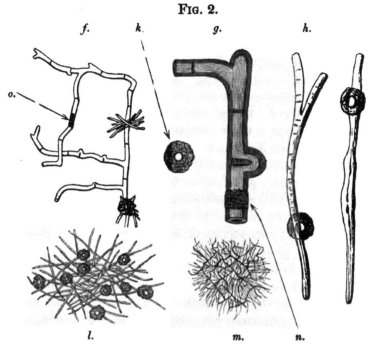

f. *Zygodesmus Berkeleyi*.   g. A portion of the same, highly magnified.   n, o. Supposed to be the adhesive portion; this appearance occasionally continues the whole length of the flocci.

h, i. *Sepedonium Wilsoni*, highly magnified.   k. One of the sporidia.   l. Portion showing the adnate habit.   m. *Sepedonium Wilsoni*, showing *Zygodesmus Berkeleyi* exterior to it.

would appear the most forcible argument, it was not on the Southport plant in any case.

Secondly.—From the examination of analogous productions, I am led to believe that the structural and habitual "uniformity of its growth" is such as to strengthen the idea that they are fungi, rather than to prove them the roots of a phenogamous plant.

Thirdly.—That although, when viewed partially dry, the substance of the rhizoma and these fibres may "*appear* perfectly continuous and identical," yet, when distended, every trace of such appearance is lost; nor does a longitudinal section of the rhizoma display any signs of an internal continuation of the fibres, which, if they were indeed the fibrillæ of the root, might be expected. Such *is* the case with the claviform branches before mentioned, (fig. 1, *a*).

Fourthly.—That in no instance, either sectional or otherwise, could any "rupture of the cuticle" be produced, except such as already explained, and shown on *f* in fig. 2.

In reference to Mr. Lees' opinion that the fibres are "suckers," I would remark, that those portions which are produced on the root of Monotropa itself, under ordinary circumstances, labour rather to maintain their connexion with it than to unite with other substances : and would refer to their structure as one not at all supporting his idea.

In conclusion, I would urge upon Mr. Luxford the more thorough examination of specimens from the locality where " the slowly decaying *leaves of the beech* " in the neighbourhood of Monotropa, " are *generally covered* with a white byssoid fungus." That such is their nature not the slightest doubt now remains in my mind ; I regard them as an interesting addition to our Cryptogamic vegetation, and hope that the following attempt to assign them " a local habitation and a name" among their fellows, undertaken as it has been at the express desire of several parties, will not be deemed an impropriety.

In proceeding with this portion of the subject I felt that there was no small difficulty to be overcome, not only on account of the intricate nature of the order, but from the fact that our knowledge of mycological productions generally, is so exceedingly imperfect. So great however was my desire lastingly to associate with the discovery the names of those gentlemen who originated or have assisted in it, that I applied to the Rev. M. J. Berkeley, enclosing him specimens and sketches. To his kind attention and assistance I am indebted for a knowledge of the genera Zygodesmus and Tuburcinia, and several other particulars, without which I could not have approximated so nearly to the truth.

EPIPHAGOS *Luxfordii*, (Ryl. MSS.) Fig. 1, *c, d, e.*

This I have applied as a provisional name to No. 1 — the species observed by Mr. Luxford on beech-leaves and Monotropa. There is much reason to believe it an undescribed genus, probably a byssoid Alga ; but owing to the total absence of fructification its position cannot be positively determined. It is to be hoped that at some future period Mr. Luxford will be fortunate enough to meet with fertile specimens.

No. 2, Mr. Berkeley informs me, is certainly a Zygodesmus (a genus of Corda, characterized as follows) : —

Tribe.—MUCEDINES.

ZYGODESMUS. Flocci creeping, branched, entangled, septate or suddenly staple-bent so as to form a geniculus (" geniculati-contracti "),* and afterwards paired by

---

* There appears to have been some mistake in the application of this phrase, at least *as regards* the species now described, though Mr. Berkeley's sketch of it exhibits

branches or geniculi developed at right angles, otherwise one floccus coupled wth another: sporidiferous branchlets erect or wart-shaped: spores acrogenous, simple, at length irregularly scattered: episporium membranaceous, smooth or hairy; nucleus solid.—*Corda, Icon. Fung.* pt. 5.

Z. *Berkeleyi,* (Ryl. MSS.) Flocci spreading, horny, brown, branched, with hemispherical swellings (" geniculi ") produced at many of the septa, from which originate the branches? Sporidia oval or ovate, smooth; nucleus simple. Fig. 2, *f, g,* and 3, *x, y,* (the sporidia).

On Monotropa roots generally.

No. 3 is also an undescribed species.

Tribe.—SEPEDONIEI.

SEPEDONIUM, *Link.* Sporidia globose (filled with sporidiola), at first covered by the flocci of the fleecy mycelium.† *Eng. Flor.* vol. v. pt. 2, p. 350.

S. *Wilsoni,* (Ryl. MSS). Flocci fleecy, interwoven or fasciculate, adnate, pellucid, white, rarely dichotomously branched: sporidia globose, tubercled or granulated, not appendiculated, but attached by the disk on their flattened surface? Fig. 2, *h, i, k, l.*

On Monotropa from Southport.

Since the former portion of this article was written, I have examined a dried specimen of Monotropa gathered at Streatley, Berks, sent because it had the flocci considered to be roots, growing on its capsules and stem; on examination I found the following: —

Tribe.—DEMATIEI.

CLADOSPORIUM, *Link.* "Sporidia arranged in" (more or less) "moniliform branchlets, at length falling off; flocci septate (above"?). *Eng. Flor.* vol. v. pt. 2, p. 338.

C. *Leesii,* (Ryl. MSS.) Flocci spreading, pellucid, straw-coloured: sporidia of various sizes, contained in branchlets, which are at length moniliform and septate? yellow, as the flocci. Fig. 3, *p, q, r.*

---

the character. The " geniculi," viewed with a power of 350 diameters (by Ross), appear to be only tumours; the outline of the floccus opposite them is perfectly continuous, and what have been considered contractions or doublings seem but to be septa left in some measure free by the swellings. Vide fig. 2, *g.*

† As this species is allied to a parasite of Monotropa already described, though not to be found in the ' English Flora,' it may be well to add the description as given by its discoverer.

TUBURCINIA, *Fries.* Sporæ rotundatæ, inæquales, opacæ; sporidiis minimis, succedentibus farctæ, plantis putridis innatæ; floccis variis, tenellis, intertextis, evanescentibus, primo adherentibus.

T. *Monotropæ.* Erumpens, sporis nigro-fuscis inæqualibus flocci evidentioribus.

Præcedenti (T. Orobanchæ) affinis, sed magis superficialis est, et erumpens, sporidia fuscescent, et flocci copiosiores adsunt.

Genus quoad fructificationem ad Sepedonium accedit.

FIG. 3.

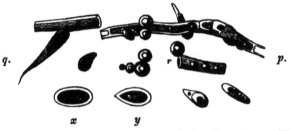

x   *y*

p. Cladosporium Leesii, showing the structure of the flocci, and a septate sporidiferous branchlet. *q.* Sporidiferous branchlet in a younger stage.   *r.* A portion where the sporidia appear to have been matured in the main portion of the floccus. The four figures not referred to are various forms of sporidia.

x, y. The sporidia of *Zygodesmus Berkeleyi*.

It is with no design to forward my own views that three of the four species are placed in genera already recognized; my feelings were for long opposed to this, but after mature examination I could find no differences sufficient whereon to found new characters. It is true, that two of the species do not *perfectly* accord with the genera in which they have been placed, such having been constructed with reference only to the kinds then known; so slight however are the discrepancies, as will be seen by the parentheses, that I have deemed it but prudent to admit them.

THOMAS G. RYLANDS.

Bewsey House, Warrington,
September 14, 1842,

ART. LXXXIX. — *Analytical Notice of the 'Transactions of the Botanical Society.'* *Vol.* i. *pt.* ii.   Edinburgh: Maclachlan, Stewart & Co.; H. Bailliére, London. 1841.

(Continued from p. 322).

VIII. *On the Botanical Characters of the British Oaks.* By R. K. GRE-VILLE, LL.D., F.L.S., &c., *Vice-President of the Botanical Society.*

FOR three years previously to the publication of this paper the author was engaged in the examination of the British oaks, with a view to ascertain the characters by which each so-called species may be distinguished from its congeners: the results of this investigation are here published. The paper is accompanied by two plates, wherein are delineated the most striking forms of leaf which have come under the author's notice, together with the peduncle and young fruit; and the author observes that "in an extensive series of specimens, these forms are united by intermediate links." In his preliminary remarks he also states : —

"When I first approached this subject with a view to ascertain the value of the characters by which the British Oaks are supposed to be distinguished, I was under a strong impression, that at least, the two forms recognised as species under the names of *Robur* and *sessiliflora* were really distinct. But I soon found that the ascribed characters were of little importance in rigid specific discrimination, and that the only one of any value at all, viz., the sessile or pedunculated fruit, must be received with so much latitude as greatly to weaken, if not to render it entirely unworthy of confidence."—p. 65.

The following remarks are made on " the characters supposed to reside in the leaves."

1. The petiole in Quercus Robur is said to be short; that is, in the form so called the petiole is supposed to be shorter than in either Q. sessiliflora or intermedia. In two of the figures "which certainly do not represent Q. Robur, they are shorter than in several of the forms which undoubtedly belong to that plant."

2. The *general* outline of the leaf. This cannot be defined "so as to distinguish any of the forms as species, as they run insensibly into each other. Nay, some of the forms in Q. Robur are absolutely repeated in Q. sessiliflora."

3. The characters supplied by the *strict* outline of the leaf and of its several parts, appear to be equally unsatisfactory. "The base varies equally in all the forms." "In Q. Robur, although generally more or less bi-auriculate, it is sometimes rounded, and even attenuated," and such is the case in Q. sessiliflora also. No greater importance is to be attached to the pinnatifid character of the leaf; the lobes of the leaf too "are sometimes rounded, when the sinuses are acute, and *vice versâ;* and sometimes both lobes and sinuses are obtuse, at others both are acute." The starry pubescence attributed by Prof. Don to Q. intermedia, the author finds to be present in the young state of almost all the forms he has figured, and it may be traced in nearly the whole series of his fully developed specimens.

"If the leaf be taken in conjunction with the peduncle, no character whatever is afforded; and I agree with Mr. Leighton, who has truly remarked that the leaves of our oaks 'vary without the least reference to the absence or presence, or relative length of the peduncle.' If the peduncle alone be taken as a guide, we shall in vain search for an immutable character, for every intermediate state is to be observed between the elongated peduncle as I have it from Cumberland, 5 inches long, and from Killin, above 4 inches long, to its total absence.

" When the peduncle becomes shorter, it is generally thickened or distorted; and when present, as it often is in Q. sessiliflora, it is very short and robust. It would appear, that in proportion as the peduncle deviates from the normal state, as we observe it in Q. Robur, and becomes more and more contracted, the acorns (no longer produced at the usual intervals) are developed in a clustered manner, till at length the peduncle becomes so short, as to render them almost, if not really sessile. The acorns of the British oaks are as sportive as the leaves; so that no characters which have been hitherto employed, taken singly or in combination, seem to be available for separating our native oaks."—p. 67.

" Were we to be guided by the result of such an examination as I have been able to institute, into the characters hitherto employed for botanically distinguishing the British oaks, we should be led to conclude that we certainly possess only one native species. At the same time, it is quite possible that characters may have been overlooked, which may really suffice to separate them. A remarkable difference in the timber has long been observed, that of Q. sessiliflora being termed Red Oak,—that of Q. Robur, White Oak, from the colour of the wood. The relative merits of the two kinds do not seem to be quite satisfactorily established; but the greatest weight of testimony seems to be in favour of red oak (Q. sessiliflora), contrary to the expressed opinion of Sir J. E. Smith.

"It may still be a question whether the superiority of the one timber over the other depends upon the specific difference of the tree. The subject is one of vast practical importance, and still requires much investigation. It has not been in my power to examine the wood of the trees from which my specimens were obtained, and therefore I am unable to throw the smallest light on this part of the subject. My only object at this time is to show, that the received botanical characters by which Q. Robur, Q. intermedia, and Q. sessiliflora are at present distinguished, pass insensibly and completely into each other, and cannot therefore be depended on, in collecting acorns for seed, and supplying the dock-yard with timber."—p. 68.

IX. *On the Vegetation and Botany of the Island of Madeira.* By JAMES MACAULAY, A.M., M.D., *Foreign Secretary of the Botanical Society.*

THE author observes that the island of Madeira "may be termed a *Transition* station between the European and the African vegetation, and intermediate between the temperate and intertropical regions of botanical geography." In character its Flora appears to be most allied to that of the northern shores of the Mediterranean, notwithstanding the situation of the island off the African coast, between 32° and 33° N. lat. Many of the plants however are species not found in Europe, although they belong to European genera; some species are peculiar to the island, while others are common to Madeira, the Canaries and the Atlantic isles.

" The scenery is of the most glorious character, both as respects the beauty of the cultivated parts and the sublimity of the mountain districts of the island. In the interior, and on the rth coast especially, the traveller meets with combinations of natural objects such as in no other part of the world can be witnessed, and which oblite.

rate every conception that had previously been formed of the grand and sublime in scenery. This is greatly the result of the geological character of the island. The mountains have not the integrity of outline and tame regularity of form that always appear in some of the formations of older geological epochs, but are composed of the most recent basalts and other igneous rocks; and, on a scale of alpine magnificence, present a scene of the wildest confusion, being everywhere deeply riven by rugged ravines, the precipitous cliffs of which are clothed to the very summit with ancient forests. Among these wild scenes there occur landscapes of the utmost loveliness, so that the scenery is altogether perhaps the finest in the world."—p. 71.

The climate and temperature of the island are quite as varied as its scenery. In sheltered spots on the south coast there is a tropical temperature, and there the climate remains more genial than an English summer; while "on the opposite coast, and in the interior of the island, the mountains are covered with snow, and all the rigour of a northern winter is experienced. Between these extremes every degree of climate is met with, and the range of vegetation varies accordingly." The author mentions some of the peculiarities arising from these varieties of climate, in connexion with the changes of the seasons.

"During the winter the residents on the coast look up from amidst their tropical vegetation and genial temperature, through every degree of climate and verdure, to the bleakest desolation on the snow-covered hills above them. In the declining months of the year, again, while on the coast the summer foliage is yet unaltered, and the influence of the sun little diminished, the upper parts of the landscape present the variegated tints and the fading foliage of autumn."—p. 72.

The author gives the following as "the most marked zones of botanical climate on the south side of the island."

"1. Region of Heaths—beyond 3500 or 4000 feet above the sea. Erica arborea is the most characteristic plant of this climate.

"2. Region of Laurels—from 3000 to 4000 feet. Laurus indica, Laurus Til, and other indigenous evergreen forest trees chiefly clothe the mountains in this zone.

"3. Region of European Trees—from 2000 to 3000 feet. Forests of chesnut and other trees introduced from Europe here principally flourish.

"4. Region of the Mediterranean Flora—from 1000 to 2000 feet. In this zone the plants of Southern Europe abound, and numbers of the trees and flowers of temperate climates have been introduced. The upper limit is marked by the vine being scarcely cultivated beyond 2000 feet above the sea.

"5. Sub-Tropical Region—from the level of the sea to 1000 feet above. The limit of this zone is well marked by a Cactus, the Opuntia Tuna of DeCandolle, which grows on the cliffs on the coast, and does not reach higher than 1000 feet.

"6. Tropical Region. In a few sheltered spots on the south coast many of the West Indian and other tropical fruits flourish."—p. 73.

The following extracts are in accordance with an opinion we have frequently ventured to express, namely, that although elevation and climate doubtless exercise considerable influence over the vegetation of a district, yet that an influence fully as powerful is exerted by the character of the soil.

" The geological structure of the island is almost wholly volcanic, basalts and basaltic conglomerates and other igneous rocks forming the whole mass of the island. At one or two places there appear beds of tertiary limestone and other non-volcanic formations, but in so very small a proportion to the whole surface, as to have no effect upon the character of the vegetation."—p. 72.

" The indigenous Flora is smaller than might be expected from the range of climate, the effect probably of the want of variety of soils, and the absence of other causes of difference of vegetation. The most conspicuous natural families are, the Filices, the Laurineæ, the Compositæ and the Labiatæ. Of the Ericeæ, there are two plants which attain to a remarkable growth in the island,—Vaccinium maderense and Erica arborea. The whortleberry forms little thickets or forests on the mountain sides, often from fifteen to twenty feet in height. The heath grows everywhere on the mountains beyond 2500 or 3000 feet above the sea. On Pico Ruivo, the highest peak in the island, it commences at about 3500, and covers most of the mountain side to nearly the summit, which is more than 6200 feet above the sea. The stems are frequently six, and sometimes eight feet in circumference.* The wood is very hard, and is used for most of their common work by the peasants. The other indigenous forest trees chiefly belong to the Laurineæ. The Laurus indica, Vinhatigo of the natives, grows to an immense size. Its wood is of a dark colour, and of excellent quality, and is employed in various articles of work, under the name of Madeira mahogany. Of foreign trees, the most conspicuous is the chesnut, Castanea vulgaris, which was introduced by the early settlers; it forms fine woods on the lower parts of the mountains, especially in the interior, and on the north coast."—p. 73.

X. *On the Specific Value of the Antherine Appendages in the Genus Viola.* By EDWARD FORBES, M.W.S., *Foreign Secretary of the Botanical Society.*

MR. FORBES observes that " form is the chief if not the only source of *specific* character in the vegetable kingdom : " the higher divisions are founded on modifications of the *internal* structure of plants, but in characterizing minor groups and species, we ought rather to attend to the modifications of external form. The reason for this will be

* Sir J. E. Smith, in his ' Tour on the Continent,' says—" I never saw Erica arborea so truly arboreous as in this place, [between Frejus and Cannes]. It was often ten feet high, with a trunk three inches in diameter, much resembling, in form and size, the trees on Box-hill in Surrey. I am informed by Dr. Lind, that it grows to a much larger size, even 18 inches in diameter, on the Serra at Madeira, 5165 feet above the sea."—i. 210, Ed. 2.

apparent when we consider that in plants, contrary to what obtains among animals, external form itself is often structure, or rather it is equivalent to the internal structure of animals: for structure being a modification of form for the performance of some function in the economy of the creature, "in animals such modification is usually an involution of form, in plants the reverse."

Naturalists are aware that in the lowest as well as in the highest groups, whether of plants or animals, there is no organ which is equally and uniformly available as affording distinctive characters. "For example, the shape of a leaf in one genus of plants may be common to all the included species, and therefore is not a specific character; in the next genus it may vary in each species, and the variation be constant in each, becoming the most evident source of distinction between the species." This depends on what Mr. Forbes terms "*the law of undulation of character,*" that is, "either on the adaptation of organization to circumstances," or on "modifications of forms &c., of no importance in the animal or vegetable economy, though of great value as marks to distinguish one original form (that is species) from another. In the last case, the character is usually also *representative,* that is, it may be the *analogue* of an organ which plays an important part in some other species, group, or type."

The author has chosen the genus Viola as affording illustrations of his positions.

"Among the irregularly-flowered genera of the family of Violarieæ, we often find certain of the stamina provided with dorsal appendages, styled by some botanists nectaries, which are lodged in the spur of the flower. In the descriptions of violets, mention is rarely made of these bodies, they being generally regarded as of generic importance only. Any one, however, who compares the nectaries of a pansy with those of a dog-violet, will see such a difference, as indicates a specific, at any rate, a sectional importance. In order to ascertain the value of the character so derived, I have compared minutely above seventy species of violets; and the results of such comparison, and their application to the elucidation of British Botany, I now lay before the Society."—p. 77.

1. *Form of Nectaries.* In the allies of V. canina they are *lancet-shaped,*—the most common form; in the pansies they are *linear;* and the rarest form is the *rotund,* found in V. palustris. These are the only three forms in the genus; the first varies in length and breadth, the linear in length.

2. *Relations of Nectary and Spur.* The spur varies in proportion according to the form of the nectaries; "it is usually thick in proportion to its length, and very blunt," when the nectaries are lancet-shaped. "The rotund nectary is associated with a very short

2 F

rounded spur, the linear usually with a long slender spur, often of great length, as in the pansies."

3. *Relation of Nectary and Colour of Flower.* Mr. Forbes is disposed to consider that greater importance is due to colour in specific distinctions than is usually assigned to it.

" In the genus Viola blue, yellow, purple and white are the colours seen. The blue may again be divided into purple-blue and sky-blue, each passing abnormally into white, but the respective whites must not be confounded. The purple-blue may also pass abnormally into rose; but the sky-blue, as far as I am aware, never does. These distinctions are of importance in the investigation of nearly allied species, such as V. canina and V. montana. The yellow passes on the one hand into white; on the other, into purple. White is rarely the normal colour of a species of violet. The lancet-shaped nectary is chiefly associated with blue flowers, sometimes with the yellows passing into white, never with the yellows which pass into purple. They have always linear nectaries. The violets, which are normally white, derived from blue, have always lanceolate or rotund appendages, never linear. The odorous violets have always purple-blue, or its derivative white flowers, and always lancet-shaped appendages, but the colour does not necessarily imply the odour. The yellows which do not pass into purple have always lancet-shaped appendagës. Among the dog-violets we find the lancet-shaped appendages lengthening in proportion to the mixture of purple in the blue, and the contrary in cerulean flowers, and their derivatives milk-white. All violets may be abnormally white, but the form of the nectary peculiar to the species, does not change with the change of colour in the individual."—p. 79.

4. *Relations of Nectary and Leaf.* The cordate leaf in violets is generally associated with a lancet-shaped nectary, which is always found when the leaf is lanceolate or truly ovate, and sometimes when through being deeply lobed a cordate or rotundo-cordate leaf becomes pinnato-palmate. " A few cordate-leaved violets have linear appendages, and some of the pinnato-palmate; but all the pansies, or violets with ovato-spathulate leaves have linear nectaries." In the stemless species with cordate leaves, " as the leaf becomes more and more rounded there is generally a tendency in the nectary to become shorter and shorter, whilst the contrary is often the case in such as have stems. Rotund appendages are associated only with reniform-cordate leaves." Mr. Forbes gives a tabular view of " the relations of leaf and nectary in seventy-one species of violets."

5. *Relation of Nectary to Bractioles and Stipules.* The relation borne by the bractiole to the nectary is one of size, not of position,— the linear nectary being generally associated with minute bractioles : the stipules " are almost always large when the nectary is linear."

6. *Relation of Nectary to Stem.* In the arborescent, and generally in the shrubby violets the nectary is lanceolate; in the more her-*baceous* stemmed species it is usually linear; while in most of the

stemless violets, including the odorous species, it is lancet-shaped, in a few only being linear.

7. *Geographical distribution of Violets, according to their Nectaries.*—

"Violets are found in most parts of the world, though their distribution is influenced materially by climate. The species from tropical countries are mostly from localities where the influence of elevation has neutralized or modified the climatal influence. The various sections of the genus have geographic centres, as may be seen in the congregations of the allies of *hirta* in North America, and of those of *tricolor* in alpine Europe. North America may be looked upon as the capital of the whole genus, since we find there representatives of all its subdivisions. In the following table the distribution of 75 species, according to the form of their nectaries, is exhibited :—

	Europe	Asia	Africa	N. America	S. America	Australia	Total
Lancet-shaped ..	19	4	1	22	1	1	48
Rotund ..	2	0	0	1	0	1	4
Linear ..	12	4	1	6	0	0	23

8. *Application of the above considerations to the arrangement of the Violets.*—

"To group the violets according to the form of the nectary, without considering the relation of that form to the other characters of the plant, would be to arrange them artificially and not naturally; for then we should have such violets as *ochroleuca* and *prionantha* associated with the pansies, and other combinations of a similarly unnatural character. But associating the form of the nectaries, with that of the leaf, with the colour, and with the geographical distribution, we obtain a very natural arrangement of the species. The odorous and hairy violets, presenting short lancet-shaped nectaries, cordate leaves, often being hairy, and frequently nearly orbicular, purple-blue flowers, no stem, and a centralization of the species in North America, form a *first* group. The same form of appendage, but usually more developed as to length, combined with a cordate or lanceolate leaf, smooth, or slightly hairy, a stem, purple-blue or cerulean flowers, and an almost equal distribution in the old and New Worlds, indicate a *second* equally natural, which may be represented by *V. canina*. *V. palustris* is the type of a *third*, and *V. biflora* of a *fourth ;* the former associating a reniform leaf with a rotund appendage and a cerulean flower, and the latter a similar leaf with an abbreviated lancet-shaped appendage and a yellow flower. The linear nectary, combined with a yellow flower and cordate leaves, forms a *fifth ;* parallel to which may be placed (*sixth*) such as have pinnate leaves, blue flowers and linear nectaries. Lastly, the pansies form a most natural group (*seventh*) of themselves, presenting us with flowers of all colours, linear nectaries, leaves peculiar to themselves, lyrate stipules, and a centralization in the mountainous countries of the western portion of the Old World."—p. 81.

XI. *An Attempt to ascertain the true Hypericum quadrangulum of Linnæus.* By CHARLES C. BABINGTON, M.A., Cantab., F.L.S., F.G.S., &c.

The author observes that the difficulty attending the determination of the Hypericum quadrangulum of Linnæus, has arisen from a belief that Linnæus himself referred to different species by that name. In the Linnæan herbarium are two specimens, on " different papers pinned together, the first of which is the *H. quadrangulum* of Eng. Bot. (t. 370), and has a number appended referring to the Sp. Pl.; the other is the *H. dubium* of Leers (Eng. Bot. t. 296), and has the same number attached to it, and also the name of " *H. quadrangulum.*" Linnæus first described his H. quadrangulum in the 'Hortus Cliffortianus,' (p. 380), where it is said to have " *Folia calycina subulata,*" and Morison (vol. ii. t. 6, f. 10) is referred to as the original authority. Morison's plant is the H. quadrangulum of Smith, " which has the *folia calycina subulata,*" while in the H. dubium of Leers and Smith they are " broad and very obtuse." Linnæus, in all his subsequently published works, refers to the ' Hortus Cliffortianus' as the authority for his H. quadrangulum; and as the plant so named in Smith's works and figured in ' English Botany' agrees with that described by Linnæus in having subulate calyx-leaves, there can be no doubt of its being the Linnæan H. quadrangulum.

Mr. Babington next shows that the plants named *H. dubium* by Leers and Smith, *H. quadrangulum* by Fries and Wahlenberg, *H. maculatum* by Crantz and Allioni, and *H. delphinense* by Villars;—" constitute two well-defined and truly distinct species," although the above names are generally considered synonymous. The paper concludes with the descriptions and synonymes of the three species, so far as the author has been able to determine the latter, " but in nearly all cases, excluding the synonymes cited by the respective authors."

1. *Hypericum quadrangulum*, Linn. Stem erect, 4-winged: leaves oval-oblong or elliptical, with pellucid dots: sepals erect, lanceolate, acute, entire; petals lanceolate. " *Linn. Hort. Cliff.* 380; *Leers, Herb.* 168; *Crantz, Aust.* 89; *Sm.! Fl. Brit.* 801; *Eng. Bot.* 370; *Gaud. Helv.* iv. 625; *Host, Aust.* ii. 78. H. tetrapterum, *Fries, Nov. Suec.* (ed. 1) 94, (ed. 2) 236; *Reich.! Excurs.* No. 5179; *Koch, Syn.* 134; *Fl. Siles.* iii. 83; *Kunth, Berol.* i. 70; *Bab.! Prim. Sarn.* 19; *Leight. Shrop.* 372. H. quadrialatum, *Wahl. Suec.* ii. 476. Androsæmum Ascyron dictum, caule quadrangulo glabro, *Morrison*, ii. 471, sect. 5. tab. 6. fig. 10."

" I have not ventured to refer to any of the other works of Linnæus, because he appears to have confounded the following species with this in all his later works."—p. 87.

2. *H. dubium*, Leers. Stem erect, obsoletely quadrangular: leaves elliptical, obtuse, with few pellucid dots: sepals reflexed, broadly elliptical, obtuse, very entire, with many black dots on the outside; petals elliptical, with many black dots beneath. " *Leers*, 169; *Sm.! Fl. Br.* 802; *Eng. Bot.* 296; *Gaud.* iv. 626; *Wallr. Sched. Crit.* 401; *Host*, ii. 79; *Boenningh. Monaster.* 227. H. quadrangulum, *Fries*, 237; *Ber-*

gius, Mat. Med. (ed. 2) p. 679; *Koch,* 134; *Fl. Siles.* iii. 83; *Sven. Bot.* t. 359; *Wahl.* ii. 478; *Sadler, Fl. Pesth.* 351; *Kunth,* ii. 70."

3. *H. maculatum,* Crantz. Stem erect, quadrangular: leaves ovate-elliptical, ob‑tuse, with few pellucid dots: sepals reflexed, ovate-lanceolate, toothed, obtuse, mucro‑nate, with pellucid striæ; petals elliptical, obtuse, with purple striæ and dots beneath. "*Crantz, St. Aust.* (ante 1769), *ed. alt.* 98; *Allioni, Fl. Pedem.* (1785) ii. 45, t. 83, f. 1, (optime). H. delphinense, *Villars!* "*Fl. Delph.* (1785) 81;" *Hist. Plant. Dauph.* (1789) iii. 497, t. 44, (male); *Reich.! Fl. Exsicc.* No. 1600. H. quadrangulum, *Leight.! Shrop.* 370.

XII. *Notes on the Distribution of British Ferns.* By H. C. WATSON, F.L.S.

MR. WATSON commences his admirable remarks by observing that "excepting some spots of small extent, whence they are banished by local peculiarities of the surface, ferns may be said to range over the whole of Britain, from south to north, from east to west, and from the shores of the sea almost to the summits of the loftiest mountains; from which latter situation they are probably absent, rather in consequence of the bleak exposure to wind, than of the diminished temperature in‑cidental to the height of any of our mountains."

The number of species of British ferns will be variously estimated, according to the views entertained by botanists regarding specific li‑mits. "The lowest estimate may be taken at 34; which is raised to 36, by the inclusion of two species now supposed to be extinct, and, perhaps, never found wild in England, namely, Asplenium fontanum and Trichomanes brevisetum." The latter number will be raised to 40, by those botanists who regard as distinct species the following plants:—Polypodium Dryopteris and calcareum; Aspidium lobatum and aculeatum; Asplenium Ruta-muraria and alternifolium; and Cis‑topteris fragilis and dentata. "And the number of 40 would be still farther augmented by the addition of four other varieties, which are sometimes accepted for species, namely,—

"Aspidium angulare, a variety of A. aculeatum or lobatum.
"Aspidium dilatatum, ... A. spinulosum.
"Aspidium dumetorum, ... A. dilatatum or spinulosum.
"Cistopteris angustata, ... C. dentata or fragilis."

Then again a few botanists would raise to the rank of species Aspi‑dium recurvum and Asplenium irriguum; while others would regard as varieties Cistopteris alpina and Woodsia hyperborea; "but since their views are not shared by many, we may hold our ferns to be esti‑mated at 36, 40, or 44."

The number of species of indigenous flowering plants would in like manner vary, according to the views of the party by whom the estimate might be made. "By rigidly excluding all species not fully recognised as indigenous, and also numerous varieties which are commonly now counted amongst species, the flowering plants of Britain will be found scarcely to exceed 1200; or, admitting doubtful species, but still excluding doubtful natives, they may be taken at 1400. To reach the number of 1636, given in the Catalogue printed by the Botanical Society of Edinburgh,* we must admit many species of foreign introduction, and a goodly list of varieties named and received as species." But in determining the proportion borne by the ferns to the flowering plants, "if we take a low estimate for one group, we must follow the same rule in the other, or their proportions will unavoidably appear wide of truth." The proportions of the two groups taken from the three estimates of the number of each given above, will be these.

"FERNS.	FLOWERS.	PROPORTIONS.
36	1200	1 to 33½
40	1400	1 to 35
44	1636	1 to 37"

The author by way of comparison next gives a table showing the relative numbers of ferns and flowering plants in eight different Floras; the proportions (omitting fractions) being as under.

Iceland......... 1 to 25	Channel Isles 1 to 51	Sweden ...... 1 to 40
Faroe .... 1 to 27	Belgium ...... 1 to 67	Lapland ...... 1 to 25
Ireland .... 1 to 30	Zealand ...... 1 to 47	

"Ferns are thus seen to bear a larger proportion relatively to flowering plants, in the northern and mountainous parts of Western Europe, than is the case with this group of plants in low countries,— such as Belgium, Zealand, and the Channel Isles, —whose latitude nearly corresponds with that of England; whilst the proportions before set down for Britain place it in the scale betwixt Faroe and Iceland, on the one hand, and Belgium and the Channel Islands on the other; the former having a relative predominance of ferns, the latter having a similar predominance of flowering plants."—p. 91.

It is then shown by a comparison of twenty local Floras, that the distribution of ferns in Britain corresponds with their distribution in the north-west of Europe generally, since both the relative and absolute number of species diminish "as we pass from the hilly districts of

*The 1st edition of the Catalogue is here referred to; the number of species (1636) given in the "Enumeration of Plants" prefixed to the Catalogue appears to include the ferns as well as flowering plants. In the 2nd edition the number of species in the two groups is distinctly given as 1594 and 55, (including Lycopodium, Isoetes, Pilularia and Equisetum).

Scotland and the north and west of England, towards the low south-eastern countries lying nearest to Belgium." In Yorkshire, according to the table given, there are 36 ferns and 1002 flowering plants; or 1 to 28: the Faversham and East Kent Flora, on the contrary, furnishes only 13 ferns and 806 flowering plants, or 1 to 62: these being the extreme proportions afforded by the twenty Floras examined. — The Midland Flora has 23 ferns and 840 flowering plants, or 1 to 37.

Mr. Watson remarks that "York is pre-eminently the county of ferns;" for although the author of the Yorkshire Flora "has multiplied species more than many other writers on local Botany," the ferns are really most numerous in that county. This the author attributes to various causes, such as its central position, and the diversified character of the country; "the climate of its low vales being sufficiently mild for the growth of species which shun the northern counties of Scotland, without being too warm or dry for the growth of boreal species, to which the hilly districts of its western border are particularly suited, as well as to the production also of the more exclusively mountain species." Then again its coast furnishes Asplenium marinum, and its various soils and rocks are adapted for those species which are attached to particular formations.

In Cambridgeshire there are 14 ferns and 847 flowering plants, or the ferns are as 1 to 60. This paucity of ferns is also explained by a reference to the character of the county. A large portion consists of low fens, with but little wood and few hedge-rows; "much of the rest is composed of gentle undulations of chalk," of inconsiderable elevation, with few trees and little water.

"Moray ranks next to Yorkshire in the high proportion of its ferns," (1 to 30). This is accounted for by "the rigid exclusion of introduced species of flowering plants" by the author of the Moray Flora.— On the other hand the number of flowering plants in Northumberland and Durham is high (1030), in consequence of "the addition of many species brought to the coasts of those counties in ship ballast." In the absolute number of species (28) the ferns of these counties rank next to Yorkshire, although, from the cause above mentioned the proportion borne by them to flowering plants is much lower, (1 to 37).

It is evident from the variation in the number of ferns in the several districts, that some species must have a partial range; accordingly it appears that "no one of the district Floras includes all the native species, whilst about half of these Floras include fewer than half of the species." Some species are so widely diffused and so abundant in individuals, that they are probably to be found in every county; others again are either confined to a few localities, or have a wide range over certain parts of the island although excluded from others.

A table is next given wherein are enumerated forty-three species, showing in how many of the twenty local Floras before mentioned, and in what number of twenty-four manuscript lists, the name of each species occurs. By this means may be gained a tolerably correct idea of the range of our native ferns; although, as the author observes, "without regard to the distinctness of the species, the dates of their first discovery, and the certainty of their nomenclature," erroneous conclusions might be drawn from the list.

Pteris aquilina, Polypodium vulgare and Aspidium Filix-mas are the only three species "so universally distributed as to be included in all the forty-four district Floras and lists." But although these are our three commonest ferns, yet to neither of them does the widest geographical range in Britain belong. "The most widely ranging of our native ferns, taking into view the three directions of latitude, longitude and elevation, are Blechnum boreale and Aspidium dilatatum (or spinulosum)."

"It has been already stated that ferns prevail chiefly in the hilly tracts towards the north and west, and that they are less numerous in the low south-eastern counties of England,—a peculiarity that is doubtless in great measure attributable to the more humid and cooler atmosphere of the former. The three circumstances on which this difference of climate depends, are those of latitude and longitude, conjointly with elevation of the surface; and the influence of these three conditions in producing the general result, will scarcely admit of divided consideration. We may, indeed, trace some agreement betwixt the range of certain species and the geographical divisions of latitude and longitude; yet this connexion (or, more strictly, this coincidence) is so much interfered with by the third condition, that of height, as to render separate investigation almost useless."—p. 97.

In proof of these positions it is remarked that about half the number of indigenous ferns are absent from the English counties lying to the east of Gloucestershire and Nottinghamshire; "whilst none of the species found growing in these south-eastern counties are wholly wanting in those to the westward of them; most of these species also being much more plentiful in the western counties." The paucity of ferns in the south-eastern counties is accounted for by the different character of the surface, owing to the absence of rocky ravines, waterfalls and mountain elevations, and the consequent dryness of atmosphere, rather than by the difference of longitude.

Nor is the northern or southern limit of a fern's range altogether to be determined by the degree of latitude, though they may be more decidedly traced than the longitudinal boundaries; yet even in this case the author states the lines of cessation to be determined more by "the hilly and broken character of the surface" than by the degree of latitude; and instances the hills of Wales as bringing "several species into a more southern latitude than it is at all likely they would have been found in, if Wales and the adjacent English counties had been as little diversified with high hills as are the counties under the same latitude on the eastern side of the island."

"The effect of the mountains, however, is probably much more decidedly shown in prolonging the southern range than in arresting the northern range of ferns; since the low coast-line, as well as small plains and valleys around, or amongst the hills, may still afford suitable localities for such ferns as are unfitted to bear the climate of the mountain summits or acclivities, although capable of growing in the climate incidental to the latitude."—p. 98.

The author's classification of ferns according to their range in Britain, and his observations, are very interesting. Those are first given which "may be considered to have a range of latitude almost through the whole of Britain."—

Cistopteris fragilis	Osmunda regalis	Aspidium dilatatum
Polypodium Phegopteris	Scolopendrium vulgare	Asplenium Filix-fœmina
vulgare	Hymenophyllum Wilsoni	Trichomanes
Pteris aquilina	Aspidium lobatum	Ruta-muraria
Blechnum boreale	Oreopteris	Adiantum-nigrum
Botrychium Lunaria	Filix-mas	marinum
Ophioglossum vulgatum	spinulosum	

Four of the above twenty species "are very rare in the south of England, namely, Botrychium Lunaria, Cistopteris fragilis, Hymenophyllum Wilsoni and Polypodium Phegopteris, especially the last; and they are not found at all in the Channel Isles." Asplenium marinum, Scolopendrium vulgare, Ophioglossum vulgatum and Osmunda regalis, are decidedly rare in the north of Scotland, "and they are not found at all in the Faroe isles, though the Ophioglossum is stated to grow in Iceland."

The following are "boreal and mountain ferns, which are unknown southward of the Thames."—

Woodsia hyperborea	Cistopteris dentata	Asplenium septentrionale
ilvensis	Polypodium Dryopteris	Aspidium Lonchitis
Cistopteris alpina	Asplenium viride	Cryptogramma crispa

Four species are given as being chiefly "confined to the middle latitudes of Britain."—

Aspidium cristatum	Asplenium alternifolium
rigidum	Polypodium calcareum

Four species "affect the southern, or southern and midland counties, being absent from the northern counties."—

Adiantum Capillus-Veneris	Aspidium Thelypteris
Asplenium lanceolatum	Grammitis Ceterach

Of four others either the range is undetermined, or their specific distinctness is questionable.—

Hymenophyllum tunbridgense	Aspidium angulare
Aspidium aculeatum	dumetorum

"Two remaining species, if ever found in England, and not now extinct, are exceedingly local in the north of England."—

Asplenium fontanum	Trichomanes brevisetum

The author observes that more complete and accurate observations are yet required "with regard to the limits of ferns in ascending the mountains;" and refers to Francis's 'Analysis of British Ferns' as containing nearly all that has been published on the subject. Five are mentioned as certainly rising "almost to the summits of the highest hills in Scotland," being found at an elevation of between 3000 and 4000 feet; the two first probably above 4000 feet.

Cryptogramma crispa	Aspidium (dilatatum ?)
Blechnum boreale	Lonchitis
Polypodium Phegopteris	

The following occur "at a lower elevation, but still probably above 2000 feet, and some perhaps above 3000 feet."—

Botrychium Lunaria	Asplenium viride	Cistopteris fragilis
Woodsia hyperborea	Cistopteris alpina	Polypodium vulgare
ilvensis	dentata	

"On quitting the mountain tracts, we leave the four following species behind us:"—

Woodsia hyperborea	Aspidium Lonchitis
ilvensis	Asplenium viride

Two others, being rare on the isolated lower hills, may be classed with the mountain ferns.—

Cryptogramma crispa	Asplenium septentrionale

Eight species, found chiefly in the hilly districts of the north and west, yet occasionally occur " so far from the mountain tracts, that they cannot be held in the character of exclusively mountain ferns."—

Cistopteris alpina	Hymenophyllum Wilsoni	Polypodium calcareum
dentata	Polypodium Phegopteris	Botrychium Lunaria
fragilis	Dryopteris	

It is next remarked that " under the combined influence of latitude and longitude, thus modified by the effect of elevation of surface, the lower limits of many ferns, equally as those of flowering plants, appear on a map like irregular lines, whose general direction runs from south-west to north-east; whilst their upper limits encircle the hills, or the hilly tracts, like zones or belts." It must not, however, be supposed that by terminal lines are to be understood any others than " artificial lines, drawn on a map, so as to connect the extreme stations for any species in either direction." Also the terms *upper* and *lower* limits are to be understood as applying " to latitude, to elevation above the sea-level, and also in some measure to the degree of proximity to the mountain tracts." In the neighbourhood of the latter a great change in the character of the Flora of a district becomes evident, although the latitude and absolute elevation remain nearly the same. Thus the upper limit of British plants will include, unless otherwise qualified, the three conditions of more northern latitude, increased elevation and greater proximity to mountain tracts. Again, by the lower limits are to be understood " the opposite conditions of southern latitude, diminished elevation, and also comparative remoteness from the mountain tracts, as centres around which the species are distributed."

These explanations being kept in view, it will be seen to be impossible at present to represent on maps the distribution of British plants with anything like accuracy, " in consequence of the upper limits of most of the species being yet so little known." Their upper limits in latitude might be traced pretty accurately ; and an approximation to their altitude above the sea-level might be arrived at; but the difficulty would be to determine their propinquity to the mountain centres. This can only be ascertained when botanists shall carefully record the places where plants of the plains are last seen by them, as they enter amongst the valleys of the mountain districts.

" As an example of such difficulties, let us take Scolopendrium vulgare, a fern widely diffused in Britain from Orkney to the Isle of Wight, and abundant in the south of England. Betwixt these extreme points, however, there are many wide spaces from which this fern is wholly absent; and one of these spaces perhaps includes nearly the whole of the Highlands. The Scolopendrium is reported to grow in the counties of Renfrew, Lanark, Edinburgh, Forfar, Nairn and Orkney, and thus seems partially to encircle the Highlands. But whether its extreme stations, or upper limits towards the mountain centres, are found in these counties, remains to be shown. Again, Pteris aquilina is exceedingly plentiful in Britain, from one extremity to the other, but it fails upon the higher hills of Wales, the north of England, and the Scottish Highlands; and may even prove to be wholly absent from a transverse belt of high moors and hills crossing Scotland about the 57th parallel of latitude. But what botanist can trace on a map of Britain those portions of the surface from which this very conspicuous fern is quite absent?"—p. 103.

"In returning from this digression respecting maps to the immediate subject of the present paper, it may be farther observed, that ferns, as a class of plants, cannot be exclusively connected with any particular local situation or quality of soil. For the most part, a shady situation, damp ground and atmosphere, and a porous or peaty soil, are suitable to ferns; whilst exposure to sun, wind, and salt spray, as well as very dry or marshy localities, are unsuitable. But some of the Asplenia grow in dry crevices of rocks and walls, as also do Grammitis Ceterach, Polypodium Dryopteris, and Polypodium calcareum, and even the less rigid ferns constituting the genus Cistopteris. On the contrary, Osmunda regalis might almost be designated a marsh fern; and other species so far approximate to the same character, as to thrive in swampy ground, that is, in watery places, where the soil is loose and spongy; for example, Aspidium Thelypteris. But ferns that grow well in swampy places will also grow well on rocks and banks where the soil is not particularly wet; for instance, Blechnum boreale and Asplenium Filix-fœmina. None are aquatics. One only is a littoral species, Asplenium marinum; and this one is occasionally seen in places many miles from the sea. Osmunda regalis so frequently occurs near the shore, within reach of the salt spray, and even at times within reach of high tides, that it might be regarded as a sub-littoral species. None perhaps require the sun's rays directly shining upon them; but some few will bear daily exposure to the sun for several hours, though the greater number thrive best on a slender allowance of sunshine. It cannot yet be stated that any species are absolutely limited to soils of a particular chemical or geognostic character; but Grammitis Ceterach, Polypodium calcareum, and the species of Cistopteris, certainly affect lime rocks, though, indeed, it is believed by some botanists, that the Polypodium calcareum is a variety of P. Dryopteris, varied in its habit through the influence of soil or exposure. If so, the only species that is limited to limestone, if even it be so limited, is Grammitis Ceterach. The harder kinds of trap and slate rocks seem favourable to Asplenium septentrionale and Woodsia ilvensis.— And, in general, the sand-stones are more productive of ferns than chalk or clays; though the difference here is probably owing more to the mechanical than to the chemical qualities of the soils."—p. 105.

(To be continued).

ART. XC.— *Varieties.*

216. *Additions to the Rarer Plants observed at Nottingham,* (Phytol. 78). I send a list of some of the rarer plants which I gathered near Nottingham during a short stay last month.

Salvia Verbenaca. Castle-rock

Lemna polyrhiza. Pools in Nottingham meadows

———— gibba. Near the rocks in the park

Crocus vernus, (in seed). Nottingham meadows

Sclerochloa distans. Ditto

———————— rigida. Colwick park-wall

Potamogeton pectinatus and perfoliatus. Canal near Lenton.

Echium vulgare. Castle-rock, &c.

Cynoglossum officinale. Rocks in Colwick park

Lysimachia Nummularia. Nottingham meadows

Hottonia palustris. Ditto

Menyanthes trifoliata. Oxton bog

Rhamnus catharticus. Near Wilford, Colwick Wood, &c.

Smyrnium Olusatrum. Castle rock

Petroselinum sativum. Rocks in Nottingham park

Ulmus suberosa. Near Beeston, &c.

Allium vineale. Rocks at Snenton, castle-rock &c.

Vaccinium Oxycoccos. Oxton bog.

Daphne Laureola. Colwick wood

Butomus umbellatus. Frequent in the Trent vale

Stellaria glauca. Nottingham meadows, and under the gypsum rock at Clifton

Sedum anglicum. Castle rock

Cerastium arvense. Nottingham forest, and banks of the Trent above Wilford

Spiræa Filipendula. Meadows nr. Beeston

Rubus Koehleri. Mapperly plains, &c.

—— Idæus. Colwick wood, &c.

Tilia parvifolia. Mapperly hills

Nuphar lutea. Pools in the meadows

Hesperis matronalis. Rocks in Colwick park

Astragalus glycyphyllos. Ditto

Trifolium subterraneum. Nottingham prk.

———————— striatum. Ditto, and near Wilford ferry

Lactuca virosa. Rocks in Colwick park

Carex stellulata. Oxton bog

—— remota and vulpina. Near Beeston

—— divulsa. Near Oxton, and in Colwick park

—— muricata. Nottingham park, &c.

—— stricta. Banks of the Trent, &c.

—— Œderi & ampullacea. Oxton bog

—— vesicaria. Banks of the Trent

—— riparia. Banks of the canal near Lenton, &c.

Quercus intermedia. Colwick wood

Salix Helix. Near Wilford, &c.

—— pentandra. Near Oxton

—— Russelliana. Nottingham meadows

—— alba. Banks of the Trent, &c.

—— acuminata. Oxton bog

—— aurita. Nottingham forest, &c.

—— cinerea. Near Wilford, &c.

—— oleifolia. Colwick wood.

—*Joseph Sidebotham ; 26, York St., Manchester, July 7,* 1842.

217. *Note on Linaria spuria and L. Elatine.* Having lately gathered some very luxuriant plants of Linaria spuria, which grew in abundance with L. Elatine, I was induced to try if I could find any character whereby to distinguish them without taking into consideration the leaves and spur; and my specimens present the following apparently constant distinctions.

L. spuria. Peduncles closely downy : calyx-segments ovate, longer than the capsule.

L. Elatine. Peduncles glabrous : calyx-segments ovate, acuminate, with a diaphanous margin reaching half way up, calyx as long as the capsule. — *Wm. Mitten ; 91, Blackman St., Boro', August 17,* 1842.

218. *New Locality for Halimus pedunculatus.* — I have much pleasure in sending you what is, I believe, an unrecorded station for the rare Halimus pedunculatus. I found it in August last, growing not very plentifully in the salt marsh about two miles below Gravesend.—*Id.; September* 17, 1842.

219. *Additional Manchester Plants.* Allow me to make the following additions to the list of Manchester plants, (Phytol. 279).

Avena fatua

Ballota ruderalis

Callitriche pedunculata, *β. sessilis*, (Bab.)

Camelina sativa

Carex angustifolia, (Sm.)

—— ampullacea, var. *longicarpa*, (mihi). This plant is found at Hale Moss; it differs from the common state of the plant, in its fruit being much longer in proportion to its breadth.

—— teretiuscula. Two varieties; one

with its fruit as figured in Leighton's Flora, the other with its fruit agreeing with Leighton's figure of the fruit of C. paniculata.

Digitaria sanguinalis. Heap bridge, near Bury

Doronicum Pardalianches. Near Littleborough

Hieracium boreale

Potamogeton oblongus

Reseda alba

Of Scleranthus I have two very different forms from the neighbourhood of Manchester, neither of them will be S. perennis, but as I have no wish at present to speak on their specific identity, but merely to call the attention of botanists to the capsules of the plants belonging to this genus, perhaps some of us will be able to make out the number of seeds contained in each. Lightfoot says " Sem. 2, calyce inclusa, (' Flora Scotica,' 225); Withering says, " Seed single, egg-shaped, inclosed by the gristly tube of the cup," (' Systematic Arrangement of British Plants,' 5th edition, vol. i. p. 240); Smith says " Seeds 2, convex at one side, flat on the other," (' English Flora,' 282); Hooker says "Capsule one-seeded, covered by the calyx," (' British Flora,' 1st ed. 188). —*Samuel Gibson ; Hebden Bridge, September* 5, 1842.

220. *Anagallis arvensis with White Flowers.* In the course of one of my rambles this month, I gathered a specimen of Anagallis arvensis, answering precisely to that described by Sir W. Hooker, viz., the " flowers pure white, with a small, well-defined, bright purplish-pink eye in the centre of every corolla,' said to have been found by Mr. John Dillwyn, at Penllegare, S. Wales. I could find but one plant, but I shall examine the field carefully next year, and earlier in the season, as my specimen had only two flowers on it, the rest having formed capsules. I should imagine that the specimen gathered by Mr. Dillwyn is the only one upon record. — *Henry Lascelles Jenner ; Chiselhurst, Kent, September* 19, 1842.

221. *Botanical chair at King's College.* We are pleased to see the name of Mr. E. Forbes advertized for the summer course on Botany.—*Ed.*

ART. XCI.—*Proceedings of Societies.*

BOTANICAL SOCIETY OF LONDON.

September 3, 1842.—John Reynolds, Esq., Treasurer, in the chair The following donations were announced :—British plants from the President, from Mr. B. D. Wardale, Mr. G. W. Francis, the Rev. T. Butler and Mr. T. Sansom. Mr. B. D. Wardale presented numerous specimens of *Lastræa cristata* (Presl), collected at Bawsey Bottom, near Lynn, Norfolk. Donations to the Library were announced from the American Academy of Sciences, Philadelphia, the Egyptian Society, Mr. G. W. Francis and Mr. S. P. Woodward.

Mr. Thomas Twining, jun., exhibited a large collection of cultivated specimens from Twickenham. A paper was read from Mr. R. S. Hill, being " An Enquiry into Vegetable Morphology."

Morphology is that division of Botany which takes cognizance of the various changes which occur in the condition of the vegetable organs, both such as are normal, as the transmutation of leaves into the several floral organs, as well as such as are abnormal, and occur only accidentally.

Taking the above as the definition of the subject, we immediately see its divisibility into two heads; the first of which treats of regular metamorphoses, which are connected with the natural structure of all vegetables: while the second includes those irregular or accidental metamorphoses, which result from an imperfect or redundant performance of the several changes noticed under the first head. These last influence particular plants, or parts of plants, and occur only in occasional instances. To this division belongs the consideration of double and other monstrous flowers.

Of regular Metamorphosis.—The great principle of regular Morphology is, that the various floral organs are but modifications of one common type, which is the leaf.

Lindley endeavours to give to Linnæus the credit of having been the author, or at least of having suggested the idea of this great fundamental principle; and in proof of this opinion quotes passages from his 'Systema Naturæ' and 'Prolepsis Plantarum,' in which the theory is imperfectly hinted at. At his suggestion the subject does not seem to have been taken up; nor was it at all until Goëthe published in 1770 his work 'On Vegetable Metamorphosis.' With a knowledge of the character of his poetical writings, it hardly need excite surprise that botanists of the day should have been sceptical on a subject so new, and at first sight so opposed to the dictates of common sense; and that his work at the time should have been considered to partake of the fanciful character of his poetry; or that they should regard it more as a poetical dream than as a sober philosophical truth, in connexion with a natural science.

Leaves are in many instances entirely wanting, or exist only in the degenerated forms of scales and spines. In these cases there is hardly any part of a plant which is not capable of being modified and rendered capable of performing the functions of leaves. For this purpose we find the excessive development of the stem which obtains throughout lactaceous plants; also the stem is furnished with leafy wings or expansions which run down its sides, as is seen in *Acacia alata*, &c. The petiole, in the form of the phyllodium, frequently takes upon itself the office of the leaf, as in most of the Acacias from New Holland. The same functions are frequently discharged by the peduncle, as in *Ruscus, Asparagus*, &c.; and this appears to be the true character of the leafy organs of Ferns, the true leaves of which exist in the degenerated form of scales known by the name of ramenta.

The evidence of the identity of bracteæ with leaves is more apparent than that of many of the other organs, for in very many plants they differ but little, except in size, from the original type, and indeed in almost every instance, even among those which are highly coloured, as in many of the genus *Salvia*, we find presented a gradual transition from the form and colour of the leaf to an almost petaloid condition; *Salvia splendens* and *Sclarea* are both good examples of this. The position of these organs on the axis shows clearly their close analogy to leaves, for they usually follow more or less the normal position of the leaves of the species, whether alternate or opposite. The calyx consists of a series of leaves arranged in a whorled manner round the axis, either distinct, or more or less combined, according to the character of the plant under consideration. Examples of partial reversion of calyx to the character of the leaf are seen in specimens of *Trifolium repens*, the *Polyanthus* of the gardens, and in cultivated roses.

In monstrous specimens we sometimes meet with the petals becoming leafy, of the occurrence of which in *Anemone nemorosa* M. DeCandolle has represented a remarkable example; and the author had seen the same condition occurring in the petals of *Papaver Argemone*.

Stamens appear to be formed from petals by the gradual narrowing of their lower part, so as to form the filament, while the anthers develope themselves on the upper margins. *Nympeæa alba* affords a beautiful example of the gradual transition of all the floral organs; and in it this transmutation of petals into stamens is clearly seen, the petals at first become narrower, then one of the margins has an abortive anther developed on it, to this another is afterwards added on the opposite margin, and finally the contraction of the petal having completely taken place, a perfect stamen is developed.

The petaloid cup which is found in the genus *Narcissus* is formed from an additional row of three stamens, as is evident from the frequent division of this organ into three pieces, which alternate with the divisions of the corolla.

The only instance with which the author was acquainted, wherein the carpellary leaf is to be found in an extended or unfolded state, naturally occurs in the order Coniferæ, where it simply covers but does not enclose the ovules. The carpellary leaf in this condition manifestly approaches a bractea.

The conversion of the pistil into a leaf is likewise frequently seen in *Trifolium repens*; and the author has a specimen of a species of *Potentilla*, which exhibits this change in a remarkable degree.

It was announced that the paper would be continued at the next meeting.

Mr. Adam White laid before the Society a selection of the plants he had lately found in a walk from Whiting Bay, Isle of Arran, to Brodick, and from Brodick to the top of Goatfell. He alluded to the strangely contrasted climates of Arran, arising partly from its insular position and its alpine mountains; he exhibited a few lichens, mosses, and phænogamic plants peculiar to alpine districts, and at the same time from the garden of Mr. Paterson of Whitehouse, Lamlash, laid before the meeting dried specimens of two species of *Leptospermum* from New Holland, one of them believed to be new, as well as of three or four other New Holland plants, the names of which were as yet unascertained. He particularly referred to the fine plants of warmer regions reared by Mr. Paterson in the open air, some of which stand the winter without shelter; as instances he referred to two species of *Salvia, Pentstemon*, &c., and exhibited luxuriant sprigs of Fuschsias, one of which was taken from a tree-like shrub, 18 feet high, and 22 feet wide, which, in its native soil, could scarcely have been more luxuriant. The damp atmosphere, he remarked, so different to their own dry climate, prevented some of the New Holland plants from flowering, but, he added, their luxuriant foliage and strong shoots nearly compensated for this. He alluded to the lists of the rarer plants found in the island, and communicated by the Rev. David Landsborough to Dr. M'Naughton, for insertion in his account of Kilbride (one of the island parishes) published in the 'New Statistical Account of Scotland,' and mentioned that he had been lately informed that Professor Gardner of Glasgow, in company with the minister of Stevenston, had found *Funaria Muhlenbergii*, a rather scarce moss, in tolerable plenty.—*G.E.D.*

| No. XVIII. | NOVEMBER, MDCCCXLII. | Price 1s. |

Art. XCII.—*Description of Equisetum hyemale, Mackaii and vari-egatum, as found on the banks and in the bed of the River Dee; with additional observations.* By J. B. Brichan.

Equisetum hyemale.

Root creeping, jointed, branched. Stems several from one branch of the root, or branched at the base, one to three feet high, or upwards, stout, erect or decumbent, articulated and fluted, occasionally throwing out catkins or very small branches near the top. Ridges or furrows fourteen to twenty-one in number, in luxuriant specimens twenty-eight; ridges grooved, and, as well as the furrows, grained like a file. Sheaths widest at top, at first pale green, with a black crenate rim; afterwards entirely black; ultimately white, with a broad black band at the base, the rim remaining black as before. The uppermost sheaths of the root *generally*, and the upper and lower of the *younger* stems *occasionally*, bear black, membranous, flexuose, deciduous teeth or bristles. The sheath of the catkin is invariably and persistently toothed. Catkins terminal, more rarely lateral, and in that case either single or in opposite pairs.

Equisetum Mackaii.

Root creeping, jointed, branched. Stems several from one branch of the root, or branched at the base, slender, often filiform, erect or decumbent, one to two and a half feet high, consisting of articulations from one to two and a half inches in length. The older stems frequently throw out long slender branches, which generally bear catkins. The stems are fluted, the ridges grooved, and both ridges and furrows grained as in E. hyemale. Number of ridges or furrows eight to twelve. Sheaths cylindrical, at first pale or yellowish green, with a narrow black band immediately under the teeth; ultimately wholly black, with the lower border of the black entire. Teeth equal in number to the ridges, membranaceous, white at the edge, long and tapering, terminating in a flexuose bristle which is generally black,

2 H

but sometimes white, in which case the white margin of the teeth is broader: they often adhere in pairs, and are decidedly persistent. — Catkins terminal, either on the stems or on the branches.

Equisetum variegatum.

Root creeping, jointed, branched. Stems many, three to twenty inches long, branched at the base and upwards, generally but not invariably decumbent and filiform, rather brittle, consisting of numerous fluted articulations, half an inch to an inch and a half long. Ridges or furrows five to nine, the former grooved, and both grained as in E. hyemale and E. Mackaii. Sheaths slightly swollen upwards, the upper half black. Lower border of the black waved or toothed, the dentations alternating with the teeth, and, like them, equal in number to the ridges. The teeth are distinct, never adhering, obtuse, somewhat ovate, black in the centre, with a broad, white, membranous margin, and tipped with a short bristle, which is either black or white, and more or less deciduous, while the teeth themselves are persistent. Catkins terminal, rather large in proportion to the size of the plant.

These descriptions are intended to apply to the plants only "as found on the banks and in the bed of the river Dee," and as seen either with the naked eye or through a small lens. I have endeavoured to exclude everything that could not with propriety be admitted as a specific distinction. The characters which I have given, and which I find to be constant, together with the various circumstances noticed in what follows, have led me to the conclusion that the three plants are well entitled to be ranked as distinct species. I conceive that the main strength of my position lies in the fact, that amidst all the varieties of size and shape which each plant presents, the distinctive characters remain the same. The species never *shade off* into one another, the smallest specimen of the largest species being readily distinguished from any specimen of the other two ; while, on the other hand, the stoutest stems of E. Mackaii and variegatum can at once be recognized as distinct from each other, and from the slenderer stems of E. hyemale. The following additional remarks apply chiefly to the situation, stems, sheaths, catkins and branching.

Situation. — The three plants are found at various parts along the course of the Dee, within the parish of Banchory, extending over a space of six or seven miles in length. There are three distinct stations for E. hyemale, four for E. Mackaii, three for E. variegatum, and several intermediate spots in which detached plants of each species or variety, especially of E. Mackaii, occur. I have no doubt whatever that both above and below the part where these stations are, there exist other localities hitherto unrecorded. In no instance have I seen E. hyemale associated with either of the other two, although E. Mackaii grows in its immediate vicinity. In two of the localities E. Mackaii and variegatum grow together, sometimes in juxta-position so close that they seem to spring from the same root. The three grow both upon the banks of the river and in the water, E. hyemale being generally highest and driest, whilst E. variegatum shows most of a tendency to *take the water*. The former seems to attain its largest growth among loose dry earth, especially where it finds shelter among large stones and trees or bushes, and insinuates its long straggling roots between and under the stones. E. Mackaii appears to prefer a locality where water, oozing from the bank, forms a green moist spot, or finds its way through a rent made by the river, or a channel worn by itself. The water where E. Mackaii thus fixes its habitat, is generally, *if not invariably*, chalybeate, being accompanied by that orange-coloured mineral or vegetable substance, and exhibiting on its surface that bluish-grey silvery scum, both of which are said to indicate the presence of iron. On the brink and in the cavities of such a spot the plant luxuriates freely, and assumes all its different forms,— short, tall and branched. E. variegatum is found in similar situations, generally running farther into the river, and the roots of both plants are frequently stained with the orange-coloured matter just mentioned. E. variegatum manifests a disposition to run its roots under large stones and grow beneath their shelter, where, and in the water, it is most luxuriant. From the stations assigned to it I am inclined to think that its natural habitat is "on the banks and in the bed" of rivers; and that, when found among sand, it has been carried down by the stream at whose mouth it grows: its greater growth in water seems to favour this conclusion. I think that when E. hyemale and E. Mackaii grow entirely in water, they are much smaller than in their peculiar situations already described.*

---

* Since writing the above I have seen another station for Equisetum Mackaii, and one for E. hyemale, both farther down the river. In these both plants grow more sparingly, — E. hyemale upon a hard, dry, grassy bank, decumbent, almost prostrate, and much more slender than usual; E. Mackaii in the immediate neighbourhood of *mineral* water, but growing partly, as it does in some of the other stations, among loose stones in the river's bed, and also nearly prostrate.

Stems. — The longest stem of E. hyemale I could find measured three feet two inches : it consisted of eighteen articulations. Several other stems measured nearly three feet in length. The longest articulation on any stem examined was about three and a half inches; occasionally a very short internode occurs among the rest, and there are sometimes two or three such in succession.

The thickest stem I observed was, in various respects, peculiar. It was composed of eleven principal articulations, the seventh of which, broken by accident, was bent at about a right angle to the lower part of the stem. The eleventh articulation was about one inch in length, and in it was inserted a sort of double stem, or rather a couple of stems, one of which, about one fourth the diameter of the original, consisted of four short articulations, having about twenty striæ, all that remained of the other being the decayed basal sheath. The broken internode and sheath were wet with an orange-coloured liquid, which, when touched, stained the hand. The cause seemed to be, that rain had lodged in the sheath and spread itself over part of the internode. Both, when dry, were partly quite black and partly of an orange colour. Does this indicate the presence of iron in the plant? I have already mentioned the fact, that I generally find E. Mackaii and E. variegatum in the vicinity of water impregnated with that mineral.

Another stem, growing in the river, had but five articulations, the upper part having decayed. At the top of the fifth articulation there were two opposite branches, the one upwards of thirteen inches long, the other rather shorter, each having twelve articulations, besides the black radical sheath. The sheaths of the main stem had fallen off, and a few setiform *black* teeth were present on the upper and lower sheaths of the two branches.

The *standard* or *normal* number of striæ both on this plant and on the other two, appears not to depend either on the length or thickness of the stem. The greatest number I have found on one of the most luxuriant stems, but not the stoutest, is twenty-eight; the smallest number, but not on the smallest stem, fourteen. On some of the most slender stems I have counted twenty-one, which I consider the *standard* number. The striæ diminish in *breadth*, but not generally in *number*, as they approach the top of the stem, nor is the greatest number always found on the thickest internode. Taking an average stem of each plant, E. hyemale appears to have the narrowest ridges; E. Mackaii the widest ridges and narrowest furrows; E. variegatum, ridges and furrows of nearly equal width, and, if there is any difference, rather broader than those of E. hyemale. From this cause stems of E. Mackaii and variegatum can at once be distinguished, at least in their recent state, by holding them between the eye and a light; E. Mackaii appears almost quite opaque, the furrows being scarcely seen, while E. variegatum and E. hyemale under the same circumstances appear semitransparent.

No stem of E. Mackaii that I have seen deserves a particular description. The longest I measured was two feet five and a half inches in length, the next longest two feet four inches. The number of articulations is about the same as in E. hyemale, the terminal one, when in fruit, being usually much longer than in the latter plant. The standard number of striæ I take to be eleven.

Of E. variegatum I have found no stem longer than about twenty inches, exclusive of the smallest portion of the root; the usual length is from three to ten. On one of the longest stems I have counted upwards of thirty articulations, the shortest has no fewer than eight; the terminal articulation, when bearing a catkin, is longer in proportion to the size of the plant than that of E. hyemale. It is likewise as stout as any on the same stem, often the stoutest of all. The standard number of striæ is eight.

The structure of the cuticle is undoubtedly the same in each species or variety; the largest plant having the narrowest striæ. In all, the furrows are more minutely grained than the ridges. It appears to me, however, that the structure of the cuticle in all the Equiseta is more or less the same, though the siliceous tubercles may not be disposed with the same regularity.

Sheaths.—These in E. hyemale have only two thirds of their length coloured differently from the internodes; in E. Mackaii two thirds or one half; in E. variegatum never above one half: the lower third or half being concolorous. In this I think they differ from some of the other Equiseta, whose sheaths are wholly coloured in a different manner from the stems. The colour changes according to the growth of the plant, as above described.

 After a good deal of searching, and examination of many specimens, I have not found the semblance of teeth upon the sheaths of E. hyemale, except on one or two of the younger stems and branches, and generally on the upper and lower sheaths. When present they are wholly black, subulate, and flexuose; they are extremely fugacious, and, from their rare occurrence, appear to me to be quite accidental. The sheath which embraces the catkin has always rather large, coarse, black, flexuose, persistent teeth, unlike the delicate membranaceous teeth of E. Mackaii and variegatum   The sheaths, having first split longitudinally along the furrows, fall off before the plant has decayed.

The sheaths of E. Mackaii invariably bear teeth such as are described above.   They remain longer on the plant than those of E. hyemale, and the teeth and bristles seldom, if ever, fall off before them.

In E. variegatum the bristle alone is usually deciduous.   The teeth remain till the sheaths decay, which they do in the same manner as in E. Mackaii.   In both the teeth of the uppermost sheath differ from the rest, only in being larger and in having shorter bristles.

The sheaths of all three turn more or less white before their final decay.   When they begin to split or fall off, the part of the internodes which they have covered is of a pale green colour, like that of the sheath in its first stage.   Soon after it begins to turn black and swell, and thus the plants in *extreme old age* become bent and geniculated. This, however, is quite different from the sinuous appearance which the stem sometimes assumes when in perfect vigour.

Catkins. — These, like the other parts of the plant, are well described by Mr. Newman.   While the catkins of E. hyemale, in one of the stations on the Dee, expand freely and shed their seed or pollen, I have not in any case observed those of E. Mackaii or variegatum to expand at all.   They appear to ripen without expanding or rising fully out of the sheath.

On the top of many stems of E. hyemale I observed a pile of small, dark brown, membranous, elastic, conical, inverted sheaths, of the same substance as the teeth of the sheath which embraces the catkin, increasing in width upwards, and so closely embracing each other that the rim only of each is seen, except the highest and largest, which gives the top of the pile a conical shape.   Although scarcely an inch in length, it resembles an *inverted abortion* of the plant.   Its lower end is at first inserted in the uppermost sheath, and surrounded by several small sheaths of a texture similar to its own, and placed within each other.   It is afterwards quite protruded, and gradually falls off, leaving a flexuose apex, somewhat like that of the catkin, whose formation I am persuaded it precedes.   On dissecting the top of several stems from which the pile of cones seemed but recently to have fallen, I found the germ of the catkin completely enclosed within its sheath, the teeth of which, not yet separate, form by their union the apex which appears when the pile falls off.   Dissection, however, when the cones are present, seems often to discover an *abortive* germ.   The number of inverted sheaths in one of these piles is about twelve ; what their peculiar function may be I cannot even conjecture : I do not find anything analogous either on E. Mackaii or on E. variegatum.*

Branching.—On what the branching depends I am unable to say. It is sometimes not easy to distinguish between branches and stems; the only distinction I can suggest seems almost a truism. Stems arise from the root, and branches from the stem. As the root is always black, they can therefore be in general distinguished, if this method of distinction is attended to. And I think it will generally be found that E. variegatum is the only plant of the three that can strictly be said to branch from the base, inasmuch as it does so in all circumstances, while the other two can scarcely be said to throw out basal branches, except when the stem has been broken off, or has decayed almost at the root. The three are occasionally branched near the top of the stem. With respect especially to E. hyemale and Mackaii, it may be said that this appears, in every instance, an effort of the plant to prolong its existence. It is always in older or broken stems that such branching takes place. E. hyemale is least subject to this manner of growth, and when branched, rarely throws out more than a single branch. I have not observed one above two or three inches long, except in the solitary instance of the stem before described. Sometimes, instead of a branch, a single catkin, and more rarely a pair of opposite catkins, appears.

E. Mackaii is, for the most part, unbranched, but of the three it is most liable to branching from the upper part of the stem. In this case the most frequent number of branches is three, and these, spring-

* Subsequent observation has enabled me to ascertain the following facts. The sheaths which in E. hyemale surround the base of the pile of cones, are the *rudiments* of several of the upper articulations of the stem. The base of the pile embraces the rim of the innermost sheath, and within these the catkin and its sheath are, in their germinal state, completely enclosed. It is only when these incipient internodes prove abortive, that they appear of a *texture similar* to that of the crowning deciduous pile. *They are entirely destitute of teeth.*

The apex of the stem of E. Mackaii, previously to the appearance of the catkin, proves, when dissected, to be composed of the converging teeth of a sheath which encloses several others disposed in a similar way, and having long, black, subulate teeth, like its own. As in E. hyemale, the germ of the catkin is found within, thus protected in its embryo state. The manner of growth in E. variegatum is exactly the same, but the teeth in every stage are somewhat white at the edge.

What length of time stems thus gradually developed require to attain their full growth, I cannot say. This, and many other particulars respecting these plants, might form very interesting subjects of observation to those by whom they are cultivated, or to those who have leisure to study them in their natural state.

ing from different parts one above another, are either horizontal, or vertical, or variously inclined. Sometimes there is but one branch, sometimes there are four, of which two are occasinally opposite. They generally, but not always, bear catkins, and that more abundantly than those of the other two varieties. The general length of the branches is from three to seven inches ; sometimes it is a foot or upwards.

E. variegatum, as already observed, *naturally* branches from the base, and, it may be added, from *all* parts of the stem. Its branches vary in length from one to fifteen inches. I know not whether the following description be worth notice, but it may perhaps illustrate the mode of growth frequently adopted by the plant, and the difficulty of distinguishing between stems and branches. On part of an ascending root five inches long, growing at the side of a large stone almost in the water, I found eight distinct ramifications. The four lowest were simple and perfect, the other four broken, or decayed, and branched. The lowest of the latter had three undivided branches ; the next, which rose from a branch of the root, had seven branches, two of which had each a branchlet : the next had one divided branch, and the highest had three branches, one of which was divided. Besides all these it threw out another root, which bore the remainder of a branched stem. A piece of the stem which formed a continuation of the primary root remained in a decayed state, and also appeared to have been branched. None of the ramifications exceeded a few inches in length.

Mr. Newman has kindly furnished me with specimens of the Dublin Canal plant, which he considers a variety distinct from that which grows in the Dee. To me they appear identical ; that from Dublin being not more luxuriant than many specimens I have gathered here.

Some of the remarks now made may appear unimportant, and even puerile ; but in so far as they contain facts, they must possess at least *some* value, and perhaps there is not too much presumption in the hope that they be the means of leading others to institute similar investigations. To myself the particulars above so minutely detailed are valuable on two accounts ; first, because they have served to convince me more of the distinctness of the plants in question as species; and secondly, because I am of opinion that the advance of botanical science, as well as of all other sciences, depends more upon the particular inquiries made by individuals than upon the general knowledge acquired without much investigation by the generality. I may be allowed to add, that although I have come to a different conclusion from Mr. Newman's, with regard to the plants I have attempted to describe, and have been most kindly encouraged by that gentleman to study the matter, and to publish the result of my inquiries, I feel all the deference due to authority so high, and therefore much diffidence in declaring myself to be of an opposite opinion.

J. B. BRICHAN.

Manse of Banchory,
October 4, 1842.

ART. XCIII.—*Additions to the List of Plants met with in the neighbourhood of Swansea*, (Phytol. 104, &c.). By T. B. FLOWER and EDWIN LEES, Esqrs. *Communicated by J. W. G. GUTCH, Esq.*

38, Foley Place, March 31, 1842.
SIR,

Through the kindness of my friends Mr. Edwin Lees, of Malvern, and Mr. T. B. Flower, I am enabled to send you the following addenda to my Swansea Flora, of plants which had escaped my notice.

I am, Sir, truly yours,

J. W. G. GUTCH.

To the Editor of ' The Phytologist.'

Ranunculus aquatilis, var. β. *pantothrix*. Frequent in watery places about Swansea.
——— *arvensis*. In Cornfields about Newton, and also near Caswell Bay.
——— *auricomus*. Frequent in the woods.
——— *circinatus*. Cromlyn Burrows and about Neath.
——— *Ficaria*. Frequent about the neighbourhood, with R. reptans.
Delphinium Consolida. I have in my herbarium two specimens of this plant gathered on the sandy shore of the bay, where it then appeared truly wild,—June, 1828; (E. Lees).
Papaver Argemone. In the cornfields, frequent.
Chelidonium majus. On old walls about Neath and Britton Ferry.
Fumaria officinalis. Common in cultivated ground and about hedges.
Cardamine pratensis. Frequent.
Draba verna. On old walls in the neighbourhood of Swansea.
Sisymbrium Sophia. On the shore on the other side of the Ferry. This is very scarce in Worcestershire; (E. Lees). Waste ground about the Infirmary, and in many other places.
——— *officinale*. Common about hedge-banks.
Viola canina. In the woods.
——— *odorata*, var. β. *imberbis*. Hedge-banks, frequent.
Polygala vulgaris, var. β. *oxyptera*. On the Downs near Pennard Castle, and frequent in other places.

Hypericum hirsutum. In thickets and hedges in the neighbourhood.

Dianthus Armeria. Banks about Britton Ferry, but not frequent.

Spergula nodosa. Frequent on the sand hills between Swansea and the Mumbles.

Sagina maritima. About Salt-house Point.

———— *apetala.* On walls and waste places, very common.

Mœhringia trinervis. In shady bushy places, common.

Stellaria aquatica. Frequent in watery places.

Tilia microphylla. You have a habitat for this I see; but it also grows in a wood between Gower Inn and Penrice, (E. Lees).

Geranium lucidum. On old walls about Newton, and other places in the neighbourhood, with G. rotundifolium.

———— *pyrenaicum.* Between Swansea Ferry and the race-course.

Chrysosplenium oppositifolium. In watery shady places.

Adoxa moschatellina. Frequent under shady hedges.

Saxifraga tridactylites. On walls and dry barren ground.

Euonymus europæus. In the hedges; also frequent about Neath.

Ononis antiquorum. In neglected pastures about the neighbourhood.

Melilotus officinalis. In the hedges and borders of fields.

———— *vulgaris.* Ballast-heaps near the west pier.

Trifolium arvense. Swansea and Sketty Burrows.

Medicago lupulina. In pastures and cultivated fields.

———— *minima.* I gathered this on the descent of a steep rock, close to the sea, eastward of the perforated rock in Oxwich Bay, 1839. It must be looked for on hands and knees, being nearly buried in sand and soil, but deserves the trouble required to attain it, from the curious structure of its legumes, (E. Lees).

Ervum tetraspermum. In the hedges and cornfields.

———— *hirsutum.* In cornfields and other cultivated ground in the neighbourhood.

Hippocrepis comosa. On dry chalky banks.

Sanguisorba officinalis. In moist meadows at the base of Craig-y-llyn-Vawr, vale of Neath, (E. Lees). Common in boggy meadows at Witch-tree bridge; and also at Neath.

Poterium Sanguisorba. On the hills, abundantly.

Geum rivale. At Cil Hepste, vale of Neath, (E. Lees).

Rosa villosa and *tomentosa.* I rather wonder you have not recorded these, nor Mr. Dillwyn either. They are eminently conspicuous and beautiful with their *deep red* flowers, in the vale of Neath, in June and July, especially about the Mellte waterfall, and at Pont Henrhyd. Rosa arvensis is very common about Swansea, (E. Lees).

Rosa systyla. This is *rare;* you have it *without* any assigned locality. I gathered it in Gower by the side of the road between Pennard and Penrice, (E. Lees).

Œnothera biennis. Naturalized in many places about Swansea and Britton Ferry.

Torilis infesta. In fields and by waysides with T. nodosa.

Pastinaca sativa. About the borders of fields.

Heracleum Sphondylium. In the hedges and meadows, common.

Æthusa Cynapium. In cultivated ground; much too common a weed.

Sium angustifolium. In ditches and rivulets.

Sison Amomum. Frequent under hedges in the neighbourhood.

Anthriscus sylvestris. In the hedges, very common.

Scandix Pecten-Veneris. In waste fields, common.

Galium palustre, var. β. *Witheringii.* In moist meadows and ditches.

———— *uliginosum.* In watery places.

Sherardia arvensis. In fields on a light soil.

Sambucus nigra. In the hedges and woods.

Hedera Helix. In woods and on old buildings.

Campanula Trachelium. In Gower, by the side of the road between Pearice and Port Eynon; numerous specimens, (E. Lees). In hedges and thickets.

———— *glomerata.* By the sides of the hills in dry open pastures.

Valeriana officinalis. In woods and marshy places.

———— *dioica.* In moist boggy meadows.

Solidago Virgaurea. In the woods, frequent.

Senecio tenuifolius. In woods and hedges and by road-sides.

Centaurea nigra, var. β. *radiata.* In pastures and by road-sides.

Sonchus arvensis. In corn-fields and hedges, frequent.

———— *asper.* Frequent, with S. oleraceus.

Lactuca muralis. On old walls, also frequent about the Mumbles Castle.

Helminthia echioides. About hedges and the borders of fields.

Picris hieracioides. On dry banks, frequent.

Symphytum officinale, with var. β. *patens.* Frequent in watery places.

Myosotis collina. Frequent on dry walls.

———— *palustris.* Common in watery places.

———— *repens.* Cromlyn Bogs and marshy places, frequent.

Verbascum nigrum. About the banks of Britton Ferry.

Solanum Dulcamara. In the hedges, frequent.

Statice Limonium. Very beautiful and luxuriant in crevices of the rocks stretching from the shore into the sea at Port Eynon, (E. Lees).

Lysimachia Nummularia. In wet meadows, and by the sides of rivulets.

Utricularia vulgaris. In a pool at Oystermouth, (E. Lees). Marshy places on Cromlyn Burrows, and also in the Neath Canal.

Veronica Beccabunga. In ditches and rivulets.

———— *scutellata.* In boggy meadows near Neath.

———— *officinalis.* Woods and pastures, frequent.

———— *montana.* Between the Lamb and Flag and Pont Nedd Vechan, vale of Neath, (E. Lees).

———— *hederifolia.* In cultivated and waste ground.

Linaria spuria. In the cornfields.

———— *Elatine.* In the corn and cultivated places.

———— *minor.* In sandy meadows.

Orobanche barbata. On ivy, on the walls of Oystermouth Castle, and at Britton Ferry.

Teucrium Scorodonia. In woods and bushy places.

Mentha rotundifolia. Very abundant about the banks of Britton Ferry.

———— *arvensis*, var. β. *acutifolia.* About the banks of Neath Canal, and also about Britton Ferry.

Rumex obtusifolius. In waste ground.

———— *conglomeratus* and *crispus.* Frequent.

———— *sanguineus.* Common by roadsides.

Polygonum Hydropiper. Frequent in watery places.
————— *mite.* About Neath and Fabian's Bay.
————— *aviculare.* Common in waste ground.
Parietaria officinalis. On old walls in the neighbourhood.
Euphorbia Peplus. In cultivated ground.
————— *amygdaloides.* In the woods.
Acorus Calamus. "Britton Ferry; Mr. Player."
Eriophorum pubescens. In great profusion and beauty on a bog called Gors Lwm, on the Banwen, Glyn Neath, (E. Lees).
Cladium Mariscus. In one spot only (as far as I saw), growing up in the hedge by the side of the Cromlyn Canal, in a marshy spot, (E. Lees).

————

The following additions and corrections have been supplied by Mr. GUTCH.

ADDITIONS.

Polystichum angulare. Hedge going from Swansea to Cromlyn Bog.
Riccia glauca. Field on the hill above Cromlyn Bog.
Jungermannia furcata, multifida, dilatata, inflata, bicuspidata, crenulata and *polyanthos.* Cromlyn Bog.
Calicium sphærocephalum. Sketty.
Opegrapha saxatilis, Verrucaria rupestris, Endocarpon Hedwigii. Rocks on the hills near the Mumbles.
Endocarpon smaragdulum. Stone walls on hills above Cromlyn Bog.
Psora cœruleo-nigricans, Squamaria crassa. Rocks on hills near the Mumbles.
Tetraspora lubrica, Entomia rotata. Water in an old quarry on hill above Cromlyn bog.
Rivularia angulosa. Cromlyn Bog.
Mouyeotia genuflexa, Tyndaridea pectinata.
Zygnema nitidum and *geminum.* Penllergare.
Cymbella ventricosa. Cromlyn Burrows.
Anabaina Jacobi. Cromlyn Bog.

CORRECTIONS.

Phytol. 105, line 18, *for* "gaur" *read* "Gam."
Senebiera didyma (Id. 106) Mr. Dillwyn informs me abounds in rubbish heaps in and about the town of Swansea.
Lepidium Draba (Id.) has been found by Mr. Moggridge; and it grows plentifully by the river side a little higher up than where gathered by Mr. Dillwyn in 1802.
Linum catharticum, (Id. 107). Mr. Dillwyn's name should have been attached to this plant, and not to L. usitatissimum.
Drosera anglica (Id.) has been found by me on the Town Hill in the boggy ground.
————— *rotundifolia* and *longifolia.* In Cromlyn Bog and Cwm Buchan.
Lavatera arborea, (Id. 117). At Paviland Cave; omitting Loughor Marshes.
Orobus sylvaticus, (Id. p. 108, line 6 from bottom), *for* "Fir Point" *read* "Fir Pont Cadley," meaning "the lands by Cadley Bridge," being the name of the tenement on which it grows.
Vicia lathyroides (Id.) Mr. Dillwyn doubts having been found.

J. W. G. GUTCH.

ART. XCIV. — *Notes on Gentiana Amarella,* (Linn.), *and Gentiana germanica,* (Willd.) By GEORGE LUXFORD, A.L.S. &c.

THESE remarks are published in the hope that others, who have had an opportunity of examining the plants to which they refer, may be induced to record the results of their investigations in 'The Phytologist.' Mr. Babington's opinion (Phytol. 310) and Mr. Brown's observations (Id. 320) have already appeared in its pages; on the other side of the question I may give the following extract from the 'Gardeners' Chronicle.' —

"*New British Plant.* — Some years ago Mr. W. Pamplin, bookseller, of Queen-street, Soho, observed a Gentian with large flowers in the neighbourhood of Tring, in Hertfordshire. Following his indication, the Rev. W. H. Coleman of Hertford has obtained specimens which have confirmed him in his suspicion that it would be found distinct from G. Amarella. In fact it proves to be the Gentiana germanica of foreign authors. He observes the following differences between it and G. Amarella : — "The plant is much smaller and less branched, while the flowers are fewer and larger. The leaves are broader — minutely but more distinctly dentate, as are also the segments of the corolla. The calyx is more rounded at the base, and its tube is equal to the segments; while G. Amarella has the segments rather longer than the tube. The segments of the corolla of such specimens of the Gentiana as he has examined, in æstivation overlap those adjacent to them on the right — while those of G. Amarella overlap towards the left. More important differences than these consist in the ovary of Gentiana germanica being stalked and the corolla widest at the throat, while in G. Amarella the ovary is sessile, and the corolla almost cylindrical. Dr. Grisebach has already referred the figures of G. Amarella in ' English Botany ' and the ' Flora Londinensis,' to G. germanica ; and there can, we think, be no doubt that it and G. Amarella are quite distinct species."—*Gardeners' Chronicle, Oct.* 9, 1841.

The perusal of this paragraph made me determine, whenever an opportunity occurred, to investigate the subject for myself, and, if possible, to satisfy my own mind as to the justness of the claims of the plant called Gentiana germanica to be considered specifically distinct from the Gentiana Amarella of Linnæus. At that time I had seen no authentic specimens of the former plant; my examination was therefore confined to the specimens of G. Amarella in my own herbarium. The results of this examination certainly did not tend to remove certain doubts which previously existed, at the same time, I was not so influenced by them but that my mind remained open to conviction should the evidence subsequently adduced tend in the opposite direction.

A few weeks since Mr. Wm. Pamplin very kindly sent me a recent specimen of a gentian, gathered by him near Streatley, Berks, together with the following remarks : —

"I have the pleasure to enclose herewith a fresh specimen of that large-flowered variety, or whatever it be, which I first gathered near Tring, in Herts, and which some have since considered, I believe, Gentiana germanica of Reichenbach and others.

"I add one or two plants of the common state of Gentiana Amarella, for your comparison."

At last, then, I had in my possession a veritable example of the wished-for plant, as well as additional materials for comparison, and I at once resolved to go to work. The first step to be taken in the enquiry was obviously that of consulting Willdenow, the original authority for Gentiana germanica. The characters assigned by him to the two plants, omitting the synonymes, are the following: —

38. *Gentiana germanica*, W. Corolla 5-cleft, salver-shaped, bearded, segments of the limb ovate, acute: leaves ovate-lanceolate: branches longer than the internodes. —' Species Plantarum,' Berolini, 1797. P. 1346.

39. *Gentiana Amarella.* Corolla 5-cleft, salver-shaped, bearded, segments of the limb lanceolate, acute: leaves lanceolate: branches shorter than the internodes.

Differing from the preceding in its whole habit, shorter branches, narrower and smaller corolla, lanceolate leaves and pale yellow root (radice *flava*).—Id. 1347.

Here, it will be observed, nothing is said about the stipitate ovarium of G. germanica, or the sessile one of G. Amarella. Willdenow's distinctive characters are founded chiefly on differences in the shape of the segments of the corolla, in the form of the leaves, and in the relative length of the branches and internodes. He also, in the remarks above quoted after G. Amarella, applies to its root the same term (*flava*) as Froelich, in the following description of his G. Amarella, applies to *its* root. Froelich's G. Amarella is quoted by Willdenow as a synonyme of G. germanica.

Froelich's description of the plant being very full, I shall give it entire, omitting only his synonymes.

33. *Gentiana Amarella.* Corolla 5-cleft, salver-shaped, acute, throat bearded; segments of the calyx sub-equal.

Hab. in the meadows of Europe. Annual. Flowers in the autumn.

Root simple, fibrous, pale yellow, (*flava*). Stem erect, obsoletely hexagonal, purplish, leafy, branched in a brachiate manner; branches opposite, single-flowered, leafy, frequently many-flowered, of the same height as the plant; frequently more simple, few-flowered, not longer than the corolla. Radical leaves numerous, lying on the ground, obovate, obtuse, narrowing into a petiole, obscurely 3-nerved, withering; stem-leaves sessile, slightly connate at the base, ovate or ovate-lanceolate, more or less acute, rough at the margin, 3-nerved; upper stem-leaves and those on the branches sub-cordate, acute. Flowers terminal and axillary, solitary or two together, erect, an inch long (shorter in small plants), peduncles erect, angular, shorter than the flower. Calyx campanulate, marked with fifteen raised lines, 5-cleft beyond the middle, about half the length of the corolla, segments lanceolate, acute, subequal. Corolla salver-shaped, purple-blue; tube widening upwards, angular, striated, transversely wrinkled between the striæ, contracted and paler at the base; limb 5-cleft, segments ovate-lanceolate, acute, entire, one-third the length of the tube: scales in the mouth of the tube five, membranaceous, white, multifid in a capillary manner at the apex. Stamina:— filaments rather longer than the tube, at the base of which they are inserted; anthers oblong, free. Pistil:—germen oblong, on a short stalk (*breviter pedicellatum*), attenuated upwards: stigmas ovate. Capsule oblong, longer than the corolla, stalked. Seeds roundish, rather compressed, smooth, brown.— Froelich, 'De Gentiana Libellus,' Erlangæ, 1796. P. 86.

Then follow the characters and an equally full description of G. pratensis, a native of Russia and Siberia, given by Willdenow as a distinct species, but quoted by Grisebach as synonymous with his G. Amarella, *α*. The following are its characters.

> 34. *Gentiana pratensis.* Corolla somewhat 5-cleft, salver-shaped, obtuse, throat bearded; segments of the calyx unequal: leaves lanceolate.

At the end of his work, Froelich gives the following interesting account of his meeting with another Gentiana Amarella, apparently distinct from the plant so called by himself and other German botanists.

> " After the preceding part was printed, the 1st fasciculus of Dickson's Dried Plants came into my hands; among them is Gentiana Amarella, No. 5, from Derbyshire, in England. The root is like that of Gent. Amarella above described, but yellow, (*lutea*). Stem as in that, also purple, but more slender; ten inches high. Branches beginning at the middle of the stem, shorter than the adjoining internodes, as in Gent. pratensis and campestris; the stem therefore is not pyramidical, as it generally is in our Amarella. Lower leaves spathulato-ovate, the next ovate-lanceolate, the upper ones lanceolate, acute, shorter than the internodes. Flowers terminal and axillary, solitary or in pairs, the lower ones seated on the branches, erect, 9 lines long, 2 lines wide, being thus much smaller than in Amarella as above described. Calyx as in that, and about half the length of the tube of the corolla, rather unequal. Corolla likewise similar, only narrower, and apparently of a paler blue. Stamina the length of the tube. Germen still more attenuated above. This seems to be a species distinct from our Amarella, and if, as appears from the title of the fasciculus, it is named on the authority of the Linnæan herbarium, it must retain the name of Amarella; whilst our plant, the Amarella of German botanists, must receive another name.—Id. 141.

Willdenow, in the following year, published the German Gentiana Amarella in his 'Species Plantarum,' under the name of G. germanica.

Turning now from Froelich and Willdenow to Grisebach, " the latest and best authority on the Gentianeæ," I must confess that I at first found myself somewhat bewildered among his characters of the " four states of each species," and the numerous synonymes assigned to each of these states. I think I cannot better illustrate the uncertainty and difficulty attending this investigation, than by giving the characters of these plants and their varieties, with other essential particulars, as I find them in Grisebach's elaborate 'Genera et Species Gentianearum.' I had intended to give the whole of the synonymes, but finding that by so doing I should considerably lengthen my paper, I determined on quoting merely the references to original authorities and the very few British works.

I may here remark that I have endeavoured to render all these translations as literal as possible, without any attempt at elegance of diction.

Section IV.—Amarella, *Gaud.* Calyx entire, forming a tube. Corolla destitute of the intermediate fold and glands, salver-shaped, internally furnished with a fringed crown, (one species excepted). Stigmata two, distinct. Anthers free. Capsule generally sessile. Testa wingless. Root annual. Style none.

† *Stem erect, branched, branches many-flowered.*

43. *Gentiana Amarella*, L. Stem slender: radical leaves oval-spathulate, upper ones ovate-lanceolate, sessile: lobes of the calyx lanceolate, subequal, shorter than the tube of the corolla: crowned corolla half an inch long, tube cylindrical: linear-oblong ovary and capsule sessile!—From a living plant.

G. Amarella, *Lin. Sp. Pl.* i. 334;* also of the Swedish botanical writers, of the Linnæan herbarium (according to Dickson), and of Smith's specimens collected near Upsal with Linnæus by Ehrhart. Dried specimens, *Dicks. Dried Pl. 5.*

β. *uliginosa*, W.! Stem short, subsimple, few-flowered: leaves lanceolate: lobes of the calyx rather unequal, about as long as the tube of the corolla: corolla slender, generally 4-cleft, 4 lines long. This appears confined to the north of Germany. Height, 1 to 4 inches.

γ. *axillaris*, Schm. Stem a foot or less in height, straight: upper leaves very acuminate: cymes axillary, densely flowered, subsessile. Transitions to var. δ. are frequently met with.

δ. *pyramidalis*, W. herb. Stem straight, much branched, the branches forming a pyramid: cymes axillary, the lower ones on long peduncles, spreading.

Description.—a. Stem slender, rather straight, 6 to 18 inches high, with upright branches, or somewhat simple. Leaves an inch long, shorter than the internodes, their margins appearing serrulated under a lens. Cyme raceme-like, sometimes few-flowered, occasionally compound and forming a slender thyrsus. Lobes of the calyx obtuse, as long as the tube, scarcely half so long as the tube of the corolla. Corolla blue; its tube obconical; the lobes elliptic-lanceolate, about half the length of the tube. Stamina about as long as the tube.

* The learned author observes, in a foot-note, that "The synonymes of the older writers had better be devoted to eternal oblivion."

Varieties. — A variable species as to the form of the leaves, the length of the calyx and the shape of its lobes ; also in the number of parts in the flowers, which are either quaternary or quinary ; in its mode of branching, and in the inflorescence, which varies from panicled to 1-flowered and sub-axillary ; also in the colour of the corolla, which varies to white. The stem occasionally sends off from its base single-flowered partially decumbent shoots, (Scotland). It is also sometimes somewhat naked, and nodding at the apex, with leaves scarcely half an inch long, rather obtuse and narrow, with few-flowered axillary cymes, rigidly spreading, (stony places in the region of Baikal). — There are scarcely any limits between the other varieties ; β. however is more constant.

Observations. — The following remarks will give some idea of the history of this species, the synonymy of which Reichenbach was the first to clear up satisfactorily. Gentiana Amarella is much more frequent in Sweden than G. germanica, if indeed the latter occur there at all, which I very much doubt, since Linnæus was either altogether unacquainted with it, or else did not distinguish it from the former, which is met with everywhere in that country. Ehrhart also collected G. Amarella with Linnæus ; and on his return to Hanover, when he first saw G. germanica, he named it G. critica. This name, according to the rule of priority, would have been substituted for Willdenow's, had not Ehrhart's specimens presented a more rare and smaller form, but still having the stipitate ovary. Ehrhart's specimens are identical ! with Smith's figure in ' English Botany,' 236, which, by mistake, is given as the G. Amarella of the Linnæan herbarium, and unnaturally represented with a sessile ovary ; * it is, however, G. germanica. Dickson, more accurate in this respect, proved the G. pratensis of Froelich to stand in the Linnæan herbarium as G. Amarella ; and Smith himself, having subsequently become sensible of his error, in his letter to Panzer written in 1822 (Flora, 1830, p. 529), says that G. germanica is distinct from the G. Amarella of Linnæus.† Sir W.

* Grisebach appears to be in error here. A very evident though short stalk is represented at the base of the separate ovary in the ' English Botany' figure.

† Sir J. E. Smith, in the second volume of his ' English Flora,' published in 1828, six years after the date of the letter above mentioned, has the following observations under Gentiana Amarella. " G. germanica, Willd. v. 1. 1346, which is G. critica of Ehrhart, Herb. 152, and, according to Swiss specimens, Haller's n. 651 (though the latter indicates many wrong synonyms, and takes it for an English plant), differs from Amarella in having *flowers* nearly twice as large, situated about the upper part of the *stem*, which is of a corymbose form of growth. It may be a good species, but has not yet been observed in England."—p. 31.

2 K

Hooker, however, did not consider the two plants to be distinct species; and has consequently again figured G. germanica under the name of G. Amarella; and in thus combining them other botanists agree, but I do not, since I have never seen intermediate forms, nor do I think that the arguments to be drawn from the different geographical stations of the two species, ought to be neglected.

Country. — Damp pastures in the plains of north-western Europe, (50° — 69°), of eastern Europe (even to the Caucasus) and Siberia. Hab. *α.* and *β.* in Scotland; *γ.* in England; *δ.* in the Highlands of Scotland; all on the authority of Hooker! Many foreign 'localities are given. Flowering in August and September.

Thus far Grisebach upon the distinctive characters of Gentiana Amarella. I cannot help suspecting that two at least of the five species which succeed it should be included among the varieties.

49. *Gentiana germanica*, W! Stem straight, robust: lower leaves spathulate, upper ones ovate-lanceolate, acuminate, sessile: lobes of the calyx subequal, ovate-lanceolate, shorter than the tube of the corolla: crowned corolla 5-cleft, tube gradually widening: linear-oblong ovary and capsule stalked.—From a living plant.

G. germanica, *Willd.! Sp. Pl.* i. 1346. Figure; *Hook. Fl. Lond.* 33.

β. *minor*, G. W. F. Meyer, (Chl. Hanov. p. 278). Stem short, subsimple: leaves subequal: corolla very large.—From a living plant. Figure; *Eng. Bot.* 236.

γ. *præcox*. Middle stem-leaves oval, lower ones obtuso-spathulate: calyx as long as the tube of the corolla. — From a living plant. A form which, according to specimens collected in the Alps, near Heiligenblut, flowers in August, has the leaves acuminate, as in *α.*, but the calyx is like that of the present variety; it is in this respect, as well as the time of flowering, a transition to the common form.

δ. *caucasica*. Stem branched: all the leaves obtuse: lobes of the calyx subulate, distant, twice the length of the tube, and as long as the tube of the corolla.— From a dried specimen. In the Carpathians it passes to *α.* Figure; *Sims, Bot. Mag.* 1038.

Description. — *α.* Stem generally a foot high, variable in ramification. Leaves two inches long, equalling the internodes, their margin appearing serrulated under a lens, point rather obtuse. Cyme truly compound, forming a panicle nearly a foot long; erect indeed, but more spreading than in G. Amarella. Lobes of the calyx acuminate, equal, as long as its tube, half the length of the tube of the corolla. — Corolla blackish-blue, varying to violet; tube obconical! lobes ovate, acute, mucronate.

Varieties. — A very variable species. Besides the forms given in the synonymes, the following are worthy of being mentioned. Under *α.* Lobes of the calyx abbreviated, rather unequal. Corolla 2 inches long, lobes elongated, elliptic-lanceolate, as long as the tube! (perhaps a monstrosity). Branches elongate, fastigiate, (approaching γ). Under β. Leaves short, obtuse, lobes of calyx abbreviated, (approaching γ). Stem slender, 1-flowered, leaves very short, sub-rotund! — Under δ. Lower leaves rosulate, spathulate, subrotund, (transition to γ). Calyx cleft on either side. Corolla 4-cleft.

Country. — Dry gravelly meadows and pastures, mountains and alpine situations at a height of from 150 to 6200 feet, in central Europe and the Caucasus. Hab. α. in England, according to a specimen collected near Ripon, (Hook.!) Flowers from August to October.

Under the name of Gentiana campestri-germanica, Grisebach also describes a plant which he considers to be a hybrid between G. germanica and G. campestris. The flower is 5-cleft, the calyx having two of its lobes much larger than the rest.

It will, I think, be considered quite unnecessary for me to give quotations from the works of any other continental botanists than those already referred to, who are all intimately connected with the subject under discussion; — Willdenow from being the original authority for G. germanica, Froelich from the full descriptions contained in his Libellus of the species of Gentiana known at the time it was written, and Grisebach, from his elaborate Monograph of the whole order, in which work he has devoted considerable space to the illustration of the plants now before us. I am, however, truly gratified in being able to close this part of the enquiry with a short extract from the *fifth* edition of Hooker's 'British Flora,' published only within the last fortnight. The two plants stand therein as distinct species; the characters assigned to each being substantially the same as Grisebach's, already given, I need not quote them: the following remarks, under G. germanica, will, however, show that although the opinions of the learned author, in deference perhaps to the high authority of Grisebach, may have been somewhat modified since the publication of the 'Flora Londinensis,' he evidently considers the question as still open to discussion.

"In the ' Flora Londinensis' I stated it as my opinion that the G. Amarella and G. germanica were not specifically different. Grisebach, Koch, and others think differently; and as the former author has examined and made his remarks on the specimens in my herbarium, I have given his characters, and would direct the attention of botanists to the subject. Mr. H. Watson is of opinion that they are but trifling varieties of each other. In all my numerous specimens of G. Amarella, the plant takes a more or less pyramidal form, and the flowers are far more numerous, crowded, and considerably smaller than in G. germanica."—Hook. Br. Fl. i. 219. Ed. 5, Oct. 1842.

Since the preceding part was written I have again examined and compared all the specimens of Gentiana at my command, that can be referred to either of the two forms. This examination has led to the following conclusions.

1. That no character can be derived from either the positive or relative size of either plants or flowers; both forms being exceedingly and equally variable in this respect. The specimen from near Streatley is much larger, more branched, and with many more flowers than the specimens of G. Amarella which accompanied it; the reverse of these conditions obtained in the plants mentioned in the extract from the 'Gardeners' Chronicle' (Phytol. 381).

2. That no dependence can be placed on the relative proportions of the leaves, branches, and internodes; these proportions frequently varying even in the same individual. In the Streatley specimen some of the branches are shorter, others are longer, than the internodes of the stem, while the leaves are shorter. The same remarks will equally apply to large-flowered continental plants, and to British specimens of G. Amarella.

3. That in form both the leaves and calyx-lobes are variable; and the proportion borne by the latter to the tube of the calyx differs in the same specimen.

4. That although the tube of the corolla may, in some specimens, be correctly described as cylindrical, and in others as gradually widening upwards, yet intermediate forms are extremely common. And Grisebach, in his descriptions, calls them both "obconical," as Mr. Brown has already remarked, (Phytol. 320).

5. That with respect to the stalk (*stipes*) of the ovary, if by that term is to be understood a stalk separate and distinct from the substance of the ovary supported by it, similar to that represented in tab. 236, 'English Botany,' then such a stalk I have not been able to find either in the Berkshire plant before mentioned, in continental specimens, or in any others that can possibly be referred to G. germanica.

6. That if, on the contrary, by the term *stipes* we are to understand such a gradual diminution of size in the base of the ovary itself, as would probably be produced by the shrinking of that part, as suggested by Mr. Brown, then such a stalk I find to be more or less evident in every specimen that I have examined, whether referrible to G. Amarella or G. germanica.

7. Moreover, in undoubted specimens of G. Amarella, gathered rather late in the season, I find some flowers with mature capsules, which contain ripe seeds, and are generally perfectly *sessile;* while in

other flowers on the same plant, sometimes even in those on the same branch, the ovary being immature when the plant was gathered, it is *stalked,* apparently in consequence of the shrinking of the lower part in the manner described.

Lastly: as a necessary consequence of the above conclusions, and in the absence of more positive evidence than any we at present possess, I think I shall be warranted in considering the numerous forms of the two plants as all belonging to one variable species.

I shall be gratified by receiving specimens of any form of these plants for further examination; and earnestly hope that the subject will be taken up by more able botanists, who will favour the public with the result of their enquiries.　　　　　　　　　Geo. Luxford.

65, Ratcliff Highway, October 25, 1842.

Art. XCV. — *Some Account of the Botanical Collections recently made by Dr. Theodore Kotschy (for the Wurtemburg Botanical Union) in Nubia and Cordofan.* * Communicated by Mr. William Pamplin, jun.

The collection of dried Nubian plants (amounting to nearly four hundred species) made by Dr. Theodore Kotschy in 1839, possesses so much interest, not only to the members of the Wurtemburg Union among whom the plants are distributed, but also to the botanical world at large, that Dr. Schnizlein, in the hope of rendering an acceptable service to botanical Geography, has kindly undertaken the task of enumerating and comparing them with the materials already known as forming part of the Floras of Egypt, Arabia, Eastern India and Western Africa. So favourable an opportunity of acquiring a knowledge of the vegetation of the eastern part of Africa, will not probably soon occur again.

In the first place, with respect to the condition of the plants, it is quite evident, on inspection, that they have been prepared with great care; of some there are specimens both in flower and fruit; all are exceedingly well preserved; and of many species the specimens are numerous. We find twenty species in the most excellent state; two hundred and fifty-six in perfectly good condition; and only about forty which are not quite complete, or, in other words, less perfect: in fact, altogether, they could not have been expected to arrive in a

* Translated and abridged from the ' Regensberg Flora oder Botanische Zeitung ' for 1842.

better state, for they are neither broken nor otherwise injured, and frequently several specimens of one species are given, and reckoned only as one : it must, however, be acknowledged, that on the whole a majority of small-sized plants occur, and a few of the larger ones, such as shrubs and trees, are cut into small pieces and occasionally divested of the root, probably with the view of economizing the space required in packing them.

Among the greatest ornaments of the collection, the following may be particularized. The two splendidly dried Utricularias—U. inflexa and U. stellaris, the new and delicate Udora cordofana. The superb grasses Fimbristylis hispidula, Isolepis prælongata, Cyperus aristatus, squarrosus, resinosus, Lappago orientalis, seven species of Aristida, Ctenium elegans, the ornamental Triachyrum cordofanum, Chloris spathacea, Pennisetum lanuginosum, Panicum Petiverii. Isnardia lythrarioides, Heliotropium pallens ; the magnificent Ipomœa repens, a bog plant, with large rose-coloured flowers ; the small but extraordinary Conomitra linearis, with its orbicular and long-pointed pod ; the conspicuous Mollugo bellidifolia ; but above all, that rare and remarkable plant the Neptunia stolonifera, whose root-stem reminds one of our Phellandria, the heads of flowers of Œnanthe, while the leaves and the fruit resemble those of a Mimosa ; Poivrea aculeata, Guiera senegalensis, six Cassias, four Bergias, the three splendid Nymphæas—cærulea, Lotus and ampla ; Striga orchidea, Acanthodium hirtum, Pedalium Caillaudii, Melhunia Kotschyi, Monsonia senegalensis, Daleschampia cordofana, four Pavonias, five Hibiscus, eight Sidas ; the fine Acacias—papyracea and seriocephala : altogether we have thirty-nine species of Papilionaceæ, among which the Requienia obovata and the two species of Alyssicarpus are worth noticing ; the remarkably well named Euphorbias—acalyphoides, covolvuloides and polycnemoides.

In the second place, the principal districts and localities in which Dr. Th. Kotschy made his collections are to be noted. They are as follow :—first in the province of CORDOFAN, the city Obied, with the neighbouring mountain Arasch Cool, the lake Tara, and the river Choor ; then the villages Abu-Gerad, Bara Chursi, Hogeli, Tejara and Uagle ; then the more southerly situated town Tekele, and the mountain Kohn : second, in the province SENNAAR, the village Wolet Medine, Gujeschab on the Nile, then the islands by the cataracts of the White Nile, at the mountain Gerri.   W. PAMPLIN, JUN.

(To be continued).

ART. XCVI.—*Analytical Notice of the 'Transactions of the Botanical Society.' Vol.* i. *pt.* ii. Edinburgh: Maclachlan, Stewart & Co.; H. Bailliére, London. 1841.

(Continued from p. 364).

XIII. *Remarks on the Structure and Morphology of Marchantia.* By GEORGE DICKIE, *Lecturer on Botany at Aberdeen.*

THE author's observations refer chiefly to Marchantia polymorpha; and they are arranged under two heads: "1. The structure of Marchantia. 2. The inferences to be drawn from the facts exposed under the first head." The paper is illustrated by figures.

In the first division of his subject Mr. Dickie describes the structure of the frond, and the fructification. In speaking of the lower surface of the frond he states that what is called by Sir W. J. Hooker a "prominent blackish midrib," "is in reality a groove in the frond, from the edges of which originate purplish scales, which, by meeting, conceal the groove; in this groove are numerous transparent filaments." Many short radicles originate from the surface of the frond at the sides of the groove.

The receptacles of both the (so-called) male and female reproductive organs, are supported by peduncles; these originate "from the grooves on the lower surface of the frond, and are grooved in a similar manner, the peduncles and their posterior grooves being continuous with the frond and the groove on its inferior surface. In each of the grooves is lodged a bundle of transparent tubular filaments, the walls of which "are covered with green granules, often arranged in a spiral manner." The lower end of the filaments is blunt and closed, the upper extremity spreads over the inferior surface of the receptacle, sending off a bundle to each lobe or ray of the former. The filaments at this part "are covered with overlapping scales, similar to those of the lower surface of the frond."

What are generally considered to be male receptacles are lobed at the margin, while the female receptacles are rayed; the rays, however, are stated to be "merely lobes bent upon themselves, from above downwards." A perpendicular section of the male receptacle shows it to consist of cellular tissue enclosed between two membranes. In the tissue "are embedded numerous flask-shaped sacs, with long necks, terminating each by an orifice on the upper surface of the receptacle.   *   *   The sacs contain the bodies called *anthers*, which vary in form according to the stage of development at which they are examined. Each of these bodies consists of a cellular sac filled with a granular matter." For a correct description of the fertile receptacles, with their involucres and capsules, the author refers to Sir W. J. Hooker's account of those parts.

Under the head of *Morphology*, the author institutes an enquiry into the true nature of the peduncles, and of the so-called male receptacles and anthers; and shows that the latter bodies bear no relation to those organs in flowering plants. He is of opinion that the lobed receptacles and the peduncles "may be considered as metamorphosed fronds," of which a longitudinal section shows them to be continuations; "the grooves of the peduncles are continuations of the grooves on the lower surface of the fronds, and they enclose the same spirally dotted filaments."

" The flat receptacle is lobed like the fronds, its structure is the same, on the lower surface, filaments in grooves covered with scales, and a cuticle with stomata on the upper surface; and the fact that one of the notches is larger than any of the others, and the concavity on one side of the peduncle, appear to lead to the inference, that this receptacle is a small frond folded horizontally upon itself. In M. conica the relation of the two is more evident, the receptacle in this species being sessile, and still more so in that variety of M. hemisphærica where the receptacles are always sessile, and embedded, as it were in the substance of the frond."—p. 110.

With respect to the bodies termed anthers, the author institutes a comparison between them as found in Marchantia and Riccia, a nearly allied genus. In the latter genus the capsules are spherical, in Marchantia the sacs " are flask-shaped, and have a long neck protruding by an orifice at the surface." The inclosed bodies also differ somewhat in shape, but in both genera they consist of "a cellular membrane enclosing minute granules." Hence the author is disposed to consider these bodies as gemmæ, or one means by which young plants are produced, but remarks that " it may appear rash to arrive at any such conclusion until they are actually seen to germinate under favourable circumstances."

XIV. *Remarks on some curious Metamorphoses of the Pistil of Salix caprea. By* the Rev. J. E. LEEFE, M.A., *Audley End, Essex.*

THE author commences his remarks by quoting Professor Lindley's observation, that the pistil is seldom found " converted *into stamens,* but it often takes upon itself the form of petals, and although *cases are very rare* of pistils bearing *pollen,* yet several instances are known *of ovules being* borne by the stamens." This rare instance of conversion occurred in the catkins of an apparently healthy plant of Salix caprea, growing on the banks of the Cam, near Audley End, Essex.

"The catkins were of a light green colour, longer, and tapering to a point, instead of being blunt, as is usual in S. caprea. In one case three apparently proceeded from the same bud. The nectary and scale were very little altered, and the change is very various in its character, and several intermediate forms occur in addition to those which I am about to mention. Those enumerated are, however, the most remarkable. I may here mention, that I do not mean that all the forms are to be found in the same catkin, and also that the greatest alteration is observable in the flowers at the apex and base of the catkins.

"1. Styles two, each bearing at the top two small pale stigmas; in other respects as usual.

"2. Scale and nectary as usual. Stalk of the ovarium extremely lengthened, resembling a filament, downy, especially towards the base, terminating in an ovarium. Style cloven, the cleft penetrating the substance of the ovarium; at the base of the cloven style, on *one face* only, appeared two yellow bodies like the lobes of an anther.

"3. Stalk greatly lengthened. Ovarium silky, not much altered at the base; stigmas almost entirely obliterated; instead of them a vertical depression or sinus, on each side of which was a yellow antheriform body, erumpent from the substance of the ovarium. These contained *perfect pollen*, and in the lower ovarium-like portion I remarked several ovules.

"4. Stalk forked or branched near the top, each fork bearing a silky body resembling an ovarium.

"5. Stalk forked; one fork filament-like, and tipped with an anther yielding pollen; the other silky, approaching an ovarium in form, and containing ovules.

"6. In this form, in which the nearest approach to a stamen was made, each fork bearing an anther containing pollen, and the ovarium being entirely obliterated, traces of the original structure are to be seen in the bifurcated filament analogous to the two stigmas."—p. 113.

All these various forms are illustrated by figures.

XV. *Descriptions of Jungermannia ulicina (Taylor), and of J. Lyoni (Taylor).* By THOMAS TAYLOR, M.D., M.R.I.A., *Dunkerron, Kenmare, Kerry.*

1. *Jungermannia ulicina,* (Taylor). Stem creeping, filiform, branched: leaves distichous, roundish-ovate, concave, bilobate, inferior lacinia much the smaller, involute; stipules ovate, bifid: fruit axillary; perichætium three-leaved, compressed, two upper leaves narrow at the base, roundish-oblong, inner margins incurved; lower leaves oblong-ovate, bifid.

On stems of Ulex europæus and Erica cinerea, "on wet banks facing the north, over Finnehy river, near Kenmare, Co. of Kerry; (T. T.) At Dolgelly, North Wales; Mr. W. Wilson."

A minute species, bearing a strong resemblance to J. minutissima, *Sm.*; from which it may be distinguished by the presence of stipules,

2 L

by the large, flattened, three-leaved perichætium, by the cauline leaves increasing in size as they approach the perichætium, by the skin being less flexuose, by the leaves being more distant, and by the paler colour of the whole plant.

Dr. Taylor refers to Hooker's ' British Jungermanniæ,' tab. lii. fig. 3, as being " a just representation of the stem, leaves, and stipules of J. ulicina ; " and observes,

" It is remarkable that on the same stems of furze in the above locality, grew every one of the minute tribe to which the present species is allied, viz., J. minutissima, J. hamatifolia, J. calyptrifolia, and J. serpyllifolia."—p. 116.

2. *Jungermannia Lyoni*, (Taylor). Stem ascending, somewhat branched : leaves distichous, alternate, somewhat quadrate, concave, recurved, trifid ; anterior lacinia rounded in front, posterior one reflexed, all acute or terminated by a single large tooth; stipules none : fruit at length lateral ; calyces oblong, obtuse, inflated, rather naked at the base, mouth fringed and plaited.

" So great is the force of individual vegetation, that within the calyx, and alongside the pistilla, a bud may sometimes be seen to arise, and at length to emerge out of the calyx, clothed with leaves. Perhaps this viviparous condition of the calyx has not yet been observed in any other species. The plant nearest in natural affinity is probably J. orcadensis, *Hook.*, equally destitute of stipules, having a similar erect, scarcely-branched stem, growing up among tufts of mosses, but differing, 1st, by the leaves being simply emarginated, 2nd, by their margins being recurved, and so assuming, when moist, a convex and tumid appearance in front."—p. 117.

Resembling J. barbata, *Schreb.* and J. incisa, *Schrad.*, but differing from the former in the presence of stipules, in its more oblong calyx, and in its less concave, subsquarrose leaves, which are also less imbricated ; and from the latter, in its larger size and ascending stems, in the leaves being anteriorly rounded and entire, with their lower lacinia reflexed and their cells smaller, in the more tumid and less plicate calyx, in its squarrose perichætial leaves, in the leaves being more distant and paler in colour.

This species is described at considerable length, and illustrated by figures.

XVI. *Extracts from the Minute-Book of the Botanical Society, from November*, 1840, *to July*, 1841.

1840, *December* 10. Read, Extracts from a letter addressed to the Secretary, by the Rev. T. B. Bell, dated December 3, 1840, containing " Observations on the specific distinctions between Asplenium Ruta-muraria, and A. alternifolium."

Mr. Bell observes, — " I am aware some botanists have remarked, that attenuated *forms of* Asplenium Ruta-muraria approach indefinitely near A. alternifolium. I believe the two species have occasionally been confounded, but I always regarded this as a mistake into which no one could fall who had *perfect* specimens before him, and who was not prepared to substitute the general aspect and habit of the plants for their specific characters. As Mr. Newman, in his recent publication on Ferns, has fallen into this mistake, and conjoined the species, I think it not out of place to communicate to the Botanical Society the following brief observations : —

" The first character is taken from the form of the frond, which is correctly stated by Sir William Hooker to be bipinnate in A. Ruta-muraria, and, in alternifolium, pinnate, the lower pinna ternate ; the pinnæ in both being alternate. Now so far from its being the tendency of attenuated or contracted forms of A. Ruta-muraria to approach the pinnate form of alternifolium—the truth of the matter is, that the more attenuated the former is, the more distinctly bipinnate does it become, or in other words, the nearer A. Ruta-muraria approaches alternifolium in its general aspect and habit, the further and more visibly does it diverge in *this* character.

" The second character is taken from the indusium, with regard to which it is hardly necessary to remark, that while that of alternifolium has a smooth even edge, the edge in all varieties of Ruta-muraria is invariably jagged or uneven, and this is quite visible to the naked eye."—p. 119.

1841, *January* 14. Dr. E. F. Kelaart made some " Observations on the cultivation &c. of Cinnamomum zeylanicum."

In these observations the cinnamon plantations of Ceylon were described at considerable length, and the distinguishing qualities of the three principal varieties of cinnamon mentioned. He also alluded to the Malabar cinnamon, " which included several sorts in one parcel, but amongst which only a few pieces are equal to the finer qualities of Ceylon cinnamon." It was also remarked that the Cassia lignea of commerce is of three kinds ; — " that from China, sometimes called Chinese cinnamon ; that from the islands bordering on China ; and that from the continent of India." The former appears to be distinct from all the varieties of cinnamon, from which it differs in its texture, taste, colour, and other characters ; " the quills of cassia are made of single rolls of bark — those of cinnamon are distinguished by being composed of several, one within the other." The different products of cinnamon were described, and observations made on the botanical characters of the plants which yield the cinnamon and cassia of commerce; allusions were also made to the difficulties attending the investigation, caused by the contradictory statements of authors. The history of the various researches into the subject was adverted to, and Dr. K. concluded his observations with a brief description of the " several species of cinnamon growing in Ceylon, Java, and the Malabar coast; as also of the Cinnamomum aromaticum, or Laurus C ı of Nees Von Esenbeck."

Prof. Christison observed that for the last twenty-fi⁓⸍

had been imported from Ceylon under the name of cassia-bark, and stated his reasons for believing the greater part of it to come from Canton. He also made some remarks on the plants from which the bark is produced.

Read, A Communication from Mr. T. Edmonston, jun., Balta Sound, Shetland, " On the native dyes of the Shetland Islands."

After some remarks on the materials formerly used for dying in the Shetland Isles, the author observes that the colours now made are only *brown, red, yellow* and *black,* to produce which the following plants are used. *Brown ;* Parmelia saxatilis, called *Scrottyie.* *Red ;* Lecanora tartarea, called *Korkalett.* *Black;* Spiræa Ulmaria. *Yellow;* Stachys palustris, the die is called *Hundie :* Galium verum is said to be used for the same colour in the parish of North Mavin. Besides these, the juice of the berries of *Empetrum nigrum* furnish a beautiful purplish-blue. The mode of using these different materials was explained.

February 11. Read, Notice of the discovery of the cones of Pinus Mughus (*Jacq.*), in peat bogs in Ireland. By Mr. Charles C. Babington, Cambridge. The specimen sent to the Society was found under " six feet of solid peat bog at Burrishoole, near Newport, Mayo." Prof. Don was of opinion that it was a "cone of Pinus Mughus, *Jacq.,* which, however, he considered a variety of P. sylvestris, but quite different from any of the varieties now native in Scotland."

" It is interesting to find that a tree which must have formed at least a portion of the native forest of that wild part of Ireland, in which a tree is now scarcely to be found, should be thus proved to belong to a form of Pinus not now native in Britain, but confined, I believe, to the Austrian Alps. The native forests of that part of Ireland have now been totally destroyed for about two hundred years, one clause in the original grants to English settlers having required their destruction and employment in the smelting of iron. Professor Don states, that these cones agree exactly with others that he has seen from the bogs of Armagh."—p. 126.

March 25. Read, a Notice of the disappearance of plants from particular localities. By Mr. J. Just, Bury, Lancashire. The first plant mentioned is Lepidium Smithii, *Hook.*, of which, in the summer of 1840, the author could not find a single specimen in a locality where it had previously abounded, although there was no apparent cause for its disappearance. Orchis maculata and the white-flowered variety of Orchis mascula are other cases mentioned. The white-flowered varieties of Myosotis sylvatica and Geranium robertianum, were introduced into a garden and allowed to shed their seeds, the plants from which show a disposition to revert to the original colour.

April 8. Mr. James Mc'Nab made a communication "On some anomalous methods of cultivating plants in hot-houses." This communication, with several illustrative wood-cuts, appeared in the 'Gardeners' Chronicle' for August 14, 1841. The plants experimented on were Ficus elastica, Polypodium aureum, Acrostichum alcicorne, Euphorbia splendens and Bilbergia nudicaulis. The roots were first denuded of the soil; the plants of Ficus were suspended from the roof in various ways, with their roots entirely exposed; the Polypodium, Acrostichum and Euphorbia had their roots covered with moss, they were then suspended from the roof in an inverted position. All the plants thus treated grew well, and the ferns had a very handsome appearance.

July 8. Dr. Douglas-Maclagan read a notice of the Chemical Constitution of the Fruit of a species of Phytelephas, commonly called " Vegetable Ivory." This fruit is the product of a South American plant, and has lately been much used as a substitute for ivory. The seed is triangular, from 1½ to 2½ inches long and from 1 to 2 broad; its substance is hard, and closely resembling ivory in its physical characters. The specimen examined, on being cut across, was found to have a cavity in the centre, the walls of which were soft and yielded to the nail. A portion of the white matter, including part of both the soft and hard substance, was analysed by the action of cold and hot water, alcohol, and subsequent incineration. The constituents of the portion examined are then stated, with their several proportions; and the results of a subsequent analysis of the ashes are also given. The notice is concluded with the statement that " from these experiments it appeared that there was nothing in the chemical constitution of the vegetable ivory, which could account for its hardness, which must be due only to a peculiar texture in the woody fibre."—p. 131.

Art. XCVII. —*Varieties.*

222. *Note on Primulas.* I have no other observation to make on the common yellow Primulæ, than that in the district haunted by me, in canton Appenzell, varying from 3000 to 8000 feet above the sea, not a specimen of P. veris (cowslip), or P. vulgaris (primrose), was to be seen, or, by the most diligent enquiry, had ever been heard of, that I could find. P. elatior (oxlip) was everywhere over the meadows, peeping up among the very earliest, on the first warm patches cleared of snow in *February*, and continuing to gladden the eye, more or less sparingly, according to the altitude and protection, until all is again covered up with snow in *November*. P. veris abounded

in the valleys below up to 1200 or 1400 feet. P. vulgaris I never saw in that part of Switzerland.—*Wm. Bennett ; 21, King William Street, Sept.* 20, 1842.

223. *Economical Use of the Dock.* As a counterpart to Mr. Lees's statement of the domestic use of the fern in the forest of Dean (Phytol. 263), I beg leave to add the following similar use of the neglected dock in this neighbourhood. In summers when food for pigs is scarce or difficult to be procured, the *dernier resort* of the housewife is the dock, which is then gathered and boiled in the manner described in the before-mentioned statement ; sometimes mixed with a few potatoes or a little barley-meal, it forms a very nice mess to help through the time of scarcity. Another use of them to the farmer's wife is as wrappers for her butter in bringing it to market, the greater portion being wrapped up in dock-leaves after they are washed clean ; it is only after these have failed, on the approach of winter, that they begin to use the cloth wrappers. —*James Bladon ; Pont-y-pool, September* 14, 1842.

224. *Enquiry respecting Bryum pyriforme.* Perhaps some of your readers would be able to say if the plant described in 'English Flora' under the name of Bryum pyriforme be the same as the one figured in 'Muscologia Britannica' under the same name. A short time ago as I was examining a few specimens of what I thought to be Bryum pyriforme, I was greatly surprised to find them so much at variance with the description of that plant by Sir W. Hooker, in 'English Flora,' vol. v. p. 60, wherein he says—" This differs from all other Brya in the remarkable shape of its *leaves*, which are almost wholly composed of *nerve*, except at the base, and there deeply serrated."— The point in which my plants differ from the description given by Sir W. Hooker, consists in the leaves being quite *smooth* on their edges. On turning to the 'Muscologia Britannica,' I find the figure in that work to represent leaves which are smooth at the base and somewhat serrated at the point. It will be evident, on examination, that the description above quoted and the figure in 'Musc. Brit.' do present characters sufficiently distinct to justify the adoption of two species. — *Samuel Gibson ; Hebden Bridge, October* 15, 1842.

225. *Note on Equisetum variegatum, var. Mackaii.* I have received the following communication from Dr. Scouler, correcting some mistakes into which he supposes I have fallen respecting the history of the Equiseta. He states that the var. Mackaii, or " E. elongatum of Hooker, was recognised as a distinct species by Mr. Whitla of Belfast, but that this gentleman's views were not adopted by any Irish botanist until the beginning of the present year. The plant was first found by Mr. Whitla in Colin Glen, near Belfast ; afterwards, in 1833, he found it in the Deer-park, near Glenarne. The remarkable variety of E. variegatum growing on the margin of the canal near Dublin, was found by Mr. Johnson, a very acute botanist." — *Edward Newman ; 2, Hanover St., Peckham, October* 20, 1842.

226. *County Lists of British Ferns.* I most earnestly solicit from those botanists who may read this notice, assistance in forming a correct and complete county list of all the British species of ferns and allied genera. The plan I wish adopted is to give the name of every species that may occur, however common, because I regard it as quite possible that even such species as Asplenium Ruta-muraria, Pteris aquilina and Lastræa Filix-mas, may not occur in every county, and their non-occurrence would thus become known and a curious fact established. All observations and localities which the writers may consider of value are also solicited, especially with regard to the rarer species. It is my intention to use these lists as data for a general work ; each, if permission be given by the writer, will be handed over to 'The Phytologist,' for inser-

tion in its pages, prior to any other use being made of it. The subjoined list of Herefordshire ferns only, as far as I have had an opportunity of observing them, will perhaps show what is desired.

List of the Ferns of Herefordshire.

Lomaria Spicant. Abundant: Dinmore Hill, Shobden-hill woods, &c.

Pteris aquilina

Allosorus crispus. Malvern Hills, (*E. Lees*).

Polypodium vulgare

    Phegopteris. Aymestree quarry, Shobden-hill woods.

    Dryopteris. Ditto,   ditto.

Polystichum angulare

    aculeatum

    lobatum

Lastræa Oreopteris. West Hope, Dinmore, Aymestree quarry.

    Filix-mas

    dilatata

Lastræa spinulosa

Athyrium Filix-fœmina

    irriguum

Asplenium Adiantum-nigrum

    Ruta-muraria

    Trichomanes

Scolopendrium vulgare

Ceterach officinarum. Abundant: garden walls in the borough of Leominster.

Ophioglossum vulgatum

Cystopteris fragilis has been recorded as growing near Ludlow, but I am neither certain as to the locality nor the authority.

Each named variety might be recorded as a species, in order to save trouble: and where a species is recorded as having been found in a county, and the writer has not himself found it, the name of the finder or recorder might be added in parentheses. —*Id.*

ART. XCVIII.— *Proceedings of Societies.*

BOTANICAL SOCIETY OF LONDON.

October 7, 1842.—Adam Gerard, Esq. in the chair. Donations to the library were announced from G. Francis, Esq. British plants were received from the Rev. A. Bloxham, Dr. Bossey, Messrs. Fordham, Bidwell, Doubleday, Holman, &c.

Mr. D. Stock, of Bungay, Suffolk, presented monstrosities, collected by him at Earsham, Norfolk, of Scolopendrium vulgare bearing two fronds, the one being barren and reniform, the other bearing sori and elongated, with the midrib spirally twisted. Also of *Aspidium lobatum*, with the rachis much abbreviated and slightly recurved, pinnæ numerous and overlapping; and of two abortive specimens of a rose from his garden, both of which produced perfectly formed and leafy branches from the axis of the flowers. Mr. Stock also presented specimens of *Thelephora caryophyllea* (new to Great Britain), discovered by him in August, 1841, in a plantation at Bungay, Suffolk. This is distinct from *Thel. terrestris* (syn. *Auricularia caryophyllea*, Bulliard), and *Thel. laciniata* (syn. *Helvella caryophyllea*, Bolton, and *Auricularia caryophyllea*, Sowerby).

Mr. John Thompson exhibited specimens of *Carex irrigua*, Sm., collected by him at Muckle Moss, near Thorngrafton, Northumberland. Mr. Thomas Twining, jun., exhibited a large collection of cultivated specimens from Twickenham.

Read, the continuation of a paper from Mr. R. S. Hill, being "An Enquiry into Vegetable Morphology."

Irregular Metamorphoses of flowers are extremely common, and usually consist either of an actual multiplication of petals, or of the transformation of stamens and pistils into petals; the effect of these changes being the formation of double flowers, the impletion of which appears to take place in different ways in different plants. In most Icosandrous and Polyandrous plants, impletion appears to result almost entirely from the conversion of the stamens, and in some instances of the pistils, into petals; in the double varieties of *Ulex europæus* it results from the same change. In Oligandrous plants we usually find an actual multiplication of petals; as may be seen in the double stocks and wallflowers of our gardens. Where the impletion is the result of this alteration of the essential floral organs, the plants are necessarily barren. Such, however, is not the case with the Dahlia, Aster, and other plants which belong to the Corymbiferous section of Compositæ; in these the impletion results, first, from the change of the tubular florets of the disk into ligulate florets, the same as those of the ray, as in the Dahlia, and, secondly, by the simple enlargement and elongation of the tubular florets, as in the many varieties of the China Aster. Such monstrosities, from the fact of the essential organs not being in any way implicated, are capable of perfectly impregnating their ovules. Thus a knowledge of the mode in which impletion occurs, is of importance to the gardener, in order that he may be enabled to calculate on the possibility of producing new varieties by seed.

Dr. Lindley says that "these changes always occur in the order of development, or from the circumference to the centre; that is to say, that the calyx is transformed into petals, petals into stamens, and stamens into ovaria; but that the reverse does not take place." In proof of this hypothesis he further says " that if the metamorphosis took place from the centre to the circumference, or in a direction inverse to the order of development, it would not be easy to show the cause of the greater beauty of double flowers than of single; because the inevitable consequence of a reversed order of transformation would be that the rich or delicate colour of the petals, upon which all flowers depend for their beauty, would be converted into the uniform green of the calyx. Such a change therefore, instead of increasing the beauty of a flower, and making it superior to its original, would tend to destroy its beauty altogether." Now were this hypothesis correct, and founded on fact, what ought to be the condition in which we find the organs in double flowers? We ought surely to find the centre of the flower filled up with an increased number of pistils. But is this the case? It is plain it is not; indeed, were it the case, the beauty of a double flower would be most effectually destroyed. This theory must therefore fall to the ground, and we must confess that we are unable to find any laws by which the order of transmutation, in such monsters, is governed.

The aim and object with the cultivators of double flowers, is to convert all the floral organs into petals, and we generally refer to cultivation as the cause of flowers becoming double; farther than this we are ignorant of the causes of their impletion.— They probably owe their origin, at first, to accidental circumstances, and afterwards the variety is carefully propagated by the methods usually adopted for that purpose.

The two classes of vegetable functions, namely, the vegetative and reproductive, notwithstanding their close connexion, appear to be performed, in some degree, in opposition the one to the other; thus any excessive development of the one class, takes place at the expense of the other.—*T. S.*

THE PHYTOLOGIST.

No. XIX. | DECEMBER, MDCCCXLII. | PRICE 1s.

ART. XCIX. — *Notes on Botanical Excursions from Glasgow during the past Summer.* By J. H. BALFOUR, M.D., Regius Professor of Botany in the University of Glasgow.

DURING last summer I made botanical excursions every week with my pupils, and the results of some of these will, I hope, not be uninteresting to the readers of 'The Phytologist.'

My first trip was to Bowling, a small village on the banks of the Clyde, about ten miles west of Glasgow. The vegetation of the trap rocks in the neighbourhood is very luxuriant, but no plants of particular interest were gathered, except Glyphomitrion Daviesii. Our visit to Campsie Glen was more productive. Here my pupil Mr. Macleod picked Equisetum Drummondii, — the first time it had been seen in the neighbourhood of Glasgow. I have no doubt that this plant is abundant in many woods, and has often been mistaken for E. arvense. I picked it in profusion afterwards near the Falls of Clyde. Among other plants met with in or near the Glen I may notice Geum intermedium, a plant which is common in all the woods in this neighbourhood, Viola lutea, Campanula latifolia, Geranium lucidum, Cardamine amara, Stellaria nemorum, Lysimachia thyrsiflora, Polypodium Phegopteris, Vaccinium Oxycoccos and Buxbaumia aphylla. Carex pauciflora was also seen in marshy ground near Strathblane, at a much lower level than usual.

The wooded banks of the Clyde at Hamilton and Cadzow furnish an ample supply of plants for the student of Botany. Here we found Carex pendula, sylvatica and pallescens, Cardamine amara and Stellaria nemorum in great luxuriance. Veronica montana, Epipactis latifolia, Arum maculatum, Doronicum Pardalianches, Trollius europæus, Ornithogalum umbellatum, Euonymus europæus, Viburnum Opulus, Ribes alpinum, Ophioglossum vulgatum and Scolopendrium vulgare. Near Bothwell Castle and Blantyre Priory we saw Carex remota, Aquilegia vulgaris, Berberis vulgaris, Galium boreale, Geranium phæum and Allium vineale. On the banks of the Clyde some alpine plants occur at an elevation considerably below that at which they usually grow. Thus, near the Falls of Corra Lin there is considerable profusion of Saxifraga oppositifolia and Asplenium viride,—alpine plants, which, however, are not found on the mountains in the vicinity.

The facilities afforded by railway and steam enabled us to visit many interesting localities at a considerable distance from Glasgow. One of these was the island of Bute, famous for the mildness of its climate. We proceeded along the shore from Rothesay towards Mount-Stuart, and returned by a road across the island. In the course of our walk we gathered Cotyledon Umbilicus in great abundance, Anagallis tenella, Saxifraga aizoides, Pinguicula lusitanica, Alisma ranunculoides, Sinapis monensis, Veronica scutellata, Anthyllis Vulneraria, Alsine peploides, Scirpus pauciflorus, Carex paniculata, Poa maritima, Osmunda regalis and Scolopendrium vulgare.

Our next excursion was made in the neighbourhood of Ratho, a village about ten miles west from Edinburgh. We examined chiefly Ravelrig Bog and Dalmahoy Hills. In the woods at Dalmahoy we picked Cephalanthera ensifolia, Pyrola minor and Listera ovata ; and on Dalmahoy Hill, Saxifraga hypnoides, Viola lutea, Trientalis europæa and Epilobium angustifolium. In Ravelrig Bog and the woods near it, we saw Corallorhiza innata, Listera cordata, Potamogeton oblongus and Linnæa borealis. Before returning to Glasgow we paid a visit to Linlithgow, and in the loch near the palace we found profusion of Ceratophyllum submersum ? Potamogeton crispus and pusillus, Nasturtium terrestre and Poa aquatica.

The kindness of my friend John Smith Esq. LL.D., enabled us on another occasion to visit the romantic glens in the neighbourhood of his residence at Crutherland, about eleven miles south-east from Glasgow. On the wooded banks we met with Carex stricta in considerable quantity, Carex sylvatica and remota, Epilobium angustifolium, Milium effusum, Impatiens Noli-me-tangere, Cistopteris fragilis, Polypodium Dryopteris and Phegopteris. On the moors in the neighbourhood grow Drosera anglica, Vaccinium Oxycoccos, Veronica scutellata, Callitriche platycarpa, Rhynchospora alba, Lycopodium Selago and Splachnum mnioides. Near Langlands I gathered Peucedanum Ostruthium* and Lamium maculatum, var. β. *lævigatum*.

The Island of Arran is interesting in a botanical as well as a geological point of view, and the facilities afforded by the Ayr railway and

* This plant was also picked subsequently near Kilwinning in Ayrshire, by my pupil Mr. James C. Murray.

the steam-boats from Ardrossan, induced me to visit it. The party spent two days on the island, making Brodick their head-quarters. Our first walk extended from Brodick, along the shore to Lamlash ; then to Whiting Bay, Dippin Point and Kildonnan Castle, whence we returned to Brodick. During this walk we gathered many interesting plants, among which may be mentioned Sinapis monensis, Steenhammera maritima (the oyster-plant as it is called), Glaux maritima, Poa maritima, Ammophila arenaria, Juncus maritimus and Gerardi, Scirpus maritimus, Blysmus rufus, Aster Tripolium, Solidago Virgaurea, Plantago maritima and Coronopus, Anagallis tenella and arvensis, Habenaria bifolia and chlorantha, Gymnadenia conopsea, Ranunculus sceleratus, Althæa officinalis, Atriplex rosea, Bromus arvensis, Carex arenaria, extensa and lævigata, Œnanthe crocata and pimpinelloides, Potamogeton oblongus, Erythræa Centaurium (a variety with very narrow leaves, the β. *compressa* of some authors), and Lycopodium selaginoides. Between Lamlash and Whiting Bay there is profusion of Hypericum dubium (Leers) and Androsæmum, Cotyledon Umbilicus, Agrimonia Eupatoria, Eupatorium cannabinum, and Isolepis setacea. Near Dippin Point, where there are fine trap cliffs, we met with Convolvulus sepium, Verbascum Thapsus, Solanum Dulcamara, Conium maculatum, Ligusticum scoticum and Vicia sylvatica.

In our second day's walk we proceeded by the sea-shore towards the northern part of the island. We visited Glen Sannox, ascended Goatfell, and returned by Glen Rosa to Brodick. Along the shore we found many of the species picked on the previous day, and besides these we noticed Drosera anglica, Lycopus europæus, Polygonum lapathifolium, Osmunda regalis, Bidens tripartita, Corydalis claviculata. Near the Inn at Brodick there is profusion of a variety of Mentha rotundifolia, called *velutina* by Mr. Babington ; and here also we picked specimens of Pastinaca sativa. In Glen Sannox we searched in vain for Avena planiculmis, which was discovered there several years ago by Mr. Stewart Murray. The ascent of Goatfell from this glen is very steep and difficult, and the granitic debris is very unproductive. Goatfell itself, although attaining an elevation of about 2700 feet, does not yield many alpine plants. This may arise partly from its insular situation, and partly from the nature of the dry granitic rocks of which it is composed. The plants picked by the party were Saxifraga aizoides and stellaris, Salix herbacea, Sedum Rhodiola, Alchimilla alpina, Armeria maritima, var. β. *alpina*, Empetrum nigrum, Juniperus communis, var. β. *nana*, Carex pauciflora and rigida, Agrostis vulgaris, var. *β. pumila*, Festuca ovina, var. *ε. vivipara*, and Lycopodium Selago. On descending into Glen Rosa we picked Drosera anglica, Lythrum Salicaria, Molinia cærulea, var. *β. alpina*, Rhynchospora alba, Galeopsis versicolor, Habenaria bifolia and chlorantha and Vaccinium Vitis-Idæa. On the sea-shore near the inn we also gathered Sagina maritima and Eryngium maritimum.

With the view of examining the Flora of the alpine districts, we made an excursion in July to the mountains on the shores of Loch Lomond. Leaving Glasgow at 4, P.M., we reached Tarbet in the evening. Here we took up our residence for two days, and made excursions in the neighbourhood. On the first day we proceeded along the shores of Loch Lomond, as far as the Sloy water; we then ascended the stream to Loch Sloy, and thence commenced our ascent of Ben Voirlich, a high hill near the head of Loch Lomond. On the shores of the Loch we picked Osmunda regalis and Hymenophyllum Tunbridgense and Wilsoni, Hypericum Androsæmum and Betula alba and glutinosa. In Loch Sloy, Lobelia Dortmanna, Littorella lacustris and Subularia aquatica were seen. On Ben Voirlich the alpine plants collected were Alchimilla alpina, Silene acaulis, Sibbaldia procumbens, Cerastium alpinum, Juncus trifidus and triglumis, Saussurea alpina, Gnaphalium supinum, Hieracium alpinum, Saxifraga aizoides, stellaris, hypnoides and nivalis, Carex rigida, Polystichum Lonchitis, Sedum Rhodiola, Polygonum viviparum, Poa alpina, and a new Poa, called by Dr. Parnell P. Balfouri, Aira alpina and var. *vivipara*, Festuca ovina, var. *ε. vivipara*, Epilobium alpinum and Thalictrum alpinum. Carex saxatilis was gathered in great profusion on the descent of the hill towards the head of Loch Lomond. On the second day Ben Lomond was visited, and on it, besides most of the alpine plants already noticed on Ben Voirlich, we saw Rubus suberectus and Chamæmorus, Scutellaria galericulata, Salix arenaria and Asplenium viride.

On the shores of Loch Lomond we particularly remarked the great quantity of fruit produced this season, by the ordinary trees, as birch, beech, oak, alder, hazel, apple, holly, &c.

J. H. BALFOUR.

Glasgow, November 11, 1842.

Art. C. — *Additions to the Phænogamic Flora of ten miles round Edinburgh.* By Thomas Edmonston, Esq.

Baltasound, Shetland Islands,
November 1, 1842.

Sir,

In the course of my botanising this season through the peculiarly rich and interesting district within ten miles of Edinburgh, having observed several plants not noted in the last List of the Flora of that district, I beg to hand you the catalogue of them, with habitats &c.; and also notes on a few of the rarer or more interesting species previously observed. If you should consider this trifling contribution to our knowledge of local Botany worthy of insertion in your valuable periodical, it is at your service.

I am, Sir,
Your very obedient Servant,
Thos. Edmonston.

To the Editor of 'The Phytologist.'

————

Bidens cernua, β. *radiata.* Lochend.

Bromus velutinus. Some dwarf specimens of the variety β. occurred near Musselburgh.

Callitriche platycarpa. Compensation pond on the Pentland hills, Lochend and Duddingston Lochs, &c. Afterwards observed in the former station by Mr. Babington, who assures me it is the true plant. It has not, I think, been previously noticed out of England.

Carex divisa. Pentland hills, scarce.

 „ *incurva.* Musselburgh and Dalmeney. I also, late in the season, picked what I believe was this plant near St. David's, on the Fife side of the Firth, but it was too far gone to be identified properly. This species is probably often passed over as C. arenaria.

Cerastium holosteoides. The C. triviale, var. β. *holosteoides*, of the Edinburgh Botanical Society's Catalogue (2nd edit.) seems a distinct plant, as I shall soon endeavour to prove. The Edinburgh specimens agree with some from Kinfauns in Perthshire, exhibited at the Botanical Society last winter. Dalkeith.

Cerastium (tomentosum ?). Hills behind Aberdour.

Dianthus glaucus, Linn. (D. deltoides, β. *glaucus*, Hook.) Left unmarked in the Edinburgh Catalogue, but abundant in the King's Park. I should be disposed to consider it distinct.

Festuca ovina, ε. *vivipara.* Pentland hills, above Collinton. It appears to me that some confusion exists with regard to the species called F. vivipara by Smith. The plant figured by that author [Eng. Bot. 1355], and abundant on all our more elevated hills, especially those of a micaceous character, seems to be distinct from the common F. ovina of our plains. It is not improbable that the alpine plant is the viviparous state of a distinct grass, which, from the great disposition of alpine grasses to assume the viviparous character, is seldom seen in its normal form, and is therefore sunk into F. ovina. The study of the alpine plant would repay any one conveniently situated for that purpose. I may notice, *en passant*, the curious tendency of alpine grasses, such as Aira alpina, Poa alpina, &c., to become viviparous. It would almost appear an effort of nature to make up for the paucity of seeds matured in such situations.

Pimpinella magna. I am not aware of this having been detected previously in Scotland. It occurs abundantly by the hedge-sides, shortly after leaving the village of Collinton, and proceeding westwards; probably merely a luxuriant variety of P. Saxifraga. The Collinton specimens agree perfectly with those from the Isle of Wight and numerous others in my collection.

Polygonum Raii. This is surely often confounded with the maritime variety of P. aviculare (I presume " P. maritimum, L." of the Edinburgh Catalogue), for the true plant occurs abundantly on both sides of the Firth of Forth. This is surely not the P. Roberti of the continental botanists, of which I possess specimens from the shores of the Adriatic, &c.

Ranunculus fluitans, Lam. Though now raised to the rank of a species, I cannot believe this to be anything more than one of the multifarious varieties of R. aquatilis; the differences merely lie in the very peculiar habit. It is sometimes two or three feet long, but this is occasioned by its growing in deeper water; it has also a curious tufted and sometimes almost geniculated appearance. Lochend.

Rubus nitidus, rhamnifolius, suberectus. Craigmillar Castle. I doubt not that many more of the reputed species of bramble may be found in this neighbourhood, and those fond of multiplying specific distinctions in this troublesome and intricate genus, will have abundant opportunity about the above-named station, the Peebles road, Dalkeith, Craigleith, and numerous other localities near Edinburgh. I will not attempt to name any but the above three, and I must confess, from my mite of experience, that I do think, without the slightest danger of confounding distinct plants, the arrangement of Koch might be followed in distinguishing nearly all our British species.

Salvia Sclarea. Armiston and Dalkeith; in all probability the outcast of a garden.

Sambucus nigra, β. laciniata. Collinton, Auchindenny and Dalhousie woods.

Saxifraga platypetala. Abundant at Habbie's How, Pentland hills. Whether this is only one of the forms of S. hypnoides, or a distinct species, seems still doubtful. The flowers are as large as, and have quite the appearance of, those of S. granulata, and the habit is more lax and elongated than in the common form.

Viola lactea. Arthur's Seat. A variety of V. canina, as I have proved by seeing every intermediate stage.

SPECIES PREVIOUSLY OBSERVED IN THE DISTRICT.

Adonis autumnalis. Fields near Comely bank; probably introduced.

Alopecurus agrestis. Preston pans. Beautiful specimens and very abundant in this station.

Arctium Bardana. This is again restored to the rank of a species, but I am unable to see permanent characters, and I must confess it seems to run into A. Lappa. What I believe to be an intermediate state, is the most common burdock about Edinburgh.

Bromus madritensis. Grange toll.

Barbarea præcox. Roslin, Currie, &c., not uncommon. The small flowers and large radical leaves of this plant are very constant in all the specimens I have seen.

Cardamine sylvatica. Very abundant, but appearing to run into C. hirsuta. In habit it combines the herbage of C. amara with the flowers of C. hirsuta, and is apparently a luxuriant variety of the latter.

Carex axillaris. Said to be found near Craigmillar Castle, but I have been unable to detect it there, and specimens I have seen under this name from that station, were varieties of C. remota.

Carex vesicaria. Duddingston Loch.

Cichorium Intybus. A rare Scottish plant, but very abundant and certainly indigenous near Granton, on the green slope on both sides of the road between the pier and Wardie.

Cochlearia danica. Abundant on both sides of the Firth.

Doronicum Pardalianches. Dalkeith, Hunter's tryste, &c. D. plantagineum is said to occur, but I cannot detect permanent characters.

Eranthis hyemalis. Craigmillar Castle.

THOS. EDMONSTON.

(To be continued).

ART. CI.—*A Flora of the Neighbourhood of Saffron Walden, Essex.*
By G. S. GIBSON, Esq.

Clematis Vitalba. Common in hedges.

Thalictrum minus. Near Linton; rare.

Anemone nemorosa. Woods and thickets, abundant.

———— *Pulsatilla.* Bartlow Hills and Hildersham.

Adonis autumnalis. Corn-fields, but rarely

Ranunculus aquatilis. Ponds and ditches, plentifully.

———— *Lingua.* At Sawston; rare.

———— *Flammula.* Marshy places, common.

———— *Ficaria.* A common weed.

———— *auricomus.* Woods &c., not uncommon.

———— *sceleratus.* Watery places, occasionally.

———— *acris, repens,* and *bulbosus.* Very common weeds.

———— *arvensis.* Corn-fields.

Helleborus fœtidus. Woods, rather unc.

Caltha palustris. Ditches &c., frequent.

Delphinium Consolida. Corn-fields, occasionally.

Berberis vulgaris. Common in hedges, though often eradicated by farmers, who imagine that it causes blight on wheat around.

Nymphæa alba. River at Audley End; perhaps planted.

Nuphar lutea. River &c. in several places

Papaver Argemone. Fields & hedge-banks occasionally.

———— *hybridum.* Once found in a field near the town.

———— *dubium.* Waste ground, frequent

———— *Rhœas.* A troublesome weed.

Chelidonium majus. Waste ground near the town, rather rare.

Fumaria officinalis. Fields and gardens, abundant.

———— *parviflora.* Cultivated fields at Littlebury.

Corydalis solida. In the Park, scarcely wild

Cheiranthus Cheiri. Park-walls, &c.

Nasturtium officinale. Ditches &c. abndt.

Barbarea vulgaris. Moist meadows, freqt.

Cardamine pratensis. Marshy ground, co.

———— *hirsuta.* A weed in gardens, &c. but local.

Draba verna. Walls, plentiful.

Thlaspi arvense. A weed in fields and gardens, though very local.

Sisymbrium officinale. Road-sides &c.fr.

———— *Sophia.* Waste ground, rare

Erysimum cheiranthoides. Walls and cultivated ground.

———— *Alliaria.* Hedges, frequent.

Camelina sativa. Cultivated fields, rare.

Coronopus Ruellii. Waste ground, not common

Capsella Bursa-pastoris. Abundant.

Brassica Napus. Waste ground, frequent

———— *Rapa.* Road-sides, occasionally.

Sinapis alba. Cultivated ground.

———— *arvensis & nigra.* Common weeds

Raphanus Raphanistrum. Ditto.

Reseda Luteola. Dry banks, not uncomn.

———— *lutea.* Chalky fields in several pl.

Helianthemum vulgare. Open banks, freq.

Viola hirta. Shady hedges, frequent.

———— *odorata* and *canina.* Abundant.

———— *tricolor.* A common weed.

Polygala vulgaris. Dry banks, not uncom.

Dianthus Armeria. Twice found in a field

———— *deltoides.* Hildersham, rare.

Saponaria officinalis. Moist meadows nr. Ickleton.

Silene inflata. Very common.

———— *noctiflora.* Corn-fields &c. occas.

———— *Armeria.* Waste ground, rare, the outcast of gardens.

Lychnis dioica (vespertina). Very common

———— ———— *(diurna).* Far less common

———— *Flos-cuculi.* Moist shady places

Agrostemma Githago. Corn-fields, freqnt.

Sagina procumbens and *apetala.* Walls & dry ground, plentiful.

Spergula arvensis. Fields occasionally.
———— *nodosa.* Meadows at Ickleton.
Stellaria media. Abundant everywhere.
———— *holostea.* Hedges, common.
———— *graminea.* At Linton.
Arenaria trinervis. Woods & shady hedges
———— *serpyllifolia.* Walls &c. common
———— *tenuifolia.* Walls at Audley End, rare.
Cerastium vulgatum. Not uncommon.
———— *viscosum* and *semidecandrum.* Fields and dry ground, common.
———— *arvense.* Dry chalky banks; Bartlow Hills, &c.
———— *aquaticum.* Sawston.
Linum usitatissimum. Cultivated fields, occasionally.
——— *catharticum.* Dry banks, common
Malva sylvestris and *rotundifolia.* Fields and roadsides, common.
——— *moschata.* Hedges in various places
Tilia europæa. Hedges &c. probably pl.
Hypericum Androsæmum. Near Newport, rare.
———— *perforatum.* Abundant.
———— *quadrangulum.* Watery places frequent.
———— *hirsutum.* Chalky banks, freq.
Parnassia palustris. Marshy meadows at Ashdon and Chesterford, rather rare
Acer campestre. Common in hedges.
——— *Pseudo-platanus.* Hedges, scarcely wild.
Geranium sanguineum. Thickets near Newport, rare.
———— *pratense.* Shortgrove Park, rare
———— *lucidum.* Old walls occasionally
———— *robertianum.* Hedges, common
———— *molle* and *dissectum.* Abundant
———— *pusillum.* Not uncommon.
Erodium cicutarium. Banks occasionally
Oxalis Acetosella. Woods, not very com.
Euonymus europæus. Hedges and groves
Rhamnus catharticus. Not rare in hedges
Ulex europæus. Barren ground, not very abundant.
——— *nanus.* Triplow, rather rare.
Genista tinctoria. Banks in several places

Cytisus Scoparius. Near Newport, not common.
Ononis arvensis. Borders of fields, plentif.
Anthyllis Vulneraria. Chalky banks, co.
Medicago lupulina. Abundant.
———— *sativa.* Cultivated ground, not wild.
Melilotus officinalis. Fields and hedges, sparingly.
Trifolium repens and *pratense.* Very com.
———— *ochroleucum.* Fields & hedges, not rare.
———— *medium.* Moist meadows in several places.
———— *scabrum.* Dry banks at Hildersham.
———— *fragiferum.* Road-sides and moist meadows.
———— *procumbens* and *minus.* Dry pastures &c. common.
———— *filiforme.* Near Hildersham &c.
Lotus corniculatus. Very common.
——— *major.* Watery places, not rare.
Astragalus glycyphyllos. Borders of fields rather sparingly.
Hippocrepis comosa. Dry chalky banks.
Onobrychis sativa. Cultivated fields and banks, frequent.
Vicia Cracca, sativa and *sepium.* Fields and hedges, common.
Ervum hirsutum & *tetraspermum.* Woods and shady hedges.
Lathyrus Aphaca. Borders of fields in several places.
——— *Nissolia.* Woods &c. but rare.
——— *pratensis.* Common.
——— *sylvestris.* Woods and hedges in several places.
——— *latifolius.* Woods &c. scarcely wild.
Prunus domestica. Hedges occasionally.
——— *insititia.* Not uncommon.
——— *spinosa.* Abundant.
——— *Cerasus.* Woods and hedges near Quendon, &c.
——— *Padus.* Woods, rare.
Spiræa Filipendula. Bartlow Hills, Newport, &c.

Spiræa Ulmaria. Watery places, com.

Geum urbanum. Freqnt. in shady hedges

Rubus rhamnifolius. Rare about Walden

—— *fruticosus* and *cæsius.* Very freqnt

—— *corylifolius.* Widdington, &c.

Fragaria vesca. Shady hedges and woods

Potentilla anserina and *reptans.* Very co.

————— *argentea.* Near Hildersham.

————— *Fragariastrum.* Woods &c. fre.

Tormentilla officinalis. Not very common

Agrimonia Eupatoria. Borders of flds. fre.

Alchimilla vulgaris. Waste grnd. very ra.

——— *arvensis.* Dry fields, common

Poterium Sanguisorba. Pastures and banks, frequent.

Rosa tomentosa. Woods and hedges in several places.

—— *rubiginosa.* Near Chesterford &c., not rare.

—— *canina, sarmentacea* and *arvensis.* Common in hedges.

Cratægus Oxyacantha. Abundant.

Pyrus communis. Woods and hedges, rather rare.

—— *Malus.* Not uncommon.

—— *aucuparia.* Woods at Debden, ra.

Epilobium angustifolium. Shortgrove Park, not common.

——— *hirsutum* and *parviflorum.* Ditches &c. very common.

——— *montanum.* Waste ground & walls, plentiful.

——— *tetragonum.* Watery places, occasionally.

Œnothera biennis. Waste ground, rare and not wild.

Circæa Lutetiana. Gardens and shady pl.

Hippuris vulgaris. River at Audley End.

Myriophyllum spicatum. Ponds, not co.

Callitriche verna. Very common.

——— *autumnalis.* In the Park &c. not rare.

Lythrum Salicaria. Ditches, frequent.

Bryonia dioica. A common weed.

Scleranthus annuus. Dry fields at Linton, sparingly.

Sempervivum tectorum. On many roofs and walls.

Sedum Telephium. Hedge-banks, rather uncommon.

—— *dasyphyllum.* On garden walls, scarcely wild.

—— *album.* Roofs, but rare.

—— *acre.* Dry banks at Linton.

—— *reflexum.* Roofs, but rare.

Ribes rubrum and *Grossularia.* Hedges occasionally.

Saxifraga granulata. At Hildersham &c.

——— *tridactylites.* Walls &c. not uncommon.

Hydrocotyle vulgaris. About Linton.

Sanicula europæa. Woods and thickets.

Conium maculatum. Road-sides, common.

Petroselinum segetum. Fields and hedge-banks, very rare.

Helosciadium nodiflorum. Common.

Sison Amomum. Borders of fields, freqt.

Ægopodium Podagraria. A troublesome weed.

Pimpinella Saxifraga. Dry pastures, ab.

——— *magna.* Shady places at Little Walden.

Sium angustifolium. Ditches, not very co.

Bupleurum rotundifolium. Corn-fields at Linton &c. sparingly.

Œnanthe fistulosa and *Phellandrium.* River at Littlebury.

Æthusa Cynapium. A common weed.

Silaus pratensis. Borders of fields, not unc.

Angelica sylvestris. Moist woods, freqnt.

Pastinaca sativa. A troublesome weed.

Heracleum Sphondylium, Daucus Carota, and *Torilis Anthriscus.* Common.

Torilis infesta. Corn-fields and road-sides, not uncommon.

—— *nodosa.* Hedge-banks, frequent.

Scandix Pecten and *Anthriscus sylvestris.* Abundant.

Chærophyllum temulentum. Fields, freqnt.

Adoxa Moschatellina. Woods, rare; Peverell's &c.

Hedera Helix. Plentiful.

Cornus sanguinea. Hedges &c. common.

Viscum album. Orchards at Debden.

Sambucus Ebulus. Moist meadows at Wenden, Bartlow, &c.

Sambucus nigra. Abundant.

Viburnum Opulus. Not rare.

———— *Lantana.* Hedges, frequent.

Lonicera Periclymenum. Very common.

———— *Xylosteum.* At Littlebury; but doubtful if wild.

Galium verum and *Mollugo.* Common.

———— *cruciatum.* Littlebury.

———— *palustre.* Watery places, very fr.

———— *saxatile.* At Audley End.

———— *tricorne.* Corn-fields at Linton and Widdington.

———— *Aparine.* Very abundant.

Sherardia arvensis. Fields, frequent.

Asperula odorata. Woods in several places

———— *cynanchica.* Dry chalky banks, plentiful.

Valeriana dioica. Marshy pl. occasionally

———— *officinalis.* Frequent.

Fedia olitoria. Corn-fields.

———— *dentata.* At Bartlow and Linton.

Dipsacus sylvestris. Road-sides, common.

———— *pilosus.* Clavering &c. but rare

Scabiosa succisa. Pastures in various parts

———— *columbaria.* Dry banks, common

Knautia arvensis. Plentiful.

Tragopogon pratensis. Pastures &c. freqt.

Helminthia echioides. Borders of fields, not rare.

Picris hieracioides. Road-sides, Little Walden, &c.

Apargia hispida, A. autumnalis, Hypochæris radicata & *Crepis virens.* Very com.

Sonchus arvensis. Frequent in fields &c.

———— *oleraceus* and *Leontodon Taraxacum.* Everywhere.

Hieracium Pilosella. Com. in dry pastures

Lapsana communis. A common weed.

Cichorium Intybus. Road-sides frequent

Arctium Lappa and *Bardana.* Woods and hedges, common.

Serratula tinctoria Borders of woods, ra.

Carduus nutans and *acanthoides.* Hedges

———— *marianus.* Hedges occasionally; Widdington, &c.

Cnicus lanceolatus and *arvensis.* Abundant

———— *palustris.* Moist meadows, frequent, and in woods.

Cnicus eriophorus. Banks &c. in several places about Walden.

———— *pratensis.* In pastures, rare.

———— *acaulis.* Dry banks and pastures, common.

Onopordum Acanthium. Road-sides, not r.

Carlina vulgaris. Dry banks, not common

Bidens tripartita. Ponds, but rare.

———— *cernua.* Pond at Ashdon.

Eupatorium cannabinum. Moist woods fr.

Tanacetum vulgare. In the Park, rather r.

Artemisia Absinthium. Road-sides, sparingly.

———— *vulgaris.* Very common.

Gnaphalium sylvaticum. Wood at Widdington, rare.

———— *uliginosum.* Moist rd-sds. &c.

———— *germanicum.* Dry pastures, frequent.

Petasites vulgaris. Mead. at Audley End.

Tussilago Farfara. Very common.

Erigeron acris. Near Littlebury, rare.

Senecio vulgaris & *Jacobæa.* Very abndnt.

———— *sylvaticus.* Near Hildersham, ra.

———— *aquaticus.* Ditches, common.

Inula Conyza. Woods, not frequent.

Pulicaria dysenterica. Common.

Doronicum Pardalianches. At Widdington, very rare.

Bellis perennis. Everywhere abundant.

Chrysanthemum Leucanthemum. Common

———— *segetum.* Fields &c., rather rare.

Pyrethrum Parthenium. Waste gr. occas.

———— *inodorum, Anthemis Cotula* & *Achillæa Millefolium.* Common.

Centaurea nigra. Abundant.

———— *Scabiosa.* Hedges, frequent.

———— *Cyanus.* Corn-flds. not uncom.

Campanula rotundifolia. Dry banks, co.

———— *Trachelium.* Hedges at Wicken, rare.

———— *glomerata.* Open chalky pastures, frequent.

———— *hybrida.* Corn-flds. not uncom.

Calluna vulgaris. Ickleton Grange, rare.

Monotropa Hypopitys. Plantation near Debden.

Ilex Aquifolium. Hedges at Newport and Quendon.

Ligustrum vulgare. Hedges in several pl.

Fraxinus excelsior. Common.

Vinca minor. Hedge-banks in several pl.

—— *major.* Shady banks, but rare.

Erythræa Centaurium. Pastures &c. occasionally.

Gentiana Amarella. Dry banks, rare.

Chlora perfoliata. Banks and ditches occasionally, not common.

Menyanthes trifoliata. Moist meadows at Ickleton.

Convolvulus arvensis. Very common.

—— *sepium.* Gardens and moist hedges, frequent.

Cuscuta europæa. Rare.

—— *Epithymum ?** On clover, but rare

Echium vulgare. Dry pastures &c. not pl.

Lithospermum officinale. Plantations at Debden.

—————— *arvense.* Waste gr. occas.

Symphytum officinale. Watery places, fr.

Borago officinalis. Waste gr. occasionally

Lycopsis arvensis. Sandy ground about Linton.

Myosotis palustris. Common.

—— *cæspitosa.* Audley End &c. not r.

—— *arvensis.* Abundant.

Cynoglossum officinale. Road-sides, not very common.

Datura Stramonium. Waste gr. very rare.

Hyoscyamus niger. Road-sides and banks occasionally.

Solanum Dulcamara. Hedges and thickets, frequent.

—— *nigrum.* Gardens and rich waste ground.

Orobanche elatior. Hedge banks, not rare.

—————— *minor.* Clover-fields, rare.

Veronica serpyllifolia, Chamædrys, hederifolia, agrestis, and *arvensis.* Waste ground, fields, &c. abundant

—— *Anagallis* and *Beccabunga.* Frequent in watery places.

Veronica officinalis. Wood at Widdington, rare.

Bartsia Odontites. Corn-fields &c. freqnt.

Euphrasia officinalis. Dry banks.

Rhinanthus Crista-galli. Moist meadows

Melampyrum cristatum. Common in many groves and shady hedges.

Pedicularis palustris. Meadows at Chesterford.

Antirrhinum majus. Garden-walls, naturalized.

Linaria Cymbalaria. Old walls &c.

—————— *spuria.* Corn-fields, not uncomn.

—————— *Elatine.* Less common than the preceding.

—————— *vulgaris.* Common.

—————— *minor.* Fields and waste ground sparingly.

Scrophularia nodosa. Woods and hedges, not uncommon.

—————— *aquatica.* Ditches, common.

—————— *vernalis.* Hedges at Hempsted.

Digitalis purpurea. Near Quendon, but very rare.

Verbascum Thapsus and *nigrum.* Banks and fields, frequent.

Lycopus europæus. Ditches & ponds, occ.

Salvia Verbenaca. Dry banks not uncom.

Mentha sylvestris. Watery places in the Park &c. rare.

—— *viridis.* At Wenden.

—— *gracilis.* Audley End, rare.

—— *hirsuta.* Ditches, common.

Thymus Serpyllum. Dry banks, common

Teucrium Scorodonia. Woods at Quendon, rare.

Origanum vulgare. Common.

Ajuga reptans. Woods, frequent.

Ballota nigra. Very common.

Galeobdolon luteum. Borders of woods, but scarce.

Galeopsis Ladanum and *Tetrahit.* Cornfields &c. not uncommon.

Lamium album & *purpureum.* Very com.

---

* The specimens sent by Mr. Gibson have the calyx nearly as long as the corolla, and the scales in the throat of the latter appear to be more deeply cut than those in C. *Epithymum,* but in dried specimens they cannot be satisfactorily examined.—*Ed.*

Lamium amplexicaule. Waste gr. occas.

—— *incisum.* Rare in waste ground.

Betonica officinalis. Woods in several pl.

Stachys sylvatica. Shady hedges, common

—— *palustris.* At Chesterford, &c.

—— *arvensis.* Corn-fields, not common

Nepeta Cataria. Road-sides in various prts.

Glechoma hederacea. Abundant.

Acinos vulgaris. Cultivated fields, occas.

Calamintha officinalis. Rare; near Chesterford.

—— *Nepeta.* Chalky banks freqt.

Clinopodium vulgare. Borders of fields, common.

Prunella vulgaris. Pastures, common.

Scutellaria galericulata. Ditches in the Park, &c.

Verbena officinalis. Hedge-banks and near houses, frequent.

Anagallis arvensis. Abundant.

—— *tenella.* At Sawston, not comn.

Lysimachia Nummularia. Moist sha. pl.

Hottonia palustris. At Sawston, rare.

Primula vulgaris. Woods, abundant.

—— *elatior.* Not rare in woods.

—— *veris.* Pastures, very abundant.

Plantago major. Borders of fields, comn.

—— *media.* Too abundant.

—— *lanceolata.* Very common.

Amaranthus Blitum. Once found on a dunghill.

Chenopodium Bonus-Henricus. Roadsides &c. common.

—— *rubrum?* and *murale.* Cultivated ground, but rare.

—— *album.* Very common.

—— *ficifolium.* Cultivated and waste ground.

Atriplex patula and *angustifolia.* Fields &c. not uncommon.

Polygonum Bistorta. Meadows near Quendon, rare.

—— *aviculare.* Very common.

—— *Fagopyrum.* Cultivated fields, not wild.

—— *Convolvulus.* Corn-fields &c. frequent.

—— *amphibium.* Ponds, not unc.

Polygonum Persicaria and *lapathifolium.* Waste ground and moist road-sides, common.

—— *Hydropiper.* On mud-heaps.

Rumex Hydrolapathum. River-side at Ickleton.

—— *crispus* and *obtusifolius.* Troublesome weeds.

—— *acutus.* Moist shady places.

—— *pulcher.* Dry banks, not uncomn.

—— *Acetosa.* Very common.

—— *Acetosella.* Dry pastures in several places.

Daphne Mezereum. Hedges occasionally, but very rare.

—— *Laureola.* Woods and groves, ra.

Thesium linophyllum. Open banks at Hildersham.

Mercurialis perennis. Common in hedges and woods.

—— *annuus.* A weed in gardens, but local.

Euphorbia helioscopia and *Peplus.* Common weeds.

—— *platyphylla.* Corn-fields, v. ra.

—— *Cyparissias.* Plantations at Audley End.

—— *exigua.* Corn-fields &c. freq.

—— *Lathyris.* Rich waste ground, occasionally.

—— *amygdaloides.* Shady hedges and groves.

Urtica dioica and *urens.* Too com. weeds

Parietaria officinalis. Under walls &c.

Humulus Lupulus. Hedges, not uncom.

Ulmus campestris, suberosa and *glabra.* Frequent in hedges.

—— *montana.* Hedges, not very comn.

Betula alba. Hedges and woods occasio.

Alnus glutinosa. Moist woods &c. freqt.

Salix Helix. At Chesterford.

—— *triandra* and *Hoffmanniana?* Osier ground.

—— *amygdalina.* Hedges at Sewers End.

—— *fragilis.* Frequent.

—— *Ruselliana.* Audley End.

—— *alba.* Very common.

—— *vitellina.* Hedges in several places.

Salix viminalis. Osier-ground, &c.

—— *cinerea, aquatica* and *oleifolia.* Hedges &c. frequent.

—— *caprea.* Hedges, not uncommon.

Populus tremula. Woods, frequent.

—— *nigra.* Banks of rivers &c. In several places.

Fagus sylvatica. Woods and hedges, oc.

Quercus Robur. Abundant.

—— *sessiliflora.* Frequent.

Corylus Avellana. Very common.

Carpinus Betulus. Woods, frequent.

Juniperus communis. Rare; (once grew at Hadstock).

Alisma Plantago. Ditches and ponds, fr.

Sagittaria sagittifolia. At Chesterford, rather rare.

Butomus umbellatus. Near Sampford, ra.

Arum maculatum. Hedges, common.

Typha latifolia. About Duxford.

Sparganium ramosum. Ditches and ponds frequent.

Lemna trisulca. Ditches, rare.

—— *minor.* Abundant.

Potamogeton densus. Pond at Audley End

————— *crispus.* Ponds in several pl.

————— *natans.* Common.

Zannichellia palustris. Pond at Audley End, rare.

Paris quadrifolia. Hales wood and other places.

Fritillaria Meleagris. Meadows near Bumpsted, rare.

Allium vineale. At Hildersham, sparingly

Ornithogalum umbellatum. Hinxton.

Hyacinthus non-scriptus. Groves and thickets, common.

Juncus glaucus, effusus and *conglomeratus.* Watery places, frequent.

—— *acutiflorus* and *obtusiflorus.* Linton, not common.

—— *lampocarpus.* Marshy places, com.

—— *compressus.* Marshes, occasionally

—— *bufonius.* Road-sides, common.

Luzula sylvatica. Woods, common.

—— *Forsteri.* Not rare in woods.

—— *campestris.* Dry pastures, freqnt.

Neottia spiralis. Near Linton, rare.

Listera ovata. Woods in several places.

—— *Nidus-avis.* Moist woods, very r.

Epipactis latifolia. At Debden, &c.

—— *grandiflora.* Shortgrove Park, &c. sparingly.

Orchis Morio. Pastures, not uncommon.

—— *conopsea.* Clay pastures, rare.

—— *mascula.* Woods &c. plentiful.

—— *ustulata.* At Linton, rare.

—— *pyramidalis.* Chalky banks, not un.

—— *latifolia.* Marshy ground.

—— *maculata.* Common.

Habenaria viridis. At Wimbush.

—— *bifolia.* Woods at Debden, &c

Aceras anthropophora. Very rare at Linton

Ophrys apifera. Pastures &c. in sev. pl.

—— *aranifera.* Dry open chalky banks near Hildersham, very sparingly.

—— *muscifera.* Woods at Widdington &c. rare.

Iris Pseudacorus. Osier ground, not very common.

— *fœtidissima.* Woods, but rare.

Crocus sativus. This plant, formerly cultivated here to so great an extent as to form the principal trade of the town, from which its name "Saffron" is derived, had been long entirely extinct till a few years ago, when some plants came up in newly trenched ground; these have however nearly disappeared again. The cultivation of it has ceased for about half a century; and the fact of its having so soon become extinct clearly demonstrates that it has but little claim to be included among our native or even naturalized plants.

Narcissus biflorus. Once found near Little Walden.

———— *Pseudo-Narcissus.* Meadows near Quendon, rare.

Galanthus nivalis. Old orchard, Littlebury

Tamus communis. Hedges, not uncomn.

Scirpus lacustris. Pond at Audley End.

—— *setaceus* and *Eleocharis palustris.* At Audley End.

Eleocharis multicaulis. At Ickleton.

Eriophorum polystachion and *angustifolium*. Ickleton and Sawston.

Carex stellulata. Near Bartlow &c.

—— *remota, intermedia* and *divulsa*. Moist meadows, at Chesterford &c.

—— *vulpina*. Ditches, common.

—— *paniculata*. In the park &c.

—— *acuta*. At Chesterford.

—— *sylvatica*. Woods, not uncommon.

—— *pendula*. Shady hedges in sev. pl.

—— *recurva*. Common.

—— *præcox*. Dry banks at Hildersham.

—— *pilulifera*. At Chesterford.

—— *hirta*. Woods and hedges, freqnt

—— *ampullacea*. Chesterford.

—— *paludosa*. Road-sides.

—— *riparia*. Osier-ground &c. comn.

—— *panicea*. At Ickleton.

Anthoxanthum odoratum and *Alopecurus pratensis*. Plentiful.

Alopecurus agrestis. Fields and road-sides, frequent.

—————— *geniculatus*. Banks of ponds, not uncommon.

Phleum pratense. Abundant.

Milium effusum. Woods, occasionally.

Calamagrostis lanceolata. Moist woods, rather rare.

Agrostis canina. Marshy ground, comn.

—— *vulgaris* and *alba*. Hedge-banks and moist meadows, common.

Catabrosa aquatica. Ditches in the Park.

Aira cristata. Dry chalky banks, freqnt.

—— *cæspitosa*. Shady places, frequent.

—— *caryophyllea*. Dry banks &c. not un.

Melica cærulea. At Ashdon, but rare.

Holcus mollis. Hedges and woods.

—————— *lanatus* and *Arrhenatherum avenaceum*. Abundant.

Poa aquatica, River at Littlebury.

— *fluitans*. Ponds and ditches, freqnt.

— *rigida* and *compressa*. Walls and dry ground, not uncommon.

— *trivialis, pratensis* & *annua*. Very fr.

— *nemoralis*. Woods in several places.

Triodia decumbens. Chesterford, rather ra.

Briza media. Meadows, frequent.

Dactylis glomerata. Abundant.

Saffron Walden, 1842.

Cynosurus cristatus. Pastures &c. freqnt.

Festuca ovina and *duriuscula*. Dry pastures &c. not uncommon.

—— *bromoides*. Linton, rather rare.

—— *loliacea*. In the Park, not comn.

—— *pratensis*. Very frequent.

—— *elatior*. Moist shady places, com.

Bromus giganteus and *asper*. Hedges and thickets in several places.

—— *sterilis*. Pastures and hedges, fr.

—— *secalinus*. Corn-fields occasionally, but rare.

—— *mollis*. Abundant.

—— *racemosus*. Meadows &c. not un.

—— *erectus*. Near Littlebury, rare.

Avena fatua. Too abundant in corn-fields.

—— *pratensis*. Near Hildersham.

—— *pubescens*. Dry banks in several pl.

—— *flavescens*. Plentiful.

Arundo Phragmites. Marshy ground, not uncommon.

Elymus europæus. Woods, but rare.

Hordeum murinum. Walls and dry ground frequent.

—————— *pratense*. Meadows, not uncom.

Triticum caninum and *repens*. Too abndt.

Brachypodium pinnatum. Open banks &c. occasionally, rather rare.

—————————— *sylvaticum*. Wds. & hdges

Lolium perenne. Very abundant.

—————— *temulentum*. Corn-fields, not unc.

—————— *arvense*. Corn-fields, rare.

Phalaris canariensis. Waste ground, oc.

—— *arundinacea*. Banks of ponds, not rare.

Polypodium vulgare. Shady hedges, occa.

Aspidium Filix-mas. Hedges &c. not un.

Asplenium Trichomanes. Walls at Wicken, rare.

—————— *Ruta-muraria*. Walls at Audley End.

Scolopendrium vulgare. Hedges &c. ra. r.

Pteris Aquilina. Not plentiful.

Botrychium Lunaria and *Ophioglossum vulgare*. Near Linton, but rare.

Equisetum fluviatile. Watery places, oc.

—————— *arvense*. Corn-fields &c. com.

—————— *palustre*. Marshy ground, fre.

G. S. Gibson.

Art. CII. — *A List of the rarer Flowering Plants and Ferns of the Neighbourhood of Dumfries; with Remarks on the Physical Conditions of the District.* By PETER GRAY, Esq.

ANNEXED is a list of the less common flowering plants and ferns, growing in the vicinity of Dumfries; and perhaps, by way of introduction, a hasty sketch of the boundaries and physical aspect of the region thus illustrated, may not be altogether inappropriate.

The Nith, in its course through the district to which it gives name, flows through three basins or valleys, which, according to the speculations of some geologists, have, at a remote period, formed the beds of as many lakes, successively drained by the river into the Solway Frith. The lowest of these, the vale of Dumfries, is enclosed on all sides, except the south, by hills chiefly composed of greywacké, and, viewed from the heights around that town, presents a natural amphitheatre of great beauty; the undulating surface characteristic of the new red sandstone formation, here the predominating one, together with the intersecting ridges that mark the outer channels of the principal river and its tributary streams, tending pleasingly to diversify the scenery of the interior.

The immediately environing hills are of no great height; but, on the north, the Moffat range, and among them Queensberry, which attains an elevation of 2259 feet, overtop these, and, although at some distance, appear continuous. On the west, the numerous ravines between them are filled, and the hills themselves belted or crowned with, plantations of modern growth, abounding in Rubus saxatilis, Vaccinium Myrtillus, Polypodium Phegopteris and Dryopteris; where bare of wood, Gymnadenia conopsea, G. albida, Habenaria viridis and Lycopodium clavatum, alpinum, Selago and selaginoides, are also met with. Criffel, a syenitic hill 1895 feet high, forms the southern terminus of these hills, and although good ground for the cryptogamist, is nowise rich in the rarer flowering plants. Loch Kinder, at its base, furnishes Lobelia Dortmanna in abundance.

On the south-east, and approaching close upon the town, lies the extensive morass of Lochar Moss, occupying many thousand acres. Although not perhaps so fruitful in botanical rarities as might be anticipated from its great extent, this moss contains a few good plants. Among these may be enumerated the pretty little Utricularia minor, in some places exceedingly abundant, Ranunculus Lingua, Drosera longifolia and anglica, Bidens tripartita, with several of lesser note.

I shall just particularize another locality, to which frequent refer-ence will be made in the catalogue. Near the suburb of Maxwelton there is, or rather has been, a system of small lochs, now all drained, with the exception of the last of the series, itself considerably reduced in size. This little district, while it may be said to present an epit-ome of the dale in its geographical details, affords also almost every species found in the valley, with others—such as Carex limosa, C. fi-liformis, Eriophorum pubescens, Lysimachia vulgaris and Lycopus europæus—that have not been observed in any other part of the latter.

These circumstances, together with its proximity to the town, ren-der it the favourite haunt of our Dumfries botanists.

Aira caryophyllea and *flexuosa*. Very com.
—— *præcox*. Glen.
Alisma ranunculoides. Maxwelton Loch, abundant; very luxuriant in ditches near Carlaverock castle.
Allium vineale. Banks of Nith near Lin-cluden abbey.
Alsine marina. Shores of the Solway.
—— *rubra*. On a wall opposite Portland place, Maxwelton.
Ammophila arenaria. Nith nr. Carsethorn.
Anchusa sempervirens. Near Rosehall; Catton's loaning. In both places in all likelihood introduced.
Anagallis tenella. Abundant near Burran point.
Andromeda polifolia. Terregles wood; about Maxwelton loch : plentiful.
Avena pratensis. Terregles, banks of the Nith, &c.
—— *pubescens*. Craigs.
Berberis vulgaris. Plntatns. nr. Lincluden
Bidens cernua. Lochar-moss; abundant.
—— *tripartita*. Lochar-moss.
Brachypodium sylvaticum. Craigs; banks of the Nith, &c.
Bromus asper. Woods & hedges, not unc.
—— *erectus*. New Abbey church-yard
—— *racemosus*. Banks of the Nith and near Terregles.
—— *secalinus*. In wheat-fields.
Calamintha Acinos. Castle-Douglas road, about a mile from Dumfries; Glen; sparingly in both localities.

Campanula latifolia. Banks of Nith near Lincluden abbey; Cluden craigs; hazel-copse near Grove.
Cardamine amara. Cluden craigs.
Carex dioica. Near Maxwelton loch; Irongray hills.
—— *filiformis*. Near Maxwelton loch.
—— *fulva*. About Maxwelton loch.
—— *muricata*. Near Lincluden abbey.
—— *Œderi*. About Maxwelton loch.
—— *pilulifera*. Irongray hills.
—— *limosa*. Terregles woods; boggy ground near Maxwelton loch.
Carum verticillatum. Very abundant about Maxwelton loch; meadow near Ma-bie-moss.
Catabrosa aquatica. Moat of Carlaverock castle.
Cerastium arvense. Bank a little beyond Portland-place; near Cluden new bridge.
Cerasus avium. Bank nr. Lincluden abbey
—— *Padus*. Pretty general in the northern parts of the district.
Cheiranthus Cheiri. Walls of Lincluden and Sweetheart abbeys; plentiful on the latter.
Circæa Lutetiana. Glen.
Drosera anglica & longifolia. Lochar-moss
Epipactis latifolia. Mavis-grove wood.
Eriophorum pubescens. Maxwelton loch.
Festuca bromoides. Fields, dykes & road-sides, common.
Gagea lutea. Hazel-copse nr. Grove, abt.

Galium boreale. Bank of Nith iu sev. pl.

Genista anglica. Terregles woods.

———— *tinctoria.* Dalscairth hills.

Gentiana campestris. Dalscairth hills; Glen.

Gymnadenia albida. Dalscairth hills, pl.

———————— *conopsea.* Dalscairth hills; Maxwelton loch.

Habenaria chlorantha. Maxwelton loch and Terregles woods.

———————— *viridis.* Dalscairth hills, spar.

Hippuris vulgaris. Lochar-moss.

Hypericum hirsutum. Glen.

Jasione montana. Not uncommon.

Lathræa squamaria. Hazel-copsc near Grove, but very sparingly.

Lepidium Smithii. Road-sides and on light soils, very common.

Linaria vulgaris, var. *Peloria.* Hedge nr. Kelton.

Littorella lacustris. Margin of the Cluden, above Lincluden abbey.

Lobelia Dortmanna. Loch Kinder, abundant; less so in Loch Lotus.

Lycopus europæus. About Maxwelton loch

Lysimachia vulgaris. About Maxwelton Loch, sparingly.

Lythrum Salicaria. Not uncommon.

Malva moschata. Banks of the Nith.

Melica uniflora. Glen ; Mavis-grove wd.

Meum Athamanticum. Banks of the Cluden, above Routing bridge.

Milium effusum. Cluden craigs; Mavisgrove woods, &c.

Molinia cærulea. About Maxwelton loch.

Myosotis repens. Boggy ground nr. Maxwelton loch.

Myrica Gale. Abundant and luxuriant in Terregles woods ; Maxwelton lo.

Nasturtium terrestre. Upon the quay at Dumfries.

Nuphar lutea. In the river Lochar.

Ornithopus perpusillus. About Maxwelton loch, and several other localities, abundant.

Orobanche major. Harleybank ; Craigs ; Cluden craigs.

Parnassia palustris. Abt. Maxwelton loch

Phalaris canariensis. Cultivated ground.

Poa aquatica. Moat of Carlaverock-castle

— *maritima.* Shores of the Solway nr. Carlaverock-castle.

Polemonium cæruleum. Catton's loaning.

Polygonum Bistorta. Banks of Nith opposite Dumfries ; Cluden craigs.

Pyrola media. Dalscairth hills.

Ranunculus Lingua. Lochar-moss ; Mabie-moss.

Ribes Grossularia. Not uncommon.

—— *nigrum.* Banks of the Nith, about a quarter of a mile below Ellisland.

———— *rubrum.* Cluden-craigs ; banks of Cluden above Routing-bridge

Rottbollia incurvata. Shores of the Solway

Rubus saxatilis. Dalscairth woods.

Sanguisorba officinalis. Field near Mabie-moss; near Kelton.

Samolus Valerandi. Maxwelton loch; ditch near Kingholm-quay ; Nith below Glencaple.

Sambucus Ebulus. Fields near Netherwood, and below the Dumfries observatory.

Saxifraga granulata. Banks of Nith and Cluden.

Scirpus pauciflorus. Lochar-moss; about Maxwelton loch.

Sedum anglicum. Road-sides near Portland-place and Dalscone ; hills near Shambelly wood ; abundant.

Silene maritima. Beach below Glencable.

Solanum Dulcamara. Plantation near Kingholm-quay ; near Lincluden-abbey.

Sparganium simplex. Maxwelton loch,

Statice Limonium. Shores of the Solway at Burran-point.

Stellaria glauca. Maxweltou loch.

———————— *nemorum.* Bank near Lincluden abbey : Cluden craigs.

Symphytum officinale. Banks of Nith nr. Albany-place.

Triodia decumbens. Maxwelton loch.

Trollius europæus. Glen : Routing-bridge, abundant.

Vaccinium Myrtillus. Dalscairth, Sham-

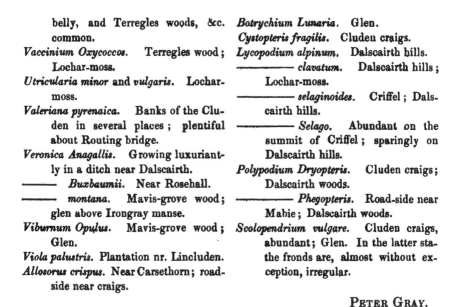

belly, and Terregles woods, &c. common.

Vaccinium Oxycoccos. Terregles wood; Lochar-moss.

Utricularia minor and *vulgaris.* Lochar-moss.

Valeriana pyrenaica. Banks of the Cluden in several places ; plentiful about Routing bridge.

Veronica Anagallis. Growing luxuriantly in a ditch near Dalscairth.

———— *Buxbaumii.* Near Rosehall.

———— *montana.* Mavis-grove wood; glen above Irongray manse.

Viburnum Opulus. Mavis-grove wood ; Glen.

Viola palustris. Plantation nr. Lincluden.

Allosorus crispus. Near Carsethorn; roadside near craigs.

Botrychium Lunaria. Glen.

Cystopteris fragilis. Cluden craigs.

Lycopodium alpinum. Dalscairth hills.

———————— *clavatum.* Dalscairth hills ; Lochar-moss.

———————— *selaginoides.* Criffel ; Dalscairth hills.

———————— *Selago.* Abundant on the summit of Criffel ; sparingly on Dalscairth hills.

Polypodium Dryopteris. Cluden craigs; Dalscairth woods.

———————— *Phegopteris.* Road-side near Mabie ; Dalscairth woods.

Scolopendrium vulgare. Cluden craigs, abundant; Glen. In the latter sta-the fronds are, almost without exception, irregular.

PETER GRAY.

Dumfries, November 12, 1842.

ART. CIII. — *Some Account of the Botanical Collections recently made by Dr. Theodore Kotschy (for the Wurtemburg Botanical Union) in Nubia and Cordofan.* Communicated by MR. WM. PAMPLIN, jun.

(Continued from p. 390)

THIRD, in the province BERBER, the city Chartum and the village Gubba, the island Tutti near Chartum; the villages Abu Haschim, Abu Hamed, with the island Mograd and Kalebsche, and the great desert of Berber. It is therefore evident that our traveller made best use of the short space of time allotted to him, in extending his excursions as much as possible through a great variety of locality, namely, mountains, hills, plains, lakes, river-sides &c., circumstances however prevented him in a great measure from taking the altitudes and noting other particulars of the mountainous districts.

In the third place, a conspectus of the natural orders and number of species in each family of which the collection consists, is here given, namely : —

ORDERS.	SPECIES.	ORDERS.	SPECIES.	ORDERS.	SPECIES.
Marsileaceæ	... 1	Brought up	...93	Brought up	...270
Alismaceæ	 3	Compositæ	28	Cassieæ	16
Hydrocharideæ	4	Portulacaceæ	... 12	Mimoseæ	 5
Gramineæ	48	Cucurbitaceæ	...10	Corniculatæ	... 2
Cyperoideæ	15	Labiatæ	 9	Aizoideæ	14
Commelinaceæ	2	Asperifoliaceæ	...12	Rosaceæ	 1
Juncaceæ	 1	Convolvulaceæ	18	Onagreæ	 5
Palmæ	 1	Polygalaceæ	... 3	Lythrarieæ	 7
Coronarieæ	 1	Personatæ	30	Tetradynamæ	... 6
Characeæ	 1	Solanaceæ	 7	Capparideæ	 6
Amentaceæ	 1	Lysimachiaceæ	2	Violaceæ	 1
Urticaceæ	 1	Asclepiadeæ	... 3	Rutaceæ	24
Nyctagineæ	 3	Contortæ	 1	Sapindaceæ	 4
Aristolochieæ	... 1	Sapotaceæ	 1	Malvaceæ	20
Laurineæ	 1	Umbelliferæ	... 1	Geraniaceæ	 4
Plumbagineæ	... 1	Terebinthaceæ	... 1	Theaceæ	 1
Rubiaceæ	 8	Papilionaceæ	...39	Tiliaceæ	 7
	93		270	Total Species,	393 ·

From the above enumeration it appears that the Gramineæ and Cyperaceæ amount to 62 species; the Papilionaceæ to 39; the Personatæ to 30; the Compositæ to 27; the Rutaceæ to 24; and the Malvaceæ to 20.　　　　　　　　　　　　　W. PAMPLIN JUN.

(To be continued).

ART. CIV.—*Notice of 'The Botanical Looker-Out among the Wild Flowers of the Fields, Woods and Mountains of England and Wales.'* By EDWIN LEES, F.L.S., &c. London: Tilt & Bogue, Fleet Street. H. Davies, Cheltenham. 1842.

WE have already given a few extracts from this work (Phytol. 301), and as our limits are very circumscribed, we cannot now do better than allow the author himself to explain the design and scope of the book before us.

" I WILL admit *in limine* that I am not here writing to instruct the professional student of Botany. Neither do I aim to surprise my brother botanists by any new arrangements in classification or discoveries in physiology. But if I take a humbler rank than the dignity of science may seem to warrant, and thus make no advances in their estimation, still I hope I may be in some degree *useful* in attracting the *many* to the pleasures afforded by an examination of plants in their wild localities, and thus, indirectly at least, subserve the cause of Natural History, by enlisting recruits, whose enthusiasm may perchance be awakened by my incitations to observation and adventure.

" In my experience as a practical collecting botanist for some years, I have invariably found that however my botanical friends might take fire at the exhibition of my specimens or the mention of their habitats, that the uninitiated in these things were unable to comprehend the sources of my pleasure, and could not understand on what principle I could experience delight in making long journeys, and taking fatiguing rambles, merely in search of plants. * * The neglect of physical and mental enjoyment lying within the reach of almost every body, appears to me to arise from a false supposition that the toils attendant upon the study of Botany would greatly counterbalance any pleasure to be derived from it. In these papers, then, I aim to show how incorrect such a conclusion is ; — and, in monthly order, my object has been to produce delineations which, even to the general eye of those unfamiliar with botanical terms, shall offer charms which may tempt the leisure of those who desire a pleasing and instructive occupation ; while I have introduced *incident* to show that the botanist, during his rambles, may still look out with all the gusto of a traveller superadded to his scientific examinations — while the stores of his collecting-book will make " a wet day at an inn " very different from " the wet day " so graphically described by Washington Irving."

* * * *

" Treating the subject thus lightly, I may hope to attract some whose attention would shrink from the study of more laboured treatises — and the ardent enquirer, if he accompanies me for *excitement*, will find abundant works before the public, where the sparks here kindled, may contribute fuel to the continuation and duration of the scientific flame he desires to nourish with increasing and perennial vigour. Even the proficient in botanical study may not be displeased with the allusions made to the habitats of some of his favourites; since as iron sharpeneth iron, so is enterprise awakened by the narration of the humblest pilgrim to the shrine that is the object of the reverence of his fraternity."—Preface.

Were we disposed to be critical, we might perchance quarrel a little with the author's style, and say that it is, in some places, rather too sentimental; but this, if it be a fault at all, is one into which a writer may readily fall, if, when he treats of flowers, they suggest to his mind other ideas than such as relate simply to their utility or scientific arrangement. However, "take it for all in all," we hail the ' Botanical Looker-Out ' as a very agreeable contribution to our stock of popular works on Botany.

ART. CV.— *Varieties.*

227. *A Catalogue of Plants found growing in the neighbourhood of Wrexham, in Denbighshire.*

Veronica serpyllifolia	Veronica Chamædrys	Circæa Lutetiana
scutellata	hederifolia	Anthoxanthum odoratum
Anagallis	agrestis	Valeriana rubra
Beccabunga	arvensis	dioica
officinalis	Pinguicula vulgaris	officinalis
montana	Lycopus europæus	Fedia olitoria

Fedia dentata
  carinata
Iris Pseud-acorus
Rhynchospora alba
Scirpus lacustris
  setaceus
  sylvaticus
Eleocharis palustris
  multicaulis
Eriophorum vaginatum
  angustifolium
Nardus stricta
Alopecurus pratensis
  geniculatus
  fulvus
Phalaris arundinacea
Phleum pratense
Milium effusum
Agrostis canina
  vulgaris
  alba
Catabrosa aquatica
Aira cæspitosa
  flexuosa
  caryophyllea
  præcox
Melica uniflora
  cærulea
Holcus mollis
  lanatus
Arrhenatherum avenaceum
Poa aquatica
  fluitans
  rigida
  trivialis
  pratensis
  annua
Triodia decumbens
Briza media
Dactylis glomerata
Cynosurus cristatus
Festuca ovina
  bromoides
  myurus
  loliacea
  pratensis
Bromus giganteus
  asper

Bromus sterilis
  secalinus
  mollis
Avena fatua
  strigosa
  pubescens
  flavescens
Arundo Phragmites
Triticum caninum
  repens
Brachypodium sylvaticum
  pinnatum
Lolium perenne
  temulentum
Montia fontana
Dipsacus pilosus
  sylvestris
Knautia arvensis
Scabiosa succisa
  columbaria
Galium verum
  cruciatum
  palustre
  saxatile
  Aparine
Asperula odorata
Sherardia arvensis
Plantago major
  lanceolata
  Coronopus
Cornus sanguinea
Alchemilla vulgaris
  arvensis
Potamogeton gramineus
  crispus
  perfoliatus
  rufescens
  natans
Sagina procumbens
  apetela
Echium vulgare
Lithospermum arvense
  officinale
  purpuro-cæruleum,
    (Denbigh)
Symphytum officinale
Lycopsis arvensis
Myosotis arvensis

Myosotis
  cæspitosa
  versicolor
Cynoglossum officinale
Anagallis arvensis
  tenella
Lysimachia nemorum
  Nummularia
Primula vulgaris
  veris
  elatior
Hottonia palustris
Menyanthes trifoliata
Erythræa Centaurium
Datura Stramonium
Atropa Belladonna
Solanum Dulcamara
  nigrum
Verbascum Thapsus
  Lychnitis (abundnt.)
  thapsiforme
  virgatum
Convolvulus arvensis
  sepium
Vinca major, (doubtfully
  wild)
  minor, (ditto).
Jasione montana
Campanula rotundifolia
  latifolia
  rapunculoides
  Trachelium
Lonicera Periclymenum
Rhamnus catharticus
Euonymus europæus
Viola hirta
  odorata
  palustris
  canina
  tricolor
  lutea
Hydrocotyle vulgaris
Sanicula europæa
Helosciadium nodiflorum
  inundatum
Sison Amomum
Ægopodium Podagraria
Bunium flexuosum

Pimpinella Saxifraga
Sium angustifolium
Œnanthe fistulosa
    peucedanifolia
    crocata
    Phellandrium
Æthusa Cynapium
Fœniculum vulgare
Angelica sylvestris
Heracleum Sphondylium
Daucus Carota
Torilis Anthriscus
    nodosa
Scandix Pecten
Anthriscus sylvestris
    vulgaris
Chærophyllum temulentum
Myrrhis odorata
Conium maculatum   [cus
Chenopodium Bonus-Henri-
    hybridum ?
    album
Viburnum Opulus
Sambucus Ebulus
    nigra
Parnassia palustris
Linum catharticum
Drosera rotundifolia
Berberis vulgaris
Peplis Portula
Galanthus nivalis
Narcissus Pseudo-narcissus
Convallaria majalis
Allium ursinum
    vineale
Ornithogalum nutans
Hyacinthus non-scriptus
Narthecium ossifragum
Juncus glaucus
    effusus
    conglomeratus
    acutiflorus
    lampocarpus
    uliginosus
    bufonius
    squarrosus
Luzula sylvatica
    pilosa

Luzula campestris
Rumex Hydrolapathum
    crispus
    sanguineus, a. and β.
    acutus
    obtusifolius
    Acetosa
    Acetosella
Triglochin palustre
Alisma Plantago
Chlora perfoliata
Erica Tetralix
    cinerea
Calluna vulgaris
Vaccinium Myrtillus
    Vitis-Idea
    Oxycoccos
Epilobium angustifolium
    hirsutum
    parviflorum
    montanum
    tetragonum
    palustre
Daphne Laureola
Polygonum Bistorta
    aviculare
    Fagopyrum
    Convolvulus
    amphibium
    Persicaria
    lapathifolium
    Hydropiper
Paris quadrifolia
Adoxa Moschatellina
Scleranthus annuus   [lium
Chrysosplenium alternifo-
    oppositifolium
Saxifraga granulata
    tridactylites
Saponaria officinalis
Silene inflata
Stellaria media
    holostea
    graminea
    uliginosa
Arenaria trinervis
    serpyllifolia
    rubra

Cotyledon Umbilicus
Sedum Telephium
    anglicum
    acre
Oxalis Acetosella
Agrostemma Githago
Lychnis Flos-cuculi
    dioica
Cerastium vulgatum
    viscosum
    aquaticum
Spergula arvensis
Lythrum Salicaria
Agrimonia Eupatoria
Reseda Luteola
Sempervivum tectorum
Prunus spinosa
    Padus
    Cerasus
Cratægus Oxyacantha
Pyrus Malus
    aucuparia
Spiræa Ulmaria
Rosa spinosissima
    villosa
    tomentosa
    canina
    systyla
    arvensis
Rubus Idæus
    suberectus
    fruticosus
    saxatilis
Fragaria vesca
    elatior
Comarum palustre
Potentilla anserina
    Fragariastrum
Tormentilla officinalis
    reptans
Geum urbanum
    rivale
Papaver Argemone
    dubium
    Rhœas
Meconopsis Cambrica
Chelidonium majus
Helianthemum vulgare

Tilia europæa
Helleborus viridis
Aquilegia vulgaris
Stratiotes aloides
Thalictrum minus
Clematis Vitalba
Anemone nemorosa
Ranunculus aquatilis
    hederaceus
    Flammula
    Ficaria
    auricomus
    sceleratus
    acris
    repens
    bulbosus
    arvensis
    parviflorus
Trollius europæus
Caltha palustris
Mentha sylvestris
    hirsuta
    gentilis
    arvensis
Thymus Serpyllum
Origanum vulgare
Teucrium Scorodonia
Ajuga reptans
Ballota nigra
Galeobdolon luteum
Galeopsis Tetrahit
    versicolor
Lamium album
    purpureum
Betonica officinalis
Stachys sylvatica
    ambigua
    palustris
    arvensis
Nepeta Cataria
Glechoma hederacea
Acinos vulgaris
Calamintha officinalis
Clinopodium vulgare
Prunella vulgaris
Bartsia Odontites
Euphrasia officinalis
Rhinanthus Crista-galli

Melampyrum pratense
Lathræa squamaria
Pedicularis palustris
    sylvatica
Antirrhinum majus
Linaria Cymbalaria
    Elatine
    vulgaris
Scrophularia nodosa
Digitalis purpurea
Verbena officinalis
Coronopus Ruellii
Capsella Bursa-pastoris
Lepidium campestre
    Smithii
Draba verna
Cardamine amara
    pratensis
    impatiens
    hirsuta
Arabis hirsuta
Barbarea vulgaris
Nasturtium officinale
    terrestre
Sisymbrium officinale
    Sophia
    Thalianum
Erysimum Alliaria
Cheiranthus Cheiri
Brassica Napus
    Rapa
Sinapis arvensis
    nigra
Raphanus Raphanistrum
Erodium cicutarium
Geranium pratense
    lucidum
    molle
    robertianum
    dissectum
    columbinum
Malva sylvestris
    rotundifolia
    moschata
Corydalis claviculata
Fumaria capreolata
    officinalis
Polygala vulgaris

Ulex europæus
Genista tinctoria
Cytisus Scoparius
Ononis arvensis
Anthyllis Vulneraria
Orobus tuberosus
Lathyrus Nissolia
    pratensis
    sylvestris
Vicia sylvatica
    Cracca
    sativa
    angustifolia
    sepium
    Bithynica
Ervum hirsutum
Astragalus glycyphyllos
Ornithopus perpusillus
Melilotus officinalis
Trifolium repens
    pratense
    arvense
    striatum
    procumbens
    filiforme
Lotus corniculatus
    major
Medicago sativa
    lupulina
    maculata
Hypericum Androsæmum
    quadrangulum
    perforatum
    humifusum
    montanum
    hirsutum
    pulchrum
    elodes
Tragopogon pratensis
Helminthia echioides
Sonchus arvensis
    oleraceus
Prenanthes muralis
Leontodon Taraxacum
Apargia hispida
Hieracium Pilosella
    Lawsoni ?
    murorum

Hieracium sylvaticum
   paludosum
   sabaudum
   umbellatum
Crepis virens
Hypochæris radicata
Lapsana communis
Cichorium Intybus
Arctium Lappa
Carduus nutans
Cnicus lanceolatus
   palustris
   pratensis
Carlina vulgaris
Bidens cernua
Eupatorium cannabinum
Tanacetum vulgare
Artemisia Absinthium
   vulgaris
Gnaphalium germanicum
   uliginosum
Conyza squarrosa
Tussilago Farfara
Petasites vulgaris
Senecio vulgaris
   sylvaticus
   tenuifolius
   Jacobæa
   aquaticus
   saracenicus
Solidago Virgaurea
Pulicaria dysenterica
Bellis perennis   [mum
Chrysanthemum Leucanthe-
   segetum
Pyrethrum Parthenium
   inodorum
Matricaria Chamomilla
Anthemis arvensis
Achillæa Ptarmica
   Millefolium
Centaurea nigra
   Cyanus

Centaurea Scabiosa
Orchis mascula
   Morio
   pyramidalis
   latifolia
   maculata
Gymnadenia conopsea
Habenaria viridis
   bifolia
Ophrys apifera
Listera ovata
   Nidus-avis
Epipactis latifolia
   palustris
Euphorbia helioscopia
   exigua
   Peplus
Callitriche verna
Zannichellia palustris
Typha latifolia
   angustifolia
Sparganium ramosum
   simplex
Carex pulicaris
   vulpina
   paniculata
   stellulata
   ovalis
   remota
   pendula
   sylvatica
   Pseudo-Cyperus
   Œderi
   binervis
   præcox
   panicea
   recurva
   cæspitosa
   stricta
   acuta
   paludosa
   vesicaria
   ampullacea

Carex hirta
Alnus glutinosa
Urtica urens
   dioica
Bryonia dioica
Myriophyllum verticillatum
Sagittaria sagittifolia
Arum maculatum
Poterium Sanguisorba
Empetrum nigrum
Ruscus aculeatus
Viscum album
Humulus Lupulus
Tamus communis
Mercurialis perennis
Taxus baccata (wild)
Atriplex angustifolia
Ceterach officinarum
Polypodium vulgare
   Phegopteris
   Dryopteris
Polystichum lobatum
   angulare
Lastræa Oreopteris
   Filix-mas
   dilatata
Cystopteris fragilis
Asplenium Trichomanes
   Ruta-muraria
   Adiantum-nigrum
Athyrium Filix-fœmina
Scolopendrium vulgare
Pteris Aquilina
Lomaria Spicant
Osmunda regalis
Botrychium Lunaria
Ophioglossum vulgatum
Equisetum fluviatile
   arvense
   sylvaticum
   limosum
   palustre
   hyemale

—*John Rowland; Queen St., Wrexham, July 9*, 1842.

228. *Note on Bryum pyriforme*, (Phytol. 398). In the ' Muscologia Britannica ' of Hooker and Taylor, it is stated that this moss " is remarkable in the shape of its leaves, of which the upper ones are the longest and most flexuose. They are com. posed, moreover, except at the very base, almost wholly of nerve; there being only a

narrow membranous margin, which, *towards the extremities*, is deeply serrated." These remarks apply only to the perichætial leaves (for the others are without serratures), and the term "deeply serrated" is somewhat too strong, even when thus limited. I have no doubt whatever that the same meaning was intended to be conveyed in the 'British Flora;' but that through an error of the transcriber (in the attempt perhaps to abbreviate the passage) the language there used has a quite different import. I have examined all the specimens in the herbarium of the author, and can therefore assure Mr. Gibson that two species have not been confounded under this name. Bryum pyriforme is indeed a peculiar species; the moss most resembling it is B. gracile (*Wilson*). In a barren state it has considerable affinity with Phascum alternifolium of Schwægrichen; but when in fruit, I know of no moss, whether British or foreign, which is likely to be mistaken for it. It is moreover one of the very few mosses which produce "anthers and pistils" within the same envelope; a circumstance which has not escaped the notice of Bruch and Schimper in their elaborate 'Bryologia Europæa.' In case this work should not be in Mr. Gibson's possession, I may mention that several new species, allied to B. cæspititium, have been introduced into it; and it may be worth while for those who have the opportunity, carefully to examine and compare such mosses as generally pass for B. capillare, B. turbinatum &c., in their various aspects and localities, with a view to ascertain the validity of the distinctions insisted upon in that work.—*W. Wilson ; Warrington, November 7, 1842.*

[The following note may be interesting to some of our readers ; it is in reply to an enquiry respecting the beautiful Hierochloe borealis.—*Ed.*]

229. *Note on Hierochloe borealis.* I fear it will be a very difficult matter to get a British specimen of Hierochloe borealis ; and I think I run no risk of contradiction when I say there is not a *living* botanist who has gathered it. For several years when residing in Angus-shire (my native county), my attention was directed to its discovery, but without success. The locality given in Hooker's 'British Flora,'—"a narrow mountain valley called *Kella*, in Angus-shire," I was never able to find. There is, about three miles from Arbroath, a place called Kelly-glen, that is a narrow valley but not a mountain one; neither do I think the plant in question is to be found there, at any rate I was never so fortunate as to find it, nor have Professors Graham and Balfour, nor any of the numerous parties which almost annually proceed from Edinburgh to the mountains of Angus-shire, ever met with it. — *James Cruickshank ; Crichton Institution, Dumfries, November 7, 1842.*

230. *Potentilla tridentata.* There are several others of the late Mr. G. Don's discoveries which have never been found except by himself, such as Potentilla tridentata; of which very rare plant I possess an original specimen, given me by Mr. James Reid, gardener to Sir James Carnegie, of Southesk. There is no locality attached to the specimen ; and Mr. Reid informed me that although an intimate friend of Mr. Don, he knew the localities of very few of his rare plants.—*Id.*

231. *Phascum axillare and patens.* About a fortnight since I found Phascum axillare in fruit, in great profusion, on the bank of a newly made ditch near the Reigate station of the Dover Railway, where the soil is particularly light. It appeared to me so very early for the fruiting of this moss (March being the time given in Hooker's Flora) that I thought the fact worth recording. The axillary capsule renders this species particularly interesting, and indeed I think it the most beautiful of our Phasca. At the same time I found Phascum patens in fruit : this was also growing on the side of a newly made ditch, near the above-named spot, the soil a stiff clay ; it was by no

means plentiful. This very minute but interesting moss differs very widely from the other species of the genus, in its patent serrated leaves, which are beautifully reticulated, with the nerve disappearing below the point. I shall be glad to supply any of your correspondents with specimens—of the former species more particularly, until my stock shall have become exhausted.—*Wm. Hanson; Reigate, November* 7, 1842.

232. *Cibotium Baromez*, (Phytol. 63). I must tell you that I again have a fine frond of Cibotium Baromez, seven feet long, sparingly in fruit. This is the third time the plant has fruited with us, which is rather singular, as there has been no particular care bestowed upon it; and why should it persist in flowering here and not elsewhere?—*David Cameron; Botanic Garden, Birmingham, November* 8, 1842.

233. *Note on Hypericum perforatum, β. angustifolium*, Koch; *H. veronense*, Schrank. As this plant has not been noticed by any British writer, perhaps it will be interesting to some of your readers to know that it has been found this season. On the 6th of July last, in company with my friend Mr. Tatham, I made a short excursion in the immediate neighbourhood of Settle, for the purpose of making myself acquainted with a few of the Carices of that part. At Giggleswick Scar I met with the above Hypericum; the plant was plentiful, but as it was only just coming into flower at that time, I did not gather much of it. As Mr. Tatham is on the spot I may perhaps, through his kindness, be able to procure specimens next season. — *Samuel Gibson: Hebden Bridge, November* 8, 1842.

[The receipt of Mr Gibson's communication reminded us of a specimen of a narrow-leaved Hypericum, received in 1839 from Mr. Cameron, of the Birmingham Botanic Garden. It is labelled " Hypericum perforatum, var. microphyllum? Ludlow: 1839." Not being able to find the letter which accompanied the specimen, we wrote to Birmingham respecting it, and have been favoured with the following reply from Mr. Westcott. Our specimen is much more branched than Hypericum perforatum generally is, the branches being nearly erect; the leaves are very narrow, and their edges revolute.—*Ed.*] *

234. *Note on the narrow-leaved Hypericum from Ludlow.* The Hypericum was first observed by Mr. Cameron and myself, in the year 1839, on the walls of Ludlow Castle, and since that time on many of the walls which surround the Castle. I have also this year found it on the rocks of Whitcliffe, and on the hedge-banks near the Angel-bank, Clee hills. In the year 1839, Mr. Cameron and myself brought roots of it from the walls of the Castle, and on our return they were planted in the Birmingham Botanic Garden, where the plant has continued to flourish, still retaining its character. I have some recollection that we sent you a specimen on our return, under the name of Hypericum perforatum, var. microphyllum; but on further observation I conceive it to be distinct from that plant in its inflorescence, and as I cannot find that such a variety is described in any author to which I have access, I have called it H. perforatum, var. linearifolium. The plant is from one foot to a foot and a half high, less branched, and in all its parts much smaller than Hypericum perforatum; branches more erect and compact, not diffuse; inflorescence more corymbose than paniculate. —*Fred. Westcott; Violet Place, Spring Street, Edgbaston, November* 21, 1842.

235. *Lapsana pusilla*, Willd. I had the pleasure to meet with this singular and interesting little plant, in considerable abundance, during the late season (1842), in the sandy meadows west of the " Broom fields," on Bexley Heath. I believe it is scarce in some districts; it is not mentioned in either the Faversham, the Yorkshire, the Manchester, or any other local Flora to which I have referred, nor do I remember

to have observed it in any station so near the metropolis as the one just given. — *Edward Edwards*; *Bexley Heath, Kent, November* 12, 1842.

236. *Cyperus fuscus*, L. It may be interesting to the readers of 'The Phytologist' to be informed that this exceedingly scarce species still exists in its only recorded British locality, where it was first noticed by the late Mr. Haworth, namely, Eel-brook meadow, a marshy pasture situate between Walham Green and Parson's Green, beyond Little Chelsea, Middlesex. On a visit to the spot in September, 1841, I was so fortunate as to obtain several good specimens, although it required some patience and diligence in order to detect them, the entire plant being of inconspicuous appearance, and only a few inches in height.—*Id.*

237. *Note on Pyrola uniflora*, L. I am not aware if any of our hand-books mention that this species seems to possess a similar property with that appertaining to our three Droseras, as observed by Sir W. J. Hooker and Mr. Wilson, (Brit. Flora, 132, 4th ed.) At least, judging from specimens in my own possession, labelled as having been collected in the woods at Scone, Perthshire, by Mr. W. Gardiner, jun. of Dundee, the single-flowered winter-green, like the British species of Drosera, retains the property of staining the paper that lies next to it in the herbarium, of a deep rusty *rose* colour, so that the form of the plant is distinctly represented through to the back of the sheet on which it is fastened, and also upon the backs of several others which, at different times, may have lain above it, and this although the specimens are perfectly dry.—*Id.*

[We can confirm our correspondent's observations on this plant; other species of Pyrola appear to possess the same property.—*Ed.*]

238. *Habitats for Petroselinum segetum*, Koch, *and Carex Pseudo-Cyperus*, L. For the use of the youthful botanists resident in or near the "great metropolis," to whom information of the whereabouts of any of the rarer species is a desideratum, I would mention that of Petroselinum segetum. Of this plant I gathered specimens a summer or two ago from Charlton lane, and about the church-yard wall, Charlton, Kent, which, if I mistake not, is an unpublished station for this pretty species. And of the beautiful Carex Pseudo-Cyperus, in June, 1842, I noticed several large tufts growing by the margins of some shallow pools, near where excavations for brick-earth were in operation, at the Highgate end of Maiden lane, Holloway, Middlesex, a short distance from the Highgate cemetery.—*Id.*

[Petroselinum segetum is not uncommon in some of the marshy ground near London, as between the Thames and the lower road to Deptford, where it sometimes grows to a considerable size.—*Ed.*]

239. *Enquiry respecting Byssus barbata*, Eng. Bot. I enclose you a specimen of a byssoid substance, or fungus, gathered by myself last week from a mass of decaying timber supporting the roof of a drift-way in an iron-mine at Clurewell Meend, Forest of Dean, Gloucestershire, 318 feet below the surface of the ground. When I first saw it in the mine by candle-light, it presented a very beautiful appearance, looking like a golden fleece with every hair glistening with moisture. It extended over a considerable surface of the roof of the drift-way, and on its sides, as far as the decaying timber extended. It no doubt belongs to the old genus Byssus of authors, which, having been abolished for sheltering too many outcasts of other genera, our present species has been turned out of doors, and from the glorious uncertainty of modern botanical nomenclature, especially in the fungoid tribes, I do not know where to find it. It certainly appears to me to bear a close resemblance to, if indeed it be not absolutely

identical with, the Byssus barbata of ' English Botany,' as drawn by Sowerby; and at p. 181 of Hooker's ' English Flora,' v. pt. ii. Fungi, it is incidentally stated that this Byssus barbata of ' English Botany' is *Ozonium auricomum.* No authority however is given for this name, and on turning to the index for *Ozonium,* no such name occurs. Under these circumstances, being dubious as to whether the Byssus barbata of Smith and Sowerby is described at all in the cryptogamic volumes of Hooker's ' British Flora,' perhaps some fungological reader of ' The Phytologist' may be able to give some information on the subject, or refer me to a recent volume where the plant is noticed under its present recognised name. I recollect no very late notice of the occurrence of this production in local Floras.—*Edwin Lees; Church-Hill Cottage, Powick, Worcestershire, November* 16, 1842.

240. *Bryum pyriforme.* Under this species the authors of ' Muscologia Britannica ' observe—" They (the leaves) are formed, moreover, almost wholly of nerve; there being only a narrow membranous margin, which, *towards the extremity,* is deeply serrated; " (Ed. 1, p. 118). I pointed out to Dr. Taylor, a few months ago, the discrepancy between this and the corresponding remarks in ' English Flora ' (quoted by Mr. Gibson), and we agreed that, if not an error of the press, it was to be referred to an inadvertence on the part of the author of the latter work. As to the actual facts, I have assuredly never seen the leaves of Bryum pyriforme serrated " at the base:" my Yorkshire specimens show faint serratures near the points of the leaves, but in a specimen now before me, gathered at Blarney, near Cork, since the publication of ' Flora Hibernica,' they are far more distinct. The figure in ' Musc. Brit.' does certainly not represent the leaf " deeply serrated," and yet I cannot doubt that the authors of that work *saw* what they described. Mr. Gibson need not be told (for no one knows better than he) that mosses, as well as other plants, are liable to variation in this respect.—*Richard Spruce; York, November* 16, 1842.

241. *Note on the discovery of Statice tatarica, near Portsmouth.* I beg to communicate to your esteemed periodical the discovery of Statice tatarica as a British plant. It has been observed by myself, as well as by Mrs. Robinson of this town, on the shore between Wicker Hard and Cams, on the borders of Portsmouth Harbour: I have also noticed it growing by the side of Fareham Creek, from half a mile to a mile below the town. I was at first undecided to what species to refer it; but on forwarding a specimen to Professor Lindley, he pronounced it to be S. tatarica. Its time of flowering is about the same as that of S. Limonium; and it is undoubtedly wild in both places. —*Wm. L. Notcutt; Fareham, November* 21, 1842.

ART. CVI. — *Proceedings of Societies.*

BOTANICAL SOCIETY OF EDINBURGH.

This Society held its first meeting for the season on Thursday the 10th instant,— Professor Graham in the chair. James Edward Winterbottom, Esq. M.A., F.L.S. &c. East Woodhay, Newbury, Berks, and Jas. Carter, Esq., M.R.C.S., Petty Cury, Cambridge, were elected non-resident Fellows; and Mr. John Thompson, Crowhall Mill, Ridley, Northumberland, an Associate of the Society. A donation to the library was announced from Thomas Brown, Esq. (per David ! ., Esq.), and numerous parcels of plants &c. were stated to have been receiv   'g.

Mr. Brand read a paper by Mr. Edmonston, jun., on the Botany of Shetland, and instituted a comparison between the number of genera and species existing in that region, and those which occur in other districts of Scotland.

"The Botany of Shetland," observes Mr. Edmonston, " though not very extensive, is interesting. Many of the less common (chiefly subalpine) plants are abundant in all situations, and many species very commonly distributed, and indeed often marked as *universal* throughout Great Britain, are very rare or altogether unknown in Shetland. Among the last may be mentioned — *Alchemilla arvensis* and *vulgaris*, *Briza media*, *Primula veris*, *Anagallis arvensis*, *Convolvulus arvensis*, *Teucrium Scorodonia*, *Geranium robertianum*, *Lapsana communis* and others of the commonest weeds. Again, *Thalictrum alpinum*, *Blysmus rufus*, and other *local* plants, are everywhere abundant, growing down to the sea level; and sylvan plants — those generally associated with woods or luxuriant pasturage, are almost entirely wanting. The Geology of Shetland is rich in interesting phenomena. The formation is almost wholly primitive—the most abundant rocks being gneiss, mica-schist, clay and chlorite slate, granite, quartz, serpentine, limestone, &c.; besides which there are amygdaloidal porphyritic rocks of different kinds. The difference of formation between Shetland and Orkney is very striking —that of the latter being as uninteresting as the former is the reverse. Orkney consists chiefly of an apparent continuation of the North coast of the Mainland, being composed of sandstone, clay-slate, and other secondary rocks, while the Shetlands may be said to belong to the oceanic series of islands. Again, the difference seems as great between the Shetland and Faroe isles—for in the latter group the rocks are all basaltic. Many of the Shetland rocks present a most remarkable degree of similarity to those of the South of England—chrome ore, native magnesia, serpentine, crystallised fluor, and several others, being common to both extremities of Great Britain, though rarely found in the intermediate space; and it is a singular fact, that some of the plants present a corresponding analogy—as for instance, *Lathyrus maritimus*, &c.

"The prevalence of peat is a very characteristic feature in the general aspect of Shetland, and proves beyond a doubt the great abundance of trees in former ages. — Judging from their remains, these seem chiefly to have belonged to Corylaceæ and Pinaceæ, as trunks and nuts of the hazel, and cones of *Abies Picea*, have repeatedly been dug out of the moors. This evidence of their existing formerly in such abundance, leads to the question whether they may still be grown. I certainly do not think that the experiment has been fairly tried, nor is it probable that it soon will be on a scale that can set the matter at rest. Indeed, many reasons seem to concur in rendering it unlikely that trees could be reared so as to render them profitable in an economical point of view. The frosts and cold weather which often occur early in autumn do not leave the plants time to form their buds for hybernation before the old leaves are nipped; and the heat of summer is by no means sufficient (as in most other northern latitudes) to compensate for the shortness of its duration. I do not attach so much importance as has sometimes been done to the influence of the sea-spray, by which, during heavy gales, Shetland is liable to be swept — for these happen generally after the sap has descended, and when therefore the plant is dormant.

I may here mention some experiments which have been carried on by my father for five or six years, in order, if possible, to settle the question. He obtained from Messrs. Lawson, of Edinburgh, all the more generally cultivated trees and shrubs, — North British, North American, and North Asiatic, — and the result has been as follows: — Among the indigenous trees of Scotland, the ash appears to stand as well as

any other, as it puts forth its leaves late and loses them early. Of the scarcely indigenous, or naturalised species, the plane-tree appears to be the hardiest ; while the birch
and Scotch fir will scarcely live a year. Again, *Pinus montana* and *Æsculus Hippo-
castanum*, comparatively tender plants, appear to thrive well ; and *Pyrus aucuparia*,
which is indigenous with us, thrives tolerably in cultivation. Almost all the willows
do well ;—*Salix Russelliana, fragilis, cinerea, viminalis* and *vitellina*, among the best.
The alder is rather too early in putting forth its leaves, but some poplars appear to do
well, especially the white Scotch, black Italian and Lombardy, and *Populus nigra* is
indigenous. Oak and beech will not thrive at all. Generally speaking, evergreens,
whether trees or shrubs, appear not to suit. *Pinus Cembra, Abies Picea,* black, white
and Norway, have all been repeatedly tried, but seldom languished a year. Even the
hardy shrubby evergreens, which are met with indigenous, or in every shrubbery on the
mainland, as *Ilex Aquifolium, Rhododendron ponticum* and *flavum, Viburnum Tinus,*
&c., die almost immediately. Among the best thriving evergreen shrubs may be mentioned *Arbutus Uva-ursi, Cotoneaster mucronata, Hedera Helix,* &c. The latter, indeed, is native, and in some situations thrives remarkably well, as it also does in Orkney.

" The climate of Orkney and Shetland are much alike, but scientific observations
have only been recorded of the former. ' Regarding it,' Mr. Clouston states, ' the high
latitude of Orkney will no doubt induce many well-informed persons even in Scotland
to suppose that our winter is much colder than that of any other country, and it may
surprise them when we say that our winter is as warm as that of Glasgow, and several
degrees warmer than that of Applegarth in Dumfriesshire, on the very southern border.
This is owing to the influence of the surrounding ocean, which elevates the temperature of winter as much as it lowers that of summer. Thus, the temperature of Orkney
in May, June and July, is 7 degrees below that of Glasgow during these months ; but
for the *whole year*, the mean annual temperature in Orkney is nearly the same as that
in Applegarth—both being about 46 degrees, or 3¼ below that of Glasgow.' "

Mr. Edmonston goes on to observe, that " the uniformity of temperature in Shetland strikes every one ; and a remarkable feature in the climate is the great and almost
constant humidity. These causes no doubt have a great influence on the vegetation,
for there is not a semblance of arctic, and scarcely (except in a very few instances) of
alpine vegetation throughout the whole islands. This is certainly rather what might
be expected than otherwise ; but there are other anomalies which cannot be altogether
referred to climate ; and the extreme scantiness of the Flora is remarkable, considering the extent of the islands, and the variety of soil, exposure and situation which they
present."

The flowering plants (including the grasses) hitherto observed in Shetland, extend
to 94 genera and 178 species ; while those found in the district of Moray amount to
333 genera and 692 species ; and even in a range of sixteen miles round Aberdeen,
there have been found 287 genera and 562 species ; and in a similar extent round Edinburgh, the numbers are 389 genera and 908 species — while the flowering plants of
Great Britain extend to 523 genera and 1594 species. The proportion of species to
genera is also least in Shetland and Aberdeenshire, being only 2 to 1, whereas in the
Edinburgh district it is 2½ to 1 ; and in Britain generally it is 3 to 1.

The statements in Mr. Edmonston's paper led to some interesting conversation—
in the course of which Professor Graham remarked, as a phenomenon which has not
hitherto received a satisfactory solution — the entire destruction or absence of wood in
many parts of Scotland where once it evidently abounded, and where the change can-

not apparently have arisen through human instrumentality; and he observed that the investigation of this subject would be attended with great interest, besides being of importance in a national point of view.

Dr. Neill said, that in his opinion the peat mosses of Scotland have generally been formed at an earlier period than is usually supposed — some of them containing trees which do not now exist in the country ; and he suggested that means should be taken to ascertain the particular species of which the mosses consist, by taking specimens of wood and seeds, or cones &c. from the successive layers, and duly noting their relative position, with all such circumstances as might tend to establish a correct theory respecting our aboriginal forest vegetation ; indeed he had once proposed that a prize should be offered by the Highland Society for the best essay on this subject ; but his proposal had not been carried into effect.

Mr. Brand remarked that in this country, as in America, the forests in many places appear to have been destroyed by fire; and he instanced some oak trees in Dalkeith Park, which seem to have been burnt down at an early period, and to have thrown out new trunks from the stumps at a later date.

Mr. Goodsir supposed that the increase of the peat might gradually render the soil unfit for the support of trees; and stated, in reference to a remark made by Professor Graham, on the approach of the alpine plants in Shetland to the sea-edge, that this peculiarity coincided with the elevation of the deep-sea invertebrate animals, to the high-water mark in the same locality.

BOTANICAL SOCIETY OF LONDON.

November 4, 1841. — Hewett Cottrell Watson, Esq., F.L.S., V.P., in the chair. Donations to the library were announced from the American Academy of Sciences Philadelphia, the Horticultural Society of Berlin, and Mr. S. P. Woodward. British plants had been received from Mr. G. W. Francis, Mr. S. P. Woodward, Mr. D. Stock, Mr. B. D. Wardale, Mr. R. Ranking and Mr. S. Gibson. The Chairman presented a specimen of *Cnicus Forsteri*, which he said corresponded exactly with the cultivated specimen of the same species preserved in Smith's herbarium. The specimen exhibited by Mr. Watson was also a cultivated one ; the root having been found near Whitemoor pond, in Surrey, in June, 1841, and flowering specimens of it exhibited before the Society last year.

The wild specimens had from two to four flowers only on each stem; whilst the cultivated specimens had ten or a dozen each. Mr. W. exhibited the specimens for the purpose of pointing out the differences between *Cnicus Forsteri* and *C. pratensis ;* branched specimens of the latter species having been in several instances mistaken for the former.

Mr. Robert Ranking, F.L.S., presented a specimen of *Plantago Coronopus*, collected at Hastings, showing the easy and natural transition from a spike to a raceme; also a specimen of *Dactylis glomerata*, in which the glumes were become foliaceous.

The commencement of a paper was read from Mr. George Clarke, of the island of Mahé (communicated by Mr. H. W. Martin), " On *Lodoicea Sechellarum*," which will be concluded at the next meeting, when a copious report will be given.—*G.E.D.*

THE PHYTOLOGIST.

No. XX.	JANUARY, MDCCCXLIII.	Price 1s.

Art. CVII. — *Notes on the Baobab Tree, (Adansonia digitata).* By George Luxford, A.L.S., &c.

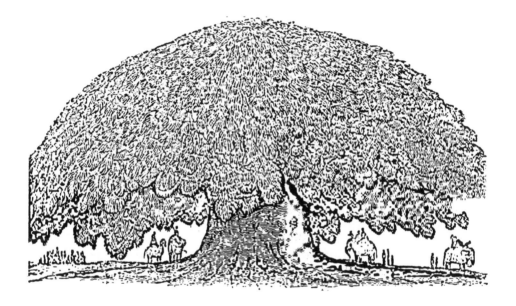

THE BAOBAB.

Egyptian Sour Gourd.   Monkey's Bread.

Baobab, Bauhin, Hist. i. 110.
Adansonia Baobab, Linn. Sp. Plant. 960.
Adansonia digitata, Linn. Syst. Veg. 620.

Group.—Syncarposæ.   Alliance.—Malvales.
Natural Order.—Sterculiaceæ.   Section.—Bombaceæ.
Linnæan Class and Order.—Monadelphia Polyandria.

Having received from a correspondent a copy of the 'Bombay Monthly Times for June, 1842,' which contains some interesting particulars relating to the Baobab-tree of Senegal, as observed in India,

2 P

it has occurred to me that a few notes on this "colossus of the vegetable kingdom," even if they contain nothing new, may not be out of
place in the pages of 'The Phytologist.'

The Baobab is a native of Senegal and other parts of the western
coast of Africa, from the Niger to Benin, "a part of the world," says
Adanson, "which has always been justly looked upon as the mother
of monsters." This celebrated French naturalist resided in Senegal
for about five years, and was probably the first botanist who had the
advantage of studying the Baobab in its native country. In 1756
M. Adanson communicated a very full account of this remarkable tree
to the Royal Academy of Sciences at Paris; his paper, together with
an admirable summary of it, were published in the Memoires of the
Academy in 1761, and appear to be the chief source whence subsequent writers have derived their knowledge of the Baobab.

A letter from Adanson to Linnæus, written four years after the return of the former from Senegal, and previously to the publication of
the memoir mentioned above, contains the characters of his new genus, and several remarks upon it; the following is an extract from this
letter, which is printed in the 'Correspondence of Linnæus and other
Naturalists,' ii. 467.

"Paris, Oct. 2, 1758.

"Among numerous new observations in natural history which I have formerly communicated to the *Académie des Sciences*, is a complete description of the *Bahobab*,
which Bernard de Jussieu has named *Adansonia*, and of which I had long ago given a
description before your letter reached me. B. de Jussieu had refrained from sending
you this description during my absence, that he might not deprive me of the opportunity of giving you pleasure. I therefore now send the essential parts of the character
which you ask for, taken from the Memoirs of the Academy intended for publication,
or rather from my own Latin manuscripts, according to the plan of your *Genera
Plantarum*, as I mean to give them to the public.

"ADANSONIA.

"*Calyx.* Perianth simple, of one leaf, cup-shaped, divided half way down into
five revolute segments, deciduous. [Fig. *c*].

"*Corolla.* Petals five,* nearly orbicular, ribbed, revolute, united by their claws to
the stamens and to each other. [Fig. *b*].

"*Stamina.* Filaments numerous (about 700), united in their lower part into a conical tube, which they crown at the top, spreading horizontally.—[Fig. *d*.] Anthers
kidney-shaped, incumbent.

"*Pistil.* Ovary nearly ovate. Style very long, tubular, variously twisted. Stigmas from 10 to 14, prismatic, shaggy, spreading from the centre.

* In Adanson's figure of the flower, of which fig. *b* at p. 436 is a fac-simile, four
petals only are shown; in his separate figure of the corolla there are five petals, which
is the normal number.

" *Pericarp.* Capsule oval, very large, woody, not bursting, internally separated into from 10 to 14 cells, filled with dry pulp and with seeds; the partitions membranous and longitudinal. [Page 437.]

" *Seeds* numerous, almost bony, kidney-shaped, lodged in friable pulp.

" Hence you may perceive how much this genus differs from the rest of the mallow tribe. First, by the calyx falling off immediately after flowering; — second, by the number and situation of the filaments at the top of a monadelphous tube; — third, by the number and form of the stigmas; — fourth, by the woody and close capsule, with its pulp and cells; — fifth, the compound fingered leaves; — and sixth, by the tree itself, which of all hitherto discovered is the most prodigious in the size of its trunk and branches, being as it were the stupendous vegetable monster of Africa. This tree is found in the country of Senegal only, from whence its fruit, with that of the *Agilhaid*, is sent every year, as an article of commerce, to Egypt. Some of its seeds having been planted there, in a garden, one or two trees were raised, which appear, from Prosper Alpinus, to have attained no remarkable size, nor perhaps to have flowered, if we may judge by the figure in that author, which in every particular, except the fruit, is erroneous. In the West Indian island of Martinico, a single tree of this kind, already full grown, bearing flowers and fruit, is carefully preserved. It was formerly sown there by the negroes. These and similar remarks are detailed, with my authorities, in the communications to the Parisian Academy."

In the above extracts allusion is made to the immense size of this tree, which has been spoken of as the largest (or rather the broadest) in the world, a title it well merits, as will be seen from the following description.*

The Baobab has more the appearance of a forest than of a single tree. It is an immense hemispherical mass of foliage, sixty or seventy feet high, and from a hundred and twenty to a hundred and fifty feet in diameter. The main trunk is very short in proportion to its size, being only about ten or twelve feet high; while it is at least twenty-five feet in diameter. Golberry, in his Travels in Africa, mentions a Baobab the trunk of which measured upwards of thirty-four feet in diameter, and was about thirty feet high. The branches are of considerable size, and fifty or sixty feet long; the central branch rises perpendicularly, the others spread around it in all directions, the low-

---

* The Norfolk-Island Pine (called *kauri* by the New Zealanders) occasionally grows to a large size. Mr. Terry, in his lately-published work on New Zealand, mentions an extraordinary individual which grows on the eastern coast, near Mercury Bay, which is the largest in New Zealand. " It is called by the natives the Father of the Kauri. Although almost incredible, it measures seventy-five feet in circumference at its base. The height is unknown, for the surrounding forest is so thick, it is impossible to ascertain it accurately. There is an arm some distance up the tree, which measures six feet in diameter at its junction with the parent trunk." It is evident that the particular tree here spoken of, far exceeds the average size of the species.

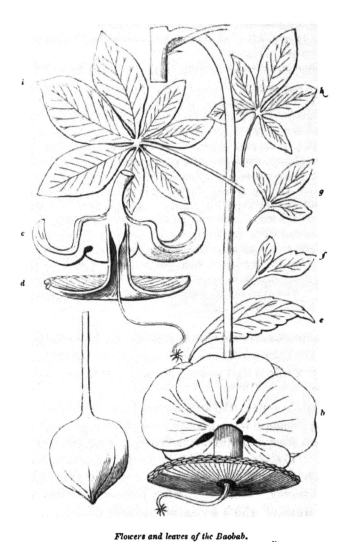

Flowers and leaves of the Baobab.

a. Flower-bud before expansion. *b.* Expanded flower. *c.* Section of the calyx. *d.* Section of the stamens. *e, f, g, h, i,* Leaves from trees of different ages.

er ones being nearly horizontal for the greater part of their length, while the extremities frequently trail on the ground, from their own weight. The roots are much longer than the branches. The central root descends perpendicularly to a great depth; the lateral ones extend horizontally, and are sometimes but a short distance below the surface of the ground. Adanson saw one of these roots, of which a great portion had been laid bare by the waters of a river; the uncovered part measured about one hundred and ten feet, and judging *from its size* he considered that forty or fifty feet might be still hidden.

The bark of the trunk and older branches is about nine lines thick, of an ash-grey colour, smooth to the touch, and having a shining appearance, as if varnished; that on the younger branches is greenish and somewhat hairy. The wood is soft, white, and extremely light, being very little heavier than cork. In the Bombay Monthly Times are given the results of carbonizing seven different sorts of wood, including that of the Adansonia. The weight per cent. of charcoal yielded by each is as follows.— Adansonia, 33; lignum vitæ, 26·4; mahogany, 25; oak, 22; beech, 19; ash, 17; Scotch fir, 16: so that the lightest wood presents the anomaly of yielding 6·6 per cent. of charcoal more than the heaviest.

The leaves are very similar in general appearance to those of the horse-chesnut, being somewhat orbicular in outline, and divided into several elliptical lobes, which are entire at the margin, and vary in number from three to seven. They are alternate, and supported by a petiole, at the base of which are two small stipules; these are said by Adanson to fall off as soon as the leaves expand; they are, however, represented in his figure. The leaves of very young trees are undivided and nearly sessile; the digitated leaves first make their appearance when the young plant is about a foot high. The figures *e—i*, on the opposite page repre-

Fruit of the Baobab.
The lower section shows the arrangement of the carpels.

sent the different forms of the leaves; the lobes of the fully developed leaves (*i*) are from four to six inches long and two or three inches wide.

The flowers of the Baobab, as might be expected from the size of the tree, are very large. The flower-bud (fig. *a*) is globose, and nearly three inches in diameter; when fully expanded the flowers are usually about six inches in diameter. There are generally two or three of these flowers on a branch, each being suspended by a peduncle which springs from the axil of one of the lower leaves, and bears a few scattered deciduous scales or bracts. The peduncle is a foot long and four lines thick. The handsome white flowers, like those of many allied plants, expand in the morning, about sun-rise, and close towards evening, thus affording an example of what Linnæus terms the sleep of plants. Golberry observes that the negroes assemble round the Baobabs to watch the expansion of their flowers; and that each flower, as it opens, is saluted with — " Good morning, beautiful lady!" An expanded flower is shown at *b*.

Omitting Adanson's minute description of the calyx and other parts of the flower, I will pass on to the fruit (p. 437), which is fromtwelve to eighteen inches long and six inches in diameter, and is suspended by a peduncle two feet long and nearly an inch thick. It is very hard and woody, and is covered with a greenish down. When cut across the fruit is found to be divided into from ten to fourteen cells by membranous dissepiments. The seeds are embedded in a spongy substance, which is whitish in fresh and healthy fruits, and of a reddish hue in those which are badly formed or very old; as it dries it becomes friable, and separates, either spontaneously or on receiving a very slight blow, into a number of irregular polyhedrons, each of which contains a single seed.

Adanson describes the structure of the seed and its mode of germination, four different stages of which are figured. The cotyledons are at first orbicular, then elliptical; on the fourth day the first true leaf is developed; at the end of a month the young tree is about a foot high, and at that time, as before stated, the digitated leaves appear; during the first summer the tree increases to about five feet in height, and is then about an inch or an inch and a half thick, whilst in France, under the most careful treatment, the author observes that within the latter period it attains no greater height than about twelve inches. A specimen in the botanic garden at Calcutta is said to have attained a circumference of eighteen feet in twenty-six years.

The Baobab comes into leaf in June, flowers in July, matures its fruit in October and November, and in the latter month it loses its leaves. It is very common both in the Island of Senegal and at the Cape de Verd, and along the sea-coast to Sierra Leone, and is even met with at Galam, which is more than a hundred leagues from the sea. M. Golberry says that in the year 1786 he " saw the greatest number of Baobabs on the isthmus of the peninsula of Cape de Verd, between the bay of Jof and that of Dakar," a space of nearly two square leagues, where there were at least sixty trees. The roots penetrate rocky soils with great difficulty, and if ever so slightly injured they decay; this decay is soon communicated to the trunk, where its progress is very rapid and the tree quickly perishes. Hence it thrives best and is most abundant in wet shifting sands, such as those which extend from Senegal to Cape de Verd, a distance of thirty leagues; while at Galam, where the soil is a hard stony clay, it occurs much less frequently and is comparatively small.

Besides a general rottenness or decay arising from injuries received by the root, this tree is occasionally subject to another disease, most probably produced by a fungus somewhat similar to that causing the dry rot, which spreads through the woody portion, and reduces it to the consistence of the pith, without either altering the colour of the wood or changing the disposition of its fibres. The bark also remains uninjured, and there is nothing in the external appearance to indicate the operations of the insidious enemy within. When thus affected the tree is frequently unable to resist the force of the wind; Adanson met with one in an island near Senegal, the trunk of which had been broken asunder in the middle during a gale. The trunk, at the time he saw it, was inhabited by an immense number of very large Coleopterous larvæ. The disease by which the tree was destroyed had most probably made considerable progress before the insects deposited their eggs in the trunk; at all events we know this to be the case with willows and other trees, which are seldom if ever attacked when in a sound and healthy state.

The rapid decay of a fine specimen of this tree, which grew at Colabah in Bombay, is doubtless to be attributed to the same disease.— This tree — one of the finest in western India — was forty-four feet in circumference. In May 1840, it was vigorous and apparently healthy; a few months after that time the large branches began to fall off and the ravages of disease proceeded with great rapidity. On examination the decayed portions were found to be perforated in all directions, like the one seen by Adanson, by the larvæ of a beetle, which were reducing the whole to a powder resembling saw-dust. Both the larva and the perfect insect are figured in the Bombay Times. The former is described as being two and a quarter inches long, and three inches in circumference at the thickest part. Some idea of the ravages of these larvæ may be formed from the statement, that a piece of the tree three feet long and eight inches in girth, apparently healthy and sound, was found to be so thoroughly perforated that scarce¹⁻ inches of solid wood could be found entire.

We have heard such astounding statements respecting the longevity of this tree, that it would not be right to pass over the subject quite without notice. In the Bombay Monthly Times before referred to, are some interesting notes furnished by the Rev. Dr. Wilson, who remarks that —

"This tree seems to be associated with absurdity among the sages of the West as well as the East. 'The Baobal-tree of Senegal,' says Lyell in his 'Principles of Geology,' ' is supposed to exceed almost any other in longevity; Adanson inferred that one which he measured, and found to be thirty feet in diameter, had attained the age of 5150 years. Having made an incision to a certain depth, he first counted three hundred rings of annual growth, and observed what thickness the tree had gained in that period. The average rate of growth of younger trees, of the same species, was then ascertained, and the calculation made according to a supposed mean rate of increase.' Now, how does the matter stand, with regard to the specimens we have before us in India? Dr. Roxburgh tells us that the tree is an *exotic* in this country — and he is quite correct. It was introduced by the Portuguese from Mozambique within the last three hundred years; and in many instances it has already attained to a growth *exceeding* that specified by Adanson and Lyell. Dr. Lindley has shown that what are called the annular [? annual] rings, are not to be depended upon in calculations as to the age of trees; and that with reference to this very extraordinary species."—Extracts from Notes of a visit to Dwaraka, by the Rev. Dr. Wilson.

There is no doubt that the Baobab lives to a very great age, as may be inferred from its enormous bulk. Adanson's observations on some trees which he met with in one of the Magdalen islands, led him to the conclusion that they were growing there at the time of the deluge, consequently that they were, at the time he saw them, upwards of five thousand years old! On these trees were carved some European names, some of which were distinctly dated in the fifteenth and sixteenth centuries, others less distinctly bore date in the fourteenth. Adanson thought it probable that the same trees were seen by Thevet when he passed these islands in 1555, on his voyage to the antarctic regions. The letters of the names were six inches high, and the names themselves occupied a length of about two feet, or somewhat less than the eighth part of the circumference of the trunk. Reasoning from these facts, and from his own observations of the rate of growth of the tree, Adanson arrived at the conclusion above stated, which is most probably an erroneous one. The same trees were seen by Golberry, thirty-six years after Adanson was on the island; he says that the words of the inscription were Dutch, and that one of the dates was 1449.

When I first read Adanson's account of the Magdalen-island Baobabs, I could not help suspecting that their size was incorrectly given, a circumference of sixteen or eighteen feet appearing to bear no proportion to the enormous age assigned to them. But on turning to the narrative of his voyage, prefixed to the 'Natural History of Senegal,' p. 66, I find that the diameter of the trees is expressly stated to be six feet. The author says that the names were deeply engraved in the bark, and that each person of the party, except himself, added his mark to those previously on the trees, but that he was satisfied with renewing two of the names which were old enough to be worth the trouble, one of them being dated in the fifteenth the other in the sixteenth century. Then follow the size of the letters, and a brief summary of the same arguments relating to the age of the trees which are afterwards given in detail in the Memoir published by the Academy. But be it observed, that neither in the Narrative nor in the Memoir does Adanson say one word about his having made an incision in the trunk and counted the number of annular layers of wood, in order to determine the age of the tree; his arguments are founded solely on the observed annual rate of increase of young trees in height and diameter, and his data are given by Sir W. J. Hooker, in the 'Botanical Magazine,' 2791-92, where the flower, fruit and leaf are beautifully figured.

The following remarks by a correspondent of the Bombay Monthly Times bear directly on the questions of the age of the tree and of its native country.

"I find you make no mention of the Adansonia digitata obtaining great perfection in the ruins of Mandoo and its environs, of which I remember once to have descanted in our evening conversations; indeed I believe these are the only localities in the upper parts of India, where the 'Khorassan Eemlee' as it is there called, is to be found in great numbers and of enormous girth.

"From Nalcha to the Delhi gate of Mandoo, a distance of six miles along the Vindyah, the road on either side is lined with the ruins of palaces, mosques and tombs, mingled with innumerable groups of the Adansonia digitata, the same extending in a long avenue from thence to the Jumna Musjid in the centre of the city, from whence they diverge to the royal parks and gardens.

"The Mahometans fondly treasure this tree as a relic of Moslem sovereignty, believing it to have been brought by the northern conquerors to embellish their imperial residences in the east; and moreover, that it languishes and dies in any Indian soil but that favoured as the abode of royalty. Sooner therefore would they lose an arm than a branch from this boasted tree, although its insidious inroads have done more to complete the ruin of Mandoo, than either the hand of time or Rajpoot bigotry; rooting itself in every crevice of the walls and roofs, and uptearing with its giant arms enormous masses of masonry. * * That the tree appears to be one to which the natives of India seem to attach much importance, is evident from both Hindoos and Mussulmen considering it in a sacred character: this probably proceeds from its prodigious size and comparative scarceness, as it is evident the Mah   dans have no claim to its importation from Khorassan, as I am credibly informed by lers from thence there is scarcely a tree in that region attains a tithe of its siz

Adanson considers the Baobab to be indigenous nowhere but in those places on the western side of Africa which have been mentioned above ; and states that the negroes, wherever they go, are in the habit of carrying with them the seeds of such plants as they make use of in cooking, or for other purposes. He enumerates many of these plants, among which are the Baobab, two kinds of cotton, the tamarind, several sorts of beans, the water-melon, &c., and observes that all these are now found in America, where they have every appearance of being indigenous, although many of them have not received American names. In support of his opinion that the Baobab is a native neither of the American continent nor of any of the West-India islands, he cites the works of Plumier, Sloane and Browne, in which it is not mentioned. He also observes that M. Thiebault de Chanvalon, an inhabitant of Martinique, speaks of a single tree growing on the island, as being the only one he had ever seen in that region. This was a young tree at the time Adanson received his information from M. Thiebault, although it had then, for some years, borne flowers and fruit. Dr. Roxburgh does not consider it to be indigenous in India, where he says it is scarce and of small size, observing that a few only have been found of any size at Allahabad, Masulipatam, on the coast of Coromandel, and in Ceylon. After reading Adanson's remarks on the custom of the negroes in transporting the seeds of the Baobab, it is not difficult to account for its introduction into Asia, and other parts of the world where it is now met with.

As a genus Adansonia seems to be chiefly distinguished from Bombax (whose habit it has, and to which Linnæus and Cavanilles say it is too nearly allied) by its deciduous calyx, its numerous stamina and its smooth shining seeds, those of Bombax being downy or woolly.— Among other and more obvious marks of agreement with other genera of the alliance Malvales, such as the extreme lightness of its wood, the large and handsome flowers, &c., may be mentioned that of the pollen grains being round and covered with minute points, as in our common Mallows. From the position of the Baobab in the system, we should expect to find its properties similar to those of its natural allies the Malvaceæ ; and the various uses made by the negroes of the different parts of the tree confirm this expectation. The mucilaginous emollient quality common to the tribe, resides principally in the bark and leaves ; these are dried in the shade, in a free current of air, then reduced to a powder of a beautiful green colour and nearly tasteless ; this powder is kept in a dry place in calico bags, and is called *lalo*. The negroes make daily use of the *lalo* in their food, for the purpose of keeping up an abundant perspiration and cooling the blood. Adanson and one of the French officers who accompanied him, made use of the *lalo* rather freely, and to it he attributes their preservation from the ardent fevers so prevalent in Senegal during September and October. The author particularly mentions the year 1751, the autumn of which year was more than usually unhealthy in Senegal, and states that himself and his friend were the only persons of the party who were able to follow their usual avocations, all the other officers being confined to their beds. The fruit of the Baobab appears to be as useful as the leaves; in a recent state its flesh is slightly acid and of an agreeable flavour, and its juice, mixed with sugar and water, forms a refreshing beverage in putrid and pestilential fevers. The fleshy envelope of the seeds, when dry, is reduced to an impalpable powder; P. Alpinus says it was sold in his time as a medicine, under the improper name of *terra sigillata,* or Lemnian earth. Monkeys are said to feed on the seeds; these are about the size of a bean, shining, and of a brownish colour, and are made into necklaces by the negroes. The shell of the fruit, and even the fruit itself when spoiled for eating, is burned, and the ley obtained from the ashes, boiled with rancid palm oil, forms an excellent soap.

Our author concludes his account of the various uses made of the Baobab, with the following singular narrative. It has been previously stated that the roots of such trees as grow in stony ground are liable to injury, and that in consequence their trunks decay and become hollow. The negroes take advantage of the cavities thus formed, and shape them regularly into chambers, or rather vast caverns, wherein they deposit the bodies of those whom they deem unworthy to receive the ordinary rites of burial. Of this class are persons called *guiriots;* these are the poets, musicians and players, of both sexes, who are hired to preside over and assist at dances and other entertainments, to which they impart much life and spirit by their buffooneries. The negroes regard these people, while living, with a kind of superstitious awe and reverence, but no sooner are they dead, than such feelings give place to horror and contempt; the natives then neither allow their bodies to be buried in the earth nor cast into the waters, imagining that if thus disposed of, the fish in the latter would be destroyed, and that the former would produce no food. By way of averting these evils the bodies of the guiriots are suspended within the hollow trunks of the Baobab, the entrances to which are closed with planks, and there, without being embalmed, they quickly dry up and become converted into a kind of mummy.

But it appears that the Baobab is not exclusively appropriated as a receptacle of the dead; the one measured by Golberry was hollow, and used as their hall of assembly by the inhabitants of the valley of Dock-Gagnack. The entrance was seventeen feet high, and faced a lake; the height of the cavity itself was twenty feet and its diameter twenty-one. The negroes had ornamented the sides of the doorway and the interior of the cavern with rude sculptures in relief. The party pitched their tents by the side of this tree, and M. Golberry was so well pleased with the chamber, that he ordered his bed to be placed within it, intending to pass the night there. This however caused so much dissatisfaction among the natives, that he abandoned his intention, although the chiefs would not have prevented him from carrying it into effect. He states that he had no occasion to repent his forbearance, having been afterwards treated with the greatest kindness by the natives.

In Dr. Wilson's notes above alluded to, it is mentioned that he visited one of these trees in India, which the Bairagees whom he found sitting in its shade told him was the only one in the world, and requested him to take off his shoes as he approached it, an honour which himself and party declined paying. He was informed that several devotees nightly took up their quarters in the hollow trunk of this tree.

It is also stated that in South America the natives hollow out the trunk of the Baobab and use it as a habitation, and that the tree thus hollowed continues to grow and flourish so long as the sap-wood and bark remain. The wild bees of Abyssinia are also reported to deposit their honey in this tree; and that the honey stored therein is the best in the country.

Dr. Alex. Gibson, in the Bombay 'Medical and Physical Transactions,' states that at Goozerat, where grow many fine specimens of the Baobab, the fishermen use the fruit as a float for their nets, and that logs of the very light wood are also employed by them as a catamaran or raft in fishing or duck-catching.

Adanson observes that in Senegal the Baobab has almost as many names as there are kingdoms. The Oualofes call the tree *Goui* and its fruit *Boui;* the French call the tree *calabassier,* and the fruit *monkeys' bread,* (*pain de singe*). Prosper Alpinus, the first botanist who wrote of this tree, says that a fruit called *Baobab* was brought from Ethiopia to Grand Cairo; from his description, and from the notes of his commentator, Wesling, there is no doubt of this fruit being that of our tree, although, as Adanson observes, the fruits seen by Alpinus must have been small and in a bad state, being probably such as are used at Senegal for no other purpose than making soap. The author considers the figure given by Alpinus to have been drawn from imagination, but remarks that Clusius contented himself with representing only what he saw. Clusius says that he received the fruit under the names of *abavo* and *abavi*, from persons who had it from English sailors returning from Ethiopia, or rather from the coast of Guinea or Senegal: he also says that the Portuguese call the fruit *calebacera*. Scaliger gives a very short description of the fruit, brought from Mozambique under the name of *guanabimus*. After stating that all the above-named authors are cited in Bauhin's Pinax, Adanson observes that M. Lippi's manuscript remarks ought not to be passed over in silence. This learned traveller, who was the victim of a voyage into Abyssinia, undertaken by order of Louis XIV. during a period of tumult and revolution in that country, gives a much more accurate description of the fruit of the Baobab, which he saw at Cairo, whither it had been brought from Upper Egypt, than any author who had preceded him. After a warm eulogium on M. Lippi, Adanson concludes his admirable memoir by stating that it is evident, from the passages quoted, that the authors cited were acquainted only with the fruit and leaves of the Baobab, but that they had no knowledge of its flowers or of the tree which bore them, the monstrous size of which presents a fact the most remarkable which the history of Botany and perhaps that of the world can furnish.

GEO. LUXFORD.

65, Ratcliff Highway,
    December 17, 1842.

ART. CVIII.—*Note on the Sands of Barry, and on Equisetum variegatum.* By MR. J. B. BRICHAN.

I HAD lately an opportunity of examining the locality named above, and as Equisetum variegatum there assumes a somewhat different appearance from that which it presents on Deeside, the few remarks I have to offer will form an appropriate sequel to my paper on the three allied Equiseta, (Phytol. 369). Whether or not the spot in which I found the plant is the same in which it was found by Don, I am unable to say; having no guide and no information, I rather stumbled upon it than searched it out, and though I went over a considerable space, I detected it only in one spot. As the locality is in various respects interesting, I may be pardoned for occupying a few sentences in particularly describing it.

A considerable part of the parish of Barry, containing from six to ten square miles, is a low sandy flat, which, at some remote period, undoubtedly lay under the sea, and is still very little elevated above its level. The links and sands, which compose the southern portion of this flat, are diversified by numerous sand-hills and knolls, which increase in size towards the south-east, and there terminate in a large sandy ridge, probably about 100 feet above the level of the sea. The smaller ones are covered with grasses, mosses, Carices &c., especially with Ammophila arenaria (*Scotticé* bent), which covers also the larger hills and part of the higher ridge already mentioned. Towards that ridge the sand-hills are rather crowded, but to the westward they decrease both in size and in number, rising however in some cases to a height of 20 or 80 feet. Of these the highest and most conspicuous hill, or group, lies close to the sea, about one mile west of *the lighthouses,* and immediately south of a large plain or meadow, where a few tall poles mark the locality of an old race-course. It is on all sides clothed with Ammophila, but is hollow and broken in the middle. A cart-road runs through the hollow, in the middle of which, and on each side of the road grows Equisetum variegatum. The spot is, on an average, not more than ten feet above the sea.

The whole district of which Barry forms a part, rests upon the Old Red Sandstone, which, according to Mr. Miller's intensely interesting work bearing the same name, derives its prevailing colour from an admixture of iron. I know not whether it is to be attributed to the same cause that much of the sand of the district is tinged with red; but in the work just alluded to I find the following. — "The oxide deposited by the chalybeate springs which pass through the lower members of the formation, would give to white sand a tinge exactly resembling the tint borne by this upper member." And it is certain that when a cut of several feet is made in almost any part of the plain of Barry, the chalybeate water immediately appears, and that its peculiar scum is seen floating along the edges of a small stream which bounds the parish on the west, and into which various chalybeate springs discharge their waters. The general appearance of the *surface* where E. variegatum grows is perhaps against the supposition that the sand has any mixture of iron; but when the sand is dug, it is found to be of a dingy brown, and the lowest stratum of the sand-hills is streaked with sand of a still deeper hue. These facts appear to confirm the remarks I lately offered on the three allied species or varieties, — E. hyemale, Mackaii and variegatum; there is, however, this difference between the two localities that I have examined, that the prevailing rocks on Deeside are granite, and, if I mistake not, granitic gneiss. The situation of E. variegatum at Barry seems to overthrow my conjecture that "the banks and bed of rivers" are its natural habitat: there is scarcely an imaginable way in which a stream could have deposited it in the spot which it now occupies.

The *roots* of the Barry plant scarcely differ from those of the Deeside variety. The *stems* are of the same variable length and number of articulations, with 4 — 10 striæ: they are completely prostrate, except in a few instances, when supported by Ammophila arenaria. When not sheltered by that or any other plant, they are brownish on the upper or exposed side and green on the under; it is, however, possible that the brown colour may be the effect of the lateness of the season. On the upper side also the bands of black upon the sheaths run farther down the stem than they do on the under side. The *teeth* are wedge-shaped, not ovate as at Banchory: the bristles are longer and apparently more persistent. The *catkins* are in general more exserted and matured, and, as well as the stems, have sometimes a reddish tinge. The plant seems to *branch* in the same manner as in the higher and moist situation on Deeside, 150 feet above the sea. When the sand is compacted by small plants which afford no shelter to the Equisetum, the latter is generally very small, slender and filiform; where the sand is loose or the plant has shelter, its growth is much stronger, and in the sheltered situation it is greener. In no case does the plant attain the same size as on Deeside. Some specimens slightly resemble E. Mackaii, but are perfectly distinct; the resemblance arises from the bristles being longer, and the amount of black upon the sheaths greater than in the usual state of the plant.

The links and sands of Barry form a very interesting botanical station. I have no doubt that a summer ramble among these hills and hollows would amply repay the researches of the botanical visitor; for although my visit took place at the end of October and the beginning of November, when they were in a state of decay, I could detect, among others, the following plants:—Astragalus hypoglottis, Elymus arenarius, Juncus balticus, and, I think, Erigeron acris. Juncus balticus is especially plentiful, being found in almost every part of the extensive links, and forming large plats in the meadows and hollows between the sand-hills. In this and similar places the intelligen observer cannot fail to be struck with the peculiar adaptation of tl..

prevailing vegetable productions to the character of the soil in which they grow. To my mind it furnishes a striking instance, not only of the "Wisdom of God in Creation," but of the wisdom of God in *providence* also. In a district of almost pure sand, while the more inland and level part is covered with common vegetation and vegetable mould, the seaward portion has been gradually overgrown with Ammophila arenaria, Carex arenaria, Elymus arenarius, and several other plants, all so well fitted by their creeping roots to bind the sand and prevent it from shifting. Even Equisetum variegatum, though confined to a small space, has its creeping root, and is calculated to serve the same end. And thus, in the first instance, an arid waste of sand, either upheaved from the bottom of the sea or exposed by the gradual retiring of its waters, and in that state utterly destitute of vegetation, has, by the agency of a powerful wind, been partly accumulated into a natural bulwark against the return of the ocean; and then, while its plains have by degrees been covered with a mould which has converted them into land capable of cultivation, its still sandy heaps which compose that bulwark have become consolidated, even to the verge of the sea, by a dense covering of plants, the most prominent and important of which are found only in such localities.   J. B. BRICHAN.

Manse of Banchory, December 1, 1842.

ART. CIX.—*County Lists of the British Ferns and their Allies.*
Compiled by EDWARD NEWMAN.

I DO not wish to conceal the fact that the perusal of Mr. Watson's admirable paper on the 'Geographical distribution of British Ferns,' has been the means of inducing me to attempt a still more rigid investigation of the produce of our counties, so far as regards this beautiful family of plants. In prosecuting this design, I find myself at the very threshold compelled to abandon my original intention of restricting each county list to the observations of an individual botanist: the kind and prompt attention with which my wish, expressed in 'The Phytologist,' has been met, enables me to make these lists far more interesting, by combining under each county the observations of many botanists. In employing the nomenclature formerly proposed by myself, and subsequently adopted in the valuable Catalogue published by the Botanical Society of Edinburgh, I conform to the usage of *the majority of* my correspondents. The authorities are arranged as nearly as possible in accordance with the dates of the letters containing the information quoted.

It appears, from carefully collating all the authorities within my reach, that the recorded species of British ferns are forty-six in number: of these one is now considered a doubtful native; twelve (printed in *italics*) are ranked by some authors as varieties; and thirty-three (printed in Roman letters) are acknowledged by every writer. By adhering to these numbers the relative proportions of each county will remain uninfluenced by the different views of botanists on this most interesting but debatable point. Species whose present existence in the county has not been verified by any of my correspondents, are marked with a dagger. †

YORKSHIRE.

Lomaria Spicant. Abundant in the neighbourhood of Sheffield; a rigid jagged variety is sometimes found, *J. Hardy :* common everywhere, *S. Gibson :* heaths, shady banks &c. abundant, *R. Spruce :* Langwith, near York, *S. Thompson.*

Pteris Aquilina. Most abundant in the neighbourhood of Sheffield, and one or two marked varieties occur, *J. Hardy :* common, *S. Gibson :* heaths, shady lanes and woods, frequent, *R. Spruce :* near York, *S. Thompson.*

Allosorus crispus. Fountain's fell, *S. Gibson :* (J. Tatham), *H. C. Watson, S. Thompson :* Cronkley-scar &c.; Teesdale; sides of Ingleborough hill; Settle, frequent; on Penhill, near the slate quarry;—*Baines's Yorkshire Flora.*

Polypodium vulgare. Abundant in the neighbourhood of Sheffield, *J. Hardy :* common in many parts of the county; var. *immersum,* mihi, this elegant variety has the sori in small dots, which are sunk in deep pits in the frond, it occurs at Whitby; — *S. Gibson :* frequent near York and Castle Howard, *R. Spruce :* near York, *S. Thompson ;* var. *serratum,* Wass-woods near Helmsley, *H. Ibbotson.*

Polypodium Phegopteris. Scarce near Sheffield, *J. Hardy :* in Cave-hole wood near Settle, *J. Tatham :* common at Hebden-bridge, Halifax, and many other parts of the county, *S. Gibson :* frequent about Halifax in rocky woods (R. Leyland), *H. C. Watson :* Scawton Howle near Helmsley, very fine, Buttercrambe-moor and Langwith lane, near York, Teesdale &c. (J. Backhouse, jun.), *R. Spruce :* Ingleborough, with P. Dryopteris, *W. Wilson :* near York, *S. Thompson :* Wensley-dale, Bell-hagg near Sheffield, Penhill, gill on Bellerby-moor, Shibden-dale, Ogden-clough &c. near Halifax, rocky woods in the vale of Todmorden frequent, near Helmsley;—*Baines's Yorkshire Flora ;* Bolton-woods, *H. Ibbotson.*

Polypodium Dryopteris. Scarce near Sheffield; Anston-rocks, fourteen miles from Sheffield, plentiful, *J. Hardy :* a very common plant in the neighbourhood of Hebden-bridge, *S. Gibson :* Bolton-abbey (Churchill Babington), common in woods about Halifax, *H. C. Watson :* Castle-Howard park, on the lower calcareous grit; Whitstoncliff near Thirsk; in various parts of the North and West Ridings, almost exclusively on the sandstone (J. Backhouse, jun.), *R. Spruce :* Ingleborough, *W. Wilson :* Brimham-rocks, Teesdale; near Pickering, near Whitby, near Richmond, near Helmsley ;—*Baines's Yorkshire Flora ;* Rievaulx woods, Luilesworth-vale, Cave-hole woods, Bolton-woods, *H. Ibbotson.*

Polypodium calcareum. Plentiful on Anston-rocks, *J. Hardy :* abundant on our hills near Settle, *J. Tatham :* neighbourhood of Settle, where the plants are smaller than those from Lancashire, *S. Gibson :* hills above Settle; this species is exceedingly distinct from Dryopteris when growing (J. Backhouse, jun.), *R. Spruce :* Ingleborough with P. Dryopteris, *W. Wilson;* Ingleborough, (W. W. Brunton), from whom I have a specimen which is eight inches from tip to tip of the lowest pinnæ, and seven and a half from the base of the lowest pinnæ to the apex of the frond, *S. Thompson :* Arncliffe and Gordale (R. B. Bowman), *H. C. Watson :* several places in Gordale, *Baines's Yorkshire Flora.*

† Woodsia Ilvensis. Richmond (J. Wood), Francis' 'Analysis of British Ferns.'

Cystopteris fragilis. Uncommon in the neighbourhood of Sheffield, *J. Hardy :* Shibden and Beeston woods near Halifax, Settle, Knaresborough, and many other places in the county. The genus Cystopteris is said to affect limestone, but I always find the varieties growing much larger, and their forms better displayed, where there is no limestone. Mr. Francis lays great stress on the length of the rachis as a character whereby to distinguish the species; if he had been a fern-collector he would have known that this character depends very much on the situation in which the plant happens to grow, if, for instance, among loose stones, the rachis will be long, if on a mortared wall, short. The same author also remarks that C. dentata is only half the size of C. fragilis, and double the size of C. alpina : my specimens of the form called fragilis vary from 2 to 18 inches in length, of that called dentata from 4 to 16 inches, and of the Low Layton plant from 1 to 8 inches; *S. Gibson:* Shibden-dale near Halifax (R. Leyland), *H. C. Watson ;* on the obelisk-bridge, Castle-Howard park; on old walls &c. in various parts of the north-eastern moors; near Helmsley, and in the long walk, Knaresborough (T. B. Powell); sparingly in Teesdale, growing in caves along with C. dentata, but keeping perfectly distinct in its habit, appearance of fronds &c. (J. Backhouse, jun.), *R. Spruce ;* near Rievaulx-abbey, Helmsley, *H. Ibbotson ;* abundant at Eggleston-bridge on the banks of the Greta, Red-scar, Applegarth ; — *Baines's Yorkshire Flora.*

Cystopteris dentata. Very common near Settle, *J. Tatham ;* Settle and other places, *S. Gibson ;* Egglestone-abbey bridge, and many other places in Teesdale, very fine (J. Backhouse, jun.), *R. Spruce.*

Cystopteris angustata. Scarce in three places, Gordale and Attermine scars and Catterick-force, *J. Tatham ;* I have found this variety growing on the very same plant with C. dentata, at Lune-bridge, Teesdale; they are no doubt the same species, (J. Backhouse, jun.), *R. Spruce:* near Aysgarth-bridge, Wensley-dale, *Baines's Yorkshire Flora.*

Cystopteris alpina or *regia.* Near Fountain's-abbey, and on wet rocks about Knaresborough, according to Teesdale in the ' Linnean Transactions,' and in Baines's Yorkshire Flora it is said to grow near Coxwold, but I have seen specimens from none of these localities, *R. Spruce.*

Polystichum aculeatum. Not common near Sheffield, *J. Hardy ;* near Triangle, four miles from Halifax, and Highgreen woods, *S. Gibson ;* woods near Halifax, (R. Leyland), *H. C. Watson ;* Whitstoncliff, near Thirsk; on the magnesian limestone at Thorpe-arch and other places (J. Backhouse jun.), *R. Spruce ;* Thorpe-arch (J. Ellis), *S. Thompson.*

Polystichum angulare. Edlington-wood near Doncaster, not common, *J. Hardy ,* in endless variety in Beeston-woods, about seven miles north-east of Halifax, Shibden and many other places near Halifax, Whitby, Richmond, &c. *S. Gibson;* Richmond, (R. B. Bowman), *H. C. Watson;* Whitstoncliff near Thirsk, on the magnesian lime-stone at Thorpe-arch and other places; I believe that angulare is a variety of lobatum and not of aculeatum (J. Backhouse jun.), *R. Spruce.*

Polystichum lobatum. Scarce in the Sheffield district, more common near Doncas-ter, *J. Hardy;* in all the woods in the neighbourhood of Hebden-bridge, *S. Gibson;* woods near Halifax (R. Leyland), a little below Catterick-force (J. Tatham), *H. C. Watson;* frequent in the stony woods about Castle-Howard, Thorpe-arch and Tad-caster, on the magnesian limestone (J. Backhouse, jun.), *R. Spruce;* near Settle (J. Tatham), *S. Thompson.*

Polystichum Lonchitis. Sparingly on the rocks above Settle, at an elevation of 1500 feet, *J. Tatham, S. Gibson;* Attermire-scar above Settle (J. Tatham), no fruit, I suspect this at least to be a young A. lobatum, *H. C. Watson;* Settle (J. Tatham), *S. Thompson,* who sends a frond, which appears to me to be the seedling, or perhaps, more properly speaking, the alpine form of acul:atum, *E. Newman;* near Malham, *W. Wilson.*

Lastræa Thelypteris. Potterie-carr, rare, *J. Hardy;* near York, in two or three lo-calities. *S. Gibson;* abundant in Askham-bogs and Heslington-fields near York, fruc-tifying in both places; Terrington-carr, amongst Bryum squarrosum and Hypnum nitens, *H. Ibbotson;* near Copgrove, near Doncaster, near Hovingham, near Settle, *Baines's Yorkshire Flora.*

Lastræa Oreopteris. Very common in the Sheffield district, *J. Hardy;* abundant above Swabeck, *J. Tatham;* very frequent in woods about Halifax (R. Leyland), *H. C. Watson;* frequent on the north-eastern moors, Castle-Howard park (very fine), wood near Earswick and Langwith-lane near York (J. Backhouse jun.), *R. Spruce.*

Lastræa Filix-mas. Abundant in the Sheffield district, *J. Hardy;* very common near York, *R. Spruce, S. Thompson.*

† Lastræa cristata. On Plumpton-rocks near Knaresborough, *Baines's Yorkshire Flora.* There is at present no other information about this species, *E.Newman.*

Lastræa rigida. Wharnside and Ingleborough, *W. Wilson;* on the rocks above Settle, at an elevation of 1500 feet, *J. Tatham;* Attermire-rocks and other places in the neighbourhood of Settle, *S. Gibson,* (J. Backhouse, jun.), *R. Spruce;* Clay-pit scar above Settle (J. Tatham), *H. C. Watson;* first gathered in 1815 on Ingleborough, near the foot of the mountain, towards the neighbouring village (W. T. Bree), *New-man's British Ferns.*

Lastræa dilatata. Abundant and very variable in the Sheffield district, *J. Hardy;* common in all the lanes and woods about Hebden-bridge &c. in endless variety, *S. Gibson;* very frequent near Halifax (R. Leyland), Black forest, Richmond (Rev. J. E. Leefe), both these appear to me very near spinulosa, *H. C. Watson;* the commonest fern in wet woods at Castle-Howard, Stockton forest, Leckby-carr &c. (J. Backhouse, jun.), *R. Spruce;* near York, *S. Thompson.*

Lastræa spinulosa. Spinulosa? Potterie-carr near Doncaster, *J. Hardy;* Scar-borough-mere, *S. Gibson;* bog near Rufforth-grange, Langwith, Stockton-forest, Thorne-moor, Leckby-carr &c. (J. Backhouse jun.), *R. Spruce;* near York, a singular variety is found at Thorne-moor, *S. Thompson.*

Lastræa dumetorum. In stony places in Castle-Howard park, *R. Spruce.*

Athyrium Filix-femina. Abundant in the Sheffield district, *J. Hardy;* in endless variety in many places, *S. Gibson;* shady banks near York, frequent, Langwith, *S. Thompson, R. Spruce.*

Athyrium irriguum. Occurs occasionally in the Sheffield district, *J. Hardy ;* Midgpool-lane near Hebden-bridge ; Mr. Francis observes that this plant is very tender and without fruit; the specimens which Sir J. E. Smith sent to his correspondents are very rigid, and bear fruit very abundantly, *S. Gibson.*

† Asplenium fontanum. I found a single root of this plant on an old wall above Skipton-castle, in July 1835 ; I took all the fronds and the plant, of course, disappeared ; and I have a specimen of the plant given to me as a Teesdale plant, but perhaps under some mistake, *S. Gibson.*

† Asplenium lanceolatum. On a wall in the village of Wharfe (Bolton), *Turner & Dillwyn*, 723.

Asplenium Adiantum-nigrum. Rather uncommon in the Sheffield district, *J. Hardy ;* near Halifax and other places, *S. Gibson ;* Old walls in Ray-wood, Castle Howard; on the moors near the Hole of Horcum ; on the banks of the Greta, and many other places in Teesdale, Whitstoncliff &c. (J. Backhouse jun.), *R. Spruce;* rocks at Ampleforth near Helmsley, walls on the moors above Pickering, *H. Ibbotson.*

Asplenium Ruta-muraria. Common in the Sheffield district, *J. Hardy ;* on the walls of Skipton-castle, near Hebden-bridge, *S. Gibson;* old walls, bridges &c. near York and Castle Howard, *R. Spruce ;* York city-walls, *S. Thompson.*

† Asplenium septentrionale. Ingleborough-hill (Mr. Tofield). It does not appear to have been found by any subsequent botanist ;—*Turner and Dillwyn*, 723. I have seen specimens from Ingleborough, and think it is probably there yet. *J. Backhouse, jun.*

Asplenium marinum. On the cliffs north of Scarborough, very rare, *S. Gibson.*

Asplenium Trichomanes. Common in the Sheffield district, *J. Hardy ;* Highgreen woods and many other places, *S. Gibson ;* old walls in Ray-wood, Castle Howard, Knaresborough, (T. B. Powell) ; on an oolitic limestone crag at Crambeck, near the Derwent; near the Hole of Horcum ; on the magnesian limestone at Thorpe-arch; Whitstoncliff near Thirsk, where I find a variety with divided fronds, (J. Backhouse), *R. Spruce;* Near Pontefract, *S. Thompson ;* a variety with the fronds deeply divided grows on the Whitstoncliff near Thirsk; *Baines's Yorkshire Flora.*

Asplenium viride and the ramose variety, very common near Settle, *J. Tatham;* common at Green-hill, Settle and other parts of Craven; var. *ramosum*, Ogden-kirk near Halifax; var. *acutifolium* mihi, with the pinnæ very long and pointed, this beautiful and very distinct variety was found near Settle by Dr. Chorley of Leeds, to whom I am indebted for specimens, *S. Gibson ;* Aislabeck, Richmond (James Ward), Gordale (R. B. Bowman), *H. C. Watson ;* specimens from near Halifax, *W. Wilson ;* Cronkley-fell and many localities in Teesdale, on the limestone (J.Backhouse jun.), *R. Spruce ;* Gilla-leys wood ; walls on the moors near Pickering, *H. Ibbotson ;* Ingleborough, Widdale-fell, Wensley-dale, Reeth-moor in Swaledale, Hill-gill near the side of the brook, *Baines's Yorkshire Flora.*

Scolopendrium vulgare. Common in the Sheffield district; var. *crispum*, plentiful near Sprotborough, two miles from Doncaster; var. *undulatum*, in Edlington-wood and on Warmsworth-cliff near Doncaster; var. *ramosum*, one plant on rocks near Sprotborough, *J. Hardy;* on the sea-coast at Scarborough, and on the hills above Settle, *S. Gibson;* Thorpe-arch (J. Backhouse, jun.); in the Long walk Castle-bank, and at Knaresborough (T. B. Powell); old walls near Castle Howard, peculiarly abundant in Mowthorpe-dell, where I have gathered many varieties ; *R. Spruce;* Plumpton-park near Harrogate (J. Richardson), *S. Thompson.*

thered it in fruit; a slender variety with the branches much attenuated is sometimes met with, *J. Hardy;* moist meadows at Ganthorpe near Castle Howard, *R. Spruce;* Goadland-dale near Whitby; near Green Hammerton, Settle, Richmond, Leeds &c.; by the brook at Hesketh-grange, near Boltby, Arncliff woods, *Baines's Yorkshire Flora;* common, *S. Gibson.*

Equisetum arvense. Too common in the Sheffield district, *J. Hardy;* moist meadows, cornfields &c. frequent, *R. Spruce;* common, *S. Gibson, H. Ibbotson.*

Equisetum fluviatile. Frequent near Castle Howard, especially in boggy woods, where I have seen branched fronds a foot high, surmounted by catkins; Langwith, Stockton and other places near York, also near Malton (J. Backhouse jun.), *R. Spruce;* in the Roche at Roche-abbey; roadside between Thornburgh and Upsall; wood on Wass-bank on the road to Helmsley; Arncliff wood, *Baines's Yorkshire Flora;* Lombard's-clough near Todmorden, *S. Gibson.*

EDWARD NEWMAN.

(To be continued).

ART. CX.—*Additions to the Phænogamic Flora of ten miles round Edinburgh.* By THOMAS EDMONSTON, jun. Esq.

(Continued from p. 407).

PLANTS NOT PREVIOUSLY OBSERVED IN THE DISTRICT.

Plantago Coronopus, var. β. *nana.* Abundant near Granton, and elsewhere.

Symphytum officinale, var. β. *patens.* Near Muttonhole, &c.

SPECIES PREVIOUSLY OBSERVED.

Geranium rotundifolium. Very fine at Preston-pans.

Geum intermedium. Abundant in many places, as at Roslin, Hawthornden, &c. Is this a species or not?

Habenaria chlorantha, Bab. After observing this plant very attentively for some time, and comparing it with the allowed H. bifol. and with specimens from the Edinburgh Botanical Society, from M. Babington, and other eminent botanists, of their H. chlor. must say that I cannot see permanent grounds of specific differ. The extreme forms appear very unlike, but the intermed so very common, that it appears to me there are sca reasons for separating the plants. I should much like Phytologist' a record of the observations of some b with the plants in a living state.

'b.

Hieracium sabaudum. One of the most co hawkweeds.

abundant on the side of Ingleborough, near the summit; Aislaby low moor near Whitby; on Sowerby, Wadsworth and Midgley moors, frequent; *Baines's Yorkshire Flora*; Cronkley-fell, Teesdale, Pen-y-ghent (R. B. Bowman), *H. C. Watson*.

Lycopodium inundatum. Plentiful in various parts of Strinsall and Towthorpe moors, as well as in Stockton-forest; Terrington-carr, scarce, *R. Spruce*; Stockton-common, *S. Thompson, S. Gibson*; in a sand-pit on the Malton-road, four and a half miles from York; Norland-moor near Halifax, *Baines's Yorkshire Flora*; Shackle-moor near Castle-Howard, *H. Ibbotson*.

Lycopodium selaginoides. Stockton-forest; Welburn-moor near Castle Howard; Teesdale, (J. Backhouse, jun.); Goldsbro'-moor near Knaresborough (J. B. Powell); Hackness near Scarborough (Mr. Peterkin), *R. Spruce*; Stockton-common, *S. Thompson*; on a part of the moor opposite the poor-houses in Wheldrake-lane, four miles east of York; on an island just above the bridge that crosses the Tees to Lower Cronkley; various places near Settle; in a marshy place on the moor north-west of the Beacon, near Richmond; on the top of Whitstoncliff near Thirsk; *Baines's Yorkshire Flora*; near the beacon, Richmond (James Ward), Towthorpe-moor (J. Storey), *H. C. Watson*.

Lycopodium Selago. High moors five miles from Sheffield, not common, *J. Hardy*; sparingly on Stockton-forest and Slingsby-moor, but not unfrequent on the north-eastern moors, *R. Spruce*; Pen-y-ghent and Ingleborough (J. Tatham), *S. Thompson*; near Settle, abundant; Penhill; moors near Halifax, Todmorden &c., *Baines's Yorkshire Flora*; Pen-y-ghent (R. B. Bowman), *H. C. Watson*; High-green woods, and many other places in the county, *S. Gibson*.

Isoetes lacustris. Castle-Howard lake; in the Foss reservoir near Coxwould, *H. Ibbotson*.

Pilularia globulifera. Stockton-common near York, Scarborough-mere &c. *S. Gibson*; margin of Gormire-pool near Thirsk, *H. Ibbotson*; Gormire, at the base of Whitstoncliff; Stockton-forest near York, (O. A. Moore); and near Richmond, *Baines's Yorkshire Flora*.

Equisetum hyemale. By the Derwent at Crambeck; near Raskelf (J. Backhouse jun.), *R. Spruce*; from Wakefield to Pontefract; Goodland-dale near Whitby; Hackness near Scarborough; near Halifax, *Baines's Yorkshire Flora*; dry woods at Castle Howard and Kirkham, not common (Teesdale), about Leeds but rare (Rev. W. Wood), Hackfall (Rev. J. Dalton) &c. *Turner & Dillwyn*; Bolton-woods in Wharfedale, *S. Gibson*; by the banks of a rivulet at Conesthorpe, *H. Ibbotson*.

Equisetum variegatum. Near Winch-bridge, and other places in Teesdale, *S. Gibson, C. C. Babington*, (J. Backhouse) *R. Spruce*; *H. Ibbotson*.

Equisetum palustre. Common in the Sheffield district, *J. Hardy*; abundant near York, particularly in ditches by the river Foss and in Hob-moor brick-ponds; not un-frequent throughout the district, *R. Spruce*; common, we have three or four varieties of this species, some of them not uncommon, *S. Gibson*.

Equisetum limosum. Plentiful in the Sheffield district, *J. Hardy*; abundant near York, particularly in ditches by the river Foss, and in Hob-moor brick-ponds; not un-frequent throughout the district, *R. Spruce*; at the bottom of Wensley-dale, *Baines's Yorkshire Flora*; common, we have two forms of this plant, perhaps as distinct as E. Mackaii from E. variegatum, one, much smaller and having fewer teeth on the sheaths, grows in the canal near Hebden-bridge, the other and larger plant grows near Selby, *S. Gibson*.

Equisetum sylvaticum. Not common in the Sheffield district; I have never gathered it in fruit; a slender variety with the branches much attenuated is sometimes met with, *J. Hardy ;* moist meadows at Ganthorpe near Castle Howard, *R. Spruce ;* Goadland-dale near Whitby; near Green Hammerton, Settle, Richmond, Leeds &c.; by the brook at Hesketh-grange, near Boltby, Arncliff woods, *Baines's Yorkshire Flora;* common, *S. Gibson.*

Equisetum arvense. Too common in the Sheffield district, *J. Hardy ;* moist mea_ dows, cornfields &c. frequent, *R. Spruce ;* common, *S. Gibson, H. Ibbotson.*

Equisetum fluviatile. Frequent near Castle Howard, especially in boggy woods, where I have seen branched fronds a foot high, surmounted by catkins; Langwith, Stockton and other places near York, also near Malton (J. Backhouse jun.), *R. Spruce;* in the Roche at Roche-abbey; roadside between Thornburgh and Upsall; wood on Wass-bank on the road to Helmsley; Arncliff wood, *Baines's Yorkshire Flora ;* Lom- bard's-clough near Todmorden, *S. Gibson.*

EDWARD NEWMAN.

(To be continued).

ART. CX.—*Additions to the Phænogamic Flora of ten miles round Edinburgh.* By THOMAS EDMONSTON, jun. Esq.

(Continued from p. 407).

PLANTS NOT PREVIOUSLY OBSERVED IN THE DISTRICT.

Plantago Coronopus, var. β. *nana.* Abundant near Granton, and elsewhere.

Symphytum officinale, var. β. *patens.* Near Muttonhole, &c.

SPECIES PREVIOUSLY OBSERVED.

Geranium rotundifolium. Very fine at Preston-pans.

Geum intermedium. Abundant in many places, as at Roslin, Haw- thornden, &c. Is this a species or not?

Habenaria chlorantha, Bab. After observing this plant very at- tentively for some time, and comparing it with the allowed H. bifolia, and with specimens from the Edinburgh Botanical Society, from Mr. Babington, and other eminent botanists, of their H. chlorantha, I must say that I cannot see permanent grounds of specific distinction. The extreme forms appear very unlike, but the intermediate ones are so very common, that it appears to me there are scarcely sufficient reasons for separating the plants. I should much like to see in 'The Phytologist' a record of the observations of some botanist familiar with the plants in a living state.

Hieracium sabaudum. One of the most common Edinburgh hawkweeds.

Hordeum pratense. Not now to be found in the district; some fine specimens occurred to Dr. Neill in the King's Park many years since.

Juncus obtusiflorus. Pentland Hills. A curious viviparous variety of J. supinus (J. uliginosus), with half-prostrate stems from a foot to two and a half feet long, occurs with the preceding in the marshes near Collinton.

Lamium Galeobdolon and *rugosum.* Dalkeith Park. L. album, *Linn.,* L. maculatum, *Linn.,* and L. rugosum, *Ait.,* appear to be correctly referred by some authors to states of the same species.

Leontodon palustre. Very abundant on the Pentland Hills and elsewhere near Edinburgh. From several years' observation of this plant in Shetland and elsewhere, I am inclined to think it is a good species, although now generally sunk into a variety of L. Taraxacum. The characters drawn from the involucre appear to be constant in all the specimens I have seen; and it is by no means improbable that small specimens of L. Taraxacum have been confounded with palustre, for I have seen them growing together, and each preserving its distinctive characters.

Leontodon autumnale. Specimens of a curious variety of this plant having a very stout scape, and covered with a very dense, long, and silky pubescence, occur near Collinton. It is exactly similar to some specimens brought from the Outer Hebrides by Dr. Balfour and Mr. Babington last year, and exhibited at the Botanical Society.

Polygala vulgaris. I have observed some curious variations in the size of the leaves and sepals, and in the habit, of some Polygalas on Arthur's Seat, Braid, Blackford and Corstorphine hills, &c., and Mr. E. Forbes has registered the same. Whether these differences may prove sufficiently constant to constitute species, I must leave to future observers.

Potamogeton. Great numbers of this intricate genus are to be met with around Edinburgh, and, I have little doubt, forms (or species) different from those described, but I confess myself perfectly unable to distinguish many of the puzzling forms; and I do think that minute differences in the shape of the fruit are not always to be depended on.

Primula veris, α. Lin. (P. veris of authors), β. Lin. (P. vulgaris) and γ. Lin. (P. elatior of British botanists, P. acaulis, β. *caulescens,* Balfour and Babington). All these forms (for it seems now fully proved that they are no more) are common about Edinburgh.

Sedum reflexum. St. David's, Fife.

Senecio aquaticus, β. *erraticus,* and other curious states, occur at Duddingston-loch.

Stratiotes aloides. Abundant in a brook which runs into Duddingston-loch. This plant, and Butomus umbellatus, are said to have been planted in the above station, but I know not on what grounds.

Thalictrum flavum. St. David's, Fife.

Trifolium incarnatum. Aberlady, Haddingtonshire, apparently indigenous.

THOS. EDMONSTON, JUN.

Baltasound, Shetland,
    November 27, 1842.

ART. CXI. — *Some Account of the Botanical Collections recently made by Dr. Theodore Kotschy (for the Wurtemburg Botanical Union) in Nubia and Cordofan.* Communicated by MR. WM. PAMPLIN, jun.

(Continued from p. 420)

WE will now proceed to give a complete enumeration of the species contained in the entire collection.

Marsileaceæ.
Marsilea nubica, *Al. Braun*
    Alismaceæ
Alisma Kotschyi, *Hochst.*
    enneandrum, *Hochst.*
Sagittaria nymphææfolia
    Hydrocharideæ.
Udora cordofana, *Hochst.*
Nymphæa cærulea, *Savi*
    ampla, *Cand.*
    Lotus, *L.*
    Gramineæ.
Cenchrus echinatus, *L.*
    longifolius, *Hochst.*
    macrostachys, *Hochst.*
Elytrophorus articulatus, *B.*
Cynodon Dactylon, *Pers.*
Digitaria ciliaris, *Koel.*
Chloris punctulata, *Hochst.*
    spathacea, *Hoc.* [*W.*
Dactyloctenium ægyptiacum

Helopus annulatus, *Nees*
Schœnefeldia gracilis, *Kth.*
Lappago occidentalis, *Nees*
    racemosa, *Schreb.*
Leptochloa arabica, *Kunth*
Aristida hordeacea, *Kunth*
    Kotschyi, *Hochst.*
    meccana, *Hochst.*
    Sieberiana, *Trin.*
    stipiformis, *Lam.*
    plumosa, *L.*
    uniglumis, *Lichst.*
Setaria imberbis, *R. S.*
    verticillata, *Beauv.*
Pennisetum lanuginosum *H.*
Gymnothrix nubica, *Hochst.*
Panicum arvense, *Kunth*
    turgidum, *Forsk.*
    subalbidum, *Kunth*
    Petiverii, *Kin.*
    Kotschyanum, *Hoch.*

Sporobolus glaucifolius, *H.*
Crypsis schœnoides, *Lam.*
Oryza sativa, *L.* [*Forsk.*
Andropogon annulatus,
    Gayanus, *Kunth*
    nervatus, *Hochst.*
Sorghum saccharatum, *Per.*
    halapense, *Pers.*
Diplachne elongata, *Hoch.*
    alba, *H.*
    pœformis, *H.*
Poa ciliaris, *L.*
Eragrostis tremula, *H.*
    pilosa, *Beauv.*
    megastachya, *Link*
Triachyrum cordofanum, *H.*
Ctenium elegans, *Kunth*
    Cyperoideæ. [*Rttb.*
Cyperus conglomeratus,
    elongatus, *Sieb.*
    aristatus, *Sieb.*

2 R

Cyperus retusus, *Nees*
squarrosus, *L.*
lepidus, *Hochst.*
Lamarckianus, *Schult.*
resinosus, *Hochst.*
pygmæus, *Rottb.*
rotundus, *L.*
vulgaris, *Kunth*
Fimbristylis hispidula, *Kth.*
dichotoma, *Vahl.*
Isolepis prælongata, *Nees*
Heleocharis monandra, *H.*
Commelinaceæ.
Commelina subaurantiaca,*H*
Forskaolii, *Vahl*
Juncaceæ.
Tenagocharis alismoides, *H.*
Palmeæ.
Cucifera thebaica, *Del.*
Coronarieæ.
Asphodelus fistulosus, *L.*
Characeæ.
Chara brachypus, var. nu-
bica, *Al. Braun*
Amentaceæ.
Salix — sine flor.
Urticaceæ.
Ficus glumosa, *Caill.*
Nyctagineæ.
Boerhaavia hirsuta, *W.*
repanda, *W.*
vulvariæfolia, *Poir.*
Aristolochieæ.
Aristolochia Kotschyi, *Hch.*
Laurineæ.
Cocculus Bakis, *A. Rich.*
Plumbagineæ.
Plumbago auriculata, *Lam.*
Rubiaceæ.
Spermacoce compacta, *Hch.*
leucodea, *H.*
Borreria radiata, *Cand.*
Mitrocarpus senegalensis, *C.*
ampliatus, *Hochst.*
Kohautia strumosa, *Hochst.*
senegalensis, *Cham.*
No. 138.
cæspitosa, *Sehnizl.*

Compositæ.
Dicoma tomentosa, *Cass.* =
Schaffnera carduoides,*Sch.*
Diplostemma acaule, *C. H.*
alatum, *Hchs* [*Schlt*
Sphæranthus angustifolius,
nubicus, *Sch.*= [*Cnd.*
Sprunera alata, *Schultz*
Vernonia paucifolia, *Less. β.*
angustifolia
Ageratum conyzoides, *L.*
Bidens bipinnata, *L.*
Ethalia gracilis, *Cand.*
Pluchea Kotschyi, *Schltz.*
Gnaphalium niliacum, *Rad.*
Cotula cinerea, *Del.*
anthemoides, *L.*
Inulaster Kotschyi, *Schultz*
Pulicaria undulata, *Cand.*
Francœuria crispa, *Cass.*
Doellia Kotschyi, *Schultz*
Pegolettia senegalensis, *Cas.*
Stengelia Kotschyana, *Hch.*
Blainvillea Gayana, *Cass.*=
Eisenmannia clandestina, *S.*
Hinterhubera Kotschyi, *Sch*
Eclipta erecta, *L.*
Sclerocarpus africanus, *Jacq*
Dipterotheca Kotschyi, *Sch.*
Microrhynchus pentaphyllus,
Sonchus cornutus,*Hoc.* [*Ho.*
Xanthium strumarium, *L.*
Cucurbitaceæ.
Bryonia fimbristipula, *Fenz*
Momordica crinocarpa, *Fnz*
Cymbalaria, *F.*
Balsamina, *L.*
Cyrtonema convolvulacea, *F*
Coniandra corallina, *F.*
Cucurbita exanthematica, *F*
Cucumis Bardana, *F.*
cognata, *F.*
ambigua, *F.*
Labiatæ et Verbenaceæ.
Moschosma polystachyum,*B*
Ocymum dichotomum, *Hch.*
lanceolatum, *Schum.*
menthæfolium, *Hchst.*

Leucas ciliata, *Bnth. β.* hirsuta
Leonotis pallida, *Benth.*
Verbena supina, *L.*
Holochiloma resinosum, *H.*
Volkameria Acerbyana, *Vis*
Asperifoliaceæ.
Echium setosum, *Del.*
Coldenia procumbens, *L.*
Heliotropium undulat. *Vhl.*
cordofanum, *Hochst.*
subulatum, *H.*
bicolor, *H.*
supinum, *L.*
pallens, *Caill.*
ovalifolium, *Forsk.*
indicum, *L.*
Cordia abyssinica, *Hochst.*
Anchusa asperrima, *Del.*
Convolvulaceæ.
Convolvulus pycnanthus,*H*
rhiniospermus, *H.*
filicaulis, *Vahl.*
lachnospermus, *H.*
microphyllus, *Sieb.*
Batatas pentaphylla, *Chois.*
auriculata, *Hochst.*
Ipomœa Kotschyiana, *Hck.*
coscinosperma, *H.*
gnaphalosperma, *H.*
coptica, *Roth*
repens, *Roth.*
palmata, *Frsk.*
cardiosepala, *H.*
pinnata, *H.*
acanthocarpa
sulphurea, *H.*
trematosperma, *H.*
Polygaleæ.
Polygala erioptera, *Cand.*
eriopt. var. pubescens
obtusata, *Del.*
Personatæ.
A.—*Rhinantheæ.*
Striga orchidea, *Hockst.*
hermontica, *Del.*
Chascanum marrubiifol. *F.*
lactenm, *F.*

B.—*Acanthaceæ.*
Acanthodium hirtum, *Hoch.*
Monechma hispidum, *H.*
  bracteosum, *H.*
Polyechma cæruleum, *Hch.*
Dipteracanthus patulus, *Nees*
Asteracantha macrurantha, *H*
Barleria Hochstetteri, *Nees*
Thunbergia annua, *Hochst.*
Thyloglossa sexangularis, *H*
  = Rostellaria sexang. *H.*
  palustris, *Hochst.* =
Gendurussa palustris, *H.*
Peristrophe bicalyculata, *N.*
Dicliptera spinulosa, *Hochst.*
Hypoëstes latifolia, *Hochst.*

Eranthemum decurrens, *H.*
C.— *Scrophulariæ.*
Macrosiphon elongatus, *H.*
  fistulosus, *Hochst.*
Chilostigma pumilum, *Hchs*
Sutera serrata, *Hochst.*
  dissecta, *Endl.*
Anticharis arabica, *Endl.*
D.—*Bignoniaceæ.* [*H.*
Ceratotheca melanosperma,
Sesamum rostratum, *Hochst.*
  orientale, *L.*
Pedalium Caillaudii, *Del.*
  Solanaceæ.
Solanum dubium, *Fres.* var.
  aculeatiss.

Solanum nigrum, *L.*
  albicaule, *Kotschy*
  hastifolium, *Hochst.*
Capsicum conicum, *Meyer*
Physalis somnifera, *L.*
  Lysimachiaceæ.
Utricularia inflexa, *Forsk.*
  stellaris, *L.*
  Asclepiadeæ.
Conomitra linearis, *Fenzl.*
Canahia Delilei, *Cand.*
Glossonema Boweanum, *C.·*
  Contortæ.
Hippion hyssopifolium, *S.*
  Sapotaceæ.
Styrax officinalis, *L.*

W. PAMPLIN, JUN.

(To be continued).

ART. CXII.— *Varieties.*

242. *On the poisonous effects of the Seeds of Hemlock.* In Lindley's ' Introduction to the Natural System,' 2nd ed. p. 22, it is stated that " the fruit of the Umbelliferæ is in no case dangerous." In the 'Pharmaceutical Journal' of this month, p. 337, Dr. Pereira mentions the case of a gentleman who suffered severely from having drank an infusion of anise, in which *some* seeds of Conium maculatum were detected. From the expressions used by the Dr. I presume that they were not numerous. It appears to me very desirable that the point should be set at rest. Dr. Maton is quoted by Paris in his Pharmacologia, as stating that the value of Extractum Conii is much increased by including the seeds in the preparation. If any confidence is to be placed in the case detailed by Dr. Pereira, which is taken by him from the ' Journal de Chimie Medicale' for August, 1842, it would appear that an infusion of the seeds is infinitely more powerful than the extract usually procured in commerce, since Dr. John Davy has administered the latter in drachm doses daily, with scarcely any untoward symptoms.—*Geo. Sparkes; Bromley, Kent, November 5, 1842.*

243. *On the folia accessoria of Hypnum filicinum, Lin.* A short time ago, whilst walking along the banks of one of our rivers, I happened to take up a tuft of that state of Hypnum filicinum which has been called H. fallax by Bridel, and on scrutinizing it with my pocket-lens, was surprised to observe, scattered here and there on the stem, several minute leaflets, scarcely one tenth so large as the true leaves, but yet resembling them in appearance. I brought the tuft home with me, and took an early opportunity of examining it with the microscope. I found these " folia accessoria " to occur chiefly towards the summit of the main stem, and more sparingly on the principal branches. In shape they are lanceolate, denticulate at the margins (*a*), nerveless (so far as I have observed), more delicate and with a wider reticulation than the true leaves. Sometimes they stand singly, as in figures *b, c, d;* not unfrequently several are found near each other, when they are smaller than usual; but most generally they stand in pairs, one leaflet partially overlying the other (*e f*). I observe them to be si-

tuated indifferently near the base or in the axil of a leaf, so that they cannot be considered stipules; but their most usual position is midway between two consecutive leaves.

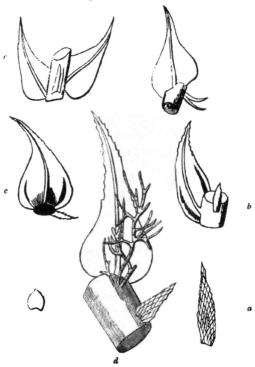

Accessory leaves of Hypnum filicinum.

a—c. Accessory leaves standing singly. *d.* Accessory leaf and radicles. *e, f.* Accessory leaves occurring in pairs. *g* Leaf of a branch bud. All the figures are magnified.

My first impression was that they were *branch-buds;* but the *true* branch-buds, which exist copiously on elongations of the stems, are first displayed as little bulbs bursting forth from the stem, and are composed of closely packed and very minute leaves, of the shape represented at fig. *g.* Afterwards, remarking that the folia accessoria were confined to the upper part of the stem, while the lower part was profusely clothed with radicles. I thought it barely possible that the former might pass into the latter; but after a very careful scrutiny I detected several instances of the two organs occurring intermixed, and each preserving its characters unaltered; thus, the radicles, in their most rudimentary state, were jointed cylinders of a deep brownish-purple hue; while the accompanying folia accessoria were pale green, cellular, foliaceous expansions: an instance of their conjunction is figured at *d.* I may add that the folia accessoria are more abundant on the upper, the radicles on the under part of the stem. These peculiar appendages I believe have been hitherto unnoticed, although they certainly exist in every state of Hypnum filicinum with which I am acquainted; it is, however, singular that they should be absent from the very nearly allied species H. commutatum, at least my specimens do not show them. It is by no means improbable that other species of Hypnum may possess folia accessoria, which, from their minuteness, have been overlooked by previous observers; and perhaps this brief account of what I have *myself notited,* may induce some of your correspondents, who have a love for muscology, to make more extensive observations with a view of ascertaining the fact.—*Rich-ard Spruce, F.B.S.E.; York, November* 16, 1842.*

244. *Information on Byssus barbata,* Eng. Bot. Your correspondent Mr. Lees requests information respecting Ozonium auricomum (Phytol. 428), the Byssus barbata of Withering and early English authors. This production was first denominated Ozonium auricomum by Link in 'Berlin Magazin,' and this generic title was confirmed by Persoon in the 'Mycologia Europæa.' It is not introduced in the second part of vol. v. of 'English Flora,' because Fries, Berkeley, and all the best mycologists of the present day, consider it best to exclude this and other doubtful productions from the catalogue of Fungi. The fructification of Ozonium is quite unknown, and this singular plant is believed to be an abnormal and barren state of some other Fungus, probably a Thelephora. I have sometimes fancied it bore some resemblance to Thelephora hirsuta, but this plant only grows on rotten wood and stumps of trees, whilst Ozonium auricomum is occasionally found on damp walls. The figure in Withering of its supposed fructification represents nothing possible. Although I fully concur in the propriety of excluding this plant, with the Rhizomorphæ and similar puzzling articles, from the body of mycological works, they are too interesting to be passed over without notice, and might, I think, be described in an Appendix, especially as most young fungists are disappointed by finding no mention made of them. Mr. Lees will find an admirable figure of Ozonium auricomum in Greville's 'Scottish Cryptogamic Flora,' iv. tab. 260; the description by Dr. G. contains all that is known concerning its structure, with the synonymy from the time of Dillenius upwards.— *H. O. Stephens; 78, Old Market-street, Bristol, December* 2, 1842.

245. *On the narrow-leaved Hypericum perforatum.* The narrow-leaved variety of Hypericum perforatum noticed in 'The Phytologist' for this month, (Phytol. 427), has been long known to me as a native of this island, where it is far from uncommon in similar situations with the ordinary form of the species, into which it passes by insensible gradations. Indeed the broad-leaved and normal state of H. perforatum is decidedly uncommon with us, though such a plant does occasionally occur, and makes a certain approach to H. dubium of Smith, in having but few pellucid dots upon the leaves, and one or more of the sepals is elliptical, oblong and obtuse, while the remainder preserve their usual acuteness of termination. I have never gathered the real dubium in flower, but I possess dried specimens in that state from others, and I found at Mucruss in October last what I take to be the genuine plant of 'English Botany' &c., but quite out of bloom, having *all* the sepals elliptical-oblong, rounded at the tip, and somewhat recurved; in habit the Irish plant was more like H. quadrangulum than H. perforatum, so similar indeed that I did not readily distinguish the two as they grew together, until more closely examined. My kind and liberal friend, Dr. Wood of Manchester, has pointed out to me since then the strongly marked pellucid reticulations of the leaves in H. dubium, when viewed by transmitted light, as affording an excellent character, in addition to those previously laid down for distin-

* Since writing the above, I have received a communication from Dr. Taylor (in answer to a letter enclosing Hypnum filicinum with folia accessoria), wherein he observes—" the name *filicinum* seems to have been applied almost prophetically to this species, which alone possesses these cauline scales that remind one of the ferns; they probably perform the same functions as the leaves, and yet app   to be universally without nerves."—*R. S.*

ART. CXIII. — *Proceedings of Societies.*

BOTANICAL SOCIETY OF EDINBURGH.

Wednesday, December 7.— Prof. Christison in the chair. The election of office-bearers for the season took place: — Dr. Neill, President; Professors Christison, Graham, Balfour and D. Stewart Esq., Vice-presidents. Various parcels of plants were announced, also donations to the library from Dr. Müller of Emmerich, Dr. Maly, &c.

Professor Christison then submitted to the Society a highly interesting communication on the Assam tea-plant, illustrated by specimens. The author stated that the different kinds of tea were produced by different modes of preparation, and showed by a series of examples of the preserved tea-leaf, that the various forms were merely varieties of the same plant. A specimen of tea, of a yellow colour, and of a remarkably strong flavour, was exhibited; also tea, in the form of small rolls, sent to this country about twenty years ago, as a present from the Emperor of China to George IV.

Mr. Goodsir then read a paper by Charles C. Babington, Esq., F.L.S., F.GS., entitled "Observations upon a few plants, respecting the claim of which to be considered as natives of Great Britain, Sir W. J. Hooker expresses doubt in the 5th edition of his ' British Flora,' with a few notes upon other species contained in that work, with reference to the Edinburgh ' Catalogue of British Plants.' The object of this paper was to show upon what evidence the authors (Professor Balfour, Mr. Babington himself, and Dr. Campbell) of the Botanical Society's ' Catalogue of British Plants ' had included in it the species concerning which Sir W. J. Hooker expresses doubt. " I cannot allow this opportunity to pass," says the author of this paper, "without expressing the great satisfaction which it gives me to see that so distinguished a botanist as Hooker has considered the catalogue deserving of quotation *throughout his work*, as I must consider it a proof that the compilers of the ' Catalogue of British Plants ' have not produced a work discreditable either to themselves or to the Society that intrusted its preparation to them."

Mr. Brand afterwards read to the Society a " Notice of the presence of Iodine in some plants growing near the sea," by G. Dickie, M.D., Lecturer on Botany in the University and King's College, Aberdeen. The author found, by chemical examination of specimens of *Statice Armeria*, from the sea-shore, and of others from the inland and higher districts of Aberdeenshire—that the former contained iodine, and that soda was more abundant in them, while potass prevailed in the latter. Iodine was also found in *Grimmia maritima*, and Mr. P. Grant of Aberdeen, has found it in *Pyrethrum maritimum*. An analysis was made of examples of *Statice Armeria, Grimmia maritima, Lichina confinis* and *Ramalina scopulorum*, all growing near the same spot, and occasionally during storms exposed to the sea spray; and all these plants, with the exception of the lichen, contained iodine. The specimens having been washed previous to analysis, the iodine could not have been derived from saline incrustation. All these vegetables were healthy, and the author of the paper has been led to conclude that marine Algæ are not the only plants which possess the power of separating from sea-water the compounds of iodine, and of condensing them in their tissues, and this without any detriment to their healthy functions.

BOTANICAL SOCIETY OF LONDON.

November 18.—Adam Gerard Esq. in the chair. Donations of British plants were announced from Miss S. B. Hawes and Miss S. K. Barnard. The continuation of the paper commenced at the last meeting, on the *Lodoicea Sechellarum*, by George Clark, Esq., was read.

The Lodoicea Sechellarum is an intertropical plant peculiar to the Sechelles Archipelago, where it grows naturally in two islands only, Praslin and Curiense. Praslin lies to the north-east of Mahè, distant twenty-one miles; Curiense to the north of Praslin, and is much smaller: a deep arm of the sea, from one to two miles in breadth, separates these two islands. They lie between 4° 15' and 4° 21' S. latitude, and 55° 39' and 55° 47' E. longitude. In the other islands of the Archipelago there are but few Lodoiceas, which have all been planted, and only two or three appear to thrive.

The trunk or stem is straight, and rises to the height of 80 or 90 feet, and is terminated by a splendid crown of winged palmated leaves; it is only from 12 to 15 inches in diameter, and so flexible that it waves to the slightest breeze. When the wind is moderately strong, the huge leaves of this giant palm are clashed together with an astonishing noise. The outside of the stem is very hard and compact, while the interior is soft and fibrous. The leaves, winged and palmated, open like a fan, and in their early growth are more than 15 feet long, without reckoning the foot-stalk, which is at least as much more. In the mature trees the leaf-stalk is not more than 8 or 10 feet long, and the whole leaf does not exceed 20 feet in length by 10 or 12 in breadth, and is entirely destitute of thorns. The nascent leaves are enveloped till the period of their expansion by a thick covering of cottony down, of a nankeen colour; but this is occasionally wanting. The unanimous testimony of the inhabitants of Praslin proves that each tree produces only one leaf a-year; and " as three leaves occupy about 8 inches of the trunk, and twenty years expire before that appears above the surface, a tree of eighty feet in height must be about four hundred years old." The flowers, about twenty in number, succeed each other one at a time, occasionally there are two together. The nuts are two-lobed, and sometimes two nuts are enclosed in one husk; three-lobed nuts are very rare, but some are met with, and it is said that specimens with five lobes have been found. The form of the nut is very singular, and cannot be compared with that of any other production of the vegetable kingdom. Two highly remarkable circumstances in the history of the Lodoicea are the duration of its blossoms and the period necessary for maturing its fruits; for the latter purpose seven or eight years are required. The Lodoicea grows in every variety of soil, but delights most in the vegetable mould of the deep gorges of the mountains. It is nevertheless found on the bare mountain tops, and forms a very conspicuous and remarkable object in such situations. It is curious that the vegetation of the nut should be prevented by its being buried, but if suffered simply to rest on the earth, in a situation not too much exposed to the sun, germination readily takes place. The purposes to which the produce of the Lodoicea is applied are numerous. The fruit, in its simple state, is an agreeable and refreshing aliment; when ripe it yields oil; its germ furnishes a very sweet food. Of the shell are made vessels of various shapes and sizes, that serve the Sechellois for nearly all domestic purposes. The entire nut is an article of commerce with India, where one of its uses is as an astringent medicine. The trunk is employed in building; split and hollowed it forms excellent gutters and paling; the leaf-stalks also are used for the latter purpose. The leaf forms a covering for roofs nearly as good as shingles, besides furnishing materials of a very superior description for hats, bonnets, wood-baskets and artificial flowers, in the manufacture of which many of the Sechelloises display great taste and skill. And lastly, the cottony down which covers the leaf previously to expansion, is a very good stuffing for pillows and mattresses. —
G. E. D.

THE PHYTOLOGIST.

| No. XXI. | FEBRUARY, MDCCCXLIII. | Price 1s. |

Art. CXIV. — *Note on Fucus Mackaii of Turner.* By Robert Kaye Greville, Esq., LL.D.

Fucus Mackaii, Turner.

Fucus Mackaii, a species established in the absence of fructification by the acute Turner, is reduced by Agardh to a mere variety of F. nodosus; and, it must be confessed, not without apparently good reason. The plant occurs in its different localities in extraordinary profusion, and strongly resembles some states of F. nodosus, so closely

2 s

indeed as scarcely to be distinguished from them. In the autumn of the present year I collected specimens of the latter plant, in Loch Ryan, which I thought conclusively established Agardh's view of the question. A few days afterwards, however, I was most agreeably surprised to observe in the collection of Mrs. Captain Maynard, at Stranraer, a specimen of Fucus Mackaii bearing receptacles; so that the doubts which have so long hung over this interesting plant, are now at an end.

The specimen (I regret to say it was a solitary one) was communicated to Mrs. Maynard by Dr. Lindsay from the Isle of Skye.

The specific character of Fucus Mackaii may now be expressed as follows. Frond coriaceous, cylindrical or subcompressed, linear, irregularly dichotomous: vesicles (when present) innate, elliptical, solitary; receptacles in pairs, ovate-oblong, terminating the lateral branches.

The annexed figure, copied from the drawing which I made at the time, of this unique specimen, will render any further description unnecessary. R. K. GREVILLE.

Edinburgh, December, 1842.

ART. CXV.—*Note on a supposed New British Cuscuta;* by CHARLES C. BABINGTON, Esq., M.A., F.L.S., F.G.S. *Communicated, with Additional Observations,* by G. S. GIBSON, Esq.*

I FULLY intended to forward C. C. Babington's notes on the Cuscuta (Phytol. 412), but have delayed doing so in the hope of obtaining clearer information as to the country from which the clover-seed came; but after considerable enquiry have not succeeded satisfactorily. The following are C. C. B.'s remarks upon it, any part of which you are of course at liberty to insert in ' The Phytologist.'

" I this morning received your letter, and immediately submitted the Cuscuta to the microscope. It certainly differs considerably from C. Epithymum, and I suspect will prove to be a new species, for 1 can find no notice of it in the continental works. I am not, however, prepared to give it as new, without more acquaintance with it than I have yet obtained. I believe that C. Epithymum is confined to shrubby plants, such as Ulex, Erica and Thymus. I add the cha-

---

* In a letter to E. Newman.

racter of the plant, according to my present ideas, and have under-lined the points in which I believe it to differ from C. Epithymum.* The provisional name that I have adopted is C. Trifolii.

"*C. Trifolii*, (Bab. MSS.)   Clusters of flowers bracteated, sessile : tube of the corolla cylindrical, *limb erect*, scales palmately cut, con-verging ; *calyx nearly or quite as long as the corolla.   Calyx* and corolla *whitish*, with acute segments.

"I had heard of a Cuscuta destroying the clover in Norfolk and Suffolk, but have never before been able to obtain specimens. I look upon it as very desirable that attention should be called to the plant, as it threatens to become a troublesome weed in fields, to which it has probably been introduced with clover-seed from the continent.— Could you ascertain if it was known on clover before recent years, and if the clover-seed was of English growth, or from what country obtained ?   I think that this query may, if it produce a decided an-swer, be of much agricultural value."

All the particulars which I have been able to ascertain, are the fol-lowing.

The clover-seed was bought by the farmer of a factor at Stortford, who distinctly states that it was foreign seed, but cannot tell from what part it came. The same farmer states that he noticed this Cus-cuta on clover in one of his fields about twelve years ago, but has not since observed it till this year.   In the adjoining part of Suffolk, it was a troublesome weed in several places, covering the clover to a considerable extent, and greatly injuring the crop ; this seed had been grown in England the previous year, and in this neighbourhood, but probably might have originally come from abroad, which I am anxious to ascertain, but fear the requisite information can scarcely now be obtained.   Perhaps if the notice of the readers of 'The Phy-tologist' be called to the subject, some more satisfactory result may be arrived at.   Should I hear any further particulars I intend to com-municate them.   G. S. GIBSON.

Saffron Walden, 3rd of 1st Month, 1843.

[Mr. Babington requests us to add that he would be much obliged to any person who would send for examination specimens of Cuscuta from clover, addressed to him at St. John's College, Cambridge.]

* These are the parts printed in *Italics.—Ed.*

ART. CXVI. — *Contributions towards a Flora of the Breadalbane Mountains.* By Wm. GARDINER, JUN., Esq.

THE plants enumerated in the following list were collected between the 28th of June and the 19th of July, 1842, on that wild and rugged mountain range which extends along the shores of Loch Tay from Kenmore to Killin, and of which the mighty mass of Ben Lawers forms the most prominent feature, rising to an altitude of 4015 feet above the sea-level. At the base of this mountain a little inn has been most conveniently placed, from which to the summit the distance is reckoned about five miles, and is usually attained with less than three hours' moderate climbing. Once at the top, the toil is amply compensated, if the mist be absent, by the magnificent panorama of the "land of the mountain and the flood" that is spread around. There are many fine sights in Breadalbane : the splendid waterfalls at Aberfeldy, Acharn, Boreland and Lawers ; —

" Th' outstretching lake, imbosom'd 'mong the hills : "

the romantic grounds of Taymouth : —

" The Tay meandering sweet in infant pride,
The palace rising on his verdant side ; "—

the quiet pastoral beauty of Glen Lochy, and the never-tiring picturesque scenery of Killin. But all these taken together fall short of producing such an interest as one feels while enjoying, in a clear sunny day, the vast prospect of alpine grandeur from the summit of Ben Lawers ; and the tourist who neglects this sight, although he has beheld the rest, misses what is most worthy of being seen. To the botanist Ben Lawers has additional charms. He seldom climbs it in the prescribed period of three hours, for almost at every step, beauty, in some form or other, beckons him to stoop and examine its charms. His pleasure is not on the summit alone, and therefore to him the ascent is not so toilsome. He walks, as it were, through a botanic garden, each compartment, as he ascends its lofty terraces, presenting him with a fresh banquet of new and lovely forms, unknown to the plains, and as he takes leisure to enjoy them, the health-giving breezes circle round him. The mountain-bee with its merry hum leads him to many a hidden gem ; and the babbling streams ever and anon are telling him of the floral treasures of their banks. He is diligent as the bee, happy as the sparkling stream. Yet no earthly happiness is without its alloy ; and rambling on Ben Lawers, whether the tour-*ist be botanist* or not, he will meet with shady moments as well as bright ones. If the sun beats strong and the air is dry, his face gets tanned and his ears blistered. If the mist comes on, and he has no compass nor guide, he runs the risk of losing himself, and descending the wrong side of the mountain. If he is overtaken by a Highland shower, he is positively certain of being thoroughly drenched to the skin; and if he has lacked the foresight to provide a spare suit to exchange for his wet one on reaching his inn, his plight will be pitiful. Moreover, he will occasionally have the comfort of getting over the ancles in a bog, of being stung by an angry wasp, or of having his blood painfully abstracted in various ways by tiny demons in the shape of insects. Yet after all, these are but trifles when we look them lightly in the face, and consider what a vast amount of pure delight is derivable from the contemplation of Nature, either in her minute details of vegetable and animal structure, or in her more general features of landscape scenery.

Thalictrum alpinum. Plentiful on Ben Lawers and the other mountains in the Breadalbane range, growing on the marshy banks of rills, as well as on wet rocks.

———— *minus, β. majus.* On the banks of Loch Tay.

Anemone nemorosa and *Trollius europæus.* These two beautiful denizens of our woods, the one with its delicately-tinted drooping flowers, and the other with its swelling globes of vegetable gold, adorned the wild and lofty rocks of Stuich-an-Lochan, and were greeted with all the rekindled warmth of early friendship. Stuich-an-Lochan is a wild rocky mountain, rising almost perpendicularly from the dark waters of Loch-na-Gat, and connected with Ben Lawers, of which it is generally considered a part, by the yet more fearful cliffs of Craig-na-Hein. It is one of the richest botanical fields in Breadalbane.

Cochlearia grænlandica. Plentiful about the summit of Ben Lawers, varying much in size, and in the colouring of the flowers and foliage.

Draba rupestris. In the crevices of rocks about the summit of Ben Lawers, but not in great quantity.

———— *incana.* More or less abundant on the rocky summits of Ben Lawers, Stuich-an-Lochan, Mael Tarmanach, Mael Greadha, and Craigalleach; varying from two inches to nearly a foot in height.

Arabis hirsuta. On the rocks of Stuich-an-Lochan, in fruit.

Silene acaulis. Abundant on all the Breadalbane mountains, at a good elevation, forming here and there a beautiful sward, glowing with its myriad blossoms of vivid red.

Lychnis diurna. On Stuich-an-Lochan, showing another example of low-ground plants seeking a high altitude.

Spergula saginoides. Rocky banks on Ben Lawers and Stuich-an-Lochan.

Alsine rubella. Very sparingly on crumbling rocks, at the summits of Ben Lawers and Mael Greadha.

Cerastium alpinum. Scattered profusely over all the range, and varying considerably in its degree of pubescence.

Cherleria sedoides. Abundant, and often forming widely extended patches.

Geranium sylvaticum and *pratense.* Plentiful in various situations, high on the mountains as well as at their bases.

Oxalis Acetosella. Stuich-an-Lochan could also boast of this graceful little gem of Flora, which recalled one from the sultry heat of July, and these perilous alpine rocks, to the soft luxury of the woodland in the merry month of May.

Anthyllis Vulneraria. Ben Lawers and Stuich-an-Lochan.

Vicia sylvatica. Den of Lawers, below the Falls.

Dryas octopetala. A single specimen without flower was picked on Stuich-an-Lochan.

Geum rivale. This, which is another species common on low grounds, occurred frequently in marshy places among the rocks of Stuich-an-Lochan.

Rubus saxatilis. On the same mountain, but not common.

—— *Chamæmorus.* Abundant on the peat-bogs and moory ground of Ben Lawers, both in flower and fruit. The latter, which are called *Avrons* by the shepherds, are esteemed for their nutritious properties as well as for their agreeable flavour. I had good reason to prefer the name of *cloud-berries*, for my specimens were culled amid the fleecy drapery of the heavens that mantled the mountain in their misty folds.

Potentilla alpestris. Not uncommon on the rocky ledges of Stuich-an-Lochan.

Sibbaldia procumbens. On the summits of all the mountains.

Alchimilla alpina. Plentiful everywhere, from the bases to the summits of the mountains, but most luxuriant where its bunches of elegant silvery foliage depend from the crevices of a rock in the immediate vicinity of one of those beautiful miniature cascades that are so numerous on an alpine rivulet.

Epilobium angustifolium. Margin of the Lochy, near Killin, imparting to the banks of that still and beautiful stream a peculiarly interesting aspect of floral grandeur, from the richness of its numerous handsome spikes of crimson blossoms.

———— *alpinum.* Marshy banks of springs and rivulets on all the mountains, generally gracefully drooping both in flower and fruit.

Circæa alpina. Moist banks near the Falls of Lawers, and in stony places shaded by trees on the banks of Loch Tay.

Sedum anglicum. On a rocky bank near Lawers, but scarce.

—— *villosum.* Growing rather sparingly on a marshy bank by the side of a stream about half way up Ben Lawers. It was also picked on a wet wall by the wayside in Glen Almond.

—— *Rhodiola.* Crevices of moist rocks on most of the mountain summits, but not so luxuriant as I have found it among the Clova mountains.

Saxifraga stellaris. Plentiful both in wet and dry places, and varying from half an inch to four or five inches in height.

———— *nivalis.* More or less abundant on the summits of Ben Lawers, Stuich-an-Lochan, Mael Tarmanach, Mael Greadha, and Craigalleach, growing in the clefts of rocks. Some of it had simple stems, others branched; and more of it was in fruit than flower.

———— *oppositifolia.* Frequent in the crevices of rocks, but its early and elegant purple blossoms were nearly gone.

———— *aizoides.* The constant ornament of all the alpine rivulets, but preferring a very moderate altitude, and even descending to the bottoms of the vallies.

———— *cernua.* This, the rarest and most interesting of the tribe, was found where

I gathered it in 1838, in the crevices of rocks and among debris, in a wild ravine near the summit of Ben Lawers, but it is now getting so scarce there, that unless picked with a sparing hand for some years to come, it will be in danger of being eradicated.

Saxifraga hypnoides. Plentiful.

Cornus suecica. Only a few specimens were picked on Ben Lawers, growing among Rubus Chamæmorus, in small boggy hollows.

Galium palustre, β. Witheringii. In marshy places by the side of Loch Tay.

———— *boreale.* Plentiful on the banks of Loch Tay in stony places, as well as in the Den of Lawers, by the side of the Lochy, and even on the rocky ledges of the lofty Stuich-an-Lochan.

Thrincia hirta. On Stuich-an-Lochan.

Crepis paludosa. About the Falls of Lawers &c.

Hieracium murorum, γ. Lawsoni. Rocks of Stuich-an-Lochan.

———— *prenanthoides* Den of Lawers, but little of it in flower.

Saussurea alpina. Frequent, though not plentiful, on the rocky shelves of Stuich-an-Lochan.

Carduus heterophyllus. Abundant in the Den of Lawers, on the banks of the Lochy, and various other places, some of its beautifully ciliato-dentate cottony leaves measuring more than a foot in length.

Gnaphalium dioicum. On the heaths, abundant.

———— *supinum.* In still greater profusion than the last, preferring bare earthy or dry rocky banks, and by no means confined to the summits of the mountains, but extending often to within a few hundred feet of their bases.

Erigeron alpinus. In tolerable plenty on the rocky ledges of Stuich-an-Lochan, generally with one flower, but occasionally with two or three.

Solidago Virgaurea, β. On the same mountain.

Vaccinium uliginosum. On boggy heaths about Loch-na-Gat, and several other places, but bearing no flowers.

Pyrola media. Picked a specimen or two about the Falls of Boreland in Glen Lochy, but past their prime.

Gentiana nivalis. A single specimen of this alpine rarity was culled on Stuich-an-Lochan, but several without flowers were left.

———— *campestris.* Plentiful on Ben Lawers.

Myosotis alpestris. On Ben Lawers and Stuich-an-Lochan in considerable abundance, gracing the rocks with its rich cærulean blossoms, than which none of our mountain flowers are more exquisitely beautiful.

Veronica serpyllifolia, β. alpina. In profusion in marshy places about springs and the sides of streams.

———— *saxatilis.* On the rocks of Stuich-an-Lochan and Craig-na-Hein, but in small quantity. This is a perfect gem, and its large, brilliant, though delicate corolla is well worth a long day's journey to look at.

Euphrasia officinalis. A variety with deep-coloured flowers and foliage was plentiful in the mountain-pastures, and in the valleys the usual pale-flowered state frequently attained a height of nearly a foot, with a more or less branched habit.

Melampyrum pratense and *sylvaticum.* In the Den of Lawers.

Digitalis purpurea. One of the most common and showy wild flowers in the Highlands.

Galeopsis versicolor. Common in the cornfields about Lawers.

Clinopodium vulgare. Banks of Loch Tay.

Armeria maritima, β. *alpina.* Summits of Ben Lawers and Stuich-an-Lochan.

Polygonum viviparum. Plentiful in the Den of Lawers, and in various places on the mountains.

Rumex alpinus. In waste places, Lawers, but without flowers this season. In 1838 I culled a specimen in the same locality, bearing flowers.

Oxyria reniformis. Abundant by the sides of the mountain streams.

Salix pentandra. In Glen Almond, with fertile catkins.

—— *reticulata.* Plentiful on the summits of all the mountains, loving best to grow on the sides of perpendicular rocks.

—— *arenaria.* By the sides of streams on Ben Lawers.

—— *cinerea.* About the falls of Boreland, the bank of Loch Tay, and by the side of Loch Frenchie in Glen Quech, varying very much in size and foliage.

—— *oleifolia.* Banks of Loch Frenchie.

—— *aurita.* Glen Quech.

—— *caprea.* Common in many places and very variable.

—— *Andersoniana.* Craigalleach, rare.

—— *tenuior.* Banks of the Lochy, near Killin, but not in great plenty.

—— *radicans* and *Weigeliana.*

—— *bicolor.* Plentiful about Loch Tay, and in Glen Quech, but without catkins.

—— *phyllyreifolia.* On the banks of Loch Tay near Lawers.

—— *herbacea.* Abundant on the rocky mountain summits, but finest on Stuich-an-Lochan.

Myrica Gale. In moory ground about the base of Ben Lawers, scenting the air with its rich and agreeable odour.

Tofieldia palustris. On Stuich-an-Lochan, Ben Lawers, Mael Tarmanach, &c.

Juncus trifidus. Near the summit of Ben Lawers, but poor starved specimens compared with those found at Clova.

—— *biglumis.* On Ben Lawers, Stuich-an-Lochan and Mael Greadha, but in very small quantity.

—— *triglumis.* Plentiful in marshy places on all the mountains, and varying with from two to four capsules.

Luzula spicata. In profusion on the rocky summits of all the Breadalbane range.

Narthecium ossifragum. In boggy places, common.

Gymnadenia conopsea. On Ben Lawers, frequent, and telling of its whereabouts by bribing the zephyrs with its aromatic breath.

—— *albida.* On grassy banks near Larich-an-Lochan.

Habenaria viridis. On Stuich-an-Lochan.

Anthoxanthum odoratum. This adventurous inhabitant of our meadows I found near the lofty summits of Ben Lawers and Craigalleach.

Phalaris arundinacea. On the banks of the Lochy.

Aira cæspitosa and *flexuosa.* Alpine states of these plants occurred on Ben Lawers and Craigalleach.

Melica uniflora. In the Den of Lawers.

Molinia cœrulea, β. *alpina.* On the banks of Loch Tay, and by the sides of streams on Ben Lawers.

Arrhenatherum avenaceum. Den of Lawers, &c.

Sesleria cœrulea. On the rocky ledges of Craigalleach.

Poa alpina. Common on all the mountains, and viviparous.

— *nemoralis.* Banks of Loch Tay.

— *cæsia.* Rocks of Stuich-an-Lochan.

Festuca ovina, ε. *vivipara.* Frequent.

Avena alpina. Rocky ledges of Stuich-an-Lochan and Ben Lawers.

Eriophorum gracile. Ben Lawers. Evidently a dwarf and slender state of E. angustifolium.

Carex pulicaris. On Ben Lawers.

—— *curta.* Ben Lawers, &c.

—— *intermedia.* Near Ben Lawers Inn, in a marshy spot.

—— *atrata.* On the ledges of Stuich-an-Lochan, not uncommon.

—— *rigida.* Rocky summits of all the mountains.

—— *saxatilis.* In marshy places on Stuich-an-Lochan and Mael Greadha.

—— *flava, pallescens and paludosa.* Den of Lawers.

—— *capillaris.* Marshy places on Stuich-an-Lochan, Mael Greadha & Craigalleach.

Polypodium Phegopteris. Abundant in various places.

———— *Dryopteris.* Den of Lawers.

Polystichum Lonchitis. In the clefts of rocks on all the mountains, but the specimens not so luxuriant as those found on the Forfarshire mountains.

———— *lobatum.* Den of Lawers, near the Falls.

Lastræa Oreopteris. Plentiful in the Den of Lawers.

Cystopteris fragilis and β. *dentata.* On moist rocks about the Falls of Lawers.

Asplenium viride. Crevices of wet rocks about the Falls of Lawers, in great profusion. This beautiful little species loves to vegetate in the immediate vicinity of waterfalls, though in 1838 I found a few plants in the shady cleft of a rock near the summit of Ben Lawers.

Hymenophyllum Wilsoni. Moist rocks about the Falls of Lawers, covering them with broad patches of its beautiful and delicate fronds.

Botrychium Lunaria. On Stuich-an-Lochan, sparingly.

Lycopodium selaginoides, alpinum and *Selago.* Plentiful on Ben Lawers and others of the range.

Equisetum palustre, β. *alpinum.* On Ben Lawers and Craigalleach.

Andræa alpina. Abundant and in fine condition about the summits of Stuich-an-Lochan, Ben Lawers, Mael Tarmanach and Mael Greadha.

—— *rupestris* and *Rothii.* Ben Lawers.

Gymnostomum lapponicum. On Ben Lawers and Mael Tarmanach.

Anyctangium ciliatum. In the Den of Lawers and many other places, common.

Diphyscium foliosum. Ben Lawers.

Splachnum sphæricum. Ben Lawers.

———— *mnioides.* On the same and all the other mountains of the range.

Conostomum boreale. Abundant and in fine perfection near the summit of Mael Tarmanach, covering a space of several yards. Sparingly on the other mountains.

Encalypta ciliata and *rhaptocarpa.* More or less diffused over the whole range.

Weissia crispula. Ben Lawers, plentiful.

—— *acuta.* Wet rocks, common.

Didymodon rigidulus. Ben Lawers.

———— *capillaceus.* In dense tufts in the clefts of rocks, abundant. Var. β. *ithyphylla.* Wet rocks in the Den of Lawers.

Didymodon longirostris. Mael Tarmanach and Mael Greadha.

Trichostomum patens. Abundant on Ben Lawers.

———————— *canescens, microcarpon, aciculare* and *fasciculare.* Ben Lawers.

———————*polyphyllum.* Rocks on the banks of Loch Tay.

Dicranum virens. Ben Lawers and Mael Gradha.

———— *falcatum.* In the greatest profusion on Ben Lawers.

———— *Starkii.* Ben Lawers, but much less common than the last.

-———— *flavescens.* Sandy deposits by the sides of Lawers-burn.

———— *scoparium, γ. fuscescens.* Mael Tarmanach &c.

———— *heteromallum* and *subulatum.* Ben Lawers.

Cinclidotus fontinaloides. On stones in Loch Tay.

Polytrichum hercynicum. Frequent on all the mountains.

———————— *juniperinum, β. gracilius, (P. strictum,* Menz.) In bogs on Ben Lawers.

———————— *septentrionale.* On wet rocks near the source of a rivulet a little below the summit of Ben Lawers, on the north side, and new, I believe, to the district.

———————— *alpinum.* The most common Polytrichum on the mountains, but most of my specimens were culled on the summit of Craigalleach, in the calm twilight of a summer's evening, after a day's laborious ramble over the whole range from Larich-an-Lochan. On the same spot (upwards of 3000 feet high) I last year had the misfortune to sprain my right ancle severely; and what I then suffered in the descent, and for two months after, would have cured many a one of botanizing for ever: but somehow or other, it only tended to increase my enthusiasm, and this prompted me in the present season to revisit that glorious mountain-altar, and there offer up to heaven grateful thanks for my recovery, and for the pure pleasure I continued to enjoy from the love of Nature's works.

Bryum trichodes. A single tuft near the summit of Ben Lawers.

————*julaceum.* Falls of Lawers and banks of the Lochy, above the bridge, but in neither place bearing fructification.

———— *crudum.* Ben Lawers &c.

———— *Zierii.* Ben Lawers and Stuich-an-Lochan.

———— *turbinatum.* About the Falls of Lawers, hill of Kenmore, and other places.

———— *nutans* and *elongatum.* Ben Lawers.

———— *ventricosum.* Falls of Lawers, and about the Falls of Boreland in Glen-Lochy.

———— *demissum.* Mael Greadha, rare.

———— *punctatum.* In many places.

Bartramia pomiformis, β. major. Den of Lawers.

———————— *ithyphylla.* On all the mountains.

———————— *gracilis.* In fructification on Ben Lawers, Stuich-an-Lochan and Mael Tarmanach.

———————— *fontana.* In bogs on most of the mountains, varying much in size. Some of the Ben Lawers specimens much elongated and elegantly slender.

Hedwigia æstiva. Plentiful but mostly barren. I last year culled it in fructification by the side of Fionlarig-burn on Craigalleach, and near a small waterfall in Glen Ogle.

Pterogonium filiforme. Abundant on the whole Breadalbane range, but barren.

Anomodon curtipendulum. Upon rocks in the woods of Glen Lochy, bearing fructification copiously.

Hypnum denticulatum, β. obtusifolium. Ben Lawers and other mountains.

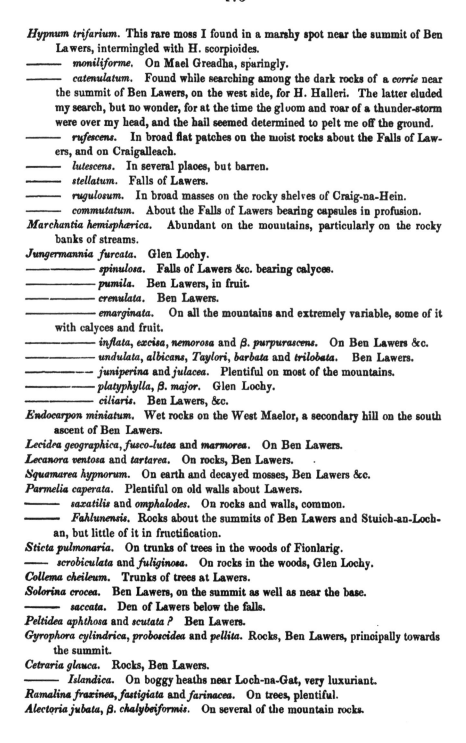

Hypnum trifarium. This rare moss I found in a marshy spot near the summit of Ben Lawers, intermingled with H. scorpioides.

——— *moniliforme.* On Mael Greadha, sparingly.

——— *catenulatum.* Found while searching among the dark rocks of a *corrie* near the summit of Ben Lawers, on the west side, for H. Halleri. The latter eluded my search, but no wonder, for at the time the gloom and roar of a thunder-storm were over my head, and the hail seemed determined to pelt me off the ground.

——— *rufescens.* In broad flat patches on the moist rocks about the Falls of Lawers, and on Craigalleach.

——— *lutescens.* In several places, but barren.

——— *stellatum.* Falls of Lawers.

——— *rugulosum.* In broad masses on the rocky shelves of Craig-na-Hein.

——— *commutatum.* About the Falls of Lawers bearing capsules in profusion.

Marchantia hemisphærica. Abundant on the mountains, particularly on the rocky banks of streams.

Jungermannia furcata. Glen Lochy.

——— *spinulosa.* Falls of Lawers &c. bearing calyces.

——— *pumila.* Ben Lawers, in fruit.

——— *crenulata.* Ben Lawers.

——— *emarginata.* On all the mountains and extremely variable, some of it with calyces and fruit.

——— *inflata, excisa, nemorosa* and *β. purpurascens.* On Ben Lawers &c.

——— *undulata, albicans, Taylori, barbata* and *trilobata.* Ben Lawers.

——— *juniperina* and *julacea.* Plentiful on most of the mountains.

——— *platyphylla, β. major.* Glen Lochy.

——— *ciliaris.* Ben Lawers, &c.

Endocarpon miniatum. Wet rocks on the West Maelor, a secondary hill on the south ascent of Ben Lawers.

Lecidea geographica, fusco-lutea and *marmorea.* On Ben Lawers.

Lecanora ventosa and *tartarea.* On rocks, Ben Lawers.

Squamarea hypnorum. On earth and decayed mosses, Ben Lawers &c.

Parmelia caperata. Plentiful on old walls about Lawers.

——— *saxatilis* and *omphalodes.* On rocks and walls, common.

——— *Fahlunensis.* Rocks about the summits of Ben Lawers and Stuich-an-Lochan, but little of it in fructification.

Sticta pulmonaria. On trunks of trees in the woods of Fionlarig.

——— *scrobiculata* and *fuliginosa.* On rocks in the woods, Glen Lochy.

Collema cheileum. Trunks of trees at Lawers.

Solorina crocea. Ben Lawers, on the summit as well as near the base.

——— *saccata.* Den of Lawers below the falls.

Peltidea aphthosa and *scutata?* Ben Lawers.

Gyrophora cylindrica, proboscidea and *pellita.* Rocks, Ben Lawers, principally towards the summit.

Cetraria glauca. Rocks, Ben Lawers.

——— *Islandica.* On boggy heaths near Loch-na-Gat, very luxuriant.

Ramalina fraxinea, fastigiata and *farinacea.* On trees, plentiful.

Alectoria jubata, β. chalybeiformis. On several of the mountain rocks.

Cornicularia aculeata, tristis and *lanata*. On Ben Lawers and Stuich-an-Lochan, the two latter species with apothecia.

Sphærophoron coralloides. In Glen Lochy and many other places, and bearing fructification on the rocks near the summit of Ben Lawers.

———————— ———— β. *cæspitosum*. On Ben Lawers.

Stereocaulon paschale, and what is usually called *botryosum*. On Ben Lawers &c., plentifully, but both so variable in form, and these forms so running into each other, that one is doubtful whether they may not after all constitute but one species.

Cladonia uncialis and β. *turgida*. Not uncommon on most of the mountains.

Scyphophorus cocciferus, gracilis, and several others, on Ben Lawers.

Pycnothelia Papillaria. On the same mountain, but rare.

WILLIAM GARDINER, JUN.

Dundee, December, 1842.

ART. CXVII.—*County Lists of the British Ferns and their Allies.*
Compiled by EDWARD NEWMAN.

(Continued from p. 455)

LANCASHIRE.

Lomaria Spicant. Common; *S. Gibson, W. Wilson, G. Pinder, J. Sidebotham, J. B. Wood :* near Liverpool, *S. Thompson :* abundant on moors near Lancaster, *S. Simpson :* common near Coniston, in wet gravelly situations, *M. Beever :* I am indebted to Miss Beever for magnificent specimens of this plant; they are of extraordinary size, and combine the characters of the fertile and barren fronds, *E. Newman.*

Pteris Aquilina. Common, *S. Gibson, G. Pinder, J. Sidebotham, J. B. Wood :* found in a very diminutive state, not exceeding three inches in length, upon the garden-wall at Knowsley, in 1840, *W. Wilson :* Penketh, *S. Thompson :* very common near Lancaster, *S. Simpson :* common near Coniston, *M. Beever.*

Allosorus crispus. Cliviger, in endless variety. Mr. Francis says that the sori are placed in lines along the transverse veins : I find them in dots at the termination of the veins, with the edge of the pinnule more or less turned over, forming a cover, sometimes however they are quite bare, and when in that state the plant might be considered a true Polypodium. Mr. Newman has observed that the character which distinguishes this plant from a true Polypodium, is that the fronds are both barren and fertile; but I find, that in common with all other Polypodia, it possesses not only barren and fertile fronds, but barren and fertile pinnæ on the same frond, and even barren and fertile pinnulæ on the same pinna, *S. Gibson :* near Lancaster, on the way to the Asylum, *W. Wilson :* Coniston, Old Man mountain, *G. Pinder :* among the rocks at Cliviger-dean, five miles from Todmorden, on the road to Burnley, (R. Leyland), *H. C. Watson :* Todmorden (J. W. G. Gutch), *H. C. Watson :* Fo-edge, near Bury, abundant, *J. Sidebotham :* abundant on moors near Lancaster, *S. Simpson :* common near Coniston, *M. Beever :* this beautiful species grows in immense profusion at Fo-edge, six miles N. W. of Bury; it also occurs on the hills near Rochdale, but I do not know the exact locality, *J. B. Wood.*

Polypodium vulgare. On old walls in many places, *S. Gibson, W. Wilson, S. Thompson :* common near Coniston, *M. Beever :* very common near Lancaster, *S. Simpson :* common on rough ditch-banks, and at the roots of trees in woods ; the variety *acutum* of Francis occurs at Sale, six miles from Manchester, *J. B. Wood :* generally distributed near Manchester, but not common; the var. *acutum* found growing with its common state, *J. Sidebotham.*

Polypodium Phegopteris. Dulesgate, near Todmorden, *S. Gibson :* near Lancaster, beyond the Asylum; Dean-church clough, near Bolton (John Martin), *W. Wilson :* Fox-clough, near Colne; Padiham-brook, near Whalley; Old Man mountain, *G. Pinder :* Mere-clough, Cotteril-clough, *J. Sidebotham :* near Lancaster, *S. Simpson:* common near Coniston, *M. Beever ;* this is not an uncommon species in the neighbourhood of Manchester, being found in several of our woods, as Mere-clough, Phillip's wood, near Prestwick, Boghart-hole clough, generally very fine and luxuriant, *J. B. Wood.*

Polypodium Dryopteris. Broad-bank, near Colne; Mr. Francis observes that the sori remain perfectly distinct, but I have found them confluent, *S. Gibson :* Dean-church clough (John Martin), *W. Wilson :* Marsden near Colne, and Padiham-brook, *G. Pinder :* Mere-clough and Cotteril-clough, *J. Sidebotham :* near Lancaster, *S. Simpson :* common near Coniston, *M. Beever :* more rare than P. Phegopteris, although found in three or four localities, as Mere-clough, Ashworth-wood &c., *J. B. Wood.*— Var. *rigidum,* mihi: Broad-bank, near Colne: this variety is sometimes taken for P. calcareum, *S. Gibson.*

Polypodium calcareum. Broad-bank and Sheden-clough. Mr. Francis has incorrectly given my name as an authority for this species being found near Lancaster; and in Mr. Watson's 'Notes on the Geographical Distribution of British Ferns,' this species is said to affect lime rocks, but the Lancashire localities are on the gravel which overlies the coal-formation, and it therefore cannot be said to be a limestone plant, *S. Gibson :* near Lancaster (— Gibson to R. B. Bowman),* Sheden-clough, three miles from Burnley (R. Leyland to Jas. Macnab), *H. C. Watson.*

Cystopteris fragilis. Broad-bank, *S. Gibson :* Wycottar-hall, walls near Colne and Clitheroe, *G. Pinder :* rare near Coniston, occasionally met with on the Old Man mountain, and the adjoining fells, more plentifully about the slate-quarries in Tibberthwaite, where it has the character of C. angustata, *M. Beever :* near Lancaster, *S. Simpson.*

Cystopteris dentata. Broad-bank, *S. Gibson :* Lymm-eye lock, between Warrington and Manchester, *W. Wilson.*

Cystopteris angustata. Cliviger, *S. Gibson.*

Cystopteris alpina. Ditto. *C. regia.* Broad-bank. All these forms of Cystopteris are common in the neighbourhood of Burnley, together with others perhaps quite as distinct: how far they may be considered species, I leave to the better judgment of others; *S. Gibson.*

Polystichum aculeatum. Sheden-clough, near Todmorden, *S. Gibson :* Fearnhead, near Warrington; Newton, *W. Wilson.*

Polystichum angulare. Marple-wood, *J. Sidebotham ;* rare near Coniston, *M. Beever.*

Polystichum lobatum. Burton-wood &c. near Warrington, *W. Wilson.* Houghend-clough, not common, *J. Sidebotham :* margin of Black-beck and above Brantwood;

---

* The discrepancy complained of by Mr. Gibson again occurs; I imagine Mr. Gibson referred to by Mr. Francis, and again by Mr. Watson, is not Mr. S. G Hebden Bridge.

M. Beever ; a plant of rare occurrence, met with however at Cotteril-clough, Bam. ford-wood, and one or two other places, *J. B. Wood.*

Lastræa Oreopteris.   Walsden, and near Todmorden, *S. Gibson :* near Warring. ton, Woolston, &c., *W. Wilson :* abundant, *G. Pinder :* generally distributed near Manchester, *J. Sidebotham :* Penketh, and many places about Liverpool, by the men employed in the Botanic Garden, *S. Thompson :* near Lancaster, *S. Simpson :* very abundant near mountains and streams in the neighbourhood of Coniston, *M. Beever.*

Lastræa Filix-mas. Common, *S. Gibson, W. Wilson, J. Sidebotham, J. B. Wood;* the variety figured in Newman's 'British Ferns,' p. 52, occurs at Rotten-clough near Todmorden, *S. Gibson :* near Liverpool, *S. Thompson :* near Lancaster, very common, *S. Simpson :* common near Coniston, also the serrated variety, *M. Beever.*

Lastræa rigida.   Top lock of Lancaster and Kendal canal (Rev. Thos. Smythes), *W. Wilson :* sparingly on the confines of the county, adjoining Westmoreland, *S. Simpson.*

Lastræa dilatata. Common in many parts of the county, *S. Gibson :* frequent, *W. Wilson :* very common, *J. Sidebotham :* near Liverpool, *S. Thompson :* near Lancaster, *S. Simpson :* near Coniston, *M. Beever :* extremely abundant, *J. B. Wood.*

Lastræa spinulosa.   Abundant on the borders of Risley-moss, near Warrington, growing in company with L. dilatata, from which it is so strikingly different in habit as to compel the opinion that it must be distinct, *W. Wilson :* Langdale in the lake district, *G. Pinder :* Lever's Hulme, the only locality I know of near Manchester, *J. Sidebotham :* near Lancaster, *S. Simpson :* this beautiful and interesting fern, which I must certainly still consider as specifically distinct from L. dilatata, is found in only one locality in this neighbourhood (Broughton, Manchester), in an old alder-swamp at Lever's Hulme ; it is intermixed with its more common ally, and when thus seen growing together the difference in their habit is remarkably striking : it approach. es most to L. cristata, *J. B. Wood.*

Lastræa dumetorum.   Cliviger, four miles from Burnley, *S. Gibson.*

Athyrium Filix-femina. Common in many places, *S. Gibson :* frequent; a varie. ty grows on Risley-moss, with more rigid habit, *W. Wilson :* common, *G. Pinder :* very common near Manchester, *J. Sidebotham :* near Coniston, *M. Beever.*

Athyrium irriguum. Sheden-clough, *S. Gibson :* near Coniston, *M. Beev·r.*

Asplenium Adiantum-nigrum. Broad-bank and Sheden-clough, *S. Gibson :* Winwick, near Warrington, *W. Wilson :* I am acquainted with but one locality, the canal wall near Marple, *J. Sidebotham :* very common near Lancaster, *S. Simpson :* very local near Coniston, and in small quantities, *M. Beever.*

Asplenium Ruta-muraria.   Broad-bank, *S. Gibson, W. Wilson :* Colne and Cli. theroe, *G. Pinder :* Hough-end Hall, common near Mottram, *J. Sidebotham :* West Derby, *S. Thompson :* very common near Lancaster, *S. Simpson :* on the Shepherd's Bridge in Yew-dale, the only locality I am acquainted with, *M. Beever :* very local, occurring on stone walls in two or three places, *J. B. Wood.*

Asplenium marinum.   In 1828 I found this species sparingly on a stone by the side of the river, about two miles from Liverpool on the way to Runcorn, *S. Gibson :* a variety still grows in Winwick Stone Delf, where it never attains any considerable size ; when fully developed by cultivation in a greenhouse, the pinnules are much shorter and rounder than those of maritime plants under the same treatment. I have not seen this species at Knot's Hole near Liverpool, but it is plentiful there according to Mr. T. B. Hall, see 'Flora of Liverpool ;' *W. Wilson :* it is still in existence in

Winwick stone-quarry, near Warrington: my friend Mr. Wilson brought me two small roots from that locality a short time since, for cultivation; it is never more than two or three inches high, and the pinnules are singularly cuneate and strongly crenate on their edges, very much resembling the Killarney plant (figured Newman's 'British Ferns,' 76); it produces seed in abundance; *J. B. Wood*: I am indebted both to Mr. Wilson and Dr. Wood for specimens from Winwick, and can bear witness to the accuracy of Dr. Wood's remark on their similarity to the Killarney specimens, *E. Newman*: the Dingle near Liverpool (T. B. Hall), *H. C. Watson*; Knot's Hole, the Dingle, *S. Thompson*; Heysham and Silverdale, *S. Simpson*.

Asplenium Trichomanes. Kant-clough, near Burnley, *S. Gibson*: near Burton, north of Lancaster, *W. Wilson*: Marsden, Clitheroe, and Stoneyhurst college-walls, *G. Pinder*: Greenfield, *J. Sidebotham*: near Lancaster, common, *S. Simpson*: common near Coniston, *M. Beever*; this is very sparingly distributed in the neighbourhood of Manchester, and is always very dwarf and poor, *J. B. Wood*. Var. *β. incisum*.— Kant-clough, near Burnley, *S. Gibson*.

Asplenium viride. Dulesgate; it is said to grow in the quarries at Staley, but I have not seen it, *J. Sidebotham*. Var *β. ramosum*; Dulesgate, Var. *γ. laciniatum*, mihi, frond deeply laciniated. Dulesgate, *S. Gibson*.

Scolopendrium vulgare. Sheden-clough, Kant-clough and other places, *S. Gibson*: Burton-wood, Fearn-head &c., near Warrington, *W. Wilson*: common, *G. Pinder*: not uncommon, Arden-hall moat, coal-pits about Arden Marple, and the variety *crispum* is found in Cotteril-clough, *J. Sidebotham*: Penketh, *S. Thompson*: only once found near Coniston in 1841, on the wall of a water-course near Tent-lodge, *M. Beever*; Bamford-wood near Manchester, *J. B. Wood*.

Ceterach officinarum. Near Kellet, north of Lancaster; I have credible information also that it grows near West Houghton, *W. Wilson*: a single plant in the Liverpool botanic garden, from Club-moor, three miles from Liverpool, *S. Thompson*: near Lancaster, *S. Simpson*.

Hymenophyllyum Wilsoni. Thevilly near Burnley, *S. Gibson*: Old Man mountain, *G Pinder*, *M. Beever*: sparingly in caves at Greenfield, *J. Sidebotham*: near Lancaster, very common, *S. Simpson*; Rake-hey Common, near Todmorden (R. Leyland), *H. C. Watson*; on hills near Bury, *J. B. Wood*.

Hymenophyllum Tunbridgense. Cliviger, *S. Gibson*: sparingly in caves at Greenfield, *J. Sidebotham*: Coniston, *S. Simpson*.

Osmunda regalis. Chilburn, near Todmorden, *S. Gibson*: Risly-moss and in a lane near Orford, one mile from Warrington, *W. Wilson*: Little Langdale, in the lake district, *G. Pinder*: sparingly at Bowden-moss, *J. Sidebotham*: near Lancaster, *S. Simpson*; near Coniston, in boggy situations, *M. Beever*; Heysham (J. Tatham to Bot. Soc. Lond.), *H. C. Watson*; not unfrequent in bogs, Chat-moss; in the neighbourhood of Southport I observed this "royal fern" to be very abundant, and attaining a most immense size, many plants exceeding ten feet in height, *J. B. Wood*.

Botrychium Lunaria. Chilburn, *S. Gibson*: near Newton, Southport (Dr. Wood), *W. Wilson*: common in the upland pastures about Oldham, very fine and luxuriant at Reddish, *J. Sidebotham*: near Lancaster, *S. Simpson*; on the common between Bootle and Crosby, ? called Seaforth Common, *H. C. Watson*; a common plant on our neighbouring hills, *J. B. Wood*.

Ophioglossum vulgatum. *S. Gibson*: Woolstone and Gate-wharf, near Warrington, *W. Wilson*: rather common in Reddish-vale, meadows near the river Tame &c.,

Worsley (T G Rylands), *S. Thompson :* near Lancaster, *S. Simpson ;* in a field near Coniston old hall, *M. Beever ;* near Crosby, *H. C. Watson ;* common near Manchester, *J. B. Wood.*

Lycopodium clavatum. Near Coniston, *M. Beever ;* Todmorden valley; hills very generally, *G. Pinder ;* Greenfield, Fo-edge, *J. Sidebotham ;* rare near Manchester; more common on the elevated moors, *J. B. Wood.*

Lycopodium alpinum. Near Coniston, *M. Beever ;* near the Holme, about five miles from Burnley, (Mr. Woodward) ; moors above the scouts near Burnley, (Rev. W. Wood); Coniston fells, (Mr. Davey); *Turner and Dillwyn,* 372 : Greenfield, *J. Sidebotham ;* Fo-edge near Bury, *J. B. Wood ;* Cliviger, at a very low elevation, barren, *W. Wilson.*

Lycopodium selaginoides. Near Coniston, *M. Beever ;* on the common between Bootle and Crosby, *H. C. Watson ;* shore at Southport, *W. Wilson.*

Lycopodium Selago. Near Coniston, *M. Beever ;* White-moor near Colne, *G. Pinder ;* Fo-edge near Bury, *J. Sidebotham ;* Fo-edge and Mottram, *J. B. Wood ;* Cliviger, *W. Wilson.*

Isoetes lacustris. Near Coniston, *M. Beever.*

Equisetum hyemale. Very rare; I believe it is to be found in small quantities in Mere-clough near Manchester; I have not seen it, but can rely on the authority of the individual who did gather it, *J. B. Wood.*

Equisetum variegatum. Bootle and Southport, *W. Wilson, S. Gibson.*

Equisetum arvense. Near Coniston, introduced? *M. Beever ;* common near Manchester, *J. Sidebotham ;* Prestwick, *H. C. Watson ;* far too common, *J. B. Wood.*

Equisetum sylvaticum. The copper-mine near Coniston, *M. Beever ;* common near Manchester, *J. Sidebotham ;* Scorton (S. Simpson to Bot. Soc. Lond.), *H. C. Watson ;* most abundant in woods, thickets and open pastures, *J. B. Wood.*

Equisetum palustre. Near Coniston, *M. Beever ;* common near Manchester, *J. Sidebotham, J. B. Wood ;* sands near Little Crosby, near Manchester, *H. C. Watson.*

Equisetum limosum. Near Manchester, *J. Sidebotham, H. C. Watson ;* many of the " pits " near Manchester are completely choked up with it, *J. B. Wood.*

Equisetum fluviatile. Common near Manchester, *J. Sidebotham ;* very abundant in swampy woods, *J. B. Wood.*

CHESHIRE.

Lomaria spicant. *W. Wilson, H. C. Watson.*

Pteris Aquilina. *W. Wilson, H. C. Watson ;* Bidston-hill &c., abundant, *S. Thompson.*

Polypodium vulgare. A serrated variety near Frodsham, *W. Wilson, H. C. Watson ;* between Woodside and Bidston, abundant, *S. Thompson.*

Polypodium Phegopteris. Werneth, *J. Sidebotham ;* Mow-cop, *G. Pinder.*

Polypodium Dryopteris. Two miles south of Warrington, attaining a very large size, *W. Wilson.*

Cystopteris fragilis. Extremely rare, I know of but one locality, Rostherne Church, *J. B. Wood, J. Sidebotham.*

Cystopteris dentata. Rostherne Church, *W. Wilson.*

Polystichum aculeatum. Very local, Marple woods, *J. B. Wood.*

Polystichum angulare. Two miles south of Warrington, in a dingle, growing in company with Polypodium Dryopteris, *W. Wilson;* plentiful in Marple wood, *J. B. Wood.*

Lastræa Thelypteris. Newchurch-bog, near Over, where it bears fruit plentifully, Pelty-pool, also near Over; Knutsford-moor and Rostherne-mere, *W. Wilson, J. B. Wood;* Rostherne-mere, where it bears fruit only in very fine seasons, as 1842, *J. Sidebotham;* Wybunbury-bog, *G. Pinder.*

Lastræa Oreopteris. *W. Wilson.*

Lastræa Filix-mas. *W. Wilson, H. C. Watson, S. Thompson.*

Lastræa cristata. Wybunbury-bog; I should be sorry should this notice lead to its extermination, *G. Pinder;* I am indebted to Mr. Pinder for specimens, they are identical with those from Norfolk, *E. Newman.*

Lastræa dilatata. *W. Wilson;* Eastherne-wood, abundant, *S. Thompson.*

Lastræa spinulosa. Newchurch-bog, Pelty-pool and Knutsford-moor, *W. Wilson;* Wybunbury-bog, *G. Pinder.*

Athyrium Filix-femina. *H. C. Watson;* attains a very great size in the Dingle, two miles south of Warrington, *W. Wilson;* lanes about Bidston, *S. Thompson.*

Asplenium Adiantum-nigrum. *H. C. Watson, W. Wilson:* Between Woodside and Oxton, very fine, *S. Thompson.*

Asplenium Ruta-muraria. *W. Wilson;* Red and Yellow Noses, *S. Thompson.*

Asplenium marinum. On the rocks called the Red Noses, at New Brighton, *W. Wilson, H. C. Watson;* on the rocks of Hilbre island at the mouth of the river Dee, on the coast of Cheshire, this species is met with in great quantities, *J. B. Wood.*

Asplenium Trichomanes. Unfrequent; I have seen it only in the Well at Becston, *W. Wilson.*

Scolopendrium vulgare. *W. Wilson, H. C. Watson;* near the toll-gate between Woodside and Bidston, *S. Thompson;* Cotterill-wood and Arden-hall, *J. B. Wood.*

Osmunda regalis. Wybunbury-bog, Smethwick, near Congleton, *G. Pinder;* sparingly at Baguley-moor, *J. Sidebotham;* Carrington-moss, *J. B. Wood.*

Botrychium Lunaria. Near Over it attains a great size, and is often branched, *W. Wilson;* Alderley-edge (Rev. Isaac Bell), *H. C. Watson:* sand-hills between Egremont and New Brighton, found abundantly in 1841 by H. E. Robson, *S. Thompson;* when botanizing near Over with my valued friend Mr. Wilson, in June, 1842, I met with some singular deviations in this interesting plant; one specimen had no less than four fertile branches and two barren ones, springing from a common stem; on several of the pinnules of the barren frond were a number of thecæ, some on their plain surface, others on their edges; many specimens had two fertile fronds generally of unequal size; they were growing in a meadow on a small declivity, in company with Habenaria chlorantha, *J. B. Wood.*

Ophioglossum vulgatum. *W. Wilson;* Davenport and Summerfield, *G. Pinder;* Alderly and Tranmere, *H. C. Watson.*

Lycopodium clavatum. Delamere forest, *W. Wilson.*

Lycopodium inundatum. Oak-mere and Baguley-moor, *W. Wilson.*

Lycopodium Selago. Risly-moss near Warrington, *W. Wilson.*

Pilularia globulifera. Bartington-heath &c. *W. Wilson.*

Equisetum hyemale. Lally's wood near Over, *W. Wilson.*

Equisetum variegatum. Sands at New Brighton, *H. C. Watson.*

Equisetum palustre. *H. C. Watson;* very abundant; the variety *polystachion* was gathered in some plenty by the sides of the embankment of the Sheffield railway near

Godley, during the last summer; in all the specimens I have seen the terminal catkin of the main stem was present, thus clearly proving that the proliferous condition is not dependent on the accidental circumstance of the top of the plant having been cropped or destroyed, an opinion very generally prevailing among botanists, *J. B. Wood.*

Equisetum limosum. *H. C. Watson.*
Equisetum arvense. *H. C. Watson.*
Equisetum sylvaticum. *H. C. Watson;* rather frequent, *W. Wilson.*
Equisetum fluviatile. *H. C. Watson.*

It is intended to give the lists for Staffordshire, Warwickshire and Worcestershire, in the March Phytologist: further information respecting these counties is particularly solicited.

EDWARD NEWMAN.

(To be continued).

ART. CXVIII. — *Some Account of the Botanical Collections recently made by Dr. Theodore Kotschy (for the Wurtemburg Botanical Union) in Nubia and Cordofan.* Communicated by MR. WM. PAMPLIN, jun.

(Continued from p. 459).

Umbelliferæ.
Coriandrum sativum, *L.*
Terebinthaceæ.
Balsamodendron Kafal, *Fors*
Papilionaceæ.
A.—*Loteæ.*
Lotus nubicus, *Hchst.*
arabicus, *L.* [*Perr.*
Cyanopsis senegalensis, *Gui.*
Trigonella hamosa, *L.*
Tephrosia leptostachys, *Can.*
uniflora, *Pers.*
anthylloides, *Hochst.*
cordofana, *H.*
Kotschyana, *Hochst.*
linearis, *Perr.*
Sesbania filiformis, *Gll. Perr*
pachycarpa, *Gll. Perr.*
punctata, *Pers.*
tetraptera, *Hochst.*
B.—*Fabaceæ.*
Kennedya arabica, *H. St.*
Rhynchosia Memnonia, *Ca.*
Dolichos angustifolius, Vhl.

Dolichos No. 218.
obliquifolius, *Schnizl.*
No. 288.
hastæfolius, *Schnizl.*
Cajanus flavus, *Cand.*
monstrosus
Clitoria Ternatea, *L.*
Indigofera diphylla, *Vert.*
deflexa, *Hochst.*
coidifolia, *Roth.*
senegalensis, *Lam.*
var. latifolia
paucifolia, *Del.*
viscosa, *Lam.*
oligosperma, *Cand.*
astragalina, *Cand.*
var. melanosperma, *C.*
argentea, *L.*
var. polyphylla
strobilifera, *Hochst.*
Anil, *L.*
var. orthocarpa, *Cand.*
semitrijuga, *Forsk.*
aspera, *Pers.*

C.—*Hedysareæ.*
Æschynomene macropoda,
[*Guill. Perr.*
Onobrychis arabica, *Hochs.*
Alyssicarpus vaginalis, *H.*
rugosus, *Hochst.*
Cassieæ.
A.—*Genisteæ.*
Requienia obovata, *Cand.*
Crotalaria macilenta, *Sm.*
lupinoides, *Hochst.*
podocarpa, *Cand.*
microcarpa, *Hochst.*
atrorubens, *Hchst.*
sphærocarpa, *Per.* var
angustifolia
thebaica, *Cand.*
B.—*Genuinæ.*
Bauhinia parvifolia, *Hoch.*
tamarindacea, *Del.*
Cassia acutifolia, *Del.*
Tora, *L.*
obovata, *Collad.*
Absus, *L.*

Cassia rhachyptera, *Hochst.*
Chamæfistula Sophora, *G.*
   Mimoseæ. [*Don.*
Neptunia stolonifera, *G. P.*
Mimosa Habbas, *Del.*
Acacia No. 294 =
  A. sericocephala, *Fenzl.*
  No. 295, =
  Inga floribunda, *Fenzl.*
  papyræcea, *Hochst.*
   Corniculatæ. .
  B.—*Saxifrageæ.*
Vahlia Weldenii, *Rchb.*
  cordofana, *Hochst.*
  Portulacaceæ.
  A.—*Paronychieæ.*
Polycarpæa glabrifolia, *Cnd.*
  rutila, *Fenzl.*
  linearifolia, *Cand.*
Mollugo bellidifolia, *Ser.*
  Cerviana, *Ser.*
Arphorsia memphitica, *Fnzl*
  B.—*Polygoneæ.*
Ceratogonum atriplicifolium
  C.—*Portulaceæ.*
Trianthema sedifolium, *Vis.*
  polyspermum, *Hochst.*
  pentandrum, *L.*
  salsoloides, *F.*
  chrystallinum, *Vahl.*
Portulaca oleracea, *L.*
  Aizoideæ.
  A.—*Atripliceæ.*
Chenopodium murale, *L.*
Limeum viscosum, *Fenzl.*
Amaranthus polygamus, *L.*
  = angustifol. *M.B.*
Celosia argentea, *L.*
  trigyna, *L.* (nec. var.)
Aërva tomentosa, *Forsk.*
  brachiata, *Mart.*
Digera arvensis, *Forsk.*
Desmochæta flavescens, *Cnd*
Achyranthes argentea, *Lam.*
Alternanthera nodiflora, *Br.*
Pongatium indicum, *Lam.*
  B.—*Genuinæ.*
Gieseckia rubella, *Hchst.*

Gieseckia pharnacioides *L.*
  Rosaceæ.
Potentilla supina, *L.*
  Onagreæ.
  A.—*Jussieuæ.*
Jussieua nubica, *Hochst.*
Isnardia lythrarioides, *H.*
  C.—*Myrobalaneæ.*
Poivrea aculeata, *Cnd.* var.
  subinermis
Guiera senegalensis, *Lam.*
Terminalia Brownei, *Fres.*
  Lythrarieæ.
  B.—*Lythreæ.*
Bergia suffruticosa, *Fenzl.*
  verticillata, *W.*
  peploides, *Guill. Per.*
  erecta, *Guill. Perr.*
Ammannia ægyptiaca, *W.*
  attenuata, *Hochst.*
Lawsonia alba, *Lam.*
  Tetradynamæ.
Senebiera nilotica, *Cand.*
Farsetia ramosissima, *Hch.*
  stenoptera, *Hochst.*
Nasturtium palustre, *Cand.*
Morettia philæana, *Cand.*
Pteroloma arabicum, *St.Hc.*
  Capparideæ.
Cleome chrysantha, *Dec.*
Polanisia orthocarpa, *Hch.*
Roscia octandra, *Hchst.*
Cadaba glandulosa, *Forsk.*
  farinosa, *F.*
Capparidea, sinc flor. et fr.
  Violaceæ.
Ionidium rhabdospermum,*H*
  Rutaceæ.
  A.—*Euphorbiaceæ.*
Euphorbia thymifolia, *Frsk.*
  granulata, *Vahl.*
  hypericifolia, *L.* var.
   angustif. et pubesc.
  acalyphoides, *Hochst.*
  convolvuloides, *H.*
  polycnemoides, *H.*
  Chamæsyce, *L.*
Dalechampia cordofana, *H.*

Mercurialis alternifolia, *H.*
Acalypha abortiva, *Hochst.*
  fimbriata, *Hochst.*
Crozophora senegalensis,,*S.*
Cephalocroton cordofanum
Croton lobatus, *L.*   [*H.*
  serratus, *Hochst.*
  obliquifolius, *Vis.*
Ricinus africanus, *Mill.*
Phyllanthus, No. 89, =
  Ph. Niruri, *L.*
  Urinaria, *L.*
  venosus, *Hochst.*
  linoides, *H.*
  B.—*Rutarieæ.*
Ruta tuberculata, *Forsk.*
  var. obovata
Moringa aptera, *Gärt.*
  Sapindaceæ.
  A.—*Tribuleæ.*
Tribulus terrestris, *L.*
Fagonia arabica, *L.*
Zypophyllum simplex, *L.*
  C.—*Paullinieæ.*
Cardiospermum Halicaca-
  Malvaceæ.   [bum
  B.—*Malveæ.*
Sida alnifolia, *L.*
  grewioides, *Gll. Perr.*
  Kotschyi, *Hchst.*
  althæifolia, *Sw.*
  heterosperma, *Hchst.*
Abutilon graveolens, *W.A.*
  ramosum, *Gll. Per.*
  asiaticum, *Guill.Per.*
Pavonia Kotschyi, *Hchst.*
  dictyocarpa, *Hochst.*
  triloba, *Hochst.*
  No. 395, =
  P. heterophylla, *Hch*
  triloba, *Guill.* var.?
  triloba, *Hochst.*
  No. 220
  hermanioides, *Fenzl.*
  heteroph. var.?
Dumreichera arabica, *H.St.*
  var. major.
  C.—*Hibisceæ.*

Lagunæa ternata, *Cav.*
Hibiscus cordatus, *Hochst.*
Trionum, *L.*
amblycarpus, *Hchst.*
Bammia, *Link*
Geraniaceæ.
A.— *Geraniæ.*
Monsonia senegalensis,*G.P.*

C.—*Byttneriæ.*
Herrmannia arabica, *Hchst.*
Waltheria indica, *L.*
Melhania Kotschyi, *Hchst.*
Theaceæ.
B.—*Celastreæ.*
Celastrus senegalensis, *Lam*
Tiliaceæ.

Antichorus depressus, *L.*
Corchorus olitorius, *L.*
brachycarpus, *Gll.P.*
tridens, *L.*
alatus, *Hochst.*
Grewia No. 281, =
G. commutata, *Cand*
echinulata, *Caill.*

W. PAMPLIN, JUN.

(To be continued).

ART. CXIX.—*Analytical Notice of the 'Transactions of the Linnean Society of London,' vol.* xix. *pt.* 1. 1842.

III. — *Some Account of* Aucklandia, *a new Genus of* Compositæ, *believed to produce the* Costus *of Dioscorides. By* HUGH FALCONAR, *M.D., Superintendent of the Hon. East India Company's Botanic Garden at Saharunpore. Communicated by* J. F. ROYLE, *M.D., F.R.S. & L.S.*

Read November 17, 1840.

THE subject of this paper is nearly allied to our own Saussurea and Carlina, from the former of which genera it seems to differ chiefly in " the rays of its feathery pappus being disposed in two rows, and cohering by twos or threes at the base." It grows in great abundance on the damp open slopes of the mountains surrounding the valley of Cashmeer, at an elevation of from 8000 to 9000 feet above the sea-level, flowering in June and maturing its fruit in October. The plant has but a slender medical reputation among the native Cashmeerians, since it is chiefly employed by them to protect bales of shawls from the attacks of moths, its odour being very pungent, and well calculated to effect this ; portions of the stem are also suspended from the necks of children, in order to protect them from the *evil eye,* and to expel worms.

The author observes that he has " frequently been asked, when in Cashmeer, where and for what purpose the immense quantities of the root, annually collected, could find a market." We give a brief summary of the commercial history of the plant, extracted from the details given in the paper. The roots are dug up in September and October; they are chopped into pieces from two to six inches long, and in this state are exported in vast quantities to the Punjab, whence the larger portion goes down to Bombay, where it is shipped for the Red Sea, *the* Persian Gulf and China; a portion finds its way into Hindoostan

Proper, whence it is taken to Calcutta, and is there bought up for the China market under the name of *Putchuk*.

One object of the author in this paper is to show that the root of Aucklandia is identical with the *Costus* of Dioscorides; and into the discussion of this question he enters pretty fully. The following are the proofs adduced in support of this opinion. 1. The correspondence of the root of Aucklandia with the descriptions of ancient authors. 2. The coincidence of names, this root being called *Koot* in Cashmeer, the Arabic *Koost* being given as synonymous. 3. At the present day the Chinese burn the root as an incense in their temples, and make use of it medicinally; the Costus was employed in the same way by the ancients. 4. The testimony of Persian authors that the Koost is not produced in Arabia, but is from the "borders of India." 5. The commercial history of the Cashmeerian root.

The author in conclusion makes some remarks on the probability that the Aucklandia, if cultivated, "would form a valuable addition to the wealth of the Hill people." He states that in consequence of the lively demand for the root in Cashmeer, there is seldom any surplus stock in hand; and the plant being a perennial, and requiring several years to mature its root, it is not likely the valley would yield any considerable increase on the quantity now collected, without ultimately reducing that quantity.

The plant is named by the author Aucklandia Costus, the generic name being given in honour of George Earl Auckland; it was met with during a journey to Cashmeer, undertaken under his Lordship's auspices while Governor-general of India. Full characters and a detailed description are given.

IV.—*Description of a new Genus of* Lineæ. *By* CHARLES C. BABING-TON, *Esq., M.A., F.L.S., F.G.S.*

Read January 19, 1841.

THE seeds of this interesting addition to the small order to which it belongs, were collected in the interior of New South Wales by Mr. Melluish, who sent them to the Cambridge Botanic Garden, where at the time the paper was written the plants raised from the seeds had flowered during three successive years.

The name given by the author is Cliococca tenuifolia; the generic name refers to the indehiscent nature of the single-seeded carpels, in which respect, and in an apparent tendency to the imperfect gynobasic structure of the Malvaceæ, the relationship between that family and the Lineæ is more fully evinced. The petals are imbricated in æstivation, and are not unguiculate; the coats of the carpels are also very thick, and the carpels themselves are perfectly closed, not even opening when they separate for the dissemination of the seed : in all which particulars the plant differs from the usual structure of Lineæ.

The paper contains full generic and specific characters, and is illustrated by a plate of details drawn by Mr. J. D. Sowerby.

V.—*On an edible Fungus from Tierra del Fuego, and an allied Chilian Species. By the Rev.* M. J. Berkeley, *M.A., F.L.S.*

Read March 16, 1841.

Mr. Darwin gives an account in his Researches of a production common on a species of beech in Tierra del Fuego, which is used by the natives as an article of food. The author's belief that these bodies were referrible to the Fungi, was confirmed by an inspection of the specimens preserved by Mr. Darwin, and by the perusal of that gentleman's original notes. With the aid of these materials he has been able to establish a new genus (which he has named Cyttaria), containing two well-defined species.

The following extracts from Mr. Darwin's rough memoranda relate to the first species — C. Darwinii.

" In the beech forests the trees are much diseased; on the rough excrescences grow vast numbers of yellow balls. They are of the colour of the yolk of an egg, and vary in size from that of a bullet to that of a small apple; in shape they are globular, but a little produced towards the point of attachment. They grow both on the branches and stems in groups. When young they contain much fluid and are tasteless, but in their older and altered state they form a very essential article of food for the Fuegian. The boys collect them, and they are eaten uncooked with the fish. When we were in Good Success Bay in December, they were then young; in this state they are externally quite smooth, turgid, and of a bright colour, with no internal cavity. The external surface was marked with white spaces, as of a membrane covering a cell. Upon keeping one in a drawer, my attention was called, after some interval, by finding it become nearly dry, the whole surface honeycombed by regular cells, with the decided smell of a Fungus, and with a slightly sweet mucous taste. In this state I have found them during January and February (1833) over the whole country. Upon dividing one, the centre is found partly hollow and filled with brown fibrous matter; this evidently merely acts as a support to the elastic semitransparent ligamentous substance which forms the base and sides of the external cells. Some of these balls remain on the trees nearly the whole year; Captain Fitzroy has seen them in June."—p. 38.

Mr. Darwin found the same species at Port Famine in February, 1834; and again, under the date of June of the same year, he describes the appearance in an older state. He found them to be much *infested* with larvæ.

The second species — C. Berteröi — is a native of Chile, and is found upon Fagus obliqua. It appears to have been first described by Bertero, in a posthumous list of Fungi published in a journal called 'Mercurio Chileno,' a translation of which by Ruschenberger was given in Silliman's Journal, xxiii. 78. The following is Bertero's account.

"*Fagus obliqua*, Mirb., *Roble*, oak, a tree common in the high mountains. In the spring is formed on the branches of this tree a great number of whitish tubercles, the parenchyma of which is spongy, though sufficiently consistent at first. I thought it a galla or excrescence, produced by the wound of some insect, as is seen on some other trees in Europe, and I gave the matter but little attention; but two days afterwards they became unglued from the branch, and I observed with surprise that the skin was broken, and the whole surface covered with pentagonal tubes precisely similar to the alveoli of a honeycomb, at first full of a gelatinous substance of the colour of milk, which disappeared with the maturation; afterwards throwing out from these cavities with some force an impalpable powder, when it was touched, exactly as is observed in the Peziza vesiculosa. At the end of two days these bodies softened, lost their expulsive property, and rotted. It perhaps forms a new genus, approximating to the Sphæ-riæ. Its vulgar name is *Dignénes*. Some persons eat them, but their insipid and styptic taste is disagreeable."—p, 39.

The following extract from Mr. Darwin's notes also refers to this species.

"Sept. 1834. On the hills near Nancagua and San Fernando, there are large woods of Roble, or the Chilian oak. I found on it a yellow fungus, very closely re-sembling the edible ones of the beech of Tierra del Fuego. Speaking from memory, the difference consists in these being paler coloured, but the inside of the cups of a darker orange. The greatest difference is, however, in the more irregular shape, in place of being spherical: they are also much larger. Many are three times as large as the largest of my Fuegian specimens. The footstalk appears longer; this is neces-sary from the roughness of the bark of the trees on which they grow. In the young state there is an internal cavity. They are occasionally eaten by the poor people. I observe that these are not infested with larvæ, like those of Tierra del Fuego."—Id.

Mr. Berkeley gives generic and specific characters, as well as a full description in English; and the illustrative plate is filled with anatomical details from his own drawings.

VII. — *On a reformed Character of the Genus* Cryptolepis, *Brown.* By HUGH FALCONER, M.D., *Superintendent of the Hon. East India Company's Botanic Garden at Saharunpore.* Communicated by J. F. ROYLE, M.D., F.R.S., F.L.S., &c.

Read June 15th, 1841.

THE genus Cryptolepis was established by Mr. Brown, who, in his Monograph published in the 'Memoirs of the Wernerian Society,' refers the genus to the Apocyneæ, placing it next to Apocynum : in this he has been followed all subsequent authors. But Dr. Falconer finds, on examination of specimens of Cryptolepis Buchanani, that it "has the whole of the accessory stigmatic apparatus of Asclepiadeæ, with granular pollen as typically developed as in Cryptostegia or any other of the Periplocæ, although in a less considerable degree of evolution; and that it must rank in that order along with them."

The author considers that the appendiculæ of the stigma, from their extreme minuteness must have eluded Mr. Brown's notice; two other points of difference lead the author to conclude that the plant examined by himself must have been different from Mr. Brown's. In the first place he does not find the five hypogynous scales mentioned in the generic definition; the same scales are wanting, so far as he has ascertained, in all the Periploceous genera allied to Cryptolepis, including even Decalepis of Wight and Arnott. Secondly, the species examined by the author has axillary instead of interpetiolar corymbs, an important character in the habit.

Many particulars relating to the reproductive organs are given, together with an amended generic character. The drawings for the illustrative plate were made by Kureem Buksh, a native artist.

IX. — *On the existence of Spiral Cells in the Seeds of* Acanthaceæ. *By* Mr. RICHARD KIPPIST, *Libr.L.S. Communicated by the Secretary.*

Read March 17th, 1840.

BOTANISTS have long been aware of the existence of spiral vessels in the envelopes of the seeds of several families of plants. Mr. Brown detected them first in the pericarps of Casuarinæ, afterwards in the testa of some Orchideæ. They were observed by Lessing in Compositæ, by Horkel and Schleiden in Labiatæ, Polemoniaceæ and Hydrocharideæ; and in the 'Botanical Register' is given an account of their appearance in the seeds of Collomia. We believe that the merit of detecting these cells in the seeds of Acanthaceæ is due to Mr. Kippist, who has in this paper given an interesting detail of his researches. The following is the author's account of his discovery.

"My attention was first directed to this subject by witnessing the very beautiful appearance under the microscope of an Acanthaceous seed, forming part of a collection brought by Mr. Holroyd from Upper Egypt, and presented by him to Professor Don. It is of a lenticular form, covered, especially towards the margin, with whitish hairs, which are closely appressed to the surface, and glued together at their extremities, so as rather to resemble corrugations of the testa than distinct hairs; on being placed in water, however, they are set at liberty, and, expanding on all sides, are seen to consist of fascicles of long, cylindrical, transparent tubes, firmly cohering for about one third of their length, and presenting all the characters of spiral vessels. These fascicles usually contain from five to twenty tubes; each tube inclosing one, two, or occasionally even three, spiral fibres, which adhere closely to the membrane. The fibre may be sometimes seen to divide into two in the upper part of a tube, the branches usually continuing distinct; sometimes, however, after making a few turns, they again coalesce. Towards the free extremity of the tube the fibre is frequently broken up into a number of distinct rings; and in other cases the spire again becomes continuous, after having been interrupted by two or three such rings. In those portions of the tubes which adhere together, the fibre is completely reticulated; towards the extremity, the coils, though quite contiguous, are usually distinct, and readily separate by the expansion of the tube; in the intermediate parts they adhere more firmly together, being connected by slender ramifications of the main fibre. The expansion of the hairs in water is accompanied by a copious discharge of mucilage, which makes its escape by distending and finally rupturing laterally the spiral tubes in which it is contained.

" The testa, which is distinctly visible in the spaces between the hairs, consists of nearly regular hexagonal cells, each containing an opake mass of grumous matter, which, not filling the entire cavity, leaves a wide transparent border. Cells similar to these, but more elongated and gradually passing into the form of tubes, immediately surround the base of each hair, which appears to be filled up by a conical mass formed of the transparent tubular portions."—p. 65.

Mr. Kippist was induced, by the peculiar appearance of these seeds, to examine those of other genera of the same natural family, with the view of ascertaining to what extent the tendency to develope spiral hairs might prevail, and whether that peculiarity might assist in characterizing genera. He finds that the presence of these cells is not universal in the Acanthaceæ, but has met with many examples, with "a considerable diversity in the structure and arrangement of the hairs which clothe the seeds." He describes the hairs of the seeds of a great number of species of Acanthaceæ, which appear to vary greatly in form and structure. We regret that our limits prevent us from selecting examples. The beautiful illustrations are from the author's own drawings.

X. *Description of a new genus of Plants from Brazil. By* J. Miers, *Esq. F.L.S*
Read March 2, 1841.

This is a minute diœcious plant found by Mr. Miers in the Organ mountains, Brazil, in February, 1838. Its texture is quite transparent, and the structure of its flowers very singular. Its position in the system is not accurately determined, but the author is disposed to place it near the Juncagineæ; in habit it resembles some Orchideæ. It is named by the author Triuris hyalina—the generic name referring to the three elongated processes of the perianth. The illustrations are from the pencil of the author.

ART. CXX.— *Varieties.*

249. *Note on Agaricus aimatochelis and A. deliciosus.* I have found in a wood near here a considerable quantity of Agaricus aimatochelis, which, when Berkeley's book was published, was not known as a native of England. My specimens had the red ring complete. Notwithstanding M. Roque's caution both I and my friends have eat. en Agaricus deliciosus, and with impunity. It was broiled with pepper, salt and but. ter, and really was delicious. I have taken a moderate-sized one at a meal.— *George Sparkes ; Bromley, Kent, November,* 1842.

250. *Note on Alaria esculenta.* In 1837 Mrs. Griffiths called my attention to the Alaria esculenta, and wished me to observe whether a new " frond was pushed out between the stem and the old one, as in Laminaria digitata and L saccharina." In con. sequence I attentively watched the plant during the three succeeding autumns, and as the result of my observations is interesting, and will serve to correct an error in Mr. Harvey's ' Manual of British Algæ,' where the Alaria is said to be annual, the follow. ing remarks may not be unacceptable to the readers of ' The Phytologist.' In spring the rocks in many parts of Mount's Bay are covered with the young fronds of the Ala. ria : the greater number of these are destroyed in the course of the summer ; those more favourably situated remain, and throw out, near the top of the stipes, a few hori. zontal leaflets. In the autumn, between these and the base of the frond, the stipes becomes elongated, bearing a new frond, which, as in Laminaria, at first has the old one attached to its summit. Numerous leaflets are thrown out from a line on each side of the new portion of the stem, and the old leaflets fall off. There is a small interval between the insertions of the old and new leaflets ; and although the marks of the former become nearly obliterated, yet by a careful examination they may still be detected, and the age of the plant ascertained. The stipes is therefore elongated whenever a new frond is formed, and this is more than once repeated, as I have frequently observed specimens which have had three sets of the horizontal leaflets.—*John Ralfs ; Penzance, December* 14, 1842.

251. *New locality for Grateloupia filicina.* I avail myself of this opportunity to notice a Welch habitat for the rare Grateloupia filicina, which Mr. Harvey, in his ' Manual of British Algæ,' mentions as having been found (in this country) only on the shores of Devonshire and Cornwall. In September last I had the pleasure of find. ing it, rather plentifully, in shallow pools on the rocks in front of the Castle Hill at Aberystwith.—*Id.*

252. *Note on the poisonous effects of Conia.* Your correspondent Mr. Sparkes, from his remarks on the poisonous effects of the seeds (or rather *fruit*) of hemlock (Phytol. 459) does not appear to be acquainted with Dr. Christison's experiments on the subject. There can be no doubt of the poisonous nature of the fruit ; Dr. C. found that it yielded a much larger quantity of the poisonous alkaloid—*conia*—than the leaves of the plant did. Conia, you are aware, is one of the most deadly poisons ; a single drop put into the eye of a rabbit, killed it in nine minutes ; three drops used in the same way, killed a strong cat in a minute and a half ; five drops poured into the throat of a small dog, began to act in thirty seconds, and in as many more, motion and respiration had entirely ceased. Conia, when injected into a vein, killed a dog *instantaneously*. *The poison has a* local irritant effect, and destroys life by causing palsy of the muscles

of respiration, being in this respect the counterpart of strychnia, which kills by caus. ing tonic contraction of the muscles of respiration. In both cases asphyxia is produ.. ced, but from different causes. The *green* fruit yields conia in large quantity. A full detail of Dr. Christison's experiments is given in the 'Transactions of the Royal So. ciety of Edinburgh, vol. xiii. As to Lindley's remark, that the fruit of the Umbelli. feræ is in no case dangerous, there surely must be some mistake. So far as I recollect he states, that while the leaves of the Umbelliferæ are always suspicious, the fruit is often aromatic. This remark holds good generally, but there are several important exceptions. Conia is easily decomposed by heat, and hence extracts of hemlock are of- ten inactive. The *dried* leaves also, according to Geiger, contain no alkaloid. These facts will account for the variable effects produced by preparations of hemlock.—*J.H. Balfour ; 11, West Regent St., Glasgow, January 4, 1843.*

[The following is the passage referred to.— " The properties of this order require to be considered under two points of view : firstly, those of the vegetation ; and, se- condly, those of the fructification. The character of the former is, generally speaking, suspicious, and often poisonous in a high degree ; as in the case of Hemlock, Fool's Parsley, and others, which are deadly poisons.   *   *   *The fruit,* vulgarly called the *seed, is in no case dangerous,* and is usually a warm and agreeable aromatic, as Ca. raway, Coriander, Dill, Anise, &c."—Lindley's 'Natural System of Botany,' Ed. 2, p. 22. Dr. Lindley also states, after Fée, that the properties of Conium maculatum are greatly affected by climate; it being inert and eatable in Russia and the Crimea, al. though it is extremely dangerous in the south of Europe. It is also stated that for medicinal purposes hemlock should be collected in June, soon after flowering, its ener. gy being much impaired if gathered later.—*Ed.*]

253. *Note on Hierochloe borealis,* (Phytol. 426 and 462). I may mention with re. gard to this beautiful and interesting grass, that I possess one or two very good spe. cimens, which I purchased along with a few others, forming part of an old herbarium bearing the date of 1805, but unfortunately I could not ascertain the name of the for- mer possessor. The handwriting is in an old style, and the following is a copy of the label :— " Holcus odoratus. Calla-glen, Angus-shire mountains ; rarissimè !" No date. Possibly these may have belonged to Mr. G. Don, but not being acquainted with his handwriting, I cannot conjecture as to the fact. The word " mountains " would seem to show that " Kelly-glen " near Arbroath could not be the spot intended. I may remark that in my herbarium there are specimens of Potentilla tridentata and opaca, collected by Mr. G. Don.— *Thos. Edmonston, jun.; Baltasound, Shetland, Ja. nuary 5, 1843.*

254. *On the Hygrometric Qualities of the Setæ of Mosses.* I do not think it has been generally noticed that many mosses possess fully as much of this quality as Fu. naria hygrometrica. Dicranum cerviculatum, varium, falcatum and strumiferum ; Orthotrichum crispum ; Bryum turbinatum, capillare, julaceum and Zierii ; Tricho. stomum polyphyllum, Tortula fallax and enervis, many Hypna, &c., I have observed to have this peculiarity to so great an extent, as to render it difficult to glue them to the paper on which they are preserved. Mosses which were dried under heavy pres- sure are comparatively destitute of this property. Thus Funaria hygrometrica, when the plants are picked and dried separately, has far less than when they had been pressed in tufts, and consequently the setæ little injured by it.—*Id.*

255. *Note on the mildness of the weather.* As an illustration of the mildness of the late season, the following notice of plants, which I gathered on the Derbyshire hills

beyond Mottram, on the 2nd of January of the present year, may prove interesting to some of the readers of ' The Phytologist.' Vaccinium Vitis-Idæa, both in flower and fruit, being the second time of its blossoms appearing within the last four months.— Empetrum nigrum, several specimens of which still bore the berries of last year; I did not see many with the flowers fully expanded, but if the weather had continued mild they would have been in full bloom in a week or two. Besides these I noticed in full flower about a dozen of the hardy spring weeds,— Stellaria media, Lamium purpureum, &c., and the following mosses were gathered in fructification.

Grimmia pulvinata	Polytrichum piliferum	Hypnum ruscifolium
Gymnostomum truncatulum	Dicranum taxifolium	Bryum pyriforme
Tortula muralis	heteromallum	argenteum
Funaria hygrometrica	Trichostomum lanuginosum	hornum
Polytrichum undulatum	Hypnum plumosum	punctatum

—*Joseph Sidebotham ; Manchester, January* 10, 1843.

256. *Note on Onoclea sensibilis.* Perhaps it will be interesting to some of the readers of ' The Phytologist' to know that Onoclea sensibilis grows in an old stone-quarry near Warrington. This fern was found in the above locality about four years ago by John Roby, Esq., of Rochdale; the plant is plentiful and grows very luxuriantly. It was also found in the north of Yorkshire a short time ago, by Mr. Baines of York.— *Samuel Gibson ; Hebden Bridge, January* 12, 1843.

257. *Note on the Poisonous Properties of the Fruit of Conium maculatum.* Professor Lindley is not the only botanist who has forgotten the qualities of the fruit of Conium maculatum, when speaking of the properties of the Umbelliferæ, for I find Dr. Willshire makes the same mistake. In his ' Principles of Botany,' he observes of the Umbelliferæ, that " the fruit is *innocuous ;* often stimulating from the essential oil it contains." It is well known that the *leaves* of Conium maculatum possess a poisonous quality, which depends on the presence of a peculiar and highly poisonous principle, called *Conia.* This principle is found in greater abundance in the seeds than in the leaves, therefore we might presume them to be more poisonous; yet I know of no case on record of poisoning by the seeds, although many unpleasant effects have been produced when small quantities have been administered. Professor Christison, in making experiments upon animals, administered about thirty grains of an extract prepared from the full-grown seeds ; it caused paralysis, convulsions and death, and this proves the fruit of Conium maculatum to possess very active and even poisonous properties. Indeed we should always be cautious how we employ the seeds of a poisonous plant, for although in some, as Papaver somniferum, they may be harmless, yet in others they possess all the active properties of the plant itself, as in the above example, to which I may also add Colchicum autumnale.— *Daniel Wheeler, M.R.C.S.L.; Reigate, January* 20, 1843.

258. *Correction of an error respecting the discovery of Statice tatarica near Portsmouth.* I regret having been the means, though inadvertently, of communicating an error to your pages, in stating Statice tatarica to have been discovered by myself near this place, (Phytol. 429). Having had some correspondence with Mr. Borrer on the subject, by whom I have been kindly furnished with specimens of the genuine S. tatarica from a foreign locality, and of S. Limonium, var. γ. (Smith), from Bosham, at the mouth of the Chichester river, I find that my plant is identical with the latter.— Its chief difference from the real S. tatarica consists in the absence of the winged stem, *which forms a* remarkable feature in that species, The very different appearance,

however, of my plant, from the normal form of S. Limonium, will I think admit of a reasonable doubt whether it is not a species distinct from that.—*W. L. Notcutt; Fareham, January 25*, 1843.

Art. CXXI. — *Proceedings of Societies.*

BOTANICAL SOCIETY OF EDINBURGH.

Thursday, January 12, 1843.—Dr. Douglas Maclagan in the chair. James Irving, Esq. was elected a resident fellow, and Baron Ludwig, — Arden Esq. and Wm. Caldwell Faure, Esq., Cape of Good Hope, non-resident fellows of the Society.

Mr. Goodsir then read two papers by John Ralfs, Esq., Penzance, on the *Diatomaceæ.* In these able papers the author described numerous species, and made some important observations on the structure and habits of these microscopic plants. Specimens were exhibited, and displayed under the microscope.

Mr. Brand read a description of two new species of British mosses, by Dr. Taylor, Dunkerron.

The next paper was entitled "Description of a new species of *Carex*, found near Hebden Bridge, Yorkshire." By C. C. Babington, Esq., M.A., F.L.S., &c., St. John's College, Cambridge. The author stated that it was now nearly two years since Mr. S. Gibson, of Hebden Bridge, had forwarded to him a *Carex* which he believed would prove to be a new species. He was now satisfied that is so, and has dedicated it to its discoverer (under the name of *Carex Gibsoni*), than whom no person can be more deserving of commemoration by means of a plant of this genus, to the careful study of which he has long and successfully applied himself.

"Remarks on the Scenery and Vegetation of Madeira." By Dr. Cleghorn, H.E.I.C.S. Dr. Cleghorn sailed from Spithead for India on the 15th of August, and reached Madeira on the 26th. His own narrative somewhat condensed is as follows.

In the evening of the 26th the vessel sailed into Funchal Bay. The sea was beautifully calm—glittering like a lake—the splash of the dolphin and the scream of the sea-bird being the only sounds. The stupendous mountains which rise behind Funchal, with their gigantic peaks, are very magnificent, and when viewed by moonlight, the general outline of these mountains resembles so much some scenes in the central Highlands, that I should have been impressed with the belief that I had been transported to Caledonia—had not the dense foliage of the plantain, the orange, and other trees fringing the coast, proved its vicinity to the tropics. From whatever quarter the island is approached, the aspect is singularly abrupt and picturesque. Next morning we went on shore. Much care is required in landing, as there is no quay, and the rocks are shelving. When the boat approaches the shore, on a signal being given, eight or ten boatmen yelling aloud, and tugging simultaneously, pull the skiff twenty yards up an inclined plane, so that passengers must *hold fast*, or they are capsized.

After a kind reception at the consul's residence, where are some splendid views of the neighbouring scenery, I sallied forth to see the lions of Funchal — the cathedral, nunneries, hospital and fortifications. Passing through the market, where were abundance of grapes, figs, dates, oranges, bananas, tomatas, &c., affording a delicious banquet to not a few of my shipmates, whose diet had consisted chiefly of salt junk, I afterwards sought the public garden or pleasure-ground, where the Portuguese were playing their national airs on the guitar and machetta, and returned to the consul's, carrying as trophies a noble specimen of *Coffea arabica*, and some ripe and unripe fruit of *Citrus Limonum* and *Limetta*.

I started early on the morning of the 27th to have a glimpse of the interior, in company with three shipmates. When entering Funchal Bay two evenings before, a striking edifice, apparently imbedded under a canopy of trellis-work and creepers, had attracted our attention. This conspicuous object, our guides now informed us, was " Nossa Senhora da Monte," or Mount Church, situated three and a half miles up the mountain, behind Funchal. As the land rises rapidly from the coast to the interior, in order to facilitate our progress, long iron spiculæ were appended to the pósterior part of our horses' shoes, presenting in an exaggerated degree the appearance of what we call in Scotland *well frosted*. This is necessary, for the angle of inclination of some roads above Funchal is certainly not less than the ascent of Arthur's Seat from the Hunter's Bog. These roads, however, though ill-suited for the progression of the tourist, are well adapted for the descent of the hogsheads of Madeira from the winepresses above the town. They are paved with small pebbles (nearly equal in size), taken from the ravines and water-courses ; and in the interstices between these stones the spiculæ mentioned are securely fixed at each step of the horse. We could not ride two abreast, as the road was only six feet broad in some places; and, speaking for myself, being mounted on a particularly fiery Pegasus, the occasional passing of a sledge, drawn by a couple of pigmy bullocks, was attended with some difficulty.

During the first part of the excursion we traversed a region of terraced vineyards, which are arranged in a singularly beautiful manner, the vines being carried on trellices over the roads, and, occasionally, this refreshing canopy is continued over some acres of rich soil—forming a lovely covering of leaves and fruit. These shady avenues are an agreeable protection from the rays of the sun, and being arched overhead, have the appearance of tunnels : in them the peasants are engaged in vine-dressing during the day—the children sport in the evening--and dogs keep watch by night.

The *Arundo donax*, which attains a height of twenty feet, is cultivated extensively for supporting the vines, and a variety of other purposes.

The soil in the vicinity of Funchal is exceedingly rich, consisting of dark vegetable mould, mixed with the debris of volcanic rock, or of beds of stiff red clay. The latter produces the best vineyards ; and a bed of this description, with traces of iron ore, extending to the depth of thirty feet, yielded as fine produce as any that we passed in the island.

As we rose from this lower region, where the mountain was clothed with vines and figs, and where flocks of canaries of a lively green flew around, we plodded on through orange bowers and festoons of fragrant jasmine (*Jasminum odoratissimum*). The proud sunflower here overtopped the modest *Heliotrope*. There were patches of *Coffea arabica* (in cultivation), *Cucurbitaceæ* in great abundance, — gourds, melons, pumpkins and cucumbers. We noticed, surrounding a merchant's villa, a stout hedge of *Fuchsiæ*, and some scattered specimens of *Passiflora edulis* and *quadrangularis*. The *Cactaceæ*, singular in appearance and various in habit, showed their spiny heads and handsome petals, their roots being sometimes firmly secured in the crevices of the granite. Here were plants of which we have no example in the north. The vegetation was of a different character from anything I had seen.

It was evident, however, that many of the rare flowers were not indigenous — but it was not easy to distinguish those that had been introduced, as the whole country at this elevation had the appearance of an ornamental garden.

There was a scene of life and motion amongst the herbage, of which our Scottish forests afford no idea — the leaves of some shrubs being covered with brilliant beetles; the abundant foliage giving plentiful nourishment to the swarms of insect tribes. — Larks in great numbers were carolling merrily, and kestrils hovering on the mountain side.

We dismounted to examine the pictures and architecture of the Mount Church, dedicated to Santa Maria, and from our now elevated position enjoyed a truly magnificent view (the platform and gateway were mantled with luxuriant shoots of *Lonicera* and *Clematis*). Below us lay the town and bay of Funchal; eastward, the singular promontory which we had rounded on entering the bay, and which is known to visitors by the name of the Brazen Nose; and to the west, beyond the tract of vineyards, is an extensive race-course, with a range of steep craggy rocks jutting out into the sea.

In pursuing the ascent, we rode along a sharp ridge leading to the Caldeira, or highest peak of the island, on which is a well-supplied ice-house. This is the range of the Pines and Spanish Chesnuts, and the timber here was not contemptible.

Closely adjoining, at a greater elevation, were many species chiefly of the *Labiatæ* and *Caryophylleæ*. We were now on either side the yawning depth of ravines, where, after a few hours' rain, most formidable torrents rush down the adjacent valley to the ocean. There were some remarkably fine specimens of *Gymnogramma Lowei* (named in honour of the resident English clergyman, a scientific naturalist) amongst the wet rocks of a narrow cleft, through which we passed in crossing a watercourse; of these specimens, some were singular varieties.

Sundry were the risks, as we slowly jogged down the mountain, keeping tight the bridle-reins, and the horses occasionally sliding on their haunches. Thus terminated a visit to Madeira, which the novelty of the scenes — the hospitality of the British consul and Portuguese merchants, and the kind attention of the house-surgeon of the hospital (an Edinburgh graduate), had rendered one of great pleasure and gratification.

BOTANICAL SOCIETY OF LONDON.

November 29, 1842.—Sixth Anniversary Meeting; J. E. Gray, Esq., F.R.S., President, in the chair. From the Report of the Council it appeared that 13 new members had been elected since the last anniversary, and that the Society consisted of 152 members.

The donations to the Library had been very considerable.

The Report of the Herbarium Committee was read, and stated that the British herbarium had been in reference order for some time, and the Committee were using their best exertions to obtain the Society's desiderata, which had lately been considerably diminished by the receipt of many rare plants from Mr. G. Francis and Mr. S. P. Woodward, the latter gentleman having presented a large series of British mints, collected by the late Mr. Sole of Bath.

Many valuable parcels of British and foreign plants had been received, and the return parcels sent to the members had given the greatest satisfaction, in many instances the return parcel having been sent within ten days after the receipt of the parcel from the member. The Committee anticipate that in future the return parcel will regularly be sent within ten days after the receipt of a parcel from any contributor.

Amongst the most valuable parcels received during the past season, may be mentioned a large collection of British plants from Mr. Hewett C. Watson, comprising upwards of 5500 specimens; also numerous Jersey plants from Mr. G. H. K. Thwaites. A large collection of Shropshire *Rubi* from Mr. H. Bidwell; 300 specimens of *Bupleurum falcatum* collected in Essex, from Mr. E. Doubleday ; and numerous specimens of *Lastræa cristata*, collected in Norfolk by Mr. B. D. Wardale, and presented by that gentleman.

The Committee pointed out the necessity of members sending two labels with each specimen, one for permanent preservation in the label-book, which the Committee anticipate will, in the course of a few years, be valuable as an authentic register of the localities of plants, and prove highly serviceable in showing the geographical distribution of the species. Many interesting monstrosities had been received from several of the members; and the Committee impressed upon the members the importance of collecting monstrosities, and their value in a public collection.

Local Herbaria.—The Society had received from Mr. Edwin Lees a herbarium of the Malvern Hills, including the Cryptogamic plants ; accompanied with many valuable remarks upon the geographical distribution of the plants of the neighbourhood, together with the geological character of the neighbouring hills : and the Committee hope next year to be able to report the receipt of other local herbaria now in course of formation for the Society.

Cryptogamous Plants.—The collection of mosses, lichens and Algæ received during the past season, had been more considerable than during any former period ; and the first three volumes of 'Algæ Danmonienses,' and part 1 of Berkeley's 'British Fungi,' had been presented to the Society by Mrs. Margaret Stovin. Mr. S. P. Woodward is now actively engaged in arranging the whole Cryptogamic collection.

Foreign Plants.—These form a valuable part of the Society's collection, and comprise plants from North and South America, British Guiana, New South Wales, Cape of Good Hope, Sierra Leone, China, and various other parts of the world.

Among the more interesting plants in this collection may be mentioned about 350 species collected by Mr. R. H. Schomburgk in British Guiana, and presented by him; 250 species collected in Natal, South Africa, by Dr. F. Krauss; many thousand specimens collected in North America by Dr. Gavin Watson, and presented by him ; and numerous species from Dr. C. F. S. De Martius, collected by him in South America.

Museum. — Numerous specimens of sections of woods, seed-vessels, barks, and several large collections of seeds, had been received, many of them purchased at the sale of the botanical museum of the late A. B. Lambert, Esq., and presented by some of the members.

The Reports of the Council and Herbarium-committee were unanimously adopted, and a ballot then took place for the Council for the ensuing year, when the chairman was re-elected President, and he nominated J. G. Children Esq., F.R.S., and Hewett C. Watson, Esq., F.L.S., Vice-Presidents ; Messrs. E. Doubleday, G. Francis, and J. G. Mitchell were elected new members of the Council, in the room of Dr. Meeson, Messrs. G. Cooper and W. H. White, who retire from the Council in accordance with the Rules of the Society. Mr. J. Reynolds, Mr. G. E. Dennes and Mr. T. Sansom were respectively re-elected Treasuror, Secretary and Librarian.—*G E. D.*

ART. CXXII. — *Notice of a new British Cerastium.*
By THOMAS EDMONSTON, jun., Esq.*

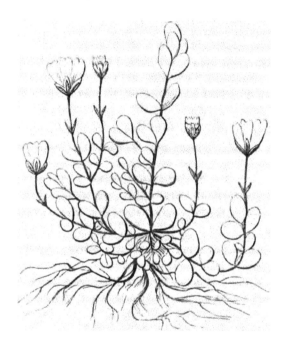

CERASTIUM LATIFOLIUM, *Linn.*

HAVING for some years entertained the opinion that the plant above
figured, although generally referred to Cerastium latifolium, was distinct from the plant called by that name in Britain, I have paid considerable attention to our alpine Cerastia, and am disposed to conclude
that my plant is truly distinct from the C. latifolium of Smith and
Hooker. I am happy to be able to say that my valued and eminent

* Communicated by Charles C. Babington, Esq., M.A., F.L S.

2 υ

friend Mr. Babington, on a careful study of the subject, concurs in this opinion, and I am delighted to fortify my own view with that of so distinguished an observer.

After attentively comparing numerous specimens, both recent and dried, of the C. alpinum and latifolium of our Scotch and Welch mountains, I am perfectly unable to detect any specific difference between them.   Sir W. Hooker, in ' Flora Scotica,' gives as the main distinction, the capsules in C. alpinum being " oblong, curved," and in C. latifolium " ovate," (Fl. Scot. 144); while in ' British Flora,' ed. 3, p. 217, he says, " I agree with Mr. Wilson in thinking that there exists scarcely any difference, either in the flower or fruit, between this (C. latifolium) and the preceding (C. alpinum) ; in both, the capsules are broadly oblong, shining, nearly twice as long as the calyx," &c., and the distinguishing characters between the two plants seem to be only in the pubescence, viz., C. alpinum being " clothed with long silky hairs," and C. latifolium " clothed with short, rigid, yellowish pubescence."   After a careful examination of numerous specimens I have drawn up the following descriptions, and shall then endeavour to establish the essential characters of the three plants, viz., the one from Shetland, and the C. alpinum and latifolium of our authors.

Cerastium latifolium, (Linn.)

Plant 2—5 inches high, branching; *stems* prostrate for about half their length, then ascending, never rooting; whole plant covered with a dense, short, glandular *pubescence; leaves* orbicular, obtuse, dark green ; *bracteas* herbaceous, lanceolate, acute; *sepals* blunt, with a membranous border half their own breadth ; *peduncles* one-flowered, equalling the flower ; *flowers* large, white, with green veins in the inside; *petals* bifid at the apex ; *capsule* ovate, shining, scarcely longer than the calyx, opening with ten valves.

This seems to be the original plant of Linnæus, and the C. latifolium of the continental botanists, for I have specimens communicated by M. Leresche from the Alpes du Vallais, exactly similar to the Shetland specimens, except in having the leaves slightly acute ; and Mr. Babington writes me that Reichenbach, in his ' Icones Floræ Germanicæ, figures the form as " C. latifolium, var. *glaciale*, of Gaudin, Fl. Helv.'

Abundant on a serpentine hill to the north of the bay of Baltasound, Unst, Shetland, extending over about a square mile of ground.   This station is interesting as being the only one hitherto known in Britain for Arenaria norvegica. I am not aware of any other British habitat for it, neither have I seen any foreign specimens.

Cerastium alpinum, Linn.

a. *Linnæanum*, (Bab. in Mag. Zool. and Bot. ii. 202). C. alpinum, Linn., Eng. Bot. 472; Hook. Fl. Scot. 144; Br. Fl. 216.

Plant 2 — 5 inches high; *stems* ascending, mostly simple; plant covered with a long, white, silky *pubescence; leaves* ovato-lanceolate, acute; *bracteas* with a narrow membranous border; *sepals* sometimes with a narrow membranous border; *flowers* one, two, or three together in a forked panicle; *peduncle* considerably longer than the flower; *flowers* as in the last; *capsule* broadly ovate, shining, twice as long as the calyx, opening with ten teeth.

β. *piloso-pubescens*, (Benth. in Lindl. Syn. ed. 1, 51). C. latifolium, Sm., Eng. Bot. 473; Hook. Fl. Scot. 144; Br. Fl. 216.

Stems generally branched, ascending, sometimes rooting at the base; *plant* with a short, rigid ("yellowish" Hook.), *pubescence; leaves* oblong, acute; *bracteas* often wanting, when present mostly with a narrow membranous margin; *peduncles* generally one-flowered, longer than the flower; *capsule* as in the last.

Of this plant Mr. Babington says "I believe that it is correctly referred by Bentham to the C. alpinum as β. *piloso-pubescens*," (Bab. in litt.)

This, with the last, grows on the more elevated mountains of Scotland, as well as on Snowdon, and I believe other Welsh mountains. β. seems less common than a., but I have specimens from several mountains of the Cairngorum range, as well as Clova, &c.

From these observations I think it will be evident not only that the Shetland plant is distinct, but that there exists no specific difference between C. alpinum and latifolium, *Sm.* and *Hook.*. The differential characters I propose are the following.

C. *alpinum*, Linn. Leaves ovate or ovato-lanceolate, acute; bracteas with a membranous border; sepals with scarcely any border; peduncle longer than the flower; capsule twice as long as the calyx.

C. *latifolium*, Linn. Leaves orbicular, obtuse; bracteas without the membranous margin; sepals with a broad border; peduncle long as a flower; capsule scarcely longer than the calyx.

500

The following figures will exemplify the many distinctive characters of these plants.

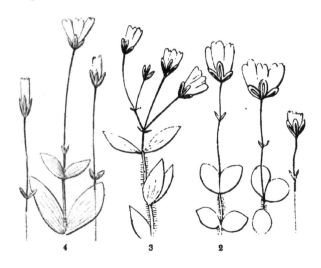

Fig. 1. End of a flowering branch and capsule of C. latifolium, *a*   Hab. Baltasound
Fig. 2. The same of C. latifolium, *glaciale* : Alps of the Vallais, Switzerland.
Fig. 3. C. alpinum, *a*. Clova, Forfarshire.   Fig. 4. C. alpinum, *β* (C. latif.): Clova.

THOS. EDMONSTON, JUN.

Baltasound, Shetland,
    January 23, 1843.

ART. CXXIII. — *Catalogue of Plants observed in the neighbourhood
of Daventry, Northamptonshire.* By MR. W. L. NOTCUTT.

IN forwarding a list of the plants of Daventry, Northamptonshire,
I have thought that a sketch of the general character and geological
formation of the neighbourhood might not be unacceptable, as I think
it of importance to trace the connexion between Geology and botanical Geography.   Daventry is a market-town seventy-two miles from
London, and is situate in the western part of the county of Northampton.   The geological formation upon which it stands is the lower
oolite, which however terminates at the distance of two or three miles
on the north-western side of the town, where the lias formation appears from beneath it : on the other side it extends over nearly the
whole county of Northampton.   In many parts of the neighbourhood
the coarse yellowish oolitic limestone lies within two or three feet of

the surface, and is quarried for mending the roads. In these spots I do not think any plants of interest have been noted. The oolite is characterized by its peculiar fossils, as pectens, Terebratulæ, Ammonites, Belemnites, &c., many species of which are found in the stonepits near the town in great abundance. On the south side of the town is Borough Hill, which is chiefly composed of sand, with nodules of sandstone, and which probably belongs to the oolitic period. On the top of this hill is a Roman entrenchment, within and on the borders of which are found many rare plants: the hill has been enclosed within the last fifty years. Three other hills on the other sides of the town, viz., Newnham Hill, Fox Hill and Welton Hill, appear to be composed of the same kind of sand as Borough Hill, but they are rather barren of interest in a botanical point of view. Two large reservoirs, formed for the purpose of supplying the Grand Junction Canal with water, are situated within a mile of the town, and bear on their margins a few aquatic and marsh plants worth notice, but from their recent formation it is not to be expected that a great variety will be found at present. Watling Street, one of the old Roman roads, runs within three miles of the town, and contains one or two very interesting species.

Many of the plants which grow around Daventry are remarkably local there, being found in only one or two spots : and this feature prevails to a greater extent than I have noticed anywhere else. I may just add that the following list is the result of a diligent investigation of the neighbourhood within a distance of three or three and a half miles from Daventry, during a residence there of between three and four years.

Anemone nemorosa. Daventry and Badby woods; Borough-hill

Ranunculus hederaceus. Rare: pool on Borough-hill

———— *aquatilis.* The reservoirs

———— *Flammula.* Old reservoir and Badby wood

———— *auricomus.* Norton & Buckby roads

———— *acris* and *bulbosus.* Common

———— *arvensis.* Borough-hill &c.

———— *repens.* Ashby road

—— —— *hirsutus.* Rare: Ashby road: Newnham

———— *Ficaria.* London road

Caltha palustris. The reservoirs

Berberis vulgaris. Rare: hedges at Staverton

Papaver dubium. Rare: Norton road

———— *Rhœas.* Corn-fields, common

———— *Argemone.* Rare: Borough-hill and Ashby road

Chelidonium majus. Rare: Everdon lane

Fumaria officinalis. Fields, common

Draba verna. Walls, common

Thlaspi Bursa-pastoris. Common

———— *arvense.* Borough-hill

Coronopus Ruellii. Local: Abbey wood

Cochlearia Armoracia. Rare: new reserv.

Cardamine pratensis and *hirsuta.* New reservoir and Newnham

Erysimum Alliaria. Ashby & London rds.

Barbarea vulgaris. Ashby & Norton rds.

Sisymbrium officinale. Roadsides, com.

Nasturtium officinale. Old reservoir; Staverton

———— *terrestre.* Local: new reserv.

———— *amphibium.* Local: canal at Braunston

Raphanus Raphanistrum. Newnham hill

Sinapis alba. Fields by Buckby road

———— *arvensis.* Common in cornfields

Brassica Napus. Field by Buckby road

Viola odorata. Weedon road; Newnham

—— *canina.* Common

—— *tricolor.* Borough-hill

—— *hirta.* Rare: Borough-hill, in the entrenchments

Polygala vulgaris. Badby wood; glen by Staverton toll-gate

Silene anglica. Rare: flds on Borough-h.

—— *inflata.* Local: Norton and Ashby roads

Agrostemma Githago. Borough-hill

Lychnis diurna. Daventry and Badby woods; Borough-hill

———— *vespertina.* Borough-hill; Welton road

———— *Flos-cuculi.* The reservoirs

Sagina procumbens. Burnt walls; Borough-hill

Spergula arvensis. Local: Borough-hill

Stellaria media. Common everywhere

———— *graminea* and *holostea.* Daventry wood and Borough-hill

Arenaria serpyllifolia. Walls in the town

———— *trinervis.* Copse at Welton-bridge

———— *rubra.* Local: Borough-hill

Cerastium viscosum. Common by road-si.

———— *arvense.* Local: Borough-hill

———— *aquaticum.* New reservoir and Norton road

Linum catharticum. Staverton glen; old reservoir; Badby

Malva sylvestris. Staverton road &c.

—— *rotundifolia.* Local: Norton and Badby

Tilia europæa. Welton road; church-yd.

Hypericum quadrangulum. London road; Borough-hill

Hypericum hirsutum. Rare: Badby road, near the bridge

———— *pulchrum.* Local: Staverton wood; Borough-hill

Acer campestre & *Pseudo-platanus.* Buckby road

Geranium robertianum. Welton and Badby roads

———— *dissectum.* New reservoir; Staverton road

———— *pratense.* Local: new reservoir

———— *molle.* Ashby rd.; Borough-h.

Oxalis Acetosella. Daventry, Staverton and Badby woods

Ononis arvensis. Norton and Badby rds.

———— *spinosa.* Ashby road

Lathyrus pratensis. New reservoir; Badby wood

———— *sylvestris.* Local; Badby wood

Vicia sepium. Badby road and wood

—— *Cracca.* Local: new reservoir; Welton canal

—— *sativa.* Borough-hill; new reservoir

—— *sylvatica.* Badby wood

Orobus tuberosus. Daventry, Staverton and Badby woods

Ulex europæus. Borough-hill; Newnham hill

—— *nanus.* Badby wood

Trifolium repens. Very common: proliferous, Norton road

———— *pratense.* Common

———— *procumbens* and *filiforme.* Borough-hill

Melilotus officinalis. Ashby road; near Braunston

Lotus corniculatus. Ashby & Norton rds.

—— *major.* Dunslade; Borough-hill; Badby road

Cytisus scoparius. Pond between Flecknoe and Drayton

Ervum hirsutum and *tetraspermum.* Borough-hill

Hedysarum Onobrychis & *Medicago lupulina.* Borough-hill

Prunus spinosa. Hedges, very common

———— *Cerasus.* Rare: Catesby and Staverton roads

Prunus domestica. Local: Borough-hill
Spiræa Ulmaria. The reservoirs
Geum urbanum. Road-sides, common
Rubus corylifolius. Staverton road
—— idæus. Borough and Newnham
hills; Staverton glen
—— fruticosus. Hedges, very common
—— cæsius. Daventry wood; Borough-hill
Fragaria vesca. Daventry & Badby wds.
Potentilla anserina, reptans and Fragari-astrum. Common
Tormentilla officinalis. Staverton and Badby woods
Agrimonia Eupatoria. Daventry wood
Alchemilla arvensis. Local: Borough-hill
Sanguisorba officinalis. Rare: new reservoir; road between Norton and Brockhall
Poterium Sanguisorba. Buckby road; Burnt walls; old reservoir
Rosa canina. Hedges, common
—— arvensis. Staverton road & Borough hill
—— spinosissima and sarmentacea. Rare: Borough-hill
Cratægus Oxyacantha. Very common
Pyrus communis. Rare: near Welton br.
—— Malus. Borough-hill; Thrupp
—— torminalis. Rare: Norton road
Epilobium hirsutum. New reservoir; London road; Badby wood
—— montanum. Badby and Staverton woods
—— parviflorum. Old reservoir; Dodford
—— tetragonum. Borough-hill; Badby
Circæa Lutetiana. Daventry and Badby woods; Watling-street road
Myriophyllum spicatum. Old reservoir
Callitriche aquatica. New reservoir; Borough-hill
Scleranthus annuus. Borough-hill
Sedum acre. Walls and roofs, common
—— reflexum. Walls and roofs, Badby
Ribes Grossularia. Staverton and Norton roads; Staverton glen

Ribes rubrum. Staverton glen
Saxifraga granulata. New reservoir
—— tridactylites. Walls
Adoxa Moschatellina. Local: Borough-hill; Staverton glen
Daucus Carota. Norton and Ashby roads
Torilis nodosa. Local: Borough-hill
—— Anthriscus. London and Norton roads
Heracleum Sphondylium. New reservoir
Pastinaca sativa. Field between Ashby and Welton road
Angelica sylvestris. Daventry and Badby woods; London road
Bunium flexuosum. Norton road; old reservoir &c.
Pimpinella Saxifraga. Dodford; Borough-hill
Sium angustifolium. Dunslade; Braunston canal
Cnidium Silaus. Old reservoir
Ægopodium Podagraria. Newnham hill; Staverton road; Badby
Æthusa Cynapium. Staverton road
Sison Amomum. Rare: Welton road
Helosciadium nodiflorum. New reservoir; Staverton glen
Chærophyllum temulentum. Near new reservoir
Anthriscus sylvestris. Norton and London roads
—— vulgaris. Norton road
Scandix Pecten-Veneris. Borough-hill; Newnham hill
Conium maculatum. Buckby road and fields adjoining
Sanicula europæa. Daventry wood; Badby
Eryngium campestre. Watling-street road near Brockhall. It is probable that this celebrated plant will not much longer exist in the above locality; it is now nearly extirpated, in consequence of the rapacity of some botanical collectors, and the fondness of cattle for it. A specimen may however still be found now and then by careful and diligent search
Hedera Helix. Badby road

Cornus sanguinea. Norton road

Sambucus nigra. Welton & Ashby roads; Borough-road

Viburnum Lantana. Staverton road.

——— *Opulus.* Staverton, Daventry and Badby woods

Lonicera Periclymenum. Norton road; Daventry and Badby woods

Galium verum. Buckby and Badby rds.

——— *saxatile.* Borough-hill; Badby wd.

——— *cruciatum.* Norton & Badby rds.

——— *mollugo.* Local: Badby wood

——— *palustre.* The reservoirs

——— *uliginosum.* Borough-hill; new reservoir

——— *Aparine.* Very common

Asperula odorata. Local: Badby wood

Sherardia arvensis. Borough-hill

Valeriana officinalis. Daventry and Badby woods

——— *dioica.* New reservoir; Staverton glen

Fedia dentata and *olitoria.* Borough-hill

Scabiosa succisa. Local: Badby wood

Knautia arvensis. Ashby road; Drayton

Dipsacus sylvestris. Watling-street road; Badby and Buckby roads [lands

——— *pilosus.* Brook at bottom of In-

Leontodon Taraxacum. Common everywh.

Sonchus arvensis. Borough-hill

——— *oleraceus.* Staverton and Badby roads: var. *asper;* Badby road

Crepis virens. Ashby road

Apargia hispida. Burnt walls; Badby rd.

——— *autumnalis.* Road between Norton and Brockhall

Hieracium Pilosella. Local: Borough-hill by Castell's farm

——— *sabaudum, umbellatum, sylvaticum* & *maculatum.* Rare: Borough hill, in the entrenchments

Tragopogon pratensis. Ashby road; field by Norton road

Lapsana communis. Norton & Badby rds.

Helminthia echioides. London road, a single plant, 1841

Serratula tinctoria. Daventry wood; Borough-hill

Arctium Lappa. London road &c.

Carduus nutans. Field by Buckby road

——— *acanthoides.* Meadow south of new reservoir

Cnicus palustris. Old reservoir

——— *arvensis.* Braunston and London roads; Borough-hill

——— *lanceolatus.* London road

——— *acaulis.* Ashby road, a single plant, 1841

Bidens tripartita. Old and new reservoirs

Tussilago Farfara. Old reservoir; Staverton glen

Petasites vulgaris. Meadow south of new reservoir

Senecio vulgaris. Very common

——— *aquaticus.* Dunslade; new reser.

——— *sylvaticus.* Local: Borough-hill; Daventry wood

——— *tenuifolius.* Dodford; Ashby rd.; Borough-hill

——— *viscosus?* Borough-hill

Bellis perennis. Everywhere

Chrysanthemum Leucanthemum. Borough hill; old reservoir

——— *segetum.* Local: Borough hill

Matricaria Chamomilla. Ashby and Welton roads

Anthemis Cotula. Fields east of new reservoir; Borough-hill

——— *arvensis.* Rare: cornfields between Newnham and Everdon

Gnaphalium uliginosum. Old reservoir; Drayton [hill

——— *germanicum.* Rare: Borough

Inula dysenterica. Dodford; Welton rd.

Solidago Virgaurea. Local: entrenchments on Borough-hill

Achillea Millefolium. Norton road; Borough hill

Centaurea nigra. Drayton

——— *Scabiosa.* Local: field by Norton road

——— *Cyanus.* Local: Borough-hill; Newnham hill

Campanula rotundifolia. Borough-hill; new reservoir

Campanula latifolia. Rare : near Thrupp farm ; Watling-street road

Jasione montana. Rare : Borough-hill

Calluna vulgaris. Local : Badby wood ; Borough-hill

Ligustrum vulgare. Staverton and Buck-by roads

Fraxinus excelsior. Norton and Buckby-roads

Menyanthes trifoliata. Rare : old reserv.

Erythræa Centaurium. Badby wood ; Welton road

Convolvulus sepium. Rare : Dodford

——— *arvensis.* Common

Myosotis arvensis. Old reservoir, &c.

——— *versicolor.* Borough-hill

——— *palustris.* New reservoir; canal at Braunston

Lithospermum arvense. Rare : Borough-hill

Symphytum officinale. Ditches near new reservoir; brook between Newnham and Badby

Hyoscyamus niger. Rare : Thrupp

Solanum Dulcamara. Norton and Welton roads

Linaria Cymbalaria. Dodford

——— *vulgaris.* Rare : Borough-hill

Scrophularia nodosa. Badby wood

——— *aquatica.* New reservoir ; Staverton

Melampyrum pratense. Daventry & Badby woods

Pedicularis sylvatica. Badby wood ; Staverton glen

——— *palustris.* Rare : old reservoir, Mr. R. H. Smith

Rhinanthus Crista-galli. Old reservoir

Euphrasia officinalis. Old reservoir; Badby wood

Bartsia Odontites. Ashby rd.; Borough-h.

Veronica officinalis. Borough-hill ; Badby wood

——— *serpyllifolia.* New reservoir ; Borough-hill

——— *Beccabunga.* New reservoir ; Welton and Badby road : a pink-flowered variety, Welton road

Veronica Anagallis. Old reservoir

——— *Chamædrys.* Welton road, &c.

——— *hederifolia.* Very common

——— *arvensis.* Dog-lane ; fields by path to Flecknoe

——— *agrestis.* Borough-hill

Lycopus europæus. Rare : the canal at Braunston

Ajuga reptans. Daventry and Badby wds. Norton road

Galeobdolon luteum. Daventry and Staverton woods

Ballota nigra. Welton and London roads

Betonica officinalis. Old reservoir; Norton road ; Badby wood

Galeopsis Tetrahit. Borough-hill, abundant ; Badby road

——— *versicolor.* Local : Cornfield by footpath to Flecknoe

Lamium album. Roadsides, common

——— *purpureum.* Staverton road &c.

——— *amplexicaule.* Rare : Borough-h

Stachys arvensis. Local : Borough-hill

——— *sylvatica.* Borough-hill ; Norton, London and Ashby roads

Glechoma hederacea. Very common

Mentha hirsuta. Old reservoir

——— *arvensis.* Local : Dunslade

——— *sylvestris.* Rare: canal, Braunston

Thymus serpyllum. Field by old reservoir

——— *Calamintha.* Local: Borough-h.

Clinopodium vulgare. Norton road ; Watling-street road   [wood

Prunella vulgaris. Old reservoir; Badby

Scutellaria galericulata. Side of canal from Braunston to Buckby locks

Primula veris. Old reservoir &c.

——— *vulgaris.* Badby, Staverton and Daventry woods

Lysimachia nemorum. Daventry and Badby woods

——— *Nummularia.* New reservoir

Anagallis arvensis. Braunston road ; Borough-hill

Plantago media and *lanceolata.* Very com.

——— *major.* Badby road

Chenopodium album. Borough-hill ; fields by Staverton road

Chenopodium rubrum. Ashby road; new reservoir

Atriplex patula. Dunslade

Rumex Acetosa. New reservoir

———— *Acetosella.* Daventry wood; Borough-hill

———— *crispus.* Fields by Welton road

———— *acutus* and *pulcher.* Borough-hill

Polygonum aviculare. Very common

———— *Bistorta.* Rare; meadow near Badby

———— *Convolvulus.* Fields on Borough-hill; Buckby road

———— *Persicaria.* Borough-hill

———— *amphibium.* Old reservoir

———— *lapathifolium.* Borough-hill; fields near Staverton wood

———— *Hydropiper.* Local: Badby wood; Borough-hill

Euphorbia exigua. Fields by Staverton toll-gate

———— *Peplus.* Cultivated ground

———— *Helioscopia.* Fields south of new reservoir

Mercurialis perennis. Ashby road; Badby wood

Urtica dioica. Everywhere

———— *urens.* Local: Welton road

Humulus Lupulus. Head of new reservoir

Ulmus campestris. London road &c.

Betula alba. Staverton road

Salix oleifolia. Staverton glen

———— *acuminata.* Staverton road; Badby wood

———— *caprea.* Ashby road; Badby wood

———— *viminalis.* New reservoir; Staverton glen

———— *alba.* New reservoir

———— *cinerea.* Staverton and Badby wds.

———— *aquatica.* Old reservoir

———— *rubra, vitellina* and *triandra.* Staverton glen

Populus alba. Road-sides

———— *nigra.* Pope-well; new reservoir

Fagus sylvatica. Badby road

Castanea vesca. Daventry wood

Quercus Robur. Borough-hill &c.

Corylus Avellana. Daventry & Badby wds.

Butomus umbellatus. Rare: the canal at Braunston

Alisma Plantago. The reservoirs; Pope-well

Sagittaria sagittifolia. Rare: canal at Braunston

Potamogeton crispus. Pope-well

———— *fluitans.* Old reservoir

———— *natans.* Pope-well; Borough-hill

———— *lucens.* New reservoir; canal at Braunston

———— *perfoliatus.* Rare: canal at Braunston

Orchis mascula. Local: Badby wood

———— *Morio.* Local: Thrupp farm

———— *maculata.* Daventry & Badby wds.

———— *latifolia.* Rare: old reservoir

Gymnadenia bifolia. Rare: Badby wood

Listera ovata. Rare: new reservoir; Badby wood

Epipactis latifolia. Rare: Daventry wd.

Iris Pseud-acorns. Rare: new reservoir

Tamus communis. Newnham hill; Badby wood

Scilla nutans. Daventry, Staverton and Badby woods; var. *flore albo*, rare, Borough-hill

Allium ursinum. Local: banks of a shady ditch at Badby

Juncus conglomeratus. Ashby road; Badby wood

———— *lampocarpus.* Old reservoir

———— *effusus.* Borough-hill

———— *obtusiflorus.* Rare: old reservoir

———— *bufonius.* Pope-well; new reservoir; Watling-street road

———— *glaucus.* Dunslade; Ashby road

———— *acutiflorus.* Rare. Badby wood

Luzula campestris. Meadows

———— *congesta.* Rare: Badby wood

———— *pilosa.* Daventry, Staverton and Badby woods

Arum maculatum. London road; Inlands

Typha latifolia and *angustifolia.* Old res.

Sparganium ramosum. New reservoir

———— *simplex.* Rare: old reservoir

Eleocharis palustris. The reservoirs

Scirpus lacustris. Old reservoir

Carex intermedia and *ovalis.* Rare: old reservoir

—— *sylvatica.* Daventry & Badby wds.

—— *recurva* and *riparia.* New reservoir

—— *cæspitosa.* Old reservoir

—— *paludosa.* Pope-well and Staverton glen

—— *hirta.* Rare: Staverton glen and old reservoir

—— *paniculata.* Rare: Staverton glen

—— *præcox.* Meadow in Staverton glen

Agrostis vulgaris. Borough-hill

Milium effusum. Local: Daventry wood

Phalaris arundinacea. The reservoirs

Phleum pratense. Meadows by old reservoir; Drayton

Alopecurus pratensis. Common

———— *geniculatus.* East side of Borough-hill; old reservoir

———— *agrestis.* Fields by Staverton toll-gate

———— *fulvus* Rare: new reservoir

Anthoxanthum odoratum. Common

Melica uniflora. Local: Lane at Badby; Daventry wood

Aira cæspitosa. Daventry wd.; new reser.

—— *flexuosa.* Rare: Daventry wood

Holcus avenaceus. London and Ashby rds. new reservoir

———— *lanatus.* London road; Borough-h

——— *mollis.* Borough-hill

Avena fatua. New reservoir

—— *flavescens.* The reservoirs; London road

Bromus mollis. Common

—— *sterilis.* Staverton and Norton rds

—— *asper.* Newnham hill; Norton & Braunston roads

—— *racemosus.* Local: Borough-hill

Festuca duriuscula. New reservoir; Borough-hill

—— *gigantea.* Norton road; Watling-street road

—— *elatior.* New reservoir; Pope-well; Norton-road

—— *myurus.* Rare: wall at Drayton

Dactylis glomerata. Common

Glyceria fluitans. Pope-well; Newnham; old reservoir

—— *aquatica.* Rare: the canal at Braunston

Poa annua. Very common

— *pratensis,* Walls; Badby wood

— *trivialis.* Common

Catabrosa aquatica. Pope-well

Briza media. Old reservoir

Cynosurus cristatus. Common

Triticum repens. London road &c.

Lolium perenne. Very common

Hordeum murinum. London road

—— *pratense.* Field by London road; old reservoir

Lemna minor. Old reservoir; Pope-well

Equisetum arvense and *palustre.* Old res.

—— *limosum.* The reservoirs

—— *sylvaticum.* Rare; Badby wd.

—— *fluviatile.* Badby wood; Staverton wood and glen

Pteris Aquilina. Borough-hill; Badby wood, &c.

Lastræa Filix-mas. Ditto

—— *dilatata.* Rare: Badby and Staverton woods

Polystichum lobatum. Rare: Newnham lane

Polypodium vulgare. Borough-hill

Ophioglossum vulgatum. Staverton, in two meadows between the toll-gate and the village; Thrapp farm

Besides the plants in the foregoing list, the following have been mentioned, in various county histories and similar publications, as growing within this district; but after diligent search I have failed to detect them. Genista anglica, Eriophorum polystachion, Agrostis stolonifera, Bromus erectus, Parnassia palustris, Allium vineale, Arenaria verna, Spergula subulata, Carduus tenuiflorus, Scirpus sylvaticus, Cotyledon Umbilicus, Cladium Mariscus, Sambucus Ebulus, Sedum Forsterianum, Verbena officinalis, Campanula Rapunculus, Juncus bulbosus, Montia fontana, and Polypodium Rhœticum, (*With.?*)

ENUMERATION.

SPECIES.		SPECIES.		SPECIES.	
Ranunculaceæ	12	Brought up	130	Brought up	294
Berberideæ	1	Grossularieæ	2	Plantagineæ	3
Papaveraceæ	4	Saxifrageæ	3	Chenopodeæ	3
Fumariaceæ	1	Umbelliferæ	21	Polygoneæ	12
Cruciferæ	17	Caprifoliaceæ	6	Euphorbiaceæ	4
Violarieæ	4	Rubiaceæ	9	Urticeæ	3
Polygaleæ	1	Dipsaceæ	4	Amentaceæ	18
Caryophylleæ	17	Valerianeæ	4	Alismaceæ	3
Lineæ	1	Compositæ	44	Potameæ	5
Malvaceæ	2	Campanulaceæ	3	Orchideæ	7
Tiliaceæ	1	Ericineæ	1	Irideæ	1
Hypericineæ	3	Jasmineæ	2	Asparageæ	1
Acerineæ	2	Gentianeæ	2	Liliaceæ	2
Geraniaceæ	4	Convolvulaceæ	2	Junceæ	10
Oxalideæ	1	Boragineæ	5	Aroideæ	1
Leguminosæ	23	Solaneæ	2	Typhaceæ	4
Rosaceæ	26	Antirrhineæ	4	Cyperaceæ	12
Onagrarieæ	5	Rhinauthaceæ	6	Gramineæ	38
Halorageæ	2	Veroniceæ	8	Lemnaceæ	1
Paronychieæ	1	Labiatæ	21	Equisetaceæ	5
Crassulaceæ	2	Primulaceæ	5	Filices	6
	130		294	Total	423

I would just call attention to the small number of ferns in this district, the proportion they bear to the flowering plants being as 1 to 70½, the smallest with which I am acquainted.   W. L. Notcutt.

Fareham, December 7, 1842.

Art. CXXIV. — *County Lists of the British Ferns and their Allies.* Compiled by Edward Newman.

(Continued from p. 482).

STAFFORDSHIRE.

Lomaria Spicant.   Harbourne, *H. C. Watson ;* woods and heaths, Madeley, *G. Pinder ;* in one or two rough places in Needwood forest, *W. L. Beynon.*

Pteris Aquilina. Woods and heaths, Madely, *G. Pinder ;* on spots not under cultivation, chiefly on the higher ground, *W. L. Beynon.*

Polypodium vulgare. Harbourne, *H. C. Watson ;* woods and heaths, Madeley, *G. Pinder ;* occasionally met with, but not in so great plenty as in the adjoining coun_ties of Warwick and Worcester, *W. L. Beynon.*

Polypodium Phegopteris. Ridge hill and Madeley manor, *G. Pinder.*

Polypodium Dryopteris. Trentham park, *G. Pinder ;* it occurs in the grounds of the Rev. Thomas Gisborne, Yoxall-lodge, where it may possibly have been introduced, as that gentleman in his youth was a collector of plants, *W. L. Beynon ;* in abun_dance on the Staffordshire side of Dove Dale, on a rock called Dove Dale Church, *Miss Beever.*

Cystopteris fragilis. Butterton park, *G. Pinder.*

Polystichum aculeatum. Sprink wood, Madeley, *G. Pinder ;* on high marly banks agreeing well with Withering's description, *W. L. Beynon.*

Polystichum lobatum. Heyley castle, *G. Pinder.*

Polystichum angulare. Heyley castle, *G. Pinder.*

Lastræa Oreopteris. Woods and heaths generally, *G. Pinder.*

Lastræa Filix-mas. Harbourne, *H. C. Watson ;* woods and heaths generally. *G. Pinder ;* in every hedge, *W. L. Beynon.*

Lastræa dilatata. Harbourne, *H. C. Watson ;* woods and heaths generally, *G. Pinder ;* in deep shade, *W. L. Beynon.*

Lastræa spinulosa. Madeley bog and elsewhere, *G. Pinder.*

Athyrium Filix-femina. Harbourne, *H. C. Watson ;* Old Manor lane and else_where, *G. Pinder ;* not unfrequent in moist shady ditches, *W. L. Beynon.*

Athyrium irriguum. Trentham park, *G. Pinder.*

Asplenium Adiantum-nigrum. Bar hill and Heyley castle, *G. Pinder ;* on one or two hedge-banks, *W. L. Beynon.*

Asplenium Trichomanes. Heyley castle, *G. Pinder ;* on old walls at Litchfield, *W. L. Beynon.*

Asplenium Ruta-muraria. Madeley village, *G. Pinder ;* on old walls at Litchfield, *W. L. Beynon.*

Scolopendrium vulgare. Heyley castle and Sprink wood, *G. Pinder ;* abundant in a deep dingle at Tatenhill, *W. L. Beynon.*

Osmunda regalis. Balterley, *G. Pinder.*

Botrychium Lunaria. Whitmore and Maer heaths, *G. Pinder.*

Ophioglossum vulgatum. Meadows at Madeley, *G. Pinder ;* in moist meadows among grass, with such plants as Orchis latifolia and Valeriana dioica, *W. L. Beynon*

Lycopodium clavatum. Swinnerton and Maer heaths, *G. Pinder ;* plentiful on dry ground, Barr common, *D. Cameron, G. Luxford ;* Perry common (Mr. Ick), *W. G. Perry.*

Lycopodium Selago. Swinnerton and Maer heaths, *G. Pinder ;* in June, 1836, Mr. Cameron picked up a single specimen, dead and withered, on nearly the highest part of Perry Bar common, no other could then be found, *G. Luxford.*

Equisetum limosum. Ditches about Trent, *W. L. Beynon.*

Equisetum palustre. Betley mere and elsewhere, *G. Pinder ;* ditches, common, *W. L. Beynon.*

Equisetum arvense. Common, *G. Pinder ;* borders of fields, *W. L. Beynon.*

Equisetum sylvaticum. Walton's wood and elsewhere, *G. Pinder.*

Equisetum fluviatile. Grafton's wood ; Madeley, *G. Pinder ;* moist coppices, *W. L. Beynon.*

WARWICKSHIRE.

Lomaria Spicant. Pretty generally distributed, *W. Southall, jun.;* abundant in many places, *D. Cameron, G. Luxford;* common in situations suitable to its growth, *W. T. Bree;* heathy places on Honiley common (*W. W. Baynes*); Haseley common, *W. G. Perry;* Coleshill heath, near the bog; very luxuriant in a lane leading down to Bannerley pool, *J. J. Murcott.*

Pteris Aquilina. Very common, *W. Southall jun., D. Cameron, W. T. Bree, W. G. Perry, J. J. Murcott.*

Polypodium vulgare. *W. Southall, jun.;* abundant both on the ground and on trunks of trees, *D. Cameron, G. Luxford;* very common, *W. T. Bree, W. G Perry.* Var. *serratum :* a very marked variety, much more nearly approaching to P. cambricum than the Irish variety, was found in a lane near Moseley, where it was pretty abundant, and though I have well hunted for another habitat I have not found one, *W. Southall, jun.* Var. β. roadside just beyond the cross at Hampton-on-the-hill, near Norton Lindsey, *W. G. Perry.*

Cystopteris fragilis. Compton Verney, near Stratford-on-Avon (*G. Cook*), *D. Cameron.*

Polystichum aculeatum. *W. Southall, jun., W. T. Bree.*

Polystichum angulare. *W. Southall, jun.;* Elmdon and near Castle-bromwich, *D. Cameron ;* rare near Warwick, *W. G. Perry ;* Radford; ditch at the top of Emscote hill, opposite the turn to Milverton, *J S. Baly.*

Polystichum lobatum. *W. Southall, jun.;* Elmdon and near Castle-bromwich, *D. Cameron;* common, *W. T. Bree;* a bank at Saltley; at Yardley; and var. *ramosum, W. G. Perry;* thicket between Huningham and Offchurch; plentiful on the road from Warwick to Henley, *J. J. Murcott;* Allesley, *J. S. Baly;* near Maxtoke, *G. Luxford.*

Lastræa Oreopteris. Coleshill heath, plentiful; Corley, *W. T. Bree;* Haseley common, *W. G. Perry.*

† Lastræa Thelypteris. Plentiful in a boggy pit in this parish [Allesley] some years ago, but the pit is drained, and the fern entirely eradicated. I never met with it elsewhere in the county. See ' Mag. Nat. Hist." iii. 166, and v. 199, for further particulars, *W. T. Bree.*

Lastræa Filix-mas. *W. Southall, jun.;* abundant, *D. Cameron, W. T. Bree, W. G. Perry, J. J. Murcott, G. Luxford.*

Lastræa dilatata. *W. Southall, jun.;* abundant in marshy places, *D. Cameron;* not uncommon, *W. T. Bree, G. Luxford ;* Coughton lane and Spernall, (Purton); Oakley wood; on rocks below Milverton by the side of the Avon, (*W. W. Baynes*); on a steep bank by the side of the horse-pond at Mr. Cook's farm-house, Woodloes, near Warwick, *W. G. Perry;* Foleshill, *J. S. Baly.*

Lastræa spinulosa. Sparingly in marshy ground near Smethwick, and lanes near Harbourne, *D. Cameron ;* the small variety of dilatata, the spinulosa of Dickson (not the var. *recurvum*), grows in woods at Allesley and on Coleshill heath, *W. T. Bree;* Chesterton wood, in a cleared part; a shady bank, Garrison-lane near Birmingham, *W. G. Perry;* Waverley wood near Weston; Coleshill heath; Frogmoor coppice near Temple Balsall, *J. J. Murcott;* Allesley, *J. S. Baly.*

Athyrium Filix-femina. *W. Southall, jun.;* common in swampy places : the red-stemmed variety is abundant in a lane near Sutton park, *D. Cameron ;* not uncommon in moist places, *W. T. Bree, G. Luxford*; between Leamington and Kenilworth, (*W. W. Baynes*); on the porch of the church at Stratford-on-Avon, *W. G. Perry*; a single plant at Waverley wood near Weston; Bannerley common and wood, also in the lane leading down to Bannerley pool, moist bank near to Stonebridge, on the Kenilworth road, *J. J. Murcott*; Allesley, *J. S. Baly*.

Athyrium irriguum. This seemingly starved variety of Filix-femina is to be found in abundance in a lane near Harbourne, growing round a small spring, which seems as if it affected the plant by some peculiarity in the water, for at the distance of ten or twelve feet it gradually merges into the more usual type of Filix-femina. Specimens of this variety were planted in the Birmingham Botanic Garden, where for two or three years they remained pretty distinct, but afterwards the distinction grew more faint, and the whole plant became more luxuriant: *D. Cameron.*

Asplenium Adiantum-nigrum. Sparingly distributed, *W. Southall, jun.*; rare; near Solly-oak, and by the side of the Warwick road near Moseley common. *D. Cameron*; partially distributed; common in the parish of Coxley; Meriden; Balsall; Allesley; *W. T. Bree:* on a high rocky bank near the river Avon below Milverton, (*W. W. Baynes*); road-side between Hampton-on-the-hill and Norton-Lindsey; Fen-end, Temple Balsall, *W. G. Perry*; on a bank a mile and a half from Henley, on the Warwick road; also a few fronds on the church at Henley, *J. J. Murcott*; a single plant on the church-yard wall at Lillington; Norton hill; Maxtoke priory, *J. S. Baly.*

Asplenium Ruta-muraria. Sparingly distributed, *W. Southall, jun.*; Aston-park wall, *D. Cameron*; Allesley; Berkenhill church; Maxtoke castle; Stonleigh, *W. T. Bree*; on an old bridge at Stonleigh; Southam church, (*W. W. Baynes*); Aston-park wall, on the side next the lane to Witton, (*Mr. Ick*); St. Mary's church-yard wall and a garden-wall in Priory-lane, Warwick, *W. G. Perry*; Tachebrooke church sparingly, and a single tuft on a stone bridge in Stonleigh deer-park, *J. J. Murcott:* vicarage and priory walls, Warwick, abundant and luxuriant; bridge between Leamington and Stonleigh; Coventry town wall, *J. S. Baly.*

Asplenium Trichomanes. Elmdon hall near Hockley, *W. Southall, jun.*; very partial, Allesley; Stonleigh, *W. T. Bree*; on a bridge near Stonleigh abbey (*W. W. Baynes*); Church-porch, Stratford-on-Avon; Coughton church, *W. G. Perry*; plentiful on a bridge in Stonleigh deer-park, *J. J. Murcott.*

Scolopendrium vulgare. Kenilworth; Knowle near Moseley, *W. Southall, jun.*; damp shady places near Elmdon, *D. Cameron*; common; frond cleft at the end and partially branched, *W. T. Bree*; rather common; boggy ground near Solihull, (*Mr. Ick*); var. fronde apice lobatâ; var. fronde profundè bipartito, laciniis incurvis, *W. G. Perry*; plentiful at Halton rock near Stratford; Kenilworth, near the ruins of the castle; bank of a pool at the Woodloes; roadside between Budbrook and Hampton, *J. J. Murcott*; luxuriant at Tachebrook; forked variety at Kenilworth castle, *J. S. Baly.*

Ceterach officinarum. On a brick wall at the back of the mansion house at Tachebrook, *W. G. Perry*; Mr. Waller's garden-wall at Tachebrook, *J. J. Murcott.*

Osmunda regalis. Found formerly at Coleshill heath and other places, but I cannot find it now; Sutton park, as I am informed, but very sparingly, *W. T. Bree.*

Botrychium Lunaria. *W. G. Perry;* on heathy ground near the upper part of Coleshill bog, on the Stonebridge side, *J. J. Murcott.*

Ophioglossum vulgatum. Near Bilsley common, *W. Southall, jun.*; abundant at Elmdon, *D. Cameron*; abundant; Maxtoke and Allesley, *W. T. Bree*; in fields near

Emscote cotton-mills (W. W. Baynes); in a coppice above half a mile beyond Saltisford common; in two fields on the left of the old park lane called Commander's fields, *W. G. Perry*; meadow at Offchurch in occupation of Mr. Coles; plantation in Warwick old park, bordering on the Woodloes; at Goodrest in Warwick old park; a single plant in Whitnash field, *J. J. Murcott*; Eastern green, *J. S. Baly*.

Lycopodium clavatum. Coleshill bog, *W. Southall, jun.*; Coleshill heath formerly, *W. T. Bree*.

Lycopodium inundatum. Coleshill heath formerly, *W. T. Bree*; near the upper end of Coleshill pool in 1842, *J. J. Murcott*.

Lycopodium Selago. Coleshill heath formerly, *W. T. Bree*; see Phytol. 61 for the Rev. Mr. Bree's observations on the Lycopodia of Coleshill, *E. Newman*.

Pilularia globulifera. Coleshill pool, *W. Southall, jun.*; abundant in Coleshill pool, both under water and on the margin, *D. Cameron*, *W. T. Bree*, *J. J. Murcott*, *G. Luxford*.

Equisetum palustre. Elmdon, *W. Southall, jun.*, *D. Cameron*; (W. T. Bree), *H. C. Watson*; on the side of a lane east of Budbrook field, Warwick, *W. G. Perry*; meadows at the Woodloes and Bubbenhall, *J. J. Murcott*; Stoke, *J. S. Baly*.

Equisetum limosum. Avern's mill-pool and other pools, *W. Southall, jun.*; Coleshill pool; Elmdon; mill-pool, Bristol-road, *D. Cameron*, *G. Luxford*; in the latter locality I have found a variety similar to the var. *polystachion* of E. palustre, *G. Luxford*; (W. T. Bree), *H. C. Watson*; river Avon; St. Nicholas' meadow, Warwick; Chesterton mill-pool, *W. G. Perry*; in a ditch near Oldham's mill, Leamington, (W. W. Baynes), *W. G. Perry*, *J. S. Baly*; Coleshill and Bannerley pools; Haseley mill-dam; several pits in Warwick old park; river Avon near Warwick, *J. J. Murcott*.

Equisetum sylvaticum. Boundary of Birmingham Botanic Garden, *W. Southall, jun.*, *D. Cameron*.

Equisetum arvense. *W. Southall, jun.*; common, *D. Cameron*, (W. T. Bree), *H. C. Watson*, *W. G. Perry*, *J. J. Murcott*, *G. Luxford*.

Equisetum fluviatile. *W. G. Perry*.

See also Mr. Perry's ' Plantæ Varvicenses Selectæ,' as further authority on Warwickshire localities.

WORCESTERSHIRE.

Lomaria Spicant. Abundant in damp woods and open wild heathy places; Bromsgrove Licky; Malvern hills; Shrawley wood; Wyre Forest, *E. Lees*; Hartlebury common, *R. J. N. Streeten*; Bromsgrove Lickey; Malvern hills and other places, *T. Westcombe*; Moseley common, *G. Luxford*, *W. G. Perry*.

Pteris Aquilina. Very common; dwarf on the Malvern hills, but in sheltered woods I have seen it full six feet high, *E. Lees*; Kempsey, in hedges; Malvern hills, *R. J. N. Streeten*; common, *T. Westcombe*.

Allosorus crispus. On the Herefordshire beacon, Malvern hills, but in one place only, so far as I have observed, and this is on the eastern or Worcestershire side of the beacon; here I have observed it for some years, but it grows very sparingly, *E. Lees*.

Polypodium vulgare. General; the variety with serrated lobes occurs on the side of a deep lane below Great Malvern, *E. Lees*; Brookend lane near Kempsey; rocks on the Malvern hills, *R. J. N. Streeten*; common, *T. Westcombe*.

Polypodium Dryopteris. On the Malvern hills, but only in one place, a stony ravine between the north and end hills north of Great Malvern; plentiful however in that locality, *E. Lees*: North hill, Malvern, *T. Westcombe*.

Cystopteris fragilis.　In fissures of the oolitic rock on the summit of Bredon hill, on the side of the precipice; near Bromsgrove Lickey, *E. Lees*.

Polystichum aculeatum, *angulare* and *lobatum*.　Of these three forms I find lobatum at least commonly occurring, and most partial to deep rocky lanes, where it sometimes attains a length of nearly three feet; it grows peculiarly fine by the travertin deposits at Eastham near Tenbury; also on the Holly-bush hill, Malvern range, *E. Lees*; P. aculeatum, Brook-end lane, near Kempsey, *R. J. N. Streeten*: near Worcester, *T. Westcombe*.

Polystichum angulare.　Near Clifton-on-Teme, *T. Westcombe*.

Polystichum lobatum.　Hedge-row in Bromhall-lane, near Worcester, *R. J. N. Streeten*; Stagbury hill, *T. Westcombe*.

Lastræa Oreopteris.　On the sides of the wet commons on the eastern side of the Malvern range; below Malvern wells, *E. Lees*; Moseley common, *W. Southall, jun.*, *D. Cameron, G. Luxford*; Malvern hills, *T. Westcombe*.

Lastræa Filix-mas.　Everywhere, *E. Lees*: Brookend cross near Kempsey, *R. J. Streeten*: common, *T. Westcombe*.

Lastræa dilatata.　Damp woods and alder-holts; very fine about Bromsgrove Lickey, *E. Lees*: Perry wood, *R. J. N. Streeten*: North hill, Malvern, very much curled and distorted; near Malvern; Blackstone rock near Bewdley, *T. Westcombe*.

Lastræa spinulosa.　A variety with overlapping pinnules and of rigid habit grows on the Malvern hills, *E. Lees*; Shrawley wood, *T. Westcombe*; I used to find it in some of the bogs on Moseley common, which I believe have since been drained, *G. Luxford*.

Athyrium Filix-femina.　*E. Lees*: Burcott, and in the wet lanes near Bromsgrove Lickey, very common (Purton's 'Midland Flora') *R. J. Streeten*: on the Malvern hills, with the broad and narrow frond, *T. Westcombe*.

Athyrium irriguum. E. Lees.

Asplenium Adiantum-nigrum.　With variegated fronds about Great Malvern, Little Malvern, and many other spots, *E. Lees*: Brook-end lane, Kempsey, *R. J. N. Streeten*: Hanbury church, and near Worcester, *T. Westcombe*.

Asplenium Ruta-muraria.　*E. Lees*: old brick wall near the Cathedral, Worcester, *R. J. N. Streeten*: Southstone rock; Stanford bridge, *T. Westcombe*.

Asplenium Trichomanes.　*E. Lees*: rocks on the north hill, Malvern, *R. J. N. Streeten*: Blackstone rock; Ham bridge; Malvern hills, *T. Westcombe*.

Asplenium viride.　Ham bridge, near Clifton-on-Teme, *E. Lees*: the reader is referred to further observations on this habitat in a preceding page (Phytol. 46); Mr. Lees obligingly sent me the specimen in question, and I beg to add my testimony to the correctness of the name, *E. Newman*: Ham bridge, very sparingly, *T. Westcombe*.

Scolopendrium vulgare.　Monstrous specimens with multilobed fronds above two feet in length grow on the conglomeratic Rosebury rock, near Knightsford bridge, *E. Lees*: Draycot near Kempsey, *R. J. N. Streeten*: near Clifton-on-Teme, *T. Westcombe*.

Ceterach officinarum.　Very sparingly on walls at Great Malvern, but not on the rocks of the hills, and I should say this fern is not at home in Worcestershire, *E. Lees*: Badsey near Evesham, *T. Westcombe*.

Osmunda regalis.　On Moseley-wake green, near Birmingham, at the northern extremity of the county; Dr. Withering records the curious appearance and disappearance of this noble fern at the above locality; it might therefore be presumed to be lost there, but a very few years since I was shown some ferns sent from this very spot to Miss Spriggs of Worcester, among which was a specimen of the Osmunda : I do not know of its growing in any other part of Worcestershire, *E. Lees ; Moseley* common, *W. Southall, jun., D. Cameron, G. Luxford,* (Mr. Ick) *W. G. Perry.*

Botrychium Lunaria.　Bredon hill (Dr. Nash); Abberley hill, (Mrs. Phipps Onslow, from whom I have received a specimen); on coal-pit banks near Stourbridge, (Mr. Waldron Hill in Withering, 2nd edition, edited by Dr. Stokes), *E. Lees.*

Ophioglossum vulgatum.　Local, though plentiful where it does occur; Grimley meadows, (A. Edmonds); near Malvern, (E. Newman); Longdon marshes, *E. Lees;* near Worcester, *T. Westcombe.*

Lycopodium clavatum.　On a sandstone cliff by the Severn, at Winterdyne near Bewdley, (T. Robinson, from whom I have a specimen), *E. Lees :* bog on Hartlebury common, *R. J. N. Streeten :* Moseley common, *W. Southall, jun.*

Lycopodium inundatum.　On a boggy part of Hartlebury common, near its termination, about a mile from Stourport, *E. Lees ;* on Hartlebury common, *R. J. N. Streeten, T. Westcombe.*

Lycopodium Selago.　Moseley common, *W. Southall, jun., D. Cameron, G. Luxford,* (Mr. Ick) *W. G. Perry.*

Equisetum hyemale.　Moseley bog (Mr. Ick), *W. G. Perry.*

Equisetum palustre.　In bogs on Hartlebury common near Stourport; by the side of the river Avon near Pershore, *E. Lees :* Fakenham bog (Purton's 'Midland Flora'); banks of the Teme near Powick, *R. J. N. Streeten :* near Worcester, *T. Westcombe.*

Equisetum limosum.　In pools and marshes generally all over the county; near Bewdley in the north, and most abundant near Chawley in the south, *E. Lees :* Severn meadows; Kempsey, *R. J. N. Streeten :* near Worcester, *T. Westcombe;* Moseley, *D. Cameron, G. Luxford.*

Equisetum sylvaticum.　In a wood in the vicinity of the Malvern hills, but very rare, and I only met with one specimen, *E. Lees :* near Clifton-on-Teme, *T. Westcombe.*

Equisetum arvense.　Excessively common, *E. Lees:* Severn meadows; Kempsey; banks of the Worcester and Birmingham canal, *R. J. N. Streeten :* common, *T. Westcombe.*

Equisetum fluviatile.　Plentiful in boggy woods near Worcester; Great and Little Malvern; indeed generally, *E. Lees :* near Worcester, *T. Westcombe.*

It is intended to give the lists for Gloucester, Wilts, Somerset, Devon and Cornwall in the April Phytologist; further information respecting these counties is particularly solicited.　In records of localities near Bristol, I shall feel much obliged by my correspondents taking great care as to the county.　Is the locality for Asplenium lanceolatum in Gloucestershire or Somersetshire, or both ?

EDWARD NEWMAN.

(To be continued).

Art. CXXIV. — *Notice of the Proceedings of the Berwickshire Naturalists' Club for 1842.* *

The perusal of this delightful summary of the proceedings of this little band of naturalists during the late glorious summer, is enough to make one "babble o' green fields," even amidst the snow-covered houses and streets of London, on this present 18th of February, 1843. There is that sparkling freshness in its details, slight though they be, which nothing but an intimate acquaintance with the manifold beauties of Nature can have imparted to them. But before we enter upon the pleasing duty of giving our readers a brief analysis of the botanical contents of this brochure, a few words on the nature and objects of the club may not be unacceptable.

From a notice in Loudon's 'Magazine of Natural History' we learn that "The Berwickshire Naturalists' Club was instituted in September, 1831, by some gentlemen who interested themselves in Natural History, and were anxious to do their best to aid one another in their pursuits, and to diffuse a taste for them among others. The club meets four times in the year, and the place of its meeting is changed every time, to afford the members an opportunity of examining in succession every part of the neighbourhood. The members meet early in the morning; they spend the forenoon in excursions, and they again assemble at dinner, after which any papers that may be laid before them are read and discussed freely."—vi. 11.

The first president was the eminent naturalist Dr. Johnston, author of the 'Flora of Berwick,' who has taken the warmest interest in the welfare of the club from its first establishment. The members held their first anniversary meeting at Coldstream, on the 19th of September, 1831. The address now before us was read at Lowick, on the 28th of September, 1843, at the eleventh annual meeting; and exhibits "the result of the labours, or rather the harvest reaped, not by bodily fatigue, but yielded to the agreeable recreation and innocent pastime of men happy to escape from the monotonous toil of their necessary occupations, to revel in all the beauty of Nature's loveliest scenes, and the thousand charms of her everchanging aspects, and to have their feelings elevated, and their minds improved, by the calm contemplation of the wonderful works of God."—*Address, p.* 1.

The anniversary meeting for 1841 was held at Kelso; the details

* We have ventured to give a title to this interesting publication : would it not be well to print it on future wrappers?

were furnished by Dr. F. Douglas, who, after describing the ramble on the banks of Tweed, states that —

"During the walk, nothing new was observed. Several plants, however, were noticed which are of rare occurrence within the limits of the club, such as Viola hirta, Thalictrum majus, Clinopodium vulgare, Epipactis latifolia, Listera cordata; and Dr. F. Douglas pointed out the habitat of a beautiful coral-like fungus — Clavaria rosea —which has not been discovered in any other locality in Great Britain. Some small specimens were gathered, but they were not in perfection, and their beauty was nearly gone."—3.

The first meeting for the year 1842 was held at Coldstream, on the 4th morning of the "merrie month of May;" and a right gladsome May-morning it must have been, as the following extract, redolent of the country, will amply testify.

"The rest of the members took a delightful walk down the banks of the Tweed by Lennel to Milnegraden, where the lovely scenery, rendered doubly alluring by the now bursting foliage of the woods spangled with heaven's own diamonds, and offering to the eye all the varied tints of ' many greens,' and the promise of future luxuriance, amply repaid all who enjoyed this walk, for the disappointment attending their search for a piece of water which was expected to afford a rich harvest of aquatic plants and insects in its ample bosom, to ' the careful and scientific explorers of its hidden treasures; ' but which the fairies or good people had either spirited away or rendered invisible to the eyes of the expectant naturalists. Be this as it may, no lake could be found, but many other very interesting objects were seen and duly appreciated. The humidity of the morning had tempted from their lurking-places several varieties of snail, and a few were gathered. The hawthorn, with its beautiful white blossoms and rich scent perfuming the air, was gathered in full blow in several situations — a proof of the forwardness of the season. The sand-martin was seen in great plenty, skimming over the waters, and excavating its simple habitation in the banks which overhang the Tweed ; and the varied sweetness of the thousand warblers trilling their songs of love, added a charm to the morning's ramble, which the denizens of our crowded and bustling towns can but rarely taste ; and if this club had no other or higher object than occasionally to give such a delightful change and peaceful recreation to those whose occupation confines them to the desk, or the close and uninteresting monotony of a town life, that object alone would render it a blessing to the neighbourhood."—p. 4.

The minutes of the next meeting, held at Gordon on the 15th of June, furnish a more purely botanical extract, which to us is doubly interesting, inasmuch as it relates principally to the charming Linnæa borealis,—that "little northern plant, long overlooked, depressed, abject, flowering early," the very type of the early fortunes of him whose name it will transmit to future ages,—which, gracefully encircling the features of our own illustrious Ray, monthly greets the readers of ' The Phytologist.'

"The chief object of the club's meeting at Gordon was to gather Linnæa borealis in the fir-woods of Lightfield, upon the Mellerstain estate, and thither, by a circuitous route over moss and muir, the eager party bent their steps. The first object which attracted special attention was an ancient and ruinous tower, situated to the west of the village, and formerly occupied by the powerful family of Setons, who were allied to the noble house of Gordon, formerly the proprietors of that district of country. Near the old ruin Chelidonium majus was found, furnishing another illustration to the opinion that this plant was introduced into horticulture at an early era. In the peaty muir on the farm of Greenknow, were gathered Stellaria glauca and Myosotis palustris, while in the nearly stagnant waters of the Eden, was observed another plant of rare occurrence in Berwickshire, viz., Sparganium natans. After leaving the moss, every fir-wood and thicket for miles around were penetrated and carefully searched for the humble little flower bearing the name of the immortal Swede. The search was, alas, in vain; and after continuing it for fully three hours, the spirits of the party flagged, and they returned disappointed to the inn, where a good dinner and excellent liquors soon dispelled any portion of vexation which might still be felt at the want of success attending the expedition. One of the members of the club, however, nothing daunted by a single failure, and anticipating better fortune in a second attempt, did not allow many days to elapse until he was again in the woods, in the hope of securing the prized Linnæa, and most fully and amply was he rewarded by beholding a large space of ground covered with the delicate shining leaves of the trailing little plant, with here and there a short flower-stalk ascending, and bearing a pair of beautiful pinkish bell-shaped flowers, bending gracefully downwards: innumerable specimens of the finest description were obtained.

"It seemed remarkable that on the first search for it, all the members had passed within five yards of the spot where the Linnæa grew. Listera cordata, Trollius europæus, Pyrola minor, were found in the woods during the course of the forenoon's walk, and a new fungus, Æcidium Pini, was added to the cryptogamic Flora of the county, the bark of the fir-tree on which it grew being totally destroyed by its ravages."—p. 5.

At this meeting was read a paper by Mr. Hardy, entitled "Butter-cups and Daisies," in which the question of the etymology of the terms goulans, gowans, gollands &c. is fully discussed, and the various plants to which these names have been applied, are pointed out. The author observes that these terms, "as applied to plants, are obviously related, and appear to derive their origin from the Anglo-Saxon *gold*; or, if we wish to consult a more remote parentage, from the Suio-Gothic *gul*, *gol*, yellow." Several dainty quotations from the old herbalists are scattered through the paper, and show that various yellow flowers, such as different species of Ranunculus, Caltha palustris, Trollius europæus and the corn-marygold, were all known as yellow gowans; while the white division includes the daisy— *the* gowan *par excellence*, which "stands at the head of its class, without a peer, the type, as it were, in which all the superior properties of the other species are blended."

Then we have "the horse-gowan, the Berwickshire name for Pyrethrum inodorum, Chrysanthemum Leucanthemum and Anthemis arvensis." But we must take leave of this subject, which we do most reluctantly, consoling ourselves as best we may with the following extract from Turner's Herbal, being a description of his "Lukken Gollande," which is believed to be no other than our Caltha palustris, though from Hodgson's 'History of Northumberland' it appears that the names "locken-gowen" and "goudie-locks" are equally applied to Trollius europæus.

'"Thys herbe useth to growe comonly about water sydes, and in watery meadowes, the proporcion of the leffe is much like unto a water-rose, otherwyse called nupefar, but the lefe is sharper and many partes lesse, and there grow many leves on one stalke, and in the toppe of the stalke is a yelow flowre like unto the kyngcuppe called ranunculus; but the leaves of the floures turne inwarde agayne, in the manner of a knoppe, or lyttell belle.' — A new Herball, &c. &c., by Wylliam Turner, Physicion unto the Duke of Somersettes Grace. Lond. 1551. Fol. k. v."

In August the club met at Abbey St Bathans, being led thither "by its retired position and celebrity for natural beauty." "In such a locality," it is remarked, "the club finds the material for forming a correct idea of the nature, extent, and composition of the ancient forests in which their forefathers may, perchance, have hunted the deer, with hound and horn, in the gallant company of a Douglas or a Percy." The following is a fancy sketch, or rather a restoration, of an ancient border forest scene, drawn from materials still existing at Abbey St. Bathans.

"In the many ravines which descended from the moor above, and in whose bottoms a runlet had cut its way amid shelving rocks, we found many springy spots occupied principally with some shrubby willows (*Salix aurita* and *cinerea*), intermingled with arching briars and wild roses. In others the alder grew predominant, while rushes and meadow-sweet and marsh thistles filled up the under ground, leaving often a middle space carpeted with mosses of yellow-green, and too moist for the growth of other plants than the willow-herbs, the forget-me-not, the ranunculus, and other semi-aquatic herblets. But the drier ground was mainly occupied with the birch, rising up from amid a bed of tall heather or of blaeberries; while a tree of oak, of the mountain ash, and of the tree willow (*Salix caprea*), grew up among the birches, marked, each of them, by its peculiar shade of green. Where, again, the streamlet had cut its channel deeper, and at a lower level, the vegetation became more free and various; the alder was more common and luxuriant; the rose and briar arched their bows with greater freedom; the rowan-tree assumed a taller habit, and by its side the hagberry grew, as if conscious Nature had pleasure in the augmented beauty which each derived from the contrast between their intermingled foliage, flowers and fruit. Here all the underground was occupied by luxuriant ferns, bending in graceful plumes over the shelving edges of the banks, with tall nodding rushes and grasses, wild geraniums, hypericums and willow-herbs, and various umbelliferous and compound syngenesious plants. Every spot is a picture, and every one so fertile in flowers, that the botanist may call there alone a richly varied herbarium, from the green moss, through whose dense mass the spring filters its waters, to the hazel and the oak that shelter the pool beneath their shade from the too hot influence of the sun.

"How different again is the wood that hangs on the sides of the hills rising from the valley on each side of the principal stream or river! It consists principally of oak, of moderate size at the base of the hill, gradually diminishing in stature as we ascend, until we find it at the summit nearly level with the surface of the ground, spreading in low, circular, leafy bushes. This troop of oaks is intermingled with a considerable quantity of birch, as various as the oak itself in size and appearance, while an ash tree rises tall above them both at distant intervals. The "bonnie broom" is frequent and tall on the lower line of this wood, while the whin occupies the line above with a denser growth than usual. The intermediate ground under and amidst the trees is full sometimes of a coarser herbage, rich in fungous growths, and where lichens make the trunks all leprous; sometimes moss predominates, and this is the habitat too for Melampyrum sylvaticum; in other places are long streams of stones and gravel, covered partially with briars, trailing roses, and with green patches of wild sage (*Teucrium Scorodonia*), or of the herb Mercury (*Mercurialis perennis*).

"Such fancy paints our ancient border forests to have been, and probably there is much of reality and truth in the picture. A wide mountainous and barren tract, intersected by a principal devious stream, having, on each side of it, an alluvial plain of some breadth that afforded good and abundant pasture for the horses, herds and flocks of the rude inhabitants. On each side there run up ravines of greater or lesser depths, every one with a burn or rivulet in its bottom; some rocky and clean — others with plashy places, while the hills are occupied with woods such as we have attempted to describe, and the plains above are brown barren moors, varied with peat-hags and covers of whins and of broom, as the depth of the soil afforded a locality for their growth. Through these forests herds of red and fallow deer were wont to roam at freedom, and were the chase which our forefathers pursued with almost savage raptures—while now the ground is occupied with new and foreign plantations, with corn, with artificial pastures, and the hills are covered with flocks of sheep, obedient to the call of the shepherd, and browsing watchless, because they know no danger."—p. 7.

Further on the writer observes it were "easy to imagine that on such a day as ours was—tempted by its sunshine and its fairness— the proud abbot of Coldingham had chosen it whereon to visit the sister abbey of St. Bathans, and was now descending, in all the state and panoply of his order, the height that overhung the hidden retreat about us!"

It is somewhat strange that in a botanical light this promising locality should possess but little interest. The only plants found worthy particular notice are said to have been Hieracium palustre, H. boreale (*Koch*); and H. sylvaticum; all three occurred "in tolerable abundance in the rocky bed of Monnie-nut burn, below Godscroft; and Melampyrum sylvaticum, which was abundant in the oak woods. — Quercus Robur was the oak principally observed, "but many specimens approached Q. sessiliflora in its peculiarities."

The last botanical notice in the number is a short but interesting separate paper entitled —

"Notice of the Smilacina bifolia (Convallaria bifolia), a British Plant. By R. EMBLETON, Surgeon.

"For my knowledge of this interesting addition to the British Flora, I am indebted to my friend, the Rev. Osd. Head, of Howick, who discovered it growing, rather sparingly, "under the shade of a wide spreading beech," in one of the woods at Howick. It has hitherto been found in France, Germany, and other parts of the continent; and I possess a specimen in my herbarium from Norway, collected by my friend Mr. R. B. Bowman, of Leadenhall St., London. It is a graceful and beautiful plant, and well deserves a share of the admiration which is so universally given to the other members of the natural family (the lily of the valley tribe) to which it belongs. It is easily recognised by its creeping roots, from which arises a delicate stem from five, to seven inches in height, with two alternate ovate leaves, and terminated by a spike of small, delicate, white flowers It flowers in July.

"Since its discovery in the woods of Howick, I am informed by Mr. Duncan, Earl Grey's gardener, that it is found in the woods at Kenwood, the seat, I believe, of the Earl of Mansfield, and from which place, through the kindness of the same individual, I possess specimens, which do not show any difference from those gathered at Howick, with the exception of their being a little more succulent. It is there found in similar situations, namely, under the shade of beech and fir trees. It may, probably, hereafter be found in many other quiet, shady spots of our native woods, and will well repay the wandering botanist for his labour of love in its search.

"Embleton, Sept. 21, 1842."

Much as we should rejoice at seeing the claims of this pretty little plant to a place among our indigenous species fully established, we must confess that at present we can only look upon it as an intruder, although one which we can hail with a hearty welcome. Loudon gives the year 1596 as the date of its introduction into Britain, so that it is perhaps only surprising that it should not have been previously met with in an apparently wild state. There are several other plants possessing an equally good title to be recognized as British, since they are at least completely naturalized, and yet botanists are not at all disposed to look upon them as true Britons: Lilium Martagon, Chelidonium majus and Ornithogalum nutans may be mentioned as examples. Linnæus, in his ' Flora Lapponica,' enters fully into the consideration of the question whether this plant should or should be considered a species of Convallaria; and comes to the conclusion that it really belongs to that genus.

We hope at no very distant period to be able to renew our acquaintance with the ' Proceedings of the Berwickshire Naturalists' Club.'

ART. CXXVI.— *Varieties.*

259. *Remedial use of a Fern called " Dail llosg y Tân."* The following notice relative to the remedial use of a fern in connexion with the late Lady Greenly, of Titley Court, Herefordshire, may perhaps be interesting, as proving the advantage of a botanical acquaintance with plants; for the worthy author of the memoir of Lady Greenly, being unable to give anything but the Welch name of the fern in question, of course it is rather uncertain what he means. I should conjecture the plant to be Ceterach officinarum.* Perhaps some Cambrian botanist may know more about it. The memoir I allude to appeared in ' The Hereford Times' of the 12th of November last, and is prefixed to a Welch *Marwnad* or death-song to her memory; composed by the Rev. J. Jones (Tegid), rector of Nevern, Pembrokeshire, and to which the prize was awarded at the ninth anniversary of the Abergavenny Cymreigyddion, Oct. 12, 1842. It is therein stated that—" She also loved flowers, but the *properties* of plants was one of her favourite studies, and a pursuit which she turned to the benefit of the poor.— She used to cultivate a variety of herbs, and administered medicine to all those who needed it in her neighbourhood. Amongst the plants for which she evinced a particular regard, was that called in Wales— ' *Dail llosg y Tân;* ' it is a species of evergreen fern, indigenous to Gwent and Morganwg ; and Lady Greenly having ascertained from her excursions among the Welch peasantry, that is was (as its name denotes) of value as a remedy for burns, she took pains to make it grow in Herefordshire, and succeeded in getting it to flourish round her favourite well at Titley."— *E. Lees ; Church Hill Cottage, Powick, near Worcester, November* 25, 1842.

260. *Note on Bryum Tozeri.* (As The Phytologist noticed Bryum Tozeri being found in a barren state this year in Kent (Phytol. 200), perhaps it is worthy of remark that Miss A. Griffiths met with the capsules of this rare moss in tolerable abundance, near Torquay, last March.†—*Torquay, December* 15, 1842.

261. *Note on Scleranthus.* Mr. Gibson (Phytol. 366) asks if these plants have one or two seeds. If he will, next summer, examine the plants carefully, he will find that the ovary has two ovules, but that

---

* In Davies's ' Welsh Botanology ' " Rhedyn y gogofau," and " Dueg-redynen feddygawl," are given as the Welsh names of Ceterach officinarum ; " Dail llosg y Tân," which we do not find in Davies, is probably the local name of some common species.—*Ed.*

† In a letter from Miss A. Griffiths to E. Newman.

2 z

only one seed usually comes to perfection. He will also observe a curious change of the dissepiment into a funiculus. At first the ovary has two cells, with one ovule in each; afterwards, as one of the ovules outgrows the other, the dissepiment gradually disappears, only a sufficient portion of it remaining to act as a funiculus, springing from the bottom of the cell and suspending the seed. No doubt occasionally two seeds are perfected, which will account for the discrepancy pointed out by Mr. Gibson.—*Charles C. Babington; St. John's College, Cambridge, January* 27, 1843.

262. *Correction of an error in Mr. Edmonston's List of Edinburgh Plants.* I shall be obliged by the insertion of the following correction of an error which somehow or other crept into my MS, on Edinburgh plants, (Phytol. 407). It is "Crichton castle," not "Craigmillan," which is the reputed station for Carex axillaris.— *Thos. Edmonston; Baltasound, Shetland, January* 28, 1843.

263. *Note on the supposed new British Cuscuta.* Having seen a notice (Phytol. 466) of the discovery of a supposed new British Cuscuta, by Mr. C. C. Babington, of Cambridge, it may be interesting to the discoverer to know that while on a visit at Ramsgate in August last, I detected a Cuscuta in one or two places in a clover-field on the right hand side of the road leading from St. Peter's to Broadstairs, and which appeared to me at the time to be a variety of C. Epithymum. I am now however inclined to think, from Mr. Babington's description, that it may turn out to be the same plant as that described by him under the name of C. Trifolii. It was by no means abundant, and it occurred in patches. I subsequently detected it in small quantity in one or two other places in the Isle of Thanet, where it appeared to have greatly injured the crops of clover. I shall take the earliest opportunity of further investigating the subject, and communicating the result in 'The Phytologist.' — *T. B. Flower; Surrey St., February* 1, 1843.

264. *List of a few Plants observed in Lincolnshire.* Lincolnshire presents one of the most attractive and prolific fields for the botanist in the east of England. The richness and variety of the soil, added to its wide extent, combine in rendering it well worthy the attention and careful investigation of every lover of Natural History. It is under this impression that I think the following list of some of the more uncommon plants observed by me during the last few years in the neighbourhood of Gainsborough, may not prove altogether unacceptable. The district to which I have confined my researches, offers perhaps a greater variety than any other part of the county; the rich

pasture land which clothes the banks of the river Trent, and the waste uncultivated tracts in the neighbourhood of Scotter and Laughton, occupy the most prominent part of it, and cannot fail to repay the researches of all those who take any interest in the botanical productions of this locality. The following then are those which have principally come under my observation, and which I consider most worthy of notice.

Pinguicula vulgaris. Abundant on Scotton common, but I am aware of no other spot in the neighbourhood in which it occurs.

Utricularia minor. Bogs near Laughton, but not common.

Menyanthes trifoliata. Extremely plentiful on Laughton common and at Scotton.

Hottonia palustris. Occurs in a pond at Knaith, but is extremely scarce.

Lysimachia vulgaris. On the banks of the Trent, not uncommon; also in tolerable abundance at Scotter.

Anagallis tenella. Occurs sparingly on Scotton common and at Knaith.

Gentiana Pneumonanthe. In abundance on Scotton common; less frequent in Scotter wood.

Drosera rotundifolia and *longifolia*. Plentiful on Scotton common

Narthecium ossifragum. Scotton common, and in the fens about Laughton, abndt.

Convallaria majalis. Lea wood; also at Manby.

Paris quadrifolia. In Gate Burton and Lee woods in abundance.

Butomus umbellatus. Ditches on the banks of the Trent and at Scotter.

Reseda Luteola. Exceedingly common on the banks of the Trent.

Geum rivale. Very common in Gate Burton wood.

Stratiotes aloides. Said to be found near Gainsborough, but this I am inclined to think is not now the case, having never met with it in my researches, nor have I heard of its being found by other collectors.

Galeopsis versicolor. Sparingly at Laughton, in cornfields.

Digitalis purpurea. Scarce in the neighbourhood, but occurring in small quantities in fields between Gate Burton and Knaith.

Eupatorium Cannabinum. Hedges near Susworth.

Petasites vulgaris. Banks of the Trent near Gainsborough.

Habenaria bifolia. Plentiful in Gate Burton wood.

Ophrys muscifera. One or two specimens have occurred at Manby wood, but it is decidedly uncommon.

Aceras anthropophora. Plentiful in Gate Burton wood, with Habenaria bifolia.

Sagittaria sagittifolia. Very common in the river Ean at Scotter.

—*T. V. Wollaston ; Cambridge, February 14, 1843.*

265. *Monstrosities in the Flowers of a Fuchsia.* The first I shall notice is interesting to the morphological enquirer, and tends strongly to support the views of Professor O. Morren of Liege (Ann. and Mag. Nat. Hist. vii. 1), that the formation of anomalous flowers is caused by a descending metamorphosis, or from the centre to the circumference, and which he sought to verify by the gradual changes observable in the double variety of the common columbine, showing that

as the cells of the anthers vanished the connective expanded, until at last it became a perfect horn-shaped nectary. In the flower alluded to above, the connective of one of the stamens shot forth beyond the anthers, and by the time they had opened and discharged their pollen, it had assumed the form of a complete ovate petal, about four lines long and three broad; the part of the connective that formed the claw or stem of the petal, was about three lines long and one in breadth; the anthers were as close together, and to all appearance as full of pollen, as those on any of the other filaments, and the colour of the petaloid expansion was the same rich purple as the petals, being in the stem part, before it reached the anthers, of the scarlet hue of the filaments. We have thus a corroborative proof of the truth of the hypothesis advanced by the Professor, — a stamen not only performing its own proper functions, but further, from its luxuriance possessing the power to form, and fully develope, one of the next outer circle of floral organs, or the very reverse of the statement of Mr. Hill (Phytol. 368), that a stamen is only a modified petal.

Another flower on the same plant exhibited a redundance in all the floral organs, with the exception of the pistil; the calyx having six segments, and there were eight petals and twelve stamens: thus departing, in its calyx and stamens, from the binary character of its family, but even in its irregularity showing the intimate connexion of the sepals and stamens.—*James Bladon; Pontypool, Feb.* 6, 1843.

266. *Remarks on the Fern-lists.* I have had much pleasure in reading the three county lists of ferns and their allies; but perhaps you will allow me, through the medium of 'The Phytologist,' to ask your Lancashire friends the two following questions: — first, whether Greenfield is situated in the county of York or Lancaster? secondly, whether Cotteril-clough is in Cheshire or Lancashire? The locality for Hymenophyllum Wilsoni at Rake-Hey common involves a twofold error, since Rake-Hey common is undoubtedly in the county of York, and the plant growing there is certainly Hymenophyllum Tunbridgense.—*R. Langthorn; Heptonstall near Halifax, Feb.* 7, 1843.

267. *Mr. Hewett Watson's grounds for connecting the name of "Mr. S. Gibson" with some botanical localities.* In the February No. (Phytol. 477) the name of *a* Mr. Gibson, but I know not whether the Mr. Gibson who writes to 'The Phytologist,' is stated to have been incorrectly connected by Mr. Francis with an alleged locality of Polypodium calcareum "near Lancaster." It is probable that the locality was communicated to Mr. Francis, on the authority of a specimen preserved in my herbarium with a memorandum to indicate that it was

received from Mr. R. B. Bowman, who in turn had received it from
— Gibson. Unluckily, I have not preserved also the label which
came to my hands along with the specimen, but there are other circumstances which will identify the donor of the specimen to Mr. Bowman. About the years 1832-4 I received numerous dried plants from
the north of England, by the kindness of that accurate botanist, but
several of them gathered by other persons. Among those plants were
a few, mostly fragmentary and ill-dried specimens, accompanied by
labels quite different from the neat labels of Mr. Bowman; and to
these labels the name of "S. Gibson" was subscribed. One of them
is now before me, the same that is referred to, on p. 652, in the 'Supplement to the New Botanist's Guide;' the name of the place cannot
be made out on this label, which was evidently written by an unedu-
cated person. From recollection I should say that the other labels,
on which the name of "S. Gibson" was written, were similar to this
one in the quality of paper and character of handwriting; and it is
thus quite possible that the writer intended some other locality, which
was misread by myself into "Lancaster." If the name of your correspondent, Mr. S. Gibson, had been attached to any locality "near
Lancaster," for Polypodium Dryopteris, in Mr. Newman's list of Lancashire ferns, I should have conjectured some discrepancy of opinion
with respect to the specific name,— for, in truth, this one, and two or
three other specimens in my herbarium, incline towards P. Dryopteris; but that not being the case, we must suppose either an error as
to the locality, or an error on the part of your correspondent in hastily
assuming himself to be the person intended. At all events, he will
know whether he formerly did send dried plants to Mr. Bowman.
The name of "S. Gibson" being the authority for other localities of
scarce plants, in the New Guide, it would be worth while to ascertain
who the person truly is, and what reliance can be placed on his
reports of localities for dubiously British plants,— for example,
Geranium nodosum, as having been gathered by him near Halifax.—
Hewett C. Watson; Thames Ditton, February 8, 1843.

268. *Enquiry respecting "Nymphæa alba minor."* In a copy of
Blackstone's 'Specimen Botanicum,' which formerly belonged to Pe-
ter Collinson, there is the following MS. note pasted in, but not in
Collinson's hand-writing.—

"I don't find the Nymphæa alba minor taken notice of in the Synopsis* at folio
368. This rare plant I have twice observed. The first was in the North Road, from

* Raii Synopsis, ed. 3tia, 1724.

York going over a stone bridge on the right hand in a river before one cames to Don-caster; it was then in flower, beginning of August.

"The second time I met with it was in going from Lyndhurst in the New Forrest to Brockenhurst; there is a watercourse at the entrance of the village, over which there are bridges, but it then being dry weather, there was no running stream, but the water stood in pools; in these pools I observed to grow both the small and great Water-Lilly; they were both in blossom together, so that the distinction was easily made and the difference was pretty remarkable.—August 10, 1739."

I shall be glad to be informed, through the medium of 'The Phytologist,' whether this plant has since been observed in the above or any other localities, and what it is.—*W. Pamplin, jun.; 45, Frith St., Soho, Feb.* 10, 1843.

269. *Supposed new British Fern.* Since the publication of a note on a supposed new British Adiantum (Phytol. 462), I have submitted the specimen to Mr. Wilson and Mr. Babington, both of whom fully concur with me in thinking it distinct. Still, as an unintentional transposition of specimens may possibly have taken place, and as I can find no Arran specimens in London, with which to compare the specimen in question, I hesitate to publish the species as new. If this should meet the eye of any botanist who can supply me with authentic specimens from Arran, even on loan, I shall esteem it a great favour.—*Edward Newman; Hanover St., Peckham, Feb.* 18, 1843.

ART. CXXVII.—*Proceedings of Societies.*

BOTANICAL SOCIETY OF EDINBURGH.

February 9, 1843.—Professor Graham in the chair. Churchill Babington, Esq., St. John's College, Cambridge, and G. G. Gibson, Esq., Saffron Walden, Essex, were elected non-resident fellows, and Alex. Paterson Esq., a resident fellow of the Society. Various donations of plants were announced.

Professor Graham then read a highly interesting account of his botanical excursion in Ross-shire during August, 1842, with a party of friends. The party left Edinburgh on the 21st of August, and met at Dingwall;—thence they walked by Garve, Auchnalt &c. for Kinlochewe. On the low hills near Garve they found a sprinkling of alpine vegetation, and Nymphæa alba, beautifully in flower, in a pool near the top of one of them, at a higher elevation than had been previously observed. The season having been remarkably dry, all the lakes were far below their usual level, and in consequence such plants as Lobelia Dortmanna, Subularia aquatica, &c., were seen, wondering at each other, in flower and fruit, on dry ground. Things, however, were now changed, for the party had scarcely a dry day during the whole of their excursion, and few such as admitted of the vegetation being carefully examined. Several days were spent among the mountains about Loch Maree, which are chiefly composed of red sandstone, with quartz tops—and by no means prolific in interesting vegetation. Cornus suecica, Saussurea alpina, Hieracium alpinum, Rubus Chamæmorus, Arbutus alpina, Azalea procumbens, Cherleria sedoides, Sibbaldia procumbens &c. were among the rarest plants observed — and, rather unusually, all the six Lycopodia were picked nearly in one spot. Tofieldia palustris, Thalictrum alpinum and Malaxis paludosa occurred at the bottom of the cliffs, and Salix herbacea was found sparingly on the red sandstone below the summit cliffs of Ben Tarshan. Opposite Applecross, in a bog which the tide could seldom reach, were picked specimens of Blysmus rufus two feet high. Here there is an extent of limestone country — easily recognised at the distance of several miles by a marked improvement in the pasturage. On it the party met with Schœnus nigricans, Gentiana Amarella, Listera ovata, and Epipactis latifolia with pale flowers, but searched in vain for Dryas octopetala, which occurred profusely in similar soils in Sutherland. In an old deserted garden between Sheildag and Jangtown, they observed Althæa officinalis, Aconitum Napellus and other introduced plants. They also saw near Janetown, Ulex europæus (a rare plant in the west of Ross-shire) growing freely and producing abundance of seed; and the elder seemed to thrive peculiarly well.

Proceeding southward, the party enjoyed one fine day at Clunie, and examined, with considerable attention some very promising mountains to the south-west of the inn. These are crumbling and micaceous, but want elevation to produce alpine plants, and the mildness of the western climate renders that all the more necessary. The only interesting vegetable feature was an immense profusion of Saussurea alpina; though in spring, before vegetation gets rank, it is not unlikely that these cliffs might be found more productive. A patch of snow observed on the south side of Maamsool, a mountain about twenty miles north of Clunie, made the party desirous of visiting it; but here again the weather baffled their intentions. The party took Ben Nevis in their route, but the same cause rendered them unable to examine, as they wished, its magnificent cliffs. They, however, picked some interesting plants, and among the rest Carex saxatilis, but only in one spot.

A letter to Professor Graham, from Mr. N. B. Ward, F.L.S., on the introduction of the Musa Cavendisii into the Navigator Islands, was read :—

" When Mr. Williams was about to leave England in 1839, for the Navigators, he was anxious to take with him some useful plants, and particularly the Musa. He enquired of me whether I thought that it would travel safely in one of the glazed cases, and, having received an answer in the affirmative, he applied to his Grace the Duke of Devonshire, who kindly gave him a healthy young plant. Mr. Williams left England on the 11th of April, 1839, and arrived at Upolu, one of the Navigator Islands, at the end of the following November. The Musa bore this long voyage well, and was transplanted into a favourable situation soon after its arrival. In May, 1840, it bore a fine cluster of fruit, exceeding 300 in number and weighing nearly a cwt. The parent plant then died, leaving behind more than thirty young ones.— These were distributed to various parts of the island, and in the following May (1841) when Mrs. Williams left the island, all of these were in a fructiferous state, and producing numerous offsets. Supposing the plants to continue to increase in the same ratio, there will be in the ensuing May (1843), more than 800,000 of them, and as the son of Mr. Williams is established as a merchant at Upolu, — is owner of two vessels constantly employed in trading between the various islands in the South Pacific, — and is moreover actuated by the same benevolent disposition which was a striking characteristic of his late father, there cannot be a doubt that in a very short time they will be common in all the islands. To estimate the importance of the introduction of this plant, we must bear in mind the great quantity of nutritious food furnished by the Banana. Humboldt has told us that he was never wearied with astonishment at the smallness of the portion of soil which, in Mexico and the adjoining provinces, would yield sustenance to a family for a year; and that the same extent of ground which, in wheat, would maintain only two persons, would yield sustenance under the Banana to fifty, although in that favoured region the return of wheat is never less than seventy, and is sometimes as much as a hundred fold. The return on an average, in Great Britain, is not more than nine for one."

Mr. Ralfs' paper on the Diatomaceæ, No. III., was then read, containing descriptions of several genera.

On the Development of Leaves. By Dr. Dickie, Lecturer on Botany, King's College, Aberdeen. — The author concludes by stating —" that it cannot be said that the forms of leaves in flowering plants have any dependence whatever on their venation, since young leaves are lobed &c. previous to the appearance of the veins. The truth appears to be, that the quantity of cellular tissue in a leaf determines the development and positions of the veins, and not the opposite."

—

BOTANICAL SOCIETY OF LONDON.

December 16, 1842.—Dr. W. H. Willshire in the chair. The following donations were announced :—British plants from the Liverpool Natural-History Society, Mr. J. Tatham, Mr. W. Baxter, Mr. J. Goodlad, jun., and Mr. W. J. West; and a small collection of plants from Sierra Leone, from Mr. Adam Gerard.

A paper was read from Dr. John Lhotsky, " On the limits of Vegetation."

January 6, 1843.—J. E. Gray, Esq., F.R.S., President, in the chair. The following donations were announced :—British plants from the Royal Horticultural Society of Cornwall, Dr. Philip B. Ayres, Mr. J. Merrick, Mr. I. Brown, Mr. T. Twining, jun., Mr. W. L. Notcutt, and Mr. T. Beesley; British Fungi from Mr. H. O. Stephens. The Rev. W. H. Coleman presented a specimen of *Carex Boennhausiana*, Weihe, found by him in Herts.

Dr. John Lhotsky read a paper " On the Sugar of Eucalyptus."

January 20, 1843.—Adam Gerard Esq. in the chair. Donations to the library were announced from the Boston Natural-History Society, the Manchester Geological Society, from the President, from Mr. W. Baxter, Mr. E. Doubleday, Mr. S. P. Woodward, Mr. Van Voorst, Mr. Lovell Reeve, Professor Meneghini, and the Academy of Natural Sciences, Philadelphia.

British plants had been received from Mr. James Buckman and Dr. J. F. Young; British mosses from Mr. W. Gardiner jun. Mr. Robert Embleton presented a specimen of *Majanthemum bifolium*, De C., (Convallaria bifolia, *Linn.*), found by him at Howick, in Northumberland.*

Mr. William Gardiner, jun., communicated a paper, being " A Notice of Localities for some of the rarer Alpine *Hypna*." The paper was accompanied with specimens.—*G. E. D.*

* See Phytol. 520.

THE PHYTOLOGIST.

Art. CXXVIII.—*A History of the British Equiseta.* By Edward
Newman. Continued from p. 340.

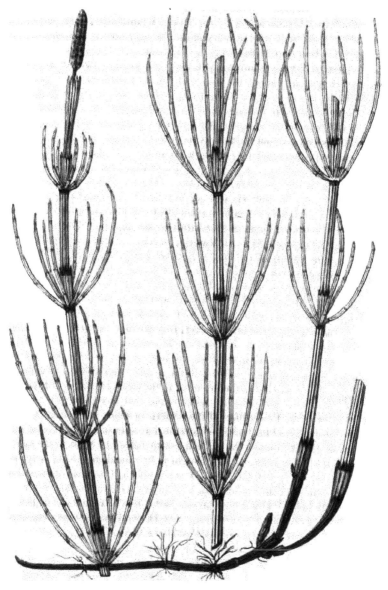

MARSH HORSE-TAIL.
Equisetum palustre of Authors.

3 a

THIS species appears to be generally distributed: it occurs in all the county lists of ferns which I have received, and is rarely mentioned either as local or uncommon. In Ireland I found it particularly abundant, especially in the north : in the vicinity of the Giant's Causeway I observed several large patches of ground densely covered with it. I have not seen it in such profusion elsewhere.

The old figures usually quoted as representing this plant must be received with considerable doubt. Those of Gerarde* and Lobel,† evidently printed from the same block, represent a plant growing in the water, and having one erect and unbranched stem, and another branched, and somewhat resembling the present species. Ray's figure ‡ represents a variety hereafter to be noticed. The modern figures of course more nearly resemble the plant.

Gerarde's description appears to comprehend more than one species. " The great thicke jointed stalk " describing Eq. limosum of Smith, while the roughness and hardness seem inapplicable to that species. I subjoin the passage as it stands in the herbal. " Water Horse-taile, that growes by the brinks of riuers and running streams, and often in the middest of the water, hath a very long root according to the depth of the water, grosse thicke and jointed, with some threds anexed thereto : from which riseth vp a great thicke jointed stalk, whereon grow long rough rushy leaues pyramide or steeple fashion. The whole plant is also rough hard and fit to shave and rub woodden things as the other."

It is not however only in these ante-Linnean works that the synonymy of this and the following species is involved in obscurity. Our modern authors, I regret to say, have hitherto done but little towards the elucidation of the nomenclature. In the hope of making the subject somewhat more clear, I have introduced some observations on the specimens in the Linnean herbarium. Unfortunately, the Linnean characters are frequently obscure, owing to the constant endeavour of their celebrated author to make them as concise as possible : in such case a reference to the specimen becomes indispensable. It is, I believe, generally known, that the Linnean herbarium was purchased by Sir J. E. Smith, and subsequently by the Linnean Society of London, in whose possession it now remains. The specimens are fixed on half sheets of foolscap paper ; they are named by Linneus himself, in his own handwriting, and have also the comments of Sir J. E. Smith

---

* Ger. Em. 1113.　　　† Lobel, 795.　　　‡ Synopsis, tab. v. fig. 3.

wherever it appeared to him necessary or useful to add an explanatory note. A few labels with MS. notes are pasted in, but I am not certain of their author. The Equiseta are comprised in a fasciculus of nine folios: the fasciculus is endorsed thus, — " 1169, Equisetum," in the handwriting of Linneus.

In the same apartment are preserved the author's own copies of the first and second editions of the 'Species Plantarum.' In the first all the species possessed by the author are distinguished by a particular mark; and the second is enriched with his own unpublished notes. I will now endeavour to combine the information obtained from these several sources, only quoting the *published* characters when requisite, and adding remarks of my own on every specimen.

Folio 1.

Linneus.—1. sylvaticum.

E. N.—A single young specimen of Eq. sylvaticum of Smith, with a very perfect catkin.

Folio 2, pinned by Linneus to the preceding.

E. N.—Two mature specimens of Eq. sylvaticum of Smith, without fructification.

Folio 3.

Linneus.—2. arvense.

Anonymous.—1061. Equisetum setis ramosis. Equisetum verticillis ad folia numerosis. Hall. Stirp. Helv. 144. Equisetum sylvaticum, tab. p. 253.

E. N.—Two specimens: right hand, a fertile specimen of Eq. arvense of Smith, with perfect catkin; left hand a mature specimen of Eq. sylvaticum of Smith, without fructification.

Folio 4, pinned by Linneus to the preceding.

Linneus.—Hispania, 713. Loeft.

E. N.—Three specimens without fructification, all of them apparently starved or distorted: they probably belong to the Eq. arvense of Smith. The Linnean MS. is on the back of the folio.

Folio 5.

Linneus.—3. palustre.

Smith.—?

Anonymous.—1060. Equisetum setis simplicibus. Equisetum minus terrestre. I. B. M. p. 730.

E. N.—Two specimens without fructification, and in a very unea-

tisfactory state of growth : right hand appears to me to be Eq. arvense
of Smith ; left hand is perhaps Eq. palustre of Smith.

FOLIO 6.

Linneus.—4. fluviatile.

Linneus [Sp. Plant.*]—Fluviatile 4. Equisetum caule striato fron-
dibus subsimplicibus ; [here follow the synonymes]. *Habitat in*
Europa *ad ripas lacuum fluviorum.* 2/.

———. [MS. addition over the word *striato*], an striato ?

———. [MS. addition on the opposite page, the copy being inter-
leaved, and in allusion to a reference to Haller]. Hoc caules prolife-
ros a sterilibus definiter profert. Hall. [The passage in Haller is this
—"Caulis floriger videtur a folioso remotus."†]

———. [MS. addition below the preceding]. Forte mera varietas
prioris [palustre] ex solo aquæ profundioris.

Smith.—limosum ? Certè.

E. N.—Four specimens, all with catkins, and identical with Eq. limosum of Smith. As there is no representative of Eq. limosum of
Linneus, and as the marked copy of the work indicates that he did
not possess it, I subjoin the character.

Linneus.—[Sp. Plant.]. Limosum, 5. Equisetum caule subnudo
lævi, [here follow the synonymes]. *Habitat in* Europæ *paludibus,*
turfosis, profundis. 2/.

———. [MS. addition on the opposite page]. Hallerus hanc facit
varietatem E. palustris.

E. N. — It is clear that Linneus trusts to Ray as the authority for
this as a distinct species, since he quotes his figure,‡ which evidently
represents the unbranched form of Eq. limosum of Smith. Hence it
seems that Eq. fluviatile of Linneus is the branched, and Eq. limosum
the unbranched form of Eq. limosum of Smith; and that Eq. fluviatile
of Smith has no representative either in the herbarium or the works
of Linneus. With respect to the observation of Linneus quoted above,
that Haller makes this species a variety of Equisetum palustre, I think
the criticism is an unjust one. Haller quotes Ray's figure 3, and, as
it seems to me, correctly, as a variety of Eq. palustre ; while Linneus
quotes Ray's figure 2, which is evidently the Eq. limosum of Smith.
A positive proof that the fluviatile of Linneus was not the fluviatile of
Smith, exists in the fact that he attempted to account for its increased
size and altered appearance by its growing from the bottom of deep
water : this is the case with Smith's limosum, but never with his flu-

---

* Sp. Plantarum, 1517. † Haller, Helv. 144. ‡ Syn. t. 5, fig. 2, *a, b.*

viatile, which, on the contrary, affects loose gravelly and sandy places unconnected with water.

FOLIO 7.

Linneus.—Œdific.: hyemale.

———. [Sp. Plant. MS. addition on the opposite page]. Equisetum caule simplici aspero vaginis non laciniatis. Hall. Helv. 143.—Caulis viridis scaber, radiis [?] vaginæ pallidæ basi marginisque denticulis obsoletis atris gibbis.

E. N.—A single specimen of Eq. hyemale of Smith. The word or abbreviation "Œdific." implying its uses, is written apart from the name.

FOLIO 8.

Linneus.—Tourelle.

Smith.—Asperrimum, Dick., variegatum, Jacq. H. B.—J. E. S

Anonymous.—Equisetum basiliense, No. 1678, haller. An species distincta apud cl. linn.? In horto cultum.

E. N.—A single specimen of Eq. variegatum of Smith. Linneus answers the question as to its being a distinct species, by giving basiliense as a synonyme of hyemale (Phytol. 338). The word "Tourelle" written by Linneus is probably a habitat.

FOLIO 9, pinned by Linneus to the preceding.

Linneus.—Suec.

E. N.—A single specimen of Equisetum variegatum of Smith, much more slender than the preceding: the Linnean MS. evidently implies the habitat, Sweden.

I think the above notes will be sufficient to show that as regards several species of Equisetum, more especially the present, generally known as palustre, the Linnean herbarium is not a sure guide. There is, however, much collateral evidence that Linneus was not only acquainted with the Equisetum palustre of modern authors, but that he referred to that species when he named the plant in question; for he expressly states that his fluviatile (Smith's limosum) may be a variety of palustre growing in deeper water. Now as he was so well acquainted with fluviatile (Smith's limosum), and possessed such good specimens, his judgment cannot be supposed so much in fault as to have referred it to arvense. The error must have arisen from a want of care in the selection of specimens for his herbarium. Moreover, the name palustre is now too universally employed to admit of its being changed, without a better reason than a discrepancy which may have originated in carelessness. The same is not the case with fluviatile: I am decidedly of opinion that the name in this instance must be changed; and I make the proposition previously to publishing the species, in order that I may be favoured with the opinions of those botanists who think otherwise. I propose restoring the Linnean name to the limosum of Smith, and sinking the name limosum to the rank of a variety; thus:—

Equisetum fluviatile, Linn. = *Equisetum limosum*, Smith.

  ,,   ,, β. *limosum*, with the stem quite simple, = *Equisetum limosum* Linn.

The species hitherto called fluviatile is already so well provided with names that it is difficult to determine which to sélect. It seems to be the Equisetum majus of Gerarde* and Ray,† the Eq. Telmateia of Ehrart ‡ and Flora Danica,§ and the Eq. eburneum of Roth.||

In a plant of which the synonymy is so imperfectly known as the marsh horse-tail, it is by no means an easy task to trace the record of medical and other properties. Haller seems to have collected together a number of wise saws from a variety of sources, and gives them under his Equisetum No. 1677: but not only do I doubt whether the whole of them were intended for any one species, but I also doubt whether his No. 1677 is the species now under consideration. The point is not worth a very rigid scrutiny. He makes out his No. 1677 to be hurtful to oxen and cows, giving them diarrhœa and making their teeth loose, but at the same time to be harmless to horses and sheep. He also speaks of the great difficulty of extirpating it from a field where it is once naturalized: of its uses as a medicine he cannot speak with certainty.¶

The roots are slender and frequently divided; they appear to spring from the joints of the rhizoma, and are generally covered with minute

---

* Ger. Em. 1113. f.   † Ray, Syn. 130.   ‡ Ehr. Beitr. ii. 159. Crypt. 31.
     § Flora Danica, tab. 1469. || Roth, Cat. i. 129.

¶ Haller, 3. 2. 1677. Hoc equisetum minus quam 1676 tamen et ipsum pecori nocet et dentium facit in bobus et vaccis vacillationem tum diarrhœam. Cum seductus pulchritudine Trifolii Equiseto 1676 inquinati, famulus, qui boum meorum curam gerebat, semel aut iterum vaccam nuper vitulam enixam hac pestilente herba aluisset, ex diarrhœa immedicabili eadem periit. Quare magnis pecuniis nostri arcanum redimerent, quo prata infaustissima herbarum liberarent. Mihi neque aratrum, neque fimus, neque alia cura profuit. Equis non nocet, neque ovibus et rangiferis. Porci nostrates recusant, cum in Suecia non detrectant. Radicibus tamen glandium simile aliquid sæpe adhæret, quod porcos credas requirere.

Vires medicas vix satis certas autumo. Aquosa planta est, parum acris: ei adstringentes vires tribuerunt in diarrhœa, in hæmoptoe efficaces, &c. &c.

fibrillæ. The rhizoma is creeping, and extends to a great length; it is of nearly the same diameter as the stem, very black and shining, and smooth to the touch: at the joints it is solid, but the internodes are more or less hollow.

The engraving at the head of this article represents a stem of marsh horse-tail of the normal size and proportions: in order to exhibit the whole at one view, the stem has been divided into three portions. — The stem is perfectly erect, about fifteen inches high, deeply furrowed and finely granulated: the furrows are eight in number, the granulations of the ridges between them feel rough to the edge of the nail if drawn along them. The stem is divided into eleven compartments by means of transverse septa; the internodes are an inch and a quarter or an inch and a half in length: the sheaths occur at the septa, and correspond in number; they considerably exceed the stem in circumference, and in consequence are loose: the ridges of the stem enter the sheaths and terminate in the teeth, which are eight in number, acute, wedge-shaped, tipped with black, and furnished with nearly transparent membranous edges. There are nine whorls of branches: these rise from the furrows of the stem, close to the base of the sheaths; they never exceed the furrows in number, and are frequently fewer; at the base of each is a short black sheath, and these form a series of black rings round the stem: they are divided into six or eight joints, of which the basal and apical are the shortest: they have five furrows, and the sheaths occurring at the joints are five-toothed; the teeth are tipped with brown: the branches vary in length in the same whorl, and still more so in different whorls.

The catkin is long and rather narrow; when mature it stands on a distinct stalk of its own length: it is terminal, and after discharging its seeds it appears to perish, the stem and branches continuing to retain their vigour. There is no apiculus, the extreme summit being composed of a scale similar to the rest: at first the scales are crowded together, forming a black mass, they afterwards separate, the peduncle supporting each becomes visible, and the catkin, increasing in length, assumes a brown colour. The catkin appears in May and June.

This species, like the preceding, is subject to extraordinary variations, the most striking of which I shall describe.

EDWARD NEWMAN.

(To be continued).

ART. CXXIX.—*Notes of a Solitary Ramble to Loch-na-gar &c., with Remarks on several species of Plants collected in the course of it, on the 14th of July,* 1842. By Mr. J. B. BRICHAN.

WEARIED with a walk of at least twenty miles, I arrived at Castleton of Braemar about 9 o'clock on the evening of Wednesday the 13th of July. For several miles, further down the Dee, I had found the pretty alpine plant, Alchemilla alpina, in situations where neither river nor torrent could have carried it from its native bed. In these situations, therefore, it is strictly wild, but at what height above the sea I am unable to say. This plant is found also within a few miles of Aberdeen, where it has undoubtedly been washed down and planted by the river.

At half-past 7 the next morning I left Castleton, and proceeded in the direction of Loch Callader, which lies to the south-west of the "dark Loch-na-gar," and the vicinity of which has been visited and explored by botanists times and ways innumerable. In my walk to this localitiy, to which one half of the road runs through Glen Callader, and through ground which is altogether *highland*, I met with no plants strictly alpine, except Alchemilla alpina, Oxyria reniformis, Saxifraga aizoides and S. stellaris. The sides of the road and ground adjacent were covered with hosts of plants, which, as they are, generally speaking, found in all situations and at all elevations, I do not consider in any sense *alpine*. Polygonum viviparum, for instance, I have frequently seen at almost no distance from the sea, and at hardly any height above its level. Oxyria reniformis, Saxifraga aizoides, and, if I mistake not, Saxifraga stellaris also, are, like Alchemilla alpina, found at a great distance from their natural habitats, by the side of the river already mentioned. I believe the same is the case with respect to many other alpine productions, and with respect to every river that has its source among the hills.

When I had arrived at Loch Callader, and had learned from the occupant of the solitary abode at its northern extremity, the best way to Loch-na-gar, I proceeded along the western margin of the lake, where, as the same obliging individual informed me, most *weeds* are to be found. I was first struck with the minute leaves of Thalictrum alpinum growing among the stones and moss close to the water. I saw them in great abundance almost to the head of the lake, but I could not detect more than half a dozen perfect plants, and these almost entirely in fruit. The beautiful Saxifraga oppositifolia next attracted my attention; there were a good many tufts or patches of it, but few specimens in flower. It is said to flower in April and May, but I am quite satisfied that in this locality it was only coming into flower at the time I gathered it: and if I recollect rightly, a dried specimen which I received some years ago, was labelled as having been gathered in July. Of Tofieldia palustris I got just two specimens, and of Juncus triglumis only a few. The four plants just named I observed in no part of the ground I traversed, except along the margin of Loch Callader; but in that station I gathered also Saxifraga stellaris, the viviparous variety of Festuca ovina, and Apargia Taraxaci, now sunk into a variety of Apargia (or Oporinia) autumnalis. The lake itself seems to contain very little that may interest the collector of "weeds." At its southern end, or head, I observed a few plants of Utricularia vulgaris, not then in flower. I may here digress so far as to remark, that this plant occurs, very sparingly, in several localities on Deeside; and that in one locality at least, a moss in the parish of Banchory, U. minor is abundant: neither of these plants, so far as I know, flowered last season. To return to Glen Callader. The end at which the lake lies presents some interesting botanical ground, which I am sorry I had not time to visit. It is surrounded by high rocks, the clefts of which contained at the time a few patches of snow.

My route now lay over the shoulder of a ridge that skirts the east side of Loch Callader, and here commenced my ascent from this interesting locality to the still more interesting mass of rock and mountain named Loch-na-gar. Instead of at once climbing the nearest height, I made a considerable circuit through a sort of hollow to the northward; and, though for some time I could not see the top of the mountain, I felt pretty sure that I was gradually winding in the direction of it. In this zigzag ascent the first plant of any interest that I met with was a solitary bush of Betula nana. As it had neither flower nor fruit, and as I confidently expected to meet with more of it, I plucked but one specimen; more, however, I did not find. Lycopodium annotinum occurred sparingly. There was, as I have invariably observed in the alpine localities I have visited, abundance of Rubus Chamæmorus not flowering, and only a patch here and there partly in flower and partly in fruit. Imbedded in moss, which was saturated with the water of a perpetual spring, and near perpetual snow, I picked three specimens of a small plant, with a decumbent rooting stem and large blue flowers, and thought I had found the rare Veronica alpina, on examination, however, it appears to be no more than the variety of V. serpyllifolia which is termed humifusa; I could not detect another specimen. Dr. Murray, "without being certain" that this plant is distinct from V. serpyllifolia, has given it as a distinct species, " as it is at least worthy of a place among a group whose members press closely upon one another." To me it certainly appears distinct, and, if I mistake not, the capsule is *ovate*, and *longer* than the style. Dr. Murray observes—" This is by no means a very alpine plant, being found at Banchory, little more elevated than the sea, and at Glentanner, not much above the common level of the district. It is even met with in the neighbourhrod of Paris, in the *Bois de Boulogne*." This seems to confirm the idea that Veronica humifusa is a veritable species. I have never seen it except where 1 found it on Loch-na-gar, at least 2000 feet above the sea. In the same spot, and in other similar places, Epilobium alpinum, and especially a small variety of Saxifraga stellaris, abounded. The latter scarcely exceeds an inch in height, but the plant, according to its elevation and exposure, grows to all intermediate sizes between one inch and eight.

I next proceeded up the course of a very small stream, which, for the length of at least one hundred feet, ran under an enormous mass of snow. Near this, I picked a specimen or two of Sibbaldia procumbens, which was plentiful, but not all in flower. I was now, although I knew it not, on the very shoulder of Loch-na-gar, but had not succeeded in getting one glimpse of its top. I therefore crossed that shoulder, and also the head of a glen which separates the mountain from another ridge, and made towards the top of the latter, which I had some idea was that of Loch-na-gar. I thought, at all events, that by gaining the height before me I should be able to see where I was. In ascending towards it I gathered Vaccinium uliginosum without either flowers or fruit; I picked also a few specimens of Carex rigida. I had by this time nearly reached the top of the ridge, when, after having several times turned to survey my ground, I had the *pleasure* of seeing the indubitable peak, from which I had wandered, considerably to the north of the height on which I stood, and which is termed Craig-dhuloch. I saw moreover that I had actually been on the shoulder of Loch-na-gar, and had no difficulty in shaping out for myself a route to the top. In order to accomplish my *right ascension*, I had of course both to descend and to retrace my steps. In this retrograde movement I first gathered some specimens of Luzula spicata, from eight to ten inches in height. I lighted also upon a patch of a single-spiked Eriophorum, which I would fain have called E. capitatum, but have not been able to *make* it any other than the more common species, E. vaginatum. In reascending Loch-na-gar I gathered Gnaphalium supinum and Azalea procumbens, the latter, of course, in fruit. Before reaching the top I passed over a long mossy level, and again gathered Sibbaldia procumbens, which was here of larger size: Luzula spicata and Carex rigida abounded. I tore up a few specimens of Salix herbacea, both male and female; and also gathered Festuca vivipara, which ascends to the very summit of the mountain, and which, I may add, finds its way down the Dee as far as Banchory, seventeen miles from the sea. My ascent ended on the top of the natural pile of huge stones that forms the peak. I thought of *rhyming* — but I remembered that this was Byron's own "dark Loch-na-gar." Indeed, my thoughts on the subject, by a sort of ingenious anticipation, had occurred long before I ascended the mountain; and when actually "at the utmost top," I was so occupied with sheltering myself from the wind, with taking care of my hat, with the contemplation of the prospect around me, and with the very interesting exercise of demolishing some eatables,—that I had at the moment far more *reason* than *rhyme* in my pate.

After seeing all I could see, I began to descend on the north side, which is excessively steep, and strewed with large blocks of stone. Both at this time, and throughout the day, the wind blew with such violence, and in gusts so sudden, that I was frequently under the necessity of throwing myself on the ground, to avoid being driven along rather faster than was agreeable. I descended in a direction which brought me to the foot of the stupendous rocks which face the north. My time did not permit me to examine this locality; and although I scrambled round the end of the small lake that lies below, my path lying partly over sloping rocks, where a very small slip might have procured me a tolerable ducking, all I found was a variety of Melampyrum pratense with very pale yellow or white flowers, having the lip externally purple and the throat yellow. It is now so black, that one can scarcely say whether it has flowers or not. I ascended once more for a short distance in an easterly direction, crossed a low ridge over which I was partly blown by the wind, and then commenced my final descent towards the water of Muick, which I reached about 6 o'clock in the evening, at the distance of about two miles from the lake whence it issues. The only other plant worth picking with which I met after I was fairly away from Loch-na-gar, was a species of Hieracium which I take to be H. Halleri.* My friend Mr. Adams informs me, that in ascending towards Loch-na-gar from the same locality to which I descended, Dr. Murray and himself gathered the species last named; I

* Apparently just coming into flower; it nearly corresponds with our own specimens of this plant from Ben-na-bourde, collected in August, 1831.—*Ed.*

could find but one specimen. The rest of my walk was of no interest to any one but myself, unless some good-natured sympathetic botanist should feel interested in the information that I continued my journey till half past 1 next morning, when I arrived, sadly way-worn, at the place whence I had started on the 13th.

The above, I fear, will be considered but a meager concern : another day, and a knowledge of the Cryptogamia, of which I am ignorant, would have done much to render even my hasty excursion more interesting. Morven, another mountain celebrated by Byron, which I visited in 1841, does not present so wide a field as Loch-na-gar; it possesses, however, considerable interest. By the side of a small stream at its base I found the beautiful Sedum villosum, which I did not detect either on Loch-na-gar or in its vicinity.

Forres, March 8, 1843.   J. B. Brichan.

Art. CXXX.— *A List of Mosses and Hepaticæ collected in Eskdale, Yorkshire.* By Richard Spruce, Esq.

York has been said by Mr. H. C. Watson to be "pre-eminently the county of ferns;" and I think I may venture to assert that it is equally unrivalled in its mosses and Hepaticæ. In proof of this I refer to the valuable list of mosses in Baines's 'Yorkshire Flora,' the result chiefly of the investigations of Messrs. Gibson, Nowell, Howarth &c. in a very small portion of the county. Since Mr. Baines's work was published, several most interesting species have been discovered, and still a considerable part of the county remains unexplored. Our dales, which afford a passage for a multitude of impetuous streams, are peculiarly prolific; and I have assigned to myself the pleasing task of exploring them, in turn, as opportunity shall offer. The vale of the Esk, whose partial examination I have now to detail, terminates at Whitby, fifty-one miles N.E. of York; the river flows nearly due east, and receives during its course several smaller streams, issuing from a like number of dales. The Whitby and Pickering railway enters Eskdale at Grosmont Bridge; and from hence to Whitby (six miles) the country is well wooded, and may be called romantic. The course of the river forms a series of almost semicircular curves, now on the right now on the left of the railway, which crosses it I suppose ten times in that distance; the south side of these curves is frequently a perpendicular cliff, rising directly from the water's edge to a height of from 30 to 150 feet; and having a northern exposure, is richly clad with mosses and Hepaticæ.   These cliffs are farther interesting from being in the line of the remarkable fault which extends from Whitby Harbour to the plain of Cleveland; Dr. Young, in the 'Geology of the Yorkshire Coast,' has estimated the amount of dislocation to be about 100 feet at the mouth of the Esk, and in some places it is probably more.   The formation is the inferior oolite, and consists of beds of alum shale* alternating with ironstone and sandstone.   Goadland beck (or the Mirk Esk) joins the Esk just above Grosmont bridge, and Lythe beck (the sweetest spot in the whole district) runs into Goadland beck about half a mile higher up.   These two becks are crossed by the great basaltic dyke; but it scarcely rises above the surface, and does not appear to produce any peculiar plants.

I devoted two days (the 30th and 31st of December, 1842) to the examination of the banks of the Esk and its tributaries above mentioned.   On the morning of my third and (as it proved) *last* day's botanizing, I proceeded per railway to Fen End, an extensive peat-moss twelve miles west of Whitby, where Goadland beck has its source;† here I encountered a snow-storm, and although I persevered in examining the rocks on the south side of Newtondale (great oolite) until they, as well as myself, were clad in white, I was eventually compelled to desist.   I regretted the necessity of this the more, as I had planned out work for other five or six days; perhaps its completion is a treat yet to come.

Note. — I have not annexed the localities to those species which occurred in every spot I visited, and which are generally distributed throughout the county.

Gymnostomum truncatulum.
———————— *curvirostrum.*   Newtondale; barren.   I gathered a moss on wet rocks
 by the Esk, almost intermediate in size and appearance between this species and
 Anictangium Hornschuchianum.   I saw the same plant last summer in Dr. Taylor's herbarium, under the MS. name of G. nimbosum; he has long known it on
 Mount Mangerton, but has never met with fruit   Mr. Ibbotson finds it on Pennyghent, and Mr. Nowell in Todmorden, but always barren.   In a dried state
 the leaves are remarkably brittle, so that on opening a package of it, I always
 find numbers of them broken off and strewed about.
Tetraphis pellucida.   Lythe beck.
——————— *Browniana.*   Newtondale; with old fruit.

---

* The alum-works near Whitby have been celebrated for centuries, and at Grosmont great quantities of iron ore are got up and sent to Newcastle to be smelted at the founderies.

† Another stream rises in this bog, which, although it eventually reaches the same sea, runs in an opposite direction.

Weissia tenuirostris, H. & T. On stones in Lythe beck ; barren.

———— *curvirostra, controversa* and *recurvata*.

———— *verticillata.* Near the waterfall in Goadland beck ; in fruit.

Grimmia apocarpa and *pulvinata*, and *Didymodon purpureus*.

Didymodon Bruntoni. Newtondale ; fruit just rising.

———— *rigidulus.* Goadland beck ; in fruit.

Trichostomum lanuginosum, heterostichum and *fasciculare*. All fruiting on an old wall in the vale of Goadland.

———————— *aciculare.* On stones by the Esk and Goadland beck; abundant.

Dicranum bryoides, adiantoides, taxifolium, flexuosum, scoparium and *heteromallum.*

———— *glaucum.* Egton moors.

———— *flavescens* and *pellucidum.* Abundant, and with fruit in a good state, by the Esk and Lythe beck. I observed a large variety of the latter, not readily to be distinguished from D. flavescens.

———— *varium.* Two or three varieties.

Tortula muralis, ruralis, subulata and *unguiculata.*

Polytrichum undulatum, commune, aloides and *nanum*, and *Funaria hygrometrica.*

Orthotrichum affine, striatum and *crispum.*

Bryum albicans. Frequent on wet rocks by the Esk; barren. This plant is remarkably fragile; two or three times I thrust my stick into the broad flakes above my head, hoping to procure a patch entire, but I only brought down a shower of mutilated stems around me.

———— *argenteum, capillare, cæspititium, ventricosum, ligulatum, punctatum* & *hornum.*

———— *marginatum.* Lythe beck ; scarce.

Bartramia pomiformis and *fontana*, and *Leucodon sciuroides.*

Daltonia heteromalla. On an apple-tree at Eskdaleside ; in fruit.

Fontinalis antipyretica. In the Esk &c.

Hookeria lucens. Goadland beck.

Hypnum trichomanoides, complanatum, undulatum and *denticulatum.*

———— *medium.* At the roots of trees by the Esk.

———— *serpens* and *populeum.* The latter rather scarce.

———— *purum.* In fruit near Lythe beck.

———— *piliferum.* Abundant, but without fruit.

———— *catenulatum.* A single tuft on a wet cliff by the Esk.

———— *plumosum.* The most frequent moss in Eskdale.

———— *Schreberi, sericeum, alopecurum, dendroides, curvatum, myosuroides, splendens, proliferum* and *prælongum.*

———— *flagellare.* Abundant by the Esk and Goadland beck, but I saw no fruit.

———— *rutabulum, velutinum, ruscifolium, cuspidatum* and *squarrosum.*

———— *confertum.* Frequent. A large variety on stones in the streams, with the leaves sub-bifariously arranged and nearly entire.

———— *loreum* and *triquetrum.* Fruiting in Newtondale.

——— *filicinum* and *commutatum.* In great profusion on wet rocks.

———— *palustre, aduncum, cupressiforme* and *molluscum.*

———— *multiflorum*, Fl. Hib. On trees in Eskdale, and on a wall in the village of Egton, but scarce.

Marchantia conica and *Jungermannia asplenioides.*

Jungermannia lanceolata, L. On wet rocks by the Esk, and with calyces on stones in

Lythe beck. The existence of this plant in Britain was formerly doubted; Sir W. J. Hooker says — " I have never seen British specimens; and I suspect the authors just named (Hudson, Withering, &c.) may have mistaken some other species for it;" (see Brit. Jung. and Eng. Flor.) I believe its claim to a place in our Flora was satisfactorily established a few years ago by Mr. Jenner, who found it at Tonbridge Wells. Its closest affinity may be with J. pumila, but in my deliberate opinion J. polyanthos is the species for which it is most likely to be passed over; I found the two plants growing intermixed, and certainly differing very little in size and general appearance. In J. lanceolata the *stems* are procumbent, and copiously furnished with strong radicles on their under side. The *leaves* are horizontal, but slightly deflexed (except the two or three terminal ones, which are vertically folded together, as we sometimes see the leaflets in leguminous plants), longer than broad by about one half, and of nearly equal breadth from the base to near the summit, where they are rounded off, or rarely somewhat retuse; their colour brownish. The *calyx* is terminal, one third longer than the perichætial leaves, curved upwards, subcylindrical, broadest at the summit, where it is remarkably depressed, with a very minute slightly elevated mouth. Jung. polyanthos is altogether of a greener hue; the stem is more slender; the leaves mostly shorter, still more quadrate in their form, and often emarginate, the reticulation is also closer. Besides these marks of distinction, there are the more important ones of the presence of *stipules* (though unusually small in proportion to the leaves), and the curious two-lipped calyx with its much exserted calyptra, which will always suffice to keep this species far apart from J. lanceolata. The plant mentioned in my list of Wharfedale mosses (Phytol. 197) as an entire-leaved variety of J. asplenioides,* is not unlike J. lanceolata in appearance, but the calyx is totally different.

Jungermannia riparia, Taylor, MS. On wet rocks by the Esk; with calyces. This species, which has very lately been distinguished by Dr. Taylor, was first detected by him at the Dargle, Co. Wicklow; and I had the pleasure of gathering it in his company last July, at Blackwater Bridge, Co. Kerry. It is intermediate in size between J. cordifolia and J. pumila, and has some points of resemblance to each, but differs from both in its obovate very obtuse calyx. I have gathered it in two other Yorkshire stations, namely, Wharfedale and Crambeck on the Derwent, and Mr. Nowell sends it from Todmorden.

———— *sphærocarpa*. Abundant, but sparingly in fruit. I observe that the young calyx has a short tubular mouth, without any trace of teeth.

———— *hyalina*. Equally common with the last, and growing intermixed with it, but distinguished by its larger size and purplish tinge, by the broader and more wavy leaves, by the vertically compressed and angulate calyx, and, above all, by the perichætial leaves growing upon, or adhering to, the calyx.

———— *emarginata*. Eskdale and Newtondale. I found a few capsuliferous specimens.

———— *inflata* and *excisa*. With calyces on Egton moors. A small variety of the latter in Newtondale, with nearly all the leaves trifid.

———— *bicuspidata*, *nemorosa* and *albicans*. With calyces.

———— *umbrosa*. Newtondale and woods round Grosmont bridge. This beau-

---

* Mr. Wilson considers this to be a distinct and undescribed species.

tiful and rare species might easily be overlooked as a young state of J. nemorosa, from which it is, however, truly distinct.

Jungermannia complanata and *scalaris*. In fruit.

———————— *polyanthos.* By the Esk and Lythe beck; with calyces.

———————— *Trichomanis* and *bidentata*.

———————— *reptans* & *platyphylla.* A solitary tuft of each observed in Newtondale

——·———— *laxifolia* & *trichophylla.* Near the waterfall in Goadland beck; calyces

———————— *dilatata, Tamarisci, pinguis, epiphylla* and *furcata*.

———————— *Lyellii* ? Near the waterfall in Goadland beck, growing upon Hypnum commutatum. This approaches nearest to the variety called J. hibernica by Hooker, but is above twice the size of any specimens in my possession under either name, and the perichætium (or outer calyx) is longer, with repeatedly laciniated segments. I found plenty of pistilla but no fruit, the season being too early.

A remarkable circumstance in the cryptogamic Flora of Eskdale is the total absence of the genera Neckera and Anomodon; whereas in other districts of similar appearance, but at a distance from the eastern coast, such as Wharfedale, Castle Howard, &c., I have found the two species of Anomodon and two of Neckera (N. pumila and crispa) distributed in tolerable plenty; and the abundance of N. viticulosum is generally characteristic. Jungermannia reptans and platyphylla also, plants which are usually so abundant in rocky situations, do not appear to exist in Eskdale; their sparing occurrence in Newtondale has been remarked above, but this is fourteen miles from the sea, and belongs to another system of drainage. RICHARD SPRUCE.

York, February 17, 1843.

———————————

Art. CXXXI.—*Remarks on the threatened extermination of rare Plants by the rapacity of Collectors.* By S. H. Haslam, Esq.

Milnthorpe, Westmoreland,
March 7th, 1843.

"Fortunati ambo, si quid mea carmina possunt."

Sir,

I hailed the first appearance of 'The Phytologist' with pleasure, as I thought it promised to supply a desideratum much wanted in our scientific periodicals, and would afford to many a plodding botanist, as yet "unknown to fame," an opportunity of recording his observations in a less formidable manner than by drawing attention to himself in a publication of higher pretensions, if, indeed, he had the chance of seeing such a book at all. I dare say there are few of us who have not within the range of our acquaintance some one or other, the result of whose labours we have often wished could be made available for the general good, but whose habits of retirement, or, it may be, the "res angusta domi," had always kept in the back ground. Now, your useful little publication seems just adapted for the very purposes I allude to, and I think I am justified in saying, that the expectations of both editor and contributor must have been very fairly realized; and I am glad to think that the pages of 'The Phytologist' have recorded many valuable hints, and put your readers in possession of much pleasing information respecting Botany and botanists.

Having premised thus much, let me draw your attention to another part of the subject, less gratifying indeed (for every medal has its reverse), but which I trust you will see in the same light as myself; and as ' The Phytologist' is the avowed organ of enlightened science and practical Botany, permit me, through the medium of its pages, to awaken a sympathetic feeling in favour of those very treasures that are the objects of our favourite pursuit. And let me, ere the botanizing season has commenced, impress upon the minds of collectors the desirableness of forbearance and moderation, when culling specimens for their herbaria, — particularly amongst what are rapidly becoming the "plantæ rariores" of our island. I frequently caution brother botanists on this head, and seldom get more than a smile in return. But if the legitimate end of Botany be a more intimate acquaintance with plants in a growing state; and if it be more delightful to feast our eyes on these gems of the earth in the garden of Nature, than to handle a dry and often disfigured specimen in the herbarium of a botanist ;— surely it is a matter of grave moment, that the war of extermination which has lately been waged against our best and rarest plants, should at length be put an end to.

Some of your readers may perhaps say that I am fighting with a shadow, or that my own alarms have pictured an exaggerated state of things; but when I see a recently-formed Society, determined, I suppose, to outstrip all others in the work of extermination, putting forth a list, including *some* of the rarest of our indigenous plants, and desiring its contributors to send no more of them, as there are enough on hand to distribute for several years ;—can it be said that my fears are altogether groundless ? What pretty picking there must have been, to amass such a lot of treasures ! And the machinery by which such a system is kept alive, is not less alarming. Tradesmen, who cannot devote the time to it themselves, or else are verily ashamed of being caught at it —" palmam qui meruit ferat "— send forth their apprentices to dig up *every* specimen they can find, of certain rare plants in their locality, and these are bundled off until the market is actually glutted.

Yet more startling facts 1 could adduce, but I trust that I have said enough to awaken attention to the subject. What can be more painful to the true and ardent lover of Botany, than to find those spots where Nature has, as it were, secreted her choicest gems, visited only for the purposes of plunder and filthy lucre? I could name many plants that, not long ago, were plentiful in this neighbourhood, but which, since the establishment of railroads and learned Societies, have become nearly extinct. The interests of science surely do not require such barbarous work, and her legitimate sons, I am equally certain, would not sanction it.

I think I may venture to assert, without fear of contradiction, that no small portion of the pleasure arising from botanical pursuits, consists in the toil and enthusiasm of a botanical ramble; and those specimens are most highly prized that have cost us most personal exertion to obtain. But it is no uncommon thing, now-a-days, to see a collection made up entirely of "contributions from friends;" and by the system I am complaining of, a herbarium may be made up "on the shortest notice," and without costing its possessor one hour's fatigue in the way of genuine Botany. All this may be well enough, if confined to the commoner plants, but I really regret to see our darling science degraded into a mere handmaid of "commerce and trade," with the additional mortification of knowing that, one by one, our rarest plants are disappearing from their long-recorded habitats : —

"Oh infelix operis summa!"

Hoping that these remarks may be taken in good part by all the *real friends* of Botany,

I remain, Mr. Editor,
Your most obedient Servant,
S. H. HASLAM.

To the Editor of 'The Phytologist.'

ART. CXXXII. — *Observations on the publication of Local Lists of Plants.* By EDWIN LANKESTER, Esq., M.D., F.L.S. & B.S.Ed.

IN prosecuting the subject of the distribution of plants, it is exceedingly desirable that correct lists of the species that grow even in small districts should be obtained. The value of lists of plants from small districts must, however, be always proportioned to the variety of soil, elevation, climate, &c., which the district may possess. Thus a list of plants from a small district of level or swampy ground, would not furnish so much valuable matter as a list from a smaller district where streams or hills produced a difference in the dryness and constitution of the soil. : Take for example the neighbourhood of Askern, a small district of Yorkshire, the centre of which is seated in about the middle portion of the great escarpment of magnesian limestone, which runs through the counties of Nottingham, York and Durham. Its extent is about four miles east and west, and two miles north and south. To the east, the magnesian limestone dips under the new red sandstone, which is covered in this district by an immense swamp. Directly through the district, in a south-easterly direction, runs the river Went, making its way along a picturesque valley, which passes at right angles through the magnesian limestone formation. The river is here about thirty feet above the level of the sea, and the highest hills above its banks are about three hundred and eighty feet. The villages of Campsall, Burghwallis and Smeaton are seated entirely on the magnesian limestone, whilst Sutton, Askern and Norton include some of the marshy districts lying above the new red sandstone.

In publishing lists of plants of particular districts, botanists have for a long time felt the inconvenience of going over a number of the names of plants which are common to almost all districts ; and yet, as no rule has been offered which would answer generally for the exclusion of common plants, all plants must be included in lists of any value.

In drawing up a Flora for my 'Account of Askern,' * I have endeavoured to obviate this difficulty ; and I am not without hope that the idea acted on may be found generally applicable, at least to Great Britain, and that it will lead to the more general publication of local Floras, which will supply material for the perfecting our knowledge of the geographical distribution of plants.

In the Flora alluded to I have distributed the plants in four tables. In the first table is a numerical statement of the genera and species of each natural order. From this it will be seen that there are 428 species of phanerogamic plants, belonging to 212 genera and 78 natural

---

* An Account of Askern and its Mineral Springs ; together with a Sketch of the Natural History, and a Brief Topography of the Immediate Neighbourhood. By Edwin Lankester, M.D., F.L.S., &c. London : Churchill, Princes St., Soho. 1842. A very neat little book, evidently got up with great care on the part of the author, and containing mnch information relative to the Antiquities, Topography and Natural History of Askern and the surrounding district.—*Ed.*

families. The second table contains a list of the plants that more particularly distinguish the district. In drawing up this table, those species have been selected that are admitted into Mr. Watson's 'New Botanist's Guide,' which contains only the plants found not to be common to eight out of twelve local Floras of Great Britain. I have followed Mr. Watson in this respect, not because I think his list the best that could be constructed, but because his book is very generally known to botanists in this country. Since the publication of Mr. Watson's book, other Floras have been published, and materials furnished for a more correct list of common plants than he has given; and I would suggest that it might be worth the while of some of our botanical societies to construct a list, which, by excluding common plants, would enable the botanist to publish only the peculiar and most interesting forms. But these two tables, even with Mr. Watson's Guide and a British Flora, would give only an incomplete view of the Flora of a particular district, since many of the common as well as of the rarer plants might be absent from it. In order to meet this defect, I have constructed a third table, including the plants which were common to the twelve counties examined by Mr. Watson, but which I did not find near Askern. In the present case I believe this table to be somewhat imperfect, but I have mentioned it in order to show the manner in which my plan may be carried out. With these three tables, a list of British plants, and Mr. Watson's Guide, we have the elements of a complete Flora of a district.

There is, however, one point in which such a Flora would be defective, and that is the comparative rarity or frequency of the excluded common plants. This is, however, always a difficult point to estimate, and little has hitherto been done towards enabling us to compute, with any proximity to the truth, the number of individual plants growing in a district. In my fourth table, by way of meeting in some measure this inconvenience, I have added the names of a few plants which are rare in the Askern district, but which are excluded as common from Mr. Watson's list.   Edwin Lankester.

19, Golden Square,
    January, 1843.

[We are greatly obliged to Dr. Lankester for his valuable suggestions; the adoption of his plan would, however, involve the necessity of printing at least *four* lists instead of *one* for every district illustrated. The subject of publishing local lists of plants is one which has often had our serious consideration. For various reasons we have hitherto preferred printing them exactly as received from the contributors, to whom our thanks are due for the pains they have taken to render their lists correct and complete. The following considerations have had great weight in inducing us to publish these lists without mutilation. Botanists confessedly require more exact information than they at present posesss, on many points relating to the general distribution of plants, common as well as rare. In all matters connected with this subject, we know of no botanist whose opinions are more worthy of regard than those of Mr. Watson; and that gentleman, at no very distant period, has well observed, that owing to the want of sufficient data we are as yet unable to say with certainty, that even such common and widely spread plants as the daisy and the dandelion are to be found in every county of the United Kingdom. Such data it is the province of local lists to supply, and we would have them include the name of every species growing in the district to which they refer. For our own part we can truly say, that if carefully drawn up, and more especially if they contain short remarks on any interesting circumstances connected with the species, our only objection to these lists arises from the fear that some of our subscribers may think that their place would be better occupied by matter more generally readable. We shall be glad to receive communications on this subject; but until some better plan be determined on, we trust we shall be excused if we "e'en gang our ain gate," printing the lists of flowering plants, mosses, &c. entire, and in future transferring localities of ferns to Mr. Newman, for publication in his county lists.—*Ed.*]

ART. CXXXIII.—*Notice of 'A Visit to the Australian Colonies.* By JAMES BACKHOUSE.' London: Hamilton, Adams & Co. 1843.

To a considerable number of our readers it is well known that the author of this work is a member of the Society of Friends, and that his visit to the Australian colonies was undertaken "for the purpose of discharging a religious duty." But our notice of this narrative will be confined to that part of it which is strict accordance with the design of 'The Phytologist,' namely, the author's observations on the vegetable productions of these colonies; and we trust that we shall be able to show that James Backhouse is an observant and accomplished botanist. To many of our readers who are acquainted with this excellent man, his profound botanical knowledge is already well known; still, even to these, our remarks may not be unacceptable, since they will collect, and somewhat condense, observations on this interesting science, which are scattered through a volume of 560 pages. The visit "occupied a period of six years, terminating with 1838." We shall divide our notice into three parts:—Van Diemen's Land, Norfolk Island and New South Wales.

VAN DIEMEN'S LAND.—As far as Botany is concerned the narrative commences at Hobart Town, where the author landed in February, 1832, and our first botanical sketch is of a hill near that town. This hill—

"Was clothed with gum trees—species of Eucalyptus—of large size, having foliage somewhat like willows, and growing among grass and small shrubs. Many trees were lying on the ground, and in various stages of decay.   Smaller trees, called here honeysuckle, she oak, cherry-tree, and wattle, were interspersed among the others, and the ground was decorated with Leptospermum scoparium, Corræa virens, Indigofera australis, and Epacris impressa; the last of which resembles heath, with white, pink, or crimson flowers.   The trees in this country often bear the name of others belonging to the northern hemisphere.   Thus the honeysuckle of the Australian regions is generally some species of Banksia, often resembling a fir in growth, but having foliage more like a holly; and the cherry-tree is an Exocarpos—a leafless, green, cypress-like bush, with small red or white fruit, bearing the stone outside!"—p. 22.

Speaking of introduced plants the author observes that the climate at Hobart town is too cold for grapes and cucumbers, but that apples, pears, quinces, mulberries and walnuts succeed better than in England.   On the basaltic hills about Hamilton, the prevailing tree is the oak — Casuarina quadrivalvis.   "It seldom grows in contact: its trunk is about 10 feet high and 5 feet round; its head is spherical, and 10 or 15 feet in diameter, and consisting of pendulous, leafless, green, jointed twigs, resembling horse-tail weed."   At Mount Wellington, Acacia Oxycedrus, 10 feet high, was in flower.   "This, along with numerous shrubs of other kinds, formed impervious thickets in some places, while, in others, Epacris impressa displayed its brilliant blossoms of crimson and rose-colour."   We pass on to the description of a fern valley, which is almost enough to make a botanist emigrate to this distant land.   In the plate accompanying this sketch the author is seen crossing the stream on the trunk of a fallen Eucalyptus; how we long to be his companion!   Loddiges and Ward, what would you not give to realize such a scene!

"The brook that supplies Hobart Town with water, flows from Mount Wellington through a valley at the foot of the mountain.   Here the bed of the brook is rocky, and so nearly flat as scarcely to deserve the name of The Cascades, by which this place is called.   Many dead trees and branches lie across the brook; by the sides of which grows Drymophila cyanocarpa — a plant allied to Solomon's seal, producing sky-blue berries on an elegantly three-branched nodding top.   Dianella cærulea—a sedgy plant — flourishes on the drier slopes: this, as well as Billardiera longiflora — a climbing shrub, that entwines itself among the bushes — was now exhibiting its violet-coloured fruit.   In damp places by the side of the brook, a princely tree-fern, Cybotium Billardieri, emerged through the surrounding foliage.   A multitude of other ferns, of large and small size, enriched the rocky margins of the stream, which I crossed upon the trunk of one of the prostrate giants of the forest, a gum-tree of large dimensions, which had been uprooted by some blast from the mountain; and in its fall had subdued many of the neighbouring bushes, and made a way where otherwise the forest would have been inaccessible.   On descending from this natural bridge, to examine a tree-fern, I found myself at the foot of one of their trunks, which was about five feet in circumference and ten feet in height.   The lower part was a mass of protruding roots, and the upper part clothed with short remains of leaf-stalks, looking rough and blackened: this was surmounted by dead leaves hanging down, and nearly obscuring the trunk

from distant view : above was the noble crest of fronds, or leaves, resembling those of Asplenium Filix-fœmina in form, but exceeding eleven feet in length, in various degrees of inclination between erect and horizontal, and of the tenderest green, rendered more delicate by the contrast with the dark verdure of the surrounding foliage. At my feet were several other ferns of large size, covering the ground, and which, through age and their favourable situation, had attained root-stocks a foot in height, crowned by circles of leaves three times that length. Other plants of tree-fern, at short distances, concealed from my view, by their spreading fronds, the foliage of the lofty evergreens that towered a hundred feet above them. The trunk of one of the tree-ferns was clothed with a Trichomanes and several species of Hymenophyllum—small membranaceous ferns of great delicacy and beauty. On a rocky bank adjoining, there were other ferns, with creeping roots, that threw up their bright green fronds at short distances from each other, decorating the ledges on which they grew. In the deepest recesses of this shade I could enjoy the novel scene—ferns above, below, around—without fear of molestation; no dangerous beasts of prey inhabiting this interesting island."
—p. 34.

On leaving Hobart Town the author sailed to Port Davey, and thence to Macquarie Harbour, where he describes the timber as being very fine. The Huon pine, valuable for ship-building &c., abounds on the eastern side; it attains a height of 100 feet, and a circumference of 25 feet; it has a pyramidal shape, and the branches are clothed with numerous slender scaly branchlets of lively green, as in the cypress and arbor vitæ. The celery-topped pine—Thalamia asplenifolia — is suitable for masts ; myrtle for keels; and the roots of light-wood — Acacia Melanoxylon — make beautiful veneers. This latter wood derives its name from swimming in the water, the other woods, pine excepted, generally sink. Hats are made of the shavings of some Acacias, " as well as from broad-leaved sedges — Lepidosperma gladiata; the leaves being first boiled and bleached." At Philips Island, in the same vicinity, we have another peep at the tree ferns.

" The huts were almost overgrown with the Macquarie Harbour vine, a luxuriant climber, bearing small acid fruit. We walked over the island, and down one of its sides, which was woody, and which exhibited the finest tree-ferns we had seen, and in great profusion. They were of two kinds, one of which we did not meet with elsewhere. Some of their larger fronds or leaves were thirteen feet long, making the diameter of the crest twenty-six feet. The stems were of all degrees of elevation, up to twenty-five or thirty feet; some of them, at the lower part, were as stout as a man's body : those of Cybotium Billardieri were covered with roots to the outside: the whole length of those of the other species—Alsophila australis— was clothed with the bases of old leaves, which were rough, like the stems of raspberries, closely tiled over each other, and pointing upwards. There was also a number of other ferns, of humble growth : two species of the beautiful genus Gleichenia had tough, wiry stems, which were used in the settlement for making bird-cages."—p. 55.

At Macquarie Harbour and Port Davey a species of Blandfordia was observed, a lily-like plant, with a crest of scarlet tubular flowers: and on the hills near the former place, was a lichen of a texture resembling net-work; this, "in the abundant rain, was distended into masses resembling cauliflowers." " In some places —— cyperaceous plants entwine themselves among the larger shrubs and ascend to their tops, and lichens hang to a great length from the boughs of some of the trees."

Returning by sea to Hobart Town, our author made an excursion on the opposite side of the river to that on which the town stands: he notices the appearance of Anguillaria dioica, a little, purple-spotted, white-blossomed, bulbous plant, which was decorating a sunny bank as one of the first harbingers of spring, (August 27th); and also comments on the strange appearance of trees in full foliage laden with snow. It is, we presume, generally known, that the trees of Australia may nearly all be regarded as a kind of evergreens, although not so strictly entitled to that name as those so commonly cultivated in our English gardens.

On the 26th of September J. Backhouse sailed for Flinder's Island, of which he records nothing botanical that is particularly worthy of notice. On the 20th of October he reached George Town, on the main land of Tasmania ; on the 30th Circular Head, and on the 31st Woolnorth. The seaweeds of this shore are of prodigious magnitude; one, " a palmate species, has a stem thicker than a man's arm, and proportionately long. The flat portion between the stem and the ribbon-like appendages is so large as to be converted by the blacks into vessels for carrying water. For this purpose they either open an oblong piece so as to form a flat bag, or run a string through holes in a circular piece, so as to form a round one." Returning to Circular Head he proceeded thence by land to Emu Bay, noticing by the way the grass-trees, to which we shall again recur, and a beautiful Blandfordia, whose stems were eighteen inches high, and supported crests of from ten to twenty pendulous red blossoms, margined with yellow, an inch and a half long, and three quarters of an inch wide-at the mouth. On some of the hills Banksia serrifolia was the prevailing tree; "it is equal to a pear-tree in size, has leaves three or four inches long and five-eighths broad, and strongly toothed : its heads of flowers are six inches long and twelve round, and the seeds are as large as almonds." After a short rest at Emu Bay,—

" We set out for the Hampshire Hills, distant 20½ miles, through one of the most magnificent of forests. For a few miles from the sea, it consists chiefly of white gum and stringy-bark, of about 200 feet in height, with straight trunks, clear of branches for from 100 to 150 feet; and resembling an assemblage of elegant columns, so irregularly placed as to intercept the view at the distance of a few hundred yards. These are elegantly crowned with branching tops of light willow-like foliage, but at an elevation too great to allow the form of the leaves to be distinguished, yet throwing a gentle shade on the ground below, which is covered with splendid tree-ferns and large shrubs, and carpeted with smaller ferns. Some of the larger stringy-barks exceed 200 feet, and rise nearly as high as ' the monument' before branching. Their trunks also will bear a comparison with that stately column, both in circumference and straightness. The bark of these trees is brown and cracked: that of the white-gums is French grey, and smooth.

" The prostrate trunks of these sylvan giants, in various stages of decay, add greatly to the interest of the scene. Some of them, lately fallen, have vast masses of the rich red earth in which they grew, still clinging to their roots; others, that have been in a state of decay before they fell, present singular ruins of shattered limbs and broken boughs; others, that seem to have been in a state of decomposition for ages, have become overgrown with various ferns and shrubs.

" As the distance from the sea increases, the Australian myrtle and sassafras, of dark dense foliage, become the prevailing trees. In these denser forests, tree-ferns form nearly the sole undergrowth, except the small, starry ferns, of low stature, of the genus Lomaria, that cover the ground thinly. Some of the tree-ferns have trunks 20 feet high. Their leaves are from 8 to 12 feet long, and the new ones, now forming, rise in the centre like elegant crosiers."—p. 111.

(To be continued).

Art. CXXXIV.—*Varieties.*

270. *Note on the occurrence of Cuscuta Epilinum and Saponaria Vaccaria in Morayshire.* In July, 1842, Mr. Wilson of Alves detect ed these two plants in a field of flax, in the parish of Alves, Morayshire, along with Camelina sativa. Mr. Babington's remark (Phytol. 250) that C. Epilinum does not make its appearance among flax raised from American and Riga seed, led me to enquire whence the seed in the present case was imported. I am informed that all the flax-seed used in this district is procured either from America or Holland; that those who are in the habit of using both, can distinguish American seed from Dutch by the *rounder* shape of the former; and that the seed respecting which I made the enquiry, was the remains of a cargo which had been cleared out at London, was obtained from the vessel on its arrival at Burghead (a small port on the Moray Frith), and was considered to be American on the ground just mentioned.— *J. B. Brichan ; Forres, February* 14, 1843.

271. *Note on a supposed new British Æcidium.* Last August I found at this place a species of Æcidium, which is not noticed in

Hooker's 'English Flora,' our text-book for British cryptogamous Botany. On referring to Gray's Natural Arrangement, I found the description of an Æcidium that in some respects answers to mine, of which I could secure but three specimens, as the season for it appeared to be nearly over. The following is Gray's description.—

" *Æcidium asperifoliarum.* Theow confluent; wine-glass shape, half immersed, pale yellow; sporidia yellowish white. *Æcidium asperifolia,* DC. Sys. 50. On the lower face of various Boragineœ. Leaves hollowed on the upper face."

My notes are as follows :—

Æcidium —— ? Spots yellowish, disfiguring and incrassating the leaves. Pseudo-peridia generally hypogynous, sometimes amphigenous, more or less confluent: sporidia orange. On Lycopsis arvensis. Thame, August 24, 1842.

I am uncertain whether the difference in the colour of the sporidia will constitute a new species; in other respects it agrees pretty well with DeCandolle's.—*Ph. B. Ayres, M.D.; Thame, Feb. 22,* 1843.

272. *Note on the Fruit of Umbelliferæ.* Before I close my note I wish to make a remark on the discussion in your journal concerning the poisonous properties of the seeds of Umbelliferæ. Lindley's statement is copied from DeCandolle's Essay on the Medical Properties of Plants, and DeCandolle is certainly in error. I think the most accurate view on this subject is, that those seeds in which vittæ are present are innocuous, while those which have no vittæ are either suspicious or poisonous. In the former, the proper juice of the plant is converted into volatile oil; in the latter, it is merely deposited in a more concentrated form.—*Id.*

273. *Note on Gigartina compressa.* My attention having been directed to some remarks on the Jusna or Ceylon moss, by M. Guibourt of Paris (Provincial Medical Journal for February, 1843, No. 128), I am induced to trouble you with some observations made by me on this plant many years since, when it occurred in considerable abundance at Sidmouth, as they agree so entirely with those of M. Guibourt, that there can be no doubt of their identity. The species in question is the Fucus lichenoides of Turner, described and figured in his 'Historia Fucorum,' ii. 124, t, 118; the Gracilaria lichenoides of Greville's 'Cryptogamic Flora,' p. 125, from specimens and a drawing of the recent plant, communicated by me from Sidmouth, Fucus lichenoides of Linn. &c. Subsequently, finding it agree with Sphærococcus compressus of Agardh, and that the term lichenoides was preoccupied and inappropriate, Dr. Greville named it Gracilaria compressa in his 'British Algæ;' and it is now Gigartina compressa of Hooker's 'British Flora,' 299, Harvey's 'Manual of British Algæ,' 74,

and Wyatt's 'Algæ Danmonienses,' iii. No. 108. *Mem.-* "Sidmouth, 1827. Turner's description of this plant is better than the figure, which is evidently from a bleached specimen. It has been abundant here from June to October, often twelve inches high, in fine fructification, capsules and granular imbedded seeds on distinct plants. The substance when fresh is cartilaginous but tender, full of moisture and brittle, breaking with the slightest touch, and shrinking to less than half its size in drying, and does not recover on subsequent immersion. It is cylindrical in the upper, compressed in the lower branches. When most plentiful I tried several experiments with it. Boiled in water it became of a most beautiful semitransparent green colour, and ate like delicate French beans, with a peculiar crispness and very agreeable taste. It did not dissolve after boiling seven hours, but lost much of its size. Vinegar nearly dissolved it, and changed the colour to yellowish brown. Boiled in syrup of preserved apricots it became a little yellowish, but retained its crispness and was extremely good: and having but little flavour of its own it might be made to taste of lemon, ginger, &c." I enclose part of two plants in the different modes of fructification. As a British species I would remark that its nearest affinity on the one hand is with Gigartina confervoides, on the other with Rhodomenia polycarpa of Greville (the Fucus Sarniensis of Turner). This last species has a much more compressed, nearly flat and broader frond, but the structure and fruit are the same, and on this account, in a future arrangement, it will doubtless be referred to its proper genus.—*Amelia W. Griffiths; Torquay, March* 1, 1843.

274. *Enquiry respecting Orchis hircina*, Scop., *and Orchis macra*, Lindl. May I enquire if any kind friend can favour me by mentioning any recently verified locality for either of these remarkable species? The first-named is said to occur in the neighbourhood of Dartford, but I am not aware of any living botanist having gathered it. Both species are noted in the books as Kentish plants.—*Edward Edwards; Bexley Heath, Kent, March* 4, 1843.

275. *Warwickshire locality for Equisetum fluviatile.* I fully expected some correspondent would have given a locality for Equisetum fluviatile, and am sorry I did not send a solitary station for the Warwickshire list of ferns. It was discovered in a damp copse near Elmdon, by James Clift; and when out there last autumn I was told of it, and saw the decaying fronds. It was scattered rather sparingly, but I should suspect it to be growing in other similar situations in that neighbourhood. Equisetum sylvaticum and palustre were growing sparingly near the same spot; and in a meadow close by, a new habi-

tat for Parnassia palustris was pointed out, where it had been in tolerable abundance last season. —*David Cameron; Botanic Garden, Birmingham, March 5, 1843.*

276. *Reply to Mr. Watson's Enquiry respecting Mr. S. Gibson.* As your correspondent Mr. Watson appears to have some doubts as to the identity of the S. Gibson mentioned by Mr. Francis in his 'Analysis of British Ferns,' in connection with Polypodium calcareum, and again in Mr. Watson's 'New Botanist's Guide, — (Phytol. 524); perhaps you will allow me to make a few remarks, which will, I think, set the question at rest. In 1830—33 I sent several parcels of dried plants to Mr. R. B. Bowman; in these parcels there was every plant mentioned in Watson's Guide, with the name of Gibson connected with them, with the exception of Geranium nodosum, and that I never sent to any correspondent, since I never had more than one specimen of the plant, and that specimen is still in my possession. If the Geranium in question be any Geranium which I sent to Mr. Bowman, it will be Geranium pyrenaicum, and the locality would be Washerlane, near Halifax: and the labels for Polypodium calcareum would be written *Sheden Clough, near Burnley, Lancashire.* If Mr. Bowman sent Mr. Watson *fragmentary* and *ill-dried* specimens, I cannot help that; but I must say that I think Mr. B. might have sent a few good ones out of the two or three hundred specimens of P. calcareum which I sent him. The locality for Meum Athamanticum should have been Ripponden, not Ripon, as in the 'Botanist's Guide,' p. 284. If Mr. Watson or any of his friends should happen to be in this neighbourhood, and should wish to see the plants growing, I shall be happy to go with them to *all* or any of the localities given on my authority, the Geranium nodosum and Asarum europæum excepted. In 1830 I could have gathered five hundred specimens of Asarum, and have left perhaps twice that number; in 1842 I visited the locality in company with Mr. Borrer, and we sought for some time before we could find a single specimen, but at length we found a few, perhaps six or seven, some of which we gathered, and the rest we left. If it do appear this season, I expect it will be the last; and then the Yorkshire Flora will lose Asarum europæum, since we have no other authentic locality for it in the county. — *Samuel Gibson; Hebden Bridge, March 15, 1843.*

277. *Note on Polypodium Dryopteris.* I am informed by my relative, the Rev. T. Gisborne, that this fern was *not* introduced at Yoxall Lodge, as suggested in a former No. (Phytol. 509). It exists in extremely small quantity, and the spot has not been altered since the enclosure of Needwood Forest.—*C. C. Babington; Cambridge, March 1843.*

ART. CXXXV.—*Proceedings of Societies.*

LINNEAN SOCIETY.*

June 7, 1842.—The Lord Bishop of Norwich, President, in the chair.

The Hon. H. Wright, of the Ceylon Civil Service, presented specimens of the fine Ceylon cinnamon of commerce, of the unusual length of eleven feet.

Read, the commencement of a paper by Mr. Clark, " On the Sea Cocoa-nut of the Seychelles, (*Lodoicea Sechellarum,* Comm. and Labill.)

June 21.—Edward Forster, Esq., V.P., in the chair.

Read, " Observations on the Growth and Reproduction of Enteromorpha intestinalis." By A. H. Hassall, Esq. The author states that in their earliest stage of development the tapering filaments of this plant consist of a single series of cells placed end to end. Each cell is afterwards bisected by a longitudinal line ; other lines subsequently make their appearance, and each original cell is thus ultimately divided into several others, each of which in its turn enlarges and is divided in the same manner. From the repetition of this process the filaments increase in size, lose their confervoid character, present a reticulated appearance and become cylindrical and hollow. The author states that in each articulation of the filaments, even when no thicker than a horse-hair, a dark central nucleus is developed, and that this is the reproductive germ, which he doubts not undergoes repeated division in the same manner as the reproductive globules of the Ulvæ. These nuclei germinate while enclosed within the cells, the filament still retaining its freshness and vigour ; and from them arise the jointed tapering filaments first described, which, after the rupture of the parent cell, and while their bases are still fixed within it, strongly resemble a parasitic Conferva. This development, division and growth of cells and reproductive bodies appears to be constantly going on, whence most specimens of the plant present examples of each stage of formation. From these observations Mr. Hassall is led to regard Enteromorpha intestinalis as bearing a relation to the Confervæ in its young articulated filaments, and to the Ulvæ in its reproductive globules. The author objects to the tautology of the specific name, and proposes that of lacustris in its place.

Read also the conclusion of Mr. Clark's paper " On the Sea Cocoa-nut of the Seychelles." For a report of a paper on this subject, subsequently read before the Botanical Society of London, see Phytol. 463; the following are additional particulars. The part of the trunk immediately above the ground forms an inverted cone, the apex of which is of an hemispherical form, with a great number of cord-like roots spreading from it in all directions, and remaining long after the destruction of the plant to which they belonged. Where the trees have disappeared from clearings, by burning or otherwise, a black circle on the surface indicates their former site. This circle is the base of the cone before mentioned, which now forms a huge bowl, often filled with decayed vegetable matter. On removing this, the interior of the bowl is found to be pierced by a number of holes, each large enough to admit the end of the fore finger. These holes are the openings of the compact, sonorous and brittle tubes into which the roots have been converted by the decay of their internal substance. Mr. Clark states that the leaves are so firmly attached to the trunk, that a man may seat himself at the end of one with perfect safety. The fibres of the leaflets are very strong and arranged in three layers, the central one being disposed transversely, the others longitudinally ; in

---

* From various causes we are sadly behind with our reports of the Linnean Society; we hope to bring up arrears in the present and next number.—*Ed.*

consequence of this arrangement, their tissue, when divested of the parenchyma, resembles coarse book muslin. Both male and female spadix are stated to pass through a fissure in the base of the accompanying leaf-stalk. The drupe is fifteen inches in length, about three feet in circumference, and weighs from thirty to forty pounds; as many as seven well-formed drupes are sometimes seen on a single spadix. The fecundation is occasionally imperfect, and then the ovary expands and lengthens, but does not assume the usual form, and at the end of two or three years it falls off. A female plant at Mahé flowered for several years without producing fruit, owing to the absence of a male plant. In 1833, a male flower was procured from an estate a few miles distant, and suspended in the tree; about two months afterwards one of the buds expanded, and the mature fruit fell from the tree at the end of 1841.

November 1.—R. Brown, Esq., V.P., in the chair.

Jonathan Pereira, M.D., F.L.S., presented specimens of the different varieties of Ceylon, Malabar and Java cardamoms, &c.

Read, " A Notice of the African Grain called Fundi or Fundungi." By Robert Clarke, Esq., Senior Assistant-Surgeon to the Colony of Sierra Leone. This grain is about the size of mignonette-seed, and is said to be cultivated in the village of Kissy and the neighbourhood of Waterloo, by individuals of the Soosoo, Foulah, Bassa and Joloff nations, by whom it is called "hungry rice." Mr. Clarke describes the mode of cultivation, and the various methods of preparing the grain for food; and he is of opinion that if imported into Europe, it might prove a valuable addition to the list of light farinaceous articles of food in use among the delicate or convalescent.

Specimens of the grass which accompanied Mr. Clarke's communication had been examined by Mr. Kippist, Libr. L.S., who added some observations on its botanical characters. It is slender, with digitate spikes, and has much of the habit of Digitaria, but on account of the absence of the small outer glume existing in that genus, it must be referred to Paspalum. Mr. Kippist regards it as an undescribed species, although specimens, collected by Afzelius at Sierra Leone, are in the herbaria of Sir J. E. Smith and Sir Joseph Banks. Mr. Kippist names it P. exile, and gives the characters.

Read also a letter from N. B. Ward, Esq. F.L.S., relative to the introduction of the Musa Cavendishii into the Navigators' Islands, (Phytol. 527).

November 15.—E. Forster, Esq., V.P., in the chair.

Mr. T. S. Ralph, A.L.S., presented numerous fruits and seeds collected in the neighbourhood of Aurungabad.

Read, a note " On the permanent varieties of Papaver orientale, *L.*" By T. Forster, M.B., F.L.S., &c. The author states that ever since the introduction of Papaver bracteatum, *Lindl.* into England, he has regarded it as a permanent variety of P. orientale. This name he retains for the species, both as being the older one, and applicable to all the varieties; of which the four following he considers as permanent:—

1. *P. orientale bracteatum;* seeds always perfect. 2. *P. orientale praecox*, the "Monkey Poppy" of the old gardeners; flowering in May with the preceding, seeds always sterile. 3. *P. orientale serotinum;* flowering in June, seeds always imperfect. 4. *P. orientale, capsulâ et floribus longioribus;* flowers in May, seeds sometimes perfect. Only met with in continental gardens. Dr. Forster states that he has been assured in the South of Europe, that P. orientale bracteatum yields the best opium, and that in the largest quantity; and as this plant seeds freely and suits the English soil, he thinks it might be advantageously substituted for P. somniferum.

Read also, a note " On Secale cornutum, the Ergot of Rye:" and " On a Species of Asplenium related to A. Trichomanes, *L.*" By A. Haro, M.D., of Metz. In the latter communication Dr. Haro calls attention to a fern discovered by himself in the well of an old castle. The well is large, four-cornered, and with a square window at the top in one of the sides. The wall opposite to the window is covered with the fern, which lies flat upon the stones, to which the fronds are attached throughout their whole length by slender roots, and adhere so firmly that it is difficult to remove them, even with a knife. A professor, to whom Dr. Haro submitted the plant, regarded it as a new species, and has named it *A. Harovii*: he has also furnished descriptive characters of the new plant, as well as of the three allied species, A. Trichomanes, viride and Petrarchæ. We give the characters of A. Harovii, which is placed between A. Trichomanes and viride.

> *A. Harovii*. Frond *decumbent*, fixed to stones by very slender fibrils, *glabrous*, unequally pinnate; stipes blackish-varnished, furnished above with an indistinct membrane, running down on each side from the insertion of the pinnules, (*appendiculatus*); middle pinnules *hastato-rhomboid, three-lobed*, upper pinnules oblong, obliquely attenuated or wedge-shaped at the base, unequally pinnatifid, all obtuse but *acutely toothed*.

December 6.—E. Forster, Esq., V.P., in the chair.

Read, a portion of " An Essay on the Distribution, Vitality, Structure, Modes of Growth and Reproduction, and Uses, of the Fresh-water Confervæ." By Arthur Hill Hassall, Esq.

December 20.—E. Forster, Esq., V.P., in the chair.

A. H. Hassall, Esq. exhibited an apple in which decay had been artificially induced by inoculating it with decayed matter from another apple containing filaments of Entophytal Fungi.

Read, a continuation of Mr. Hassall's memoir on the fresh-water Confervæ.

Read also, " Some further Observations on the Nature of the Ergot of Grasses." By Edwin John Quekett, Esq., F.L.S.

BOTANICAL SOCIETY OF LONDON.

February 17, 1843.—J. E. Gray, Esq., F.R.S. &c., President, in the chair. The following donations were announced. British plants from Dr. Streeten, and foreign plants from Mr. Samuel Simpson. Donations to the library were announced from Mr. H. C. Watson and Mr. E. Newman. Mr. T. Clarke, jun. presented specimens of a very large variety of Lastræa Filix-mas, found by him at King's Cliff Valley, near Bridgewater.

Mr. G. H. K. Thwaites read a paper, being a Notice of the discovery of Grimmia orbicularis, a moss new to Britain, which was found by him upon St. Vincent's Rocks, Bristol. The foliage is not distinguishable from that of Grimmia pulvinata; the capsule however is abundantly distinct, being globose instead of ovate, and having a conical instead of a rostrate operculum. Both species grow upon St. Vincent's Rocks, and are sometimes intermingled, but each retains its peculiar characteristics, so that Grimmia orbicularis cannot be considered a variety of G. pulvinata. Specimens of the former species accompanied the paper.

Read also, a paper from Mr. T. Beesley, being " Additions to the List of Plants found in the neighbourhood of Banbury, Oxfordshire, in 1842."

March 17.—J. E. Gray, Esq., F.R.S., &c., President, in the chair. Mr. David Moore, of the Royal Botanic Garden, Dublin, presented a specimen of Carex paradoxa

(*Willd.*) found by him in Ladiston Woods, Mullingar, Westmeath, Ireland, in July last. Mr. Robert Castle presented a specimen of Araucaria excelsa. Col. Jackson presented an interesting collection of foreign plants. The President presented the 1st fasciculus of Leefe's British Willows : and British plants had been received from Mr. T. B. Hall, Dr. Ayres, Mr. Henfrey and Miss Beever. Donations to the library were announced from Professor Meneghini, Mr. W. M. Chatterley and the American Philosophical Society.

Mr. Arthur Henfrey read a paper " On the British Species of Statice." * Specimens of British and foreign species in the Society's collection were exhibited.—*G.E.D.*

MICROSCOPICAL SOCIETY OF LONDON.

March 15, 1843.—George Loddiges, Esq. in the chair.

Read, a paper from the Rev. J. B. Reade, entitled " Microscopic ¦Chemistry, No. II." The paper was headed "On the existence of Ammonia in Vegetable Substances described as containing Nitrogen." After stating that very minute portions of sulphate of lime in snow may be rendered manifest by means of the microscope, and also that the almost inappreciable quantitiy of ammonia mentioned by Liebig as existing in the atmosphere would be capable of detection by the same means ; the author proceeded to show the existence of ammonia in the seeds of plants, which he stated may be rendered apparent by burning the common field bean in a spirit-lamp, until flame and smoke entirely cease. The gas given off is to be received on slips of glass moistened with pure hydrochloric acid. The salt thus obtained he describes as a salt of ammonia, which he considers to be produced by the decomposition of an ammoniacal salt previously existing in the bean, and not by the destructive distillation of an organic body in contact with the atmosphere. This presence of ammonia the author looks upon as proved in various ways : — 1. By the before-mentioned production of crystals of hydrochlorate of ammonia on slips of glass, when the gas from the bean is exposed to the vapour of volatile hydrochloric acid. 2. By the odour of this gas when received into an eight or ten ounce bottle, being clearly that of ammonia. 3. By the production of crystals of bi-tartrate of ammonia on the addition of a little tartaric acid to the hydrochlorate. 4. By the action of the supposed ammoniacal gas on testpapers, furnishing a proof of the presence of volatile alkali. And lastly, by an experiment in which he sublimed over hydrochlorate of ammonia, in an unchanged state, into a drop of distilled water. The acid with which the ammonia is combined he supposes in some instances at least to be silicic acid. In answer to an objection made to these views, that the ammonia is chemically formed by the destructive distillation of the vegetable compound in contact with the atmosphere, the author adduced what, in his opinion, must be considered both negative and positive evidence : the former being founded on the known reluctance of nitrogen to enter into combination with all other substances; the latter principally from the evolution of ammonia from bean-meal, heated in a glass tube with the mouth inserted into hydrochloric acid, thus preventing contact with the atmosphere. The author concluded by describing a method of readily obtaining as a standard of measurement, a minute quantity of hydrochlorate of ammonia, equal to about the $\frac{1}{10000}$ of a grain.

---

* This paper will most probably appear in our next number.—*Ed.*

THE PHYTOLOGIST.

No. XXIV. | MAY, MDCCCXLIII. | PRICE 1s.

ART. CXXXVI.—*Description of a Species and Variety of the Genus Statice, known to British Botanists as the Limonium Anglicum minus of Ray's Synopsis.* By ARTHUR HENFREY, Esq., A.L.S., M. Mic. Soc., Curator to the Botanical Society of London.

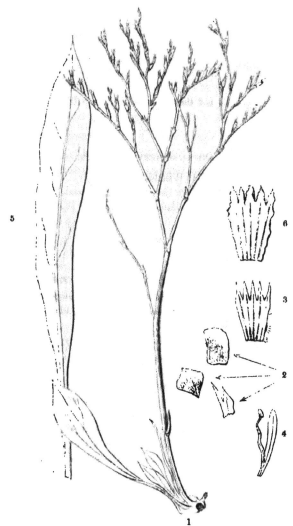

1. *Statice rariflora*, Drejer, half the natural size.   2. Scales of the involucre, magnified.   3. Calyx, ditto.   4. Petal with stamen, ditto.   5. Leaf of *Statice Limonium*, β.   6. Calyx, magnified.   The figures exhibit the relative sizes of the calyces, being taken from plants of equal size.

THE following descriptions were read before the Botanical Society of London. At that time I imagined the first of these plants to be undescribed; but by the kindness of Mr. H. C. Watson, who suggested the reference, I have been able to identify it with Drejer's description of Statice rariflora.

STATICE *rariflora*, Drejer.

Caule erecto, angulare, ramoso, spicis diffusis, elongatis; floribus paucis, distantibus, attenuatis; calyce angusto, implicato, lobis 5, acutissimis, petalis angustissimis; foliis parvis, spathulato-lanceolatis, mucronatisque (sub-triplinervis). *S. rariflora*, Drejer, Flora Excursoria Haffniensis, 121. *Limonium Anglicum minus*, Ray, Syn. 202. *S. Limonium*, γ. Sm. Fl. Brit. i. 341.

Scape 8—10 inches high, erect, somewhat angular, slightly furrowed, with few branches; spikes diffuse, attenuate, few-flowered; flowers distant, small, elongated, calyx with a long narrow tube, and five very acute lobes (occasionally with intermediate teeth), not plaited, petals narrow; leaves small, spathulato-lanceolate, obscurely triple-nerved, mucronate; footstalks bordered.

I have drawn up the above description from a specimen presented to the Botanical Society of London, by Mr. W. L. Notcutt, who gathered it at Fareham.* It is a much more delicate-looking plant than any other British Statice I have seen, and has a very peculiar starved appearance. The leaves are small and coriaceous, the only very distinct vein being the central one, which is very prominent at the back. The most characteristic point I can find is the form of the calyx, which, as will be seen on a reference to the figures (3 and 6 at page 561), differs materially from that of S. Limonium; and I have ventured, from this circumstance, strengthened by the form of the leaves and the general character of the plant, to continue Drejer's specific distinction. As the 'Flora Haffniensis' is a rather uncommon book in this country, it may not be amiss to give here the extract containing the description of this plant: for this also I am indebted to Mr. Watson.

" 349. *St. rariflora :* ramis inflorescentiæ adrectis dissitifloris, bract. oblique truncatis muticis exteriore latiore inferiorem amplectente, omn. florigeris. *Limonium anglicum minus, flor. in spicis rarius sitis*, Ray, Syn. 202. *Lim. humile*, Mill Ed. germ.

* See Mr. Notcutt's notes, Phytol. 429 and 492.—*Ed.*

"Omni parte tenuior, minus ramosa, in ramis exterior. fl. unilaterales solitarii interdum gemini, in interioribus in axi flexuosæ distinctius alterni. Fol. juniora lanceol. in mucronem attenuata, adultiora obovato-spathulata mucrone (elongato recto v. brevissime curvato) sub apice emergente. Panic. deflorata ob bract. coloratas max. membranac. fusco-variegata. Huc St. Lim. Valensienes, Fr. Mant. p. 10 (ex specim. ad J. Agardh datis) quæ sola magnitudine a nostra differt. 7—8. 2/. In littoribus.— P. 121.

Mr. Notcutt has been good enough to furnish me with the following particulars respecting the habitat, which would tend to disprove the idea of its being a starved specimen of S. Limonium. "Found in salt marshes on both sides of the town," (Fareham). "On the side between Fareham and Portchester the common form of S. Limonium grows in profusion, but I could not perceive any specimens which presented any intermediate grade between my plant and it, though they both grow near each other."

STATICE *Limonium,* Linn., var. β. *longifolia.*

Spicis elongatis, floribus minus confertis; foliis lanceolatis, angustis, acuminatisque, submucronatis. *Limonium Anglicum minus,* Ray, Syn. 202. *Statice Limonium* γ. Sm. Fl. Brit. i. 341.

Spikes elongated, flowers more distant, leaves lanceolate, narrow, sometimes linear-lanceolate (young leaves occasionally resembling the normal form in S. Limonium), generally with long bordered footstalks and a weak mucro, formed by the cartilaginous margins of the leaves.

This description is taken from a number of specimens in the herbarium of the Botanical Society of London, from two localities in Scotland; one being St. Mary's Isle (the specimens are from the Botanical Society of Edinburgh); and the other Garlieston, Wigtonshire, (the specimens collected by Professor Balfour and communicated by Mr. Watson).

The calyx of this variety exactly resembles that of Statice Limonium, *a.*; and 1 have examined a number of specimens presenting every form of leaf between elliptic and linear-lanceolate, but very few approach to spathulate; the spikes are very numerous, bearing many flowers, but are elongated, and each flower is perfectly distinct. On this account the variety appears somewhat like S. rariflora, but is distinguished by the shape and relative size of the calyx and the form of the leaves. The whole plant, when dried, has a reddish brown tint.

London, March 25, 1843. ARTHUR HENFREY.

ART. CXXXVII. — *Sketch of Botanical Rambles in the Vicinity of Bristol.* By LEO. H. GRINDON, Esq.

THE sketches of botanical rambles which have occasionally appeared in the pleasant pages of ' The Phytologist,' embolden me to offer you the following notes of two or three excursions made last summer in the vicinity of Bristol, a part of our island well known to be remarkably rich in rare and beautiful plants.

The lovely morning of July 3rd, 1842, found me approaching my native city by the Gloucester road ; and although the luxuriant and picturesque scenery upon either side afforded of itself sufficient delight and employment for the eye, I found time to remark profusion of Galium Mollugo, Pastinaca sativa, Malva sylvestris and Melilotus officinalis by the way-sides, while the hedges wore a thick mantle of our elegant southern climber, Clematis Vitalba. Here and there was a field literally purple with the bloom of the cultivated teasel, Dipsacus Fullonum.

Early in the forenoon of the following day (Monday) I rambled away towards Horfield. In the suburb of the town was abundance of Mercurialis annua (a very common weed), Diplotaxis muralis, Coronopus Ruellii, Plantago media and Convolvulus arvensis ; the hedges being formed chiefly of Rubus fruticosus and Cornus sanguinea. A little further in the country Centaurea Jacea, Agrimonia Eupatoria, Hypericum hirsutum and Hordeum pratense became plentiful. The hedges were here completely enveloped in luxuriant Tamus communis and Bryonia dioica, occasionally relieved by a mass of wild roses, (R. canina and systyla). Viburnum Lantana and Opulus formed a considerable portion of many of the fences, as we left the smoke of the town behind us, and these, together with the Cornus, were in green fruit. At Ashley was plenty of Rhamnus catharticus in the hedges, likewise in green fruit. Asplenium Ruta-muraria was common upon the old walls, and Hordeum murinum by the road-sides. On Brandon hill, in the afternoon of this day, I noticed Plantago Coronopus, *β. nanus*, in fruit, but was too late for Trifolium subterraneum, which likewise abounds there. The evening was devoted to a walk to Redland : on the walls near the green, Hieracium murorum and Linaria Cymbalaria were blooming in great luxuriance ; Ceterach officinarum was also in perfection. I visited the Lamium longiflorum* habitat, but, as anticipated, found very little remaining in flower. This is a most beau-

---

* L. album, β. Hooker, Br. Fl. ed. 4 ; L. maculatum, β. *lævigatum*, Ed. Cat. ?

tiful plant, perfectly distinct from L. album, and from all the cultivated states of L. maculatum I have ever seen. It occurs with both pink and white flowers.

The afternoon of Tuesday the 5th found me again at Redland, whither I went for the purpose of collecting Bromus erectus, which is the principal grass in many meadows both here and at Horfield : but the scythe had swept all away. To make amends, and not go home empty handed, I pursued my walk until my vasculum was filled with Acer campestre in fruit, Poa compressa, Festuca myurus and Arenaria serpyllifolia.

On Wednesday, July 6, we made an excursion to Clevedon, a lovely and highly picturesque watering place, twelve miles S.W. of Bristol. Our road lay first through Ashton, where, on the old shaded wall ascending the hill towards Failand, six species of ferns at once presented themselves;—Ceterach officinarum, Asplenium Adiantum-nigrum, A. Trichomanes, A. Ruta-muraria, Scolopendrium vulgare and Polypodium vulgare : Lastræa Filix-mas and Pteris Aquilina grow upon the bank above, only a few feet distant. It was upon this identical wall that, ten years ago, I first gathered them; and although the interest then excited by their novelty could not be renewed to me, their original beauty was still present, and filled my mind with a thousand pleasant memories.

Further up the hill, Cnicus acaulis β. *caulescens* was growing among the gravel by the road-side. Thenceforward, until we approached Tickenham, nothing remarkable presented itself, that is, nothing of *botanical* interest, a circumstance which none could regret, when compensated for by so glorious a prospect as that here enjoyed. On descending the hill to the village of Tickenham, I found Linum angustifolium in flower and fruit, Phleum arenarium, Verbena officinalis, and Salvia verbenaca in fruit; while the high banks, even down to the very carriage-way, were adorned with a profusion of flowers such as I have never beheld, except in that one favoured spot. Papaver Rhœas, Cichorium Intybus, Galium verum, Malva moschata, M. sylvestris, wild roses, and a multitude of others no less showy, blended their bright hues beneath the unclouded sun more beautifully than pen can describe. Thence, all the way to Clevedon, the old walls and dry hedge-banks were clothed with an infinite quantity of Ceterach officinarum and Cotyledon Umbilicus; the handsome branched variety of the latter being equally abundant with the ordinary simplestemmed state of the plant. Here and there in the hedges Campanula Trachelium showed itself; and occasionally, on walls, Sedum acre, S.

dasyphyllum, S. reflexum and Valeriana rubra.   Near the Bristol Hotel was Torilis nodosa, very large.

We reached the shore of the Bristol Channel about mid-day.   For a considerable distance the coast is here very precipitous and romantic, being formed of black, uncouth masses of rock, which appear to have fallen into their present position through the undermining influence of the sea.   Inland the hills are steep, and, on their seaward slopes, comparatively barren, affording little besides pasturage for sheep.   Filago germanica was, however, abundant, and in a little natural shrubbery of furze and brambles, I found Senecio sylvaticus and Calamagrostis Epigejos.   The path winds along the extreme edge of the cliffs, forming, as it were, the line of demarcation between earth and the oceanic territory.   Extending from the path down to high-water mark, except where interrupted by the rocks, is a steep grassy bank, abounding with beautiful flowers.   Here I gathered Iris fœtidissima, Orchis pyramidalis in profusion, Orobanche minor, Euphorbia amygdaloides in fruit, Chlora perfoliata in profusion, Silene maritima, Festuca elatior, Scolopendrium vulgare, Daucus Carota, and many others.   The vegetation by the sides of the path consists, in a great measure, of rather stunted Anthyllis Vulneraria.   Near Walton, the rocks are less rugged, and allow of walking upon and between them. Here it was that in a low curious cave I met with Asplenium marinum, being the first time I had ever seen living plants.   Some of the finest specimens were growing many yards from the entrance of the cave, where little light could enter, and so shut in by the sloping roof, that I had to creep till I was almost prostrate before they were accessible. In the crevices of the rocks was abundance of Thrincia hirta, and of a curious and striking variety of Plantago Coronopus, having succulent, densely hairy leaves, and exceedingly numerous flower-scapes.   Further on, towards Portishead, Hypericum Androsæmum was in bloom, still retaining the ripe fruit of last year.   A few diminutive specimens of Samolus Valerandi were visible in a little cave ; and by the side of a fresh-water spring, which bubbles forth from amid the bosom of the rocks, Schœnus nigricans, brown and muddy from the tide washing over it.   After collecting these, together with some shells and specimens of the different Fuci, with which the rocks are thickly tapestried, we returned to Bristol ; and though the *botanical* value of the day's gatherings was not of the very first order, the delightful influences and associations under which they were collected have given them a deep and unfading interest.

During the 7th and 8th of July torrents of rain prevented all botanizing. Saturday the 9th was little better, but we then lost all patience, defied the weather, and sallied forth for the purpose of exploring Leigh woods. On the banks of the Avon, near Rownham, Trifolium fragiferum was in flower and fruit abundantly. We likewise met with Cochlearia anglica, Plantago maritima, Scirpus maritimus, Apium graveolens, &c. It was desperately wet in the woods, and after a succession of shower-baths from above, and drenchings from around and below, we were glad to retreat, in possession, however, of luxuriantly beautiful Cystopteris fragilis, Polypodium vulgare (eighteen inches long), Asplenium Trichomanes, Quercus sessiliflora, Pyrus Aria in green fruit, Tilia parvifolia, Acer campestre in green fruit, &c.

Tuesday the 12th was brilliantly fine, and being the last day of my stay at Bristol, it was spent in part upon St. Vincent's Rocks; but the fatigue of a boisterous voyage to South Wales the previous day had quite unnerved me, and I was only tempted to summon up my remaining strength by the prospect of obtaining Orobanche barbata, Veronica hybrida, Rubia peregrina, Centaurea Scabiosa, Petroselinum sativum, Bromus diandrus, &c.; with regard to all which, and many others of no less interest, I perfectly succeeded. And thus ended my week's botanizing at Bristol. Leo. H. Grindon.

Manchester, March 13, 1843.

Art. CXXXVIII.—*Plants observed in the neighbourhood of Ludlow, Shropshire.* By Frederick Westcott, Esq., A.L.S., &c.

Spring St., Edgbaston, December 3, 1842.

Sir,

I send you an enumeration of plants which I gathered or observed during a short stay at Ludlow in October last. I regret that I was too late for the grasses, Carices and Orchideæ, and had no means of ascertaining the Fungi, which appear to be numerous: all these deficiencies I hope to supply next summer, when it is probable I may revisit the neighbourhood.

I send the list, not because it will be found to contain any very remarkable plants, but in the hope that persons in the neighbourhood may be induced to pay more attention to the Botany of their district; for I have no doubt that many interesting discoveries may be made there, especially among the mosses, the lichens and the Fungi. The Clee Hills, more particularly, would be found a rich locality.

Hoping that some one on the spot will follow up this subject in good earnest, and communicate the result to the interesting pages of your journal, I am, Sir,

Yours very truly,

FRED. WESTCOTT.

To the Editor of 'The Phytologist.'

Ranunculus acris, bulbosus, and repens, banks of the castle walk
—— fluviatilis, abundant in the Teme
—— hederaceus
Caltha palustris
Trollius europæus
Aquilegia vulgaris, ruins of Richard's castle
Berberis vulgaris, Ludford
Diplotaxis tenuifolia and Cheiranthus Cheiri, rocks about the castle and banks of the castle-walk
Cardamine hirsuta
Hesperis inodora, ruins within the castle
Sinapis nigra, banks of the castle walk
Erysimum Alliaria
Montia fontana
Cerastium vulgatum and arvense
Arenaria serpyllifolia, walls about the cas.
Sagina procumbens
Stellaria media, uliginosa and holostea
Lychnis dioica and Flos-cuculi
Dianthus plumarius, on the walls of the castle, on the right hand side on entering the door, by the keep.
Malva sylvestris and moschata
Geranium robertianum, molle, dissectum and lucidum. The last is abundant on the walls of the castle, where its shining leaves have a very pleasing appearance.
Epilobium montanum and palustre
Circæa Lutetiana, banks of the Teme
Dipsacus sylvestris, hedge near Ludlow
Viburnum Lantana
Hedera Helix and Sambucus nigra, about the castle
Lonicera Periclymenum, Whitcliff coppice
Cornus sanguinea
Anthriscus sylvestris
Bunium flexuosum

Pimpinella Saxifraga
Chærophyllum temulum
Conium maculatum
Ægopodium Podagraria
Plantago major, media and lanceolata
Acer campestre
Hypericum pulchrum, humifusum, and hirsutum, banks of castle walk
—— perforatum, β. angustifolium, walls of Ludlow castle; rocks of Whitcliffe; hedge-bank near the Angel bank, Clee-hills, (Phytol. 427 and 461).
Geum urbanum
Agrimonia Eupatoria
Potentilla Fragariastrum, reptans and anserina
Prunus spinosa
Fragaria vesca, ruins of the castle and Whitcliffe coppice
Rosa tomentosa, canina, and canina β. sarmentacea
Rubus fruticosus, cæsius, Idæus, rhamnifolius & carpinifolius, Whitcliffe coppice and hedges
Ribes Grossularia
Viola canina, odorata, palustris and lutea. The last grows in great abundance in moist places on the top of the Clee hills
Oxalis Acetosella
Galium cruciatum, saxatile & uliginosum
Linum catharticum, Whitcliffe
Cotyledon Umbilicus, Ludford
Sedum reflexum and dasyphyllum, rocks of castle walk
—— acre, on the walls
—— rupestre, Clee hills
—— Telephium, var. alpinum. In habit this plant is nearly prostrate, slen-

der, and weak in all its parts. The leaves are also frequently opposite, thinner, and of a reddish green colour; it is also destitute of the leafy bracteal appendages which are present on Sedum Telephium. It was gathered by Mr. Cameron and myself on the Clee hills, in 1839, when we considered that its different appearance arose from growing at an elevation of from 14—15,000 feet above the level of the sea; it has however been cultivated in the Birmingham Botanic Garden, for two years, next to the true S. Telephium, without suffering the slightest change, and the difference between the two plants remains distinctly marked.

Saxifraga tridactylites and granulata, on the banks of the castle walk

—— hypnoides, on the stones of that part of the Clee hills called the Hoar edge, abundant

Chrysosplenium oppositifolium, abundant among the stones under the dripping rocks of Whitcliffe

Trifolium arvense, pratense and repens, banks of the castle walk

Medicago lupulina, walls of the Castle

Ulex europæus, Whitcliffe

Lotus corniculatus and major, ditto

Spartium Scoparium, Whitcliffe coppice

Vicia sylvatica, ditto

Euonymus europæus

Achillæa Millefolium

Apargia autumnalis and hirta

Sonchus oleraceus

Cirsium lanceolatum

Cnicus palustris

Leontodon Taraxacum

Tanacetum vulgare

Hieracium Pilosella and murorum

Carlina vulgaris

Pyrethrum Parthenium

Senecio Jacobæa

Petasites vulgaris

Conyza squarrosa

Eupatorium Cannabinum

Lapsana communis

Prenanthes muralis

Tussilago Farfara

Lappa glabra

Artemisia vulgaris

Erigeron acris

Campanula rotundifolia

Jasione montana, bank of the castle walk and Whitcliffe

Vaccinium Myrtillus, Whitcliffe coppice

Pyrola media, ditto

Glechoma hederacea

Ballota fœtida

Thymus Serpyllum

Galeopsis Tetrahit

Teucrium Scorodonia

Stachys sylvatica

—— Betonica, Whitcliffe coppice

Prunella vulgaris

Origanum vulgare, castle walk

Melissa Calamintha, banks of the castle walk, near Mortimer's tower

Primula veris and vulgaris

Lysimachia nemorum

Lathræa squamaria, near Steventon

Scrophularia nodosa and aquatica

Bartsia Odontites

Digitalis purpurea, Whitcliffe

Verbascum Thapsus, ditto

Veronica polita, Chamædrys and officinalis, banks of the castle walk

Linaria vulgaris, ditto

—— Cymbalaria, walls about the town

Melampyrum pratense, Whitcliffe coppice

Verbena officinalis, banks of castle walk

Myosotis palustris

Lithospermum officinale, abundant on the banks of the castle walk

Rumex Acetosa and sanguineus, var. with green veins, ditto

Polygonum Bistorta, Persicaria and Hydropiper

Parietaria officinalis, walls about Ludlow

Urtica dioica

Mercurialis perennis

Euphorbia amygdaloides and helioscopia, abundant in Whitcliffe coppice

Typha latifolia, Oakley park pool
parganium ramosum, banks of the Teme
iris Pseudacorus, ditto, abundant
Epipactis latifolia, Whitcliffe, in great
plenty, and very fine, some speci-
mens being from 2 to 3 feet high
Spiranthes autumnalis, fields adjoining
Ludlow
Listera ovata
Juncus conglomeratus, Whitcliffe coppice
Luzula pilosa, ditto
Butomus umbellatus, in the river between
the new bridge and the Mill-street
weir
Aira flexuosa
Milium effusum, Whitcliffe coppice
Holcus mollis, common
Bromus sterilis, walls about the castle,
and castle ditch
Dicranum scoparium, Whitcliffe coppice
—— heteromallum & bryoides, moist bks.
Bryum ventricosum, rks. under Whitcliffe
—— capillare, rocks of castle walk
—— palustre
—— hornum, Whitcliffe coppice
—— pyriforme, walls about castle-walk
Hypnum stellatum, splendens, cuspidatum
and purum, Whitcliffe
—— myosuroides, complanatum and denticulatum, Whitcliffe coppice
—— molluscum, on the stones among the
ruins of the castle
—— triquetrum, Whitcliffe

Hypnum prælongum & confertum, banks
Bartramia pomiformis
—— fontana. This moss grows in great
abundance among the wet rocks
under Whitcliffe, where I found
it in a beautiful state of fructifi-
cation, which to me is of rare oc-
currence
Gymnostomum microstomum
Grimmia pulvinata, Titterstone
Orthotrichum crispum, upon the trees in
Whitcliffe coppice, abundant
Polytrichum alpinum, undulatum and
commune, Whitcliffe
Sphagnum obtusifolium and acutifolium,
wet places near the river
Marchantia polymorpha
Jungermannia asplenioides
Borrera ciliaris, trees in Whitcliffe coppice
Cetraria glauca
Parmelia parietina
Peltidea canina
Scyphophorus pyxidatus
Cladonia rangiferina   [the stones
Sphærophoron coralloides, Titterstone, on
Lecanora Hæmatomma, Titterstone
Enteromorpha intestinalis, abundant in
the river, below the bridge
Protonema Orthotrichi, on trees in Whit-
cliffe coppice, with Orthotrichum
crispum
Boletus luteus
Agaricus pratensis, campestris & procerus

Art. CXXXIX.—*Notice of 'A Visit to the Australian Colonies.* By
James Backhouse.' London: Hamilton, Adams, & Co. 1843.

(Continued from p. 553)

During his stay at the Hampshire hills, J. Backhouse made frequent excursions in the neighbouring country: in one of these he
noticed Telopea truncata, or Van Diemen's Land tulip-tree, a laurel-
like shrub bearing heads, four inches across, of brilliant, scarlet, wiry
flowers; an upright Phebalium, with silvery leaves and small white
flowers; and a white-flowered sorrel — Oxalis lactea; a Telopea, the

flowers of which abound in honey, which our author found it easy to extract by means of the slender tubular stems of grass ; and a shrubby Aster, with toothed leaves, so profusely loaded with pure white blossoms as to bend gracefully in all directions.   We now quote a passage that will give some idea of the denseness of the forests in this island.

"On an old road called the Lopham-road, a few miles from the Bay, we measured some stringy-bark trees, taking their circumference at about 5 feet from the ground. One of these, which was rather hollow at the bottom, and broken at the top, was 49 feet round; another that was solid, and supposed to be 200 feet high, was 41 feet round ; and a third, supposed to be 250 feet high, was 55⅓ feet round.   As this tree spread much at the base, it would be nearly 70 feet in circumference at the surface of the ground.   My companions spoke to each other, when at the opposite side of this tree to myself, and their voices sounded so distant that I concluded they had inadvertently left me, to see some other object, and immediately called to them.   They, in answer, remarked the distant sound of my voice, and asked if I were behind the tree ! When the road through this forest was forming, a man, who had only about 200 yards to go, from one company of the work-people to another, lost himself: he called, and was repeatedly answered; but getting further astray, his voice became more indistinct, till it ceased to be heard, and he perished.   The largest trees do not always carry up their width in proportion to their height, but many that are mere spars are 200 feet high.

" The following measurement and enumeration of trees growing on two separate acres of ground in the Emu Bay forest, made by the late Henry Hellyer, the Surveyor to the V. D. Land Company, may give some idea of its density.

"FIRST ACRE.

500	Trees under			12 inches in girth.
992	do.	...	1 to 2 feet	do.
716	do.	...	2 to 3 do.	do.
56	do.	...	3 to 6 do.	do.
20	do.	...	6 to 12 do.	do.
12	do.	...	12 to 21 do.	do.
4	do.	...	30 do.	do.
84	Tree Ferns.			

2,384 Total.

SECOND ACRE.

704	Trees under			12 inches in girth.
880	do.	...	1 to 2 feet	do.
148	do.	...	2 to 3 do.	do.
56	do.	...	3 to 6 do.	do.
32	do.	...	6 to 12 do.	do.
28	do.	...	12 to 21 do.	do.
8	do.	...	21 to 30 do.	do.
8	do.	...	30 feet and upwards.	
112	Tree Ferns.			

1,976 Total."—p. 115.

The measurement of individual trees seems really enormous. We have a prostrate tree measuring 200 feet to the first branch; a second cut into rails each 180 feet long; a third *so large that it could not be cut into lengths for splitting*, and a shed had been erected against it, the tree serving for a back. The following dimensions are given of ten standing trees, which occurred within half a mile: their circumference was taken at four feet from the ground.

No. 1,—45 feet; No. 2,—37½; No. 3,—35; No. 4,—38; No. 5,—28; No. 6,—30; No. 7,—32; No. 8,—55; No. 9,—40½; No. 10,—48.

On the banks of the Emu river was a laurel-like shrub of great beauty, with clusters of white blossoms half an inch across (Anoptera glandulosus). In the same vicinity occur three edible plants; the first a fungus which grows on the myrtle, and is known in the colony by the name of "punk:" the second is also a fungus, produced in clusters from swollen portions of the branches of the same shrub, and varying in size from that of a nut to that of a walnut; its taste is like cold cow-heel: the third is "Gastrodium sesamoides, a plant of the orchis tribe, which is brown, leafless, and 1½ foot high, with dingy, whitish, tubular flowers. It grows amongst decaying vegetable matter, and has a root like a series of kidney potatoes, terminating in a branched thick mass of coral-like fibres. It is eaten by the Aborigines, and is sometimes called Native Potato, but the tubers are watery and insipid."

In returning over the island to Hobart Town, there appear to have been but few plants that attracted much notice. In the vicinity of this place a species of Conospermum, with narrow strap-shaped leaves and small flowers, was noticed. In October, 1833, the travellers ascended Mount Wellington.

" At the base, sandstone and limestone form low hills; further up, compact argillaceous rock rises into higher hills, which abound in marine fossils. The height of the mountain is 4000 feet. Near the top, basalt shows itself in some places, in columnar cliffs. The trees, for two-thirds of its height, are stringy-bark, white and blue gum, peppermint, &c. A species of Eucalyptus, unknown in the lower part of the forest, is frequent at an elevation of 3000 feet. Another is found on the top of the mountain. The different species of Eucalyptus are very common, and form at least seven-eighths of the vast forests of Tasmania. In the middle region of the mountain, the climate and soil are humid. The Tasmanian myrtle—Fagus Cunninghamii, here forms trees of moderate size; the Australian pepper-tree, — Tasmania fragrans, is frequent; the broad-leaved grass-tree—Richea Dracophylla, forms a striking object; it is very abundant, and on an average, from ten to fifteen feet high; it is much branched, and has broad, grassy foliage. The branches are terminated by spike-like panicles of white flowers, intermingled with broad, bracteal leaves, tinged with pink. Culcitium salicifolium, Hakea lissosperma, Telopea truncata, Corrœa ferruginea, Gaultheria hispida,

Prostanthera lasianthos, Friesia peduncularis, and many other shrubs, are met with in the middle region of the mountain. For a considerable part of the way up, we availed ourselves of a path that is nearly obliterated, which was used by the workmen, when laying a watercourse from the breast of the mountain, for the purpose of supplying Hobart Town with water. This path led through a forest of tree-ferns, surmounted by myrtle, &c. Nearer the top, we had to pass a large tract of tumbled basalt. The upper parts of many of the stones were split off, probably by the alternations of frost and heat. A few patches of snow were still remaining.

" The top of the mountain is rather hollow, sloping toward Birches Bay, in the direction of which, a stream of excellent water flows. The ground is swampy, with rocks and stony hills. Astelia alpina, Gleichenia alpina, Drosera arcturi, several remarkable shrubby Asters, a prostrate species of Leptospermum, Exocarpos humifusus, a dense bushy Richea, and several mountain shrubs, of the Epacris tribe, are scattered in the swamps, and among the rocks."—p. 159.

In the narrative of a second visit to Flinders Island, in December, we find a more detailed account of the grass-trees than any that has previously been given. Their stems are five to seven feet in height, and as many in circumference; the crest or summit consists of a number of grass-like leaves, three or four feet in length, and from the centre of these rises a single erect flower-spike, varying from five to ten feet in height: this is thickly clothed with hard scales, and small, white, star-like flowers, except for about eighteen inches, at the base, which is bare. The trunks of these grass-trees are charred with continual burnings of the scrub; and abundance of red resin, capable of being used in making sealing-wax and French polish, exudes from them: this resin fills the place left by the decay of the flower-stalk, and is abundant at the base of the stem, protecting this part from excess of moisture. The head of a grass-tree that has not thrown out a flower-stem is pleasant eating, and has a nutty flavour. Accompanying the description is a plate, representing these extraordinary plants.

In May, 1834, our travellers being at Hobart Town, visited a small settlement on the Derwent, called Brown's River, and noticed in their walk Sprengelia incarnata, a heath-like shrub, which was in flower in some marshy ground by the way: also Plagianthus discolor, one of the Malvaceæ, bearing clusters of white blossoms: the species of this genus are called Currijong, in common with others whose bark is sufficiently tenacious for making cordage.

NORFOLK ISLAND.—James Backhouse landed on Norfolk Island in the beginning of March, 1835: he thus describes it.

" Norfolk Island is about seven miles long and four broad. A small portion of its southern side is limestone; to the east of this there is a still smaller portion, of coarse, siliceous sandstone. The remainder of the island is basaltic, and rises into hills, covered with grass and forest. The highest hill is Mount Pitt, which is on the north side of the island, and about 1,200 feet above the level of the sea. The upper portions of the valleys, and the higher parts of the hills, are covered with wood. The Norfolk Island pine, Altingia excelsa, towers a hundred feet above the rest of the forest; it also grows in clumps, and singly, on the grassy parts of the island, to the very verge, where its roots are washed by the sea, in high tides. In figure, this tree resembles the Norway spruce, but the tiers of its branches are more distant. Its appearance is remarkably different, in its native soil, from what it is in the fine collection of trees at Kew; where it nevertheless exhibits many of its striking and beautiful features.— Where the wood of Norfolk Island merges into open grassy valley, a remarkable tree-fern, Alsophila excelsa, exhibits its rich crests among the surrounding verdure. The fronds are from seven to twelve feet long; they resemble those of Aspidium Filix-mas, and are produced in such a quantity, as to make this noble fern excel the princely palm-tree in beauty. It usually has its root near the course of some rain-stream, but as its trunk rises to fifty feet in height, and its top does not affect the shade, like many of its congeners, it forms a striking object in the landscape.

" Much of the land was formerly cultivated, but this is now overrun with the apple-fruited guava, and the lemon, which were introduced many years ago, when the island was settled, with a view to its becoming a granary to New South Wales. Grape vines, figs, and some other fruits, have also become naturalized. In the garden at Orange Vale, coffee, bananas, guavas, grapes, figs, olives, pomegranates, strawberries, loquats and melons, are cultivated successfully. Apples are also grown here, but they are poor and will not keep."—p. 251.

" One of the remarkable vegetable productions of this island is Freycinetia Baueriana, or the N. I. grass-tree. It belongs to the tribe of Pandaneæ, or screw pines. Its stem is marked by rings, where the old leaves have fallen off, and is an inch and a half in diameter; it lies on the ground, or climbs like ivy, or winds round the trunks of trees. The branches are crowned with crests of broad, sedge-like leaves. From the centre of these, arise clusters of three or four oblong, red, pulpy fruit, four inches in length, and as much in circumference. When the plant is in flower, the centre leaves are scarlet, giving a splendid appearance to the plant, which sometimes is seen twining round the trunk of the princely tree-fern. The New Zealand flax, Phorminm tenax, a large, handsome plant, with sedgy leaves, covers the steep declivities of many parts of this island, particularly at the tops of the cliffs of the coast. It is suffered to grow to waste, except a little that is converted into small nets and cordage, by the prisoners, for their own use. Two New Zealanders were once introduced, to teach the prisoners to prepare it; but their process was so tedious, that the scheme was abandoned."—p. 256.

On the 16th our traveller rode with Major Anderson to Anson's Bay, on the northern side of the island.

" The road was chiefly through thick forest, overrun with luxuriant climbers. Among them was a Wistaria, with pea-flowers, of purple and green, and leaves something like those of the Ash. It hangs in festoons of twenty or thirty feet, from the limbs of the trees that support it. One of the most beautiful climbers of the island is Ipomœa pendula, which has handsome fingered foliage, and flowers like those of the major convolvulus, but of a rosy pink, with a darker tube. The remains of two pines, which were noted for their magnitude, and were blown down in a storm, were lying by the side of the road. These were called ' The Sisters ; ' they were nearly two hundred feet in height."—p. 258.

The author informs us he frequently took a walk before breakfast, and explored the thickly wooded hills and valleys. On the borders of the woods there was a great variety of beautiful shrubs ; among these is —

" The slender jasmine, Jasminum gracile, known in England as a delicate green-house plant. Here it climbs over the bushes, or with twisted stems, as thick as a man's wrist, reaches the branches of lofty trees, at fifty feet from the ground, and climbs in their heads. In these cases, it has probably grown up with the trees, the lower branch-es of which have progressively died away, and left the wreathed stems of the jasmine like ropes, hanging from the upper boughs. Scattered on the grassy hills is Hibiscus or Lagunea Patersonii, which forms a spreading tree of forty feet in height : it is called here the white oak : its leaves are of a whitish green, and its flowers pink, fading to white, the size of a wine-glass. It is perhaps the largest plant known to exist, belong-ing to the mallow tribe.* In a thick wood, I met with it eighty feet high, and with a trunk sixteen and a half feet round."—p. 258.

On the 28th of March James Backhouse, accompanied by the agri-cultural superintendant, walked to a stock-station, called Cheeses Gully, on the north side of the island. He here observed two re-markable arches of rock, one of them connecting the columnar basalt of the cliff with a little inaccessible islet, inhabited by gannets and tropic birds. He noticed many of the old timber roads, grown up with the Cape gooseberry, Physalis edulis,—

" Which produces abundance of pleasant, small, round fruit, in a bladder-like ca-lyx. This is eaten by the prisoners, who also collect and cook the berries of the black nightshade, Solanum nigrum. These berries are accounted virulently poisonous in England, but their character may possibly be changed by the warmer climate of Nor-folk Island.

" In the woody gullies, the Norfolk-Island cabbage-tree, Areca sapida, abounds. It is a handsome palm, with a trunk about twenty feet in height, and from one and a half to two feet in circumference, green and smooth, with annular scars, left by the fallen leaves. The leaves or fronds form a princely crest, at the top of this elegant column ; they are pectinate, or formed like a feather, and are sometimes nineteen feet in length; they vary from nine to fifteen in number. The apex of the trunk is en-closed in the sheathing bases of the leaf-stalks, along with the flower-buds, and young leaves. When the leaves fall they discover double compressed sheaths, pointed at the upper extremity, which split open indiscriminately, on the upper or under side, and fall off, leaving a branched spadix, or flower-stem, which is the colour of ivory, and at-tached by a broad base to the trunk. The flowers are produced upon this spadix : they are very small, and are succeeded by round seeds, red externally, but white, and

* Except the Baobab, Phytol. 433.—Ed.

as hard as horn, internally. As the seeds advance towards maturity, the spadix becomes green. The young, unfolded leaves of this cabbage-tree, rise perpendicularly,
in the centre of the crest. In this state, they are used for making brooms ; those still
unprotruded, and remaining enclosed within the sheaths of the older leaves, form a
white mass, as thick as a man's arm ; they are eaten raw, boiled or pickled. In a raw
state, they taste like a nut, and boiled, they resemble artichoke-bottoms. The seeds
furnish food for the wood-quest, a large species of pigeon, which has a bronzed head
and breast, and is white underneath, and principally slate-coloured on the back and
wings. This bird is so unconscious of danger, as to sit till taken by a noose at the end
of a stick ; when one is shot, another will sometimes remain on the same bough, till
itself also is fired at. We measured a Norfolk Island pine, twenty-three feet, and another twenty-seven feet, in circumference. Some of them are nearly two hundred feet
high. The timber is not of good quality, but is used in building ; it soon perishes
when exposed to the weather. This is said to be the case with all the other kinds of
wood on the island. Norfolk Island iron-wood, Olea apetala, is the only other sort reputed to be worth using. No fences of wood are expected to stand above three years.
Vegetation is rapid in this fine climate, but decay is rapid also. There are very few
dead logs lying in the bush."—p. 264.

On the 2nd of April our traveller explored a gully on the north side
of the island, and found it " shaded by forest and abounding in ferns
and young palms ; " he also observed four orchideous epiphytes on the
upper branches of the trees. Peperomias and ferns were plentiful ;
the former " are spreading green plants allied to pepper ; " they were
growing on moist rocks, " on the dark sides of which Trichomanes
Bauerianum, a membranaceous fern of great beauty, forms tufts exceeding a foot in height." On the rocks of the south coast he found
Asplenium difforme, a fern resembling Asplenium marinum : a little
way inland the leaves of this fern are more divided, and it varies
through every intervening form, until, in the woods in the interior of
the island, the leaves are separated into such narrow segments, that
the fructification becomes marginal, and in this state the plant is called Cœnopteris odontites. On the 4th, after visiting a gang of invalids
employed in stone-breaking, J. Backhouse explored a place called the
Cascade, fringed in places with copses and straggling tree-ferns. A
little brook winds from the woody hills to an open valley, formerly inhabited by settlers, whose chimneys were still standing, and whose
orchards, now run wild, have spread grape-vines, lemons, figs and
guavas all around.

"Their sugar-canes have also become naturalized, and border the streamlet thickly, till it falls over a basaltic rock, about twenty feet high, decorated with ferns, and a
variety of other plants. Here the brook is again narrowed by woody hills, and margined by luxuriant plants of the broad, sedgy-leafed New Zealand flax, and water-
cress, till it emerges on an open, flat, basaltic promontory, from the very point of which
it falls, about twenty feet, to the sea beach, where it is lost among the large, rounded,

tumbled stones. Among the sugar-cane and scrub at this point, a beautiful convolvulus-like plant, Ipomœa cataractæ, is entwined, and exhibits its large, purple flowers, shot with red. It was named from this place, by Bauer, a celebrated botanist, who accompanied one of the earliest navigators of these seas, and whose 'Flora of Norfolk Island' has lately been published by a person named Endlicher.

"Ipomœa carinata, a large plant of the Convolvulus tribe, having white flowers, with long tubes, that open at night, climbs among the trees in the borders of the woods. Among the bushes there are two pretty species of passion-flower, Disemma adiantifolia and D. Baueriana, with copper-coloured blossoms."—p. 268.

(To be continued).

ART. CXL.—*Rarer Plants found near Castle-Howard, Yorkshire.* By H. IBBOTSON, Esq.

Thalictrum flavum. Banks of the Derwent and Areyholme beck.
Ranunculus Lingua. Bogs near Kirkham.
Trollius europæus. Terrington North Carr and in Holly-hill bogs.
Helleborus viridis. About Mowthorpe and Conesthorpe.
Aquilegia vulgaris. In most of the Castle-Howard Woods.
Actæa spicata. Kitscrew wood.
Berberis vulgaris. Hedges near Slingsby.
Corydalis claviculata. Bulmer Hagg.
Fumaria capreolata. In the park.
Nasturtium terrestre. Ponds near Bulmer and Terrington.
——— *amphibium.* Banks of the Derwent.
Arabis hirsuta. Fields at Baxtonholme, and on rocks in Mowthorpe dale.
Cardamine amara. Banks of the Derwent.
Thlaspi arvense. Mowthorpe fields.
Erysimum cheiranthoides. Fields near Terrington.
Viola palustris. Terrington Carr and Holly-hill bogs.
——— *hirta.* Gilla Leys wood.
Drosera longifolia. Slingsby moor.
——— *anglica.* Terrington carr, very ra.
Dianthus deltoides. In the Coom near Terrington.
Silene Otites. East Moors, now probably extinct.

Silene noctiflora. Cornfields, not unfreq.
Sagina apetala. Garden walks, Ganthorpe
Stellaria nemorum. Oxcar's wood.
——— *glauca.* Boggy ground near the Derwent.
Cerastium arvense. About Terrington and Conesthorpe.
Malva moschata. Fields at Mowthorpe.
Hypericum elodes. Slingsby moor.
——— *montanum.* Gilla Leys wood.
Geranium pyrenaicum. East moors, and near Conesthorpe.
Euonymus europæus. Kitscrew wood.
Rhamnus Frangula. Broat's plantation, near Ganthorpe.
Astragalus Hypoglottis. Welburn moor.
Vicia sylvatica. Frequent in the woods.
Cerasus Padus. Hedges at Ganthorpe.
Spiræa Filipendula. Welburn Moor.
Rubus suberectus. Potichar bank wood.
——— *corylifolius.* Ray-wood; Cross hill, Ganthorpe.
——— *Koëhleri.* Raywood; Ganthorpe Broats plantation.
——— ——— *β. fusco-ater.* Raywood.
——— *rudis.* Raywood.
——— *rhamnifolius.* Cross hill, Ganthorpe
Sanguisorba officinalis. Meadows, com.
Rosa spinosissima. Common.
——— *villosa.* Banks of the Derwent.
——— *canina β. sarmentacea.* Hedges near Baxtonholme &c.

3 E

Rosa arvensis. Ganthorpe moor.

Myriophyllum verticillatum. Pools and ditches near Crambeck.

Callitriche autumnalis. Terrington Carr.

Hippuris vulgaris. In the Derwent at Crambeck.

Peplis Portula. Ganthorpe Broats.

Bryonia dioica. Hedges at Welburn.

Sedum Telephium. Near Welburn.

—— *dasyphyllum.* Walls at Terrington

Saxifraga granulata. Pasture at Howthorpe.

Chrysosplenium alternifolium. Boggy ground near Dalby.

Sium latifolium. Bogs near Crambeck.

—— *angustifolium.* Terrington Carr.

Œnanthe Phellandrium. By the Derwent at Crambeck.

Torilis nodosa. Fields near Terrington and Conesthorpe.

Myrrhis odorata. Plentiful at Crambeck, Baxtonholme and Mowthorpe.

Fedia dentata. Conesthorpe fields.

Inula Helenium. Mowthorpe dale.

Bidens cernua. Ponds in the park.

Chrysanthemum segetum. Fields at Baxtonholme.

Antennaria dioica. Slingsby moor.

Cirsium eriophorum. Roughills plantation, near Ganthorpe.

Carlina vulgaris. Ganthorpe moor.

Serratula tinctoria. Fields near Ganthorpe and in Head Hagg wood.

Picris hieracioides. Mowthorpe dale.

Jasione montana. Terrington Broats.

Campanula glomerata. Meadows, freqnt.

Specularia hybrida. Fields near Hovingham and Conesthorpe.

Oxycoccos palustris. Terrington Carr.

Pyrola minor. In several of the Castle-Howard woods.

Villarsia nymphæoides. Lakes in the park.

Gentiana Pneumonanthe. Terrington Carr

Lithospermum officinale. Oxcar's wood and hedges near Welburn.

Verbascum Thapsus. Mowthorpe dale.

Linaria minor. Bulmer fields.

Rhinanthus major. Cornfields nr. Welburn

Veronica polita. Conesthorpe fields.

Lycopus europæus. By the Derwent at Crambeck.

Melissa Acinos. Fields at Baxtonholme.

—— *Calamintha.* Roadside between Hovingham and Slingsby.

Nepeta Cataria. Hedges near Fryton.

Lamium amplexicaule. Fields near Terrington.

Galeopsis Ladanum. Flds. nr. Hovingham

Verbena officinalis. About Ganthorpe and Welburn.

Utricularia vulgaris. Ditches nr. Crambeck

—— *minor.* Terrington Carr.

Hottonia palustris. Ditches nr. Crambeck

Lysimachia vulgaris. Banks of the Derwent

Polygonum Bistorta. Meadows near Ganthorpe

Daphne Laureola. Gatherley mills farm.

Orchis ustulata. St. Ann's meadow.

Gymnadenia conopsea. Meadows, not unfr.

Habenaria viridis. Welburn moor; Ganthorpe town's pasture.

—— *bifolia.* Cum Hagg wood, and other places.

Ophrys muscifera. Oxcar's wood.

Spiranthes autumnalis. Ganthorpe moor.

Listera cordata. Ganthorpe Broats plantation.

—— *Nidus-avis.* Cum Hagg wood, Thortle wood, &c.

Epipactis latifolia. Thortle wood.

—— *palustris.* Terrington N. Carr.

Paris quadrifolia. Woods, very common.

Convallaria majalis. Cum Hagg wood, Slingsby wood, &c.

Gagea lutea. Oxcar's wood.

Ornithogalum umbellatum. Terrington Broats.

Colchicum autumnale. St. Ann's meadow.

Sagittaria Sagittifolia. In the Derwent.

Butomus umbellatus. Ditto.

Lemna trisulca. Ditches near Crambeck.

—— *polyrhiza.* Castle-Howard ponds.

Isolepis fluitans. Terrington Carr.

Blysmus compressus. Welburn moor, rare.

Scirpus sylvaticus. By the Derwent near Crambeck.

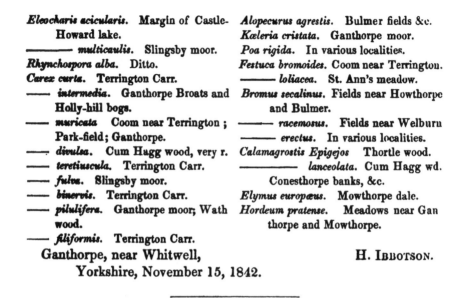

Eleocharis acicularis. Margin of Castle-Howard lake.

———— *multicaulis.* Slingsby moor.

Rhynchospora alba. Ditto.

Carex curta. Terrington Carr.

—— *intermedia.* Ganthorpe Broats and Holly-hill bogs.

—— *muricata* Coom near Terrington ; Park-field; Ganthorpe.

——, *divulsa.* Cum Hagg wood, very r.

—— *teretiuscula.* Terrington Carr.

—— *fulva.* Slingsby moor.

—— *binervis.* Terrington Carr.

—— *pilulifera.* Ganthorpe moor; Wath wood.

—— *filiformis.* Terrington Carr.

Alopecurus agrestis. Bulmer fields &c.

Kœleria cristata. Ganthorpe moor.

Poa rigida. In various localities.

Festuca bromoides. Coom near Terrington.

———— *loliacea.* St. Ann's meadow.

Bromus secalinus. Fields near Howthorpe and Bulmer.

———— *racemosus.* Fields near Welburn

———— *erectus.* In various localities.

Calamagrostis Epigejos Thortle wood.

———————— *lanceolata.* Cum Hagg wd. Conesthorpe banks, &c.

Elymus europæus. Mowthorpe dale.

Hordeum pratense. Meadows near Ganthorpe and Mowthorpe.

Ganthorpe, near Whitwell,
  Yorkshire, November 15, 1842.

H. IBBOTSON.

ART. CXLI. — *Varieties.*

278. *Note on Adiantum Capillus-Veneris.* About sixteen years ago I found Adiantum Capillus-Veneris on the Clee hill, Titterstone. It was growing among the stones, on the ascent to the group of rocks called the Giant's Chair. I plucked a piece of it as a specimen, and placed it in my book, leaving the root. This specimen I kept by me for some time, but at last it was lost, and of the loss I took no notice, not doubting that the next time I visited the spot, I should again find the plant. However, I have hitherto been unsuccessful in my researches, but it would be well if some one would diligently search for it, and perhaps it may again be discovered.—*Fred. Westcott; Spring St., Edgbaston, December,* 1842.

279. *Note on Convallaria bifolia,* Linn., *as a reputed British Species.* In connexion with the notice of the recent discovery of this beautiful little plant in England (Phytol. 520), I beg to add that it is mentioned as indigenous in the woods at Hampstead, Middlesex, in the list of wild plants in the 'History of Hampstead' by Park, published, I believe, thirty or forty years ago: and in 1835 I detected several patches of the plant, apparently well established and really wild, under the shade of fir-trees, growing near the highest parts of Caen wood, the property of the Earl of Mansfield, between Hampstead and Highgate. A year or two before that time, I had also observed it under fir-trees in Aspley wood, Bedfordshire. The village of Aspley is situate at the distance of a short walk from the town of Woburn : I have no means of ascertaining if specimens may still be obtained from thence. Convallaria majalis was plentiful all over Aspley wood, but past flowering when C. bifolia was found. This wood is one of the most picturesque and delightful that can be imagined. It was a favourite resort of the late amiable poet, Wiffen, when domiciled at Woburn Abbey ; and many of the charming graphic descriptions of woodland scenery appearing in his works, may well be supposed to have been depicted from the originals of this delightful locality.—*Edward Edwards ; Bexley Heath, Kent, March 4*, 1843.

280. *Note on the Surrey locality for Fritillaria Meleagris.* In the ' Naturalists' Almanack ' for 1843, it is said that " this very beautiful and local plant flowers profusely in some meadows at Mortlake," (p. 9). This species, called by the country folk " snake's head," used to flower in *a* meadow at Mortlake, Surrey, known from that circumstance as " the Snake's-head Meadow," but of late years it has become very scarce, if not altogether eradicated by the ruthless hands of the village children, by whom the early showy plant was coveted as an ornament to their May garlands. The meadow is at the Thames side, beyond the brewhouses, and about midway between the village and Kew bridge. On visiting the spot at the proper time, during several seasons within the last five years, I was not able to obtain more than a single specimen. I am not aware of any other recorded station for the Fritillary in the immediate environs of the metropolis.—*Id.*

281. *Lithospermum purpureo-cæruleum.* " The purple gromwell, a local and very beautiful plant, found in Darenth wood, in Kent,"— (Nat. Alm. 11). To the best of my belief, this species does not *now* occur in Darenth wood. I cannot learn that any of my friends have detected it there during recent years ; neither have I, after numerous diligent searches, been able to meet with it. I possess specimens from Babbicombe, Devon.—*Id.*

282. *Pæonia corallina.* " The peony, a plant now only found in some small islands of the Bristol Channel, and even in these it is becoming year after year less abundant, and will perhaps before long cease to exist in Britain in a state of nature,"—(Nat. Alm. 13). May I venture to enquire, through the medium of ' The Phytologist,' if the above statement can be verified by any reader of that useful periodical ; and if it is within possibility to hope to obtain a specimen of so great a desideratum to our herbaria ?—*Id.*

283. *Scilla autumnalis.* " There are several spots on Blackheath where it is abundant,"—(Nat. Alm. 19). Within recollection this plant

was tolerably plentiful on Blackheath, but I fear it is now well nigh lost in that locality. Certainly it has been far from abundant for many years, owing to the heath having become a well-trodden promenade, and the frequent resort of cricket-players &c., which formerly was not the case. I noticed the plant, in small quantity, near the clump of trees on the heath, near the highway to Eltham, a few autumns ago; last year not a single specimen appeared. I believe it may still be met with in abundance at the Warren, at Shorne in this county.—*Id.*

284. *Habitat for Tordylium officinale*, Linn. (Eng. Bot. 1st edit. t. 2440). I used to meet with a plant which I believe to have been this species, about Swanscombe, in bushy places between the church and the entrance to Swanscombe wood, in passing from the village to the wood. It was to be found there in 1839; not having visited the locality since that time, I cannot affirm that it yet exists there. This station for it is not given in any list of localities with which I am acquainted. The plant seems to be now erased from our Flora; it does not appear in the Edinburgh Society's Catalogue.—*Id.*

285. *Note on Dicranum adiantoides and taxifolium.* Allow me, through the medium of your valuable periodical, to offer a few remarks on Dicranum adiantoides and taxifolium. Since the commencement of the present season, my attention has been particularly directed to the consideration of these two mosses, and every observation tends to confirm my opinion that they are varieties of the same species, though their extreme forms are widely different. The nearly allied species, D. bryoides, is very variable in form and size. I have luxuriant specimens before me from one of our peat bogs, two inches high, which preserve their character of terminal fruit-stalks, and render that species truly distinct: but the characters of the *lateral* and *radicular* fruit-stalks of D. adiantoides and D. taxifolium are not always to be depended on as specific distinctions. A few days ago I gathered both growing within a yard of each other; D. taxifolium covered the bank with its beautiful green foliage, and D. adiantoides flourished on the stump of a tree, intermingled with several other mosses. Some specimens of the latter were very fine and characteristic, bearing several lateral fruit-stalks, while others were small, producing them also from the base. D. taxifolium occurred mostly with radicular footstalks, but I detected several producing them also laterally, which differed not materially from the smaller specimens of adiantoides. The character of flexuose fruit-stalks is not peculiar to any of the species, as I have observed it occasionally in all three. I hope these remarks may elicit further information on the subject from those who are more

competent to treat on it than myself.—*Joseph Sidebotham; Manchester, March 6*, 1843.

286. *Note on Vegetable Morphology.* I cannot see how the luxuriant growth of a stamen, causing it to produce a petaloid expansion at its upper extremity, in addition to those parts necessary for its peculiar functions, can be regarded as a proof of a descending metamorphosis, (Phytol. 523); by which I understand a dwindling away as it were of the vital energy of the plant, preventing the development of the elementary structures into the highest forms of which they are capable. This theory of a descending metamorphosis appears to me to be unphilosophical in the extreme, for surely, if we can trace the same type through a series of organs, the simplest and first developed of which can, if necessary, perform the collective functions of the whole, we cannot hesitate to take this simplest form as primary. I consider that the monstrosities we see in Dahlias &c. are caused, immediately, by an excess of nutriment afforded to the plant, which necessitates a great development of the organs of digestion and respiration, viz., the leaves: this, of course, diminishing the power of the plant to perfect its floral organs, and thus causing what may be termed an arrest of development, whereby stamens remain petals, &c. This does not explain the fact mentioned by Mr. Bladon; but if the anther *was* perfect, which he says it appeared to be, there is nothing very extraordinary in the production of a small petaloid expansion from the stamen, arising from a redundancy of vital action, when we consider how closely the two parts in question are allied.—*Arthur Henfrey, M. Mic. Soc., Curator to the Bot. Soc. Lond.; March* 8, 1843.

287. *Note on the " Daill llosg y Tân."* In answer to Mr. Lees' enquiry respecting the above plant (Phytol. 521), I have enclosed a few leaves and a young plant of the species known by that name in this part of Gwent,* and applied to the same purposes. As I have never examined the inflorescence of this plant I cannot give its name, but it is evidently monocotyledonous, and not a fern. In the summer the leaves are considerably larger, some of them being an inch or an inch and a quarter in breadth. They never rise above the water, but at that season lie incumbent on the surface. At the present time those with the longest stems lie horizontally, about half an inch below the surface, while the shorter-stemmed ones are as nearly upright as those of the generality of plants. Whether the circumstance of their sinking below the surface is owing to the late frosts or not, I am

* Gwent, the northern and western districts of Monmouthshire.

unable to say, but I had to break the ice to procure the specimens sent. The manner of using the leaves is to lay a number of them on the burn, and as they dry to replace them by fresh ones. A friend of mine, a native of Morganwg,* informs me that in that part the leaves are mixed up with lard, so as to form an ointment; he thought that the leaves known by the above Cambro-British name were very much larger than the size stated; perhaps some other species may be used by the Glamorganshire people for the same purpose, as the name only indicates the "leaf for a burn by fire."—*James Bladon; Pont-y-Pool, March* 8, 1843.

[The plant sent by our correspondent as the " Daill llosg y Tân" of Gwent, we believe to be the common pondweed,—Potamogeton natans.—*Ed.*]

288. *On the influence of Light in producing the Green Colour of Plants.* About Christmas, 1841, I was searching in a wood, chiefly oak, for some lichens to decorate the perches in a glazed case, intended for the reception of some ornithological specimens. I happened to turn over with my foot a piece of oak bark, about fifteen inches long: the side next the ground (the external part of the bark when *in situ*) was covered with lichens of the most vivid green, quite as bright as that of any leaves in early summer, not the pale colour of young shooting leaves, but of those arrived at mature growth. From the appearance of the grass under and on each side of the place where the bark had been, it had evidently lain there at least all the previous summer: yet I have never seen any lichens of the same or any other species, exposed to the full light of " day's garish eye," in the least approaching the vividness of the colour in the specimens alluded to above.—*Id.*

[The following passage relating to the Algæ, which are nearly allied to the Lichens, occurs in the Introduction to Harvey's 'Manual of British Algæ.' After mentioning the three principal varieties of colour among the Algæ, namely, grass-green, olivaceous-brown or olive-green, and red, the author states that the first of these colours is characteristic chiefly of such species as are " found in fresh water, or in very shallow parts of the sea, along the shores, and generally above half-tide level," the great mass of the green Algæ being inconsiderably submerged. " The olivaceous-brown or olive-green is almost entirely confined to marine species; * * the red also is almost exclusively marine, and reaches its maximum in deep water. * * How far below low-water-mark the red species extend has not been ascertained, but those from the extreme depths of the sea are of the olive series in its darkest form. For the colours of these last it has puzzled botanists not a little to account. It is well known that *light* is ab-

solutely necessary to the growth of land-plants, and that the green colour of their foliage altogether depends upon its supply: and if they be placed in even partial darkness, the green quickly acquires a sickly yellowish hue, and finally becomes whitish.
But with Algæ it is different. At enormous depths, to which the luminous rays, it is
known, do not penetrate, species exist as fully coloured as those along the shore. They
therefore, in this respect, either differ from all other plants (Fungi included), or perhaps, what are called the *chemical* rays, in which seem to reside the most active principles of solar light, may be those which cause colour among vegetables, and these *may*
penetrate to depths to which luminous rays do not reach. But this is mere supposition. Lamouroux suggests that ' the particles of light, or its elementary molecules,
combined or mixed with the water,' suffice for this purpose. However this may be, it
is worth remarking that this property among Algæ, of producing vigorous growth and
strong colour without the agency of light, affords another link between them and the
animal kingdom, among the lower tribes of which, light is by no means essential to
growth and the most brilliant colour."—p. ix. This passage is interesting in itself,
and in some measure applies to the subject of the preceding communication.—*Ed.*]

289. *Note on Viviparous Grasses.* It appears to me that some
misapprehension exists with regard to what are called " viviparous "
grasses. I have several times been deceived by the term, and should
like to see the matter cleared up in the pages of 'The Phytologist.'
In works of authority we are told that " in wet seasons the seeds of
grasses frequently germinate before they fall from the husks, and that
a crop of young plants at the summit of the parent stem is the consequence," or words to that effect. Now in the greater part, or all the
cases of viviparous grasses which have come under my observation,
the plants *have never flowered at all,* and of course *produced no seed.*
How far, *en passant,* may this circumstance be considered as illus-
trating the morphological doctrine, that every flower is but a stunted
branch ? Festuca ovina, β. *vivipara,* growing in my garden, produces
heads of young plants in the above manner every year, but never flowers at all. That such is the case with wild specimens of this and
other species that are found viviparous, we are by no means led to
suppose; in fact we are informed just the opposite. Do the seeds of
pasture grasses *ever* germinate in the husk, like wheat, when it is said
to "sprit"? The following are the grasses which I possess or have
heard of as being occasionally viviparous. — Nardus stricta, Alopecurus pratensis, Agrostis vulgaris and alba, Aira cæspitosa and alpina,
Glyceria fluitans, Poa alpina, Dactylis glomerata, Cynosurus cristatus,
Festuca ovina and duriuscula, and ? Lolium perenne.— *Leo. H. Grindon; 32, Higher Temple St., Manchester, March* 13, 1843.

290. *Note on Polygonum Convolvulus.* We have a beautiful variety of Polygonum Convolvulus growing in many places about Manchester, *with winged fruit,* and so exceedingly luxuriant in growth

that it was mistaken for P. dumetorum by two or three botanists last autumn. But the wing shrivels so much in drying, that the error can only be made with recent specimens, and even then the roughness of the testa would of course remove all doubt as to the species.—*Id*.

[We were informed by the late Professor Don, that it was this winged variety of Polygonum Convolvulus which led to the insertion of P. dumetorum (under the name of Fagopyrum membranaceum, *Mœnch*.) in Gray's ' Natural Arrangement of British Plants,' previous to the discovery of the latter as an indigenous species.—*Ed*.]

291. *On the arrangement of a Herbarium.* In reply to the enquiry on the wrapper of the last No. of ' The Phytologist,' respecting the best method of arranging a herbarium, I beg to offer a description of the plan of my own, which is both compact, neat and easy of reference, and answers the end for which it was designed admirably well. In the first place I have six guard-books, made of *blue demy* paper, three quires in each, but this being rather too little, I would recommend three and a quarter: they must be at least 4½ inches wide in the binding. I have them labelled, " Herbarium Britannicum, No. 1, 2," &c. and also " Linnæan System," with the names of the classes contained in each volume. No. 1 has the first four classes; No. 2 has one — Pentandria; No. 3 the next seven; and No. 4, 5 and 6, four classes each. Dividing the system in this manner renders the contents of each volume as nearly equal as possible. At the beginning of each volume I have an index to the classes, orders and genera contained in it, referring to the same pasted on the corner of the left hand page, where each commences, one or more leaves being allotted to each genus. The whole is arranged after the fourth edition of Hooker's ' British Flora.' I have also one of Francis's lists of species, which is taken from the above work, cut up and pasted at the ends of the volumes. On this I mark off the species as I get them; so that by turning to the lists I can see at once both what each volume contains and what are desiderata. In the next place I have the specimens fastened down on half sheets of *printing demy*, with very narrow strips of blue paper; for this purpose I use common paste. When this is done, and the paste is dry, to prevent the attacks of insects, I lay on a little weak solution of corrosive sublimate in spirits of turpentine. In case of small species, such as Veronicas, violets, saxifrages, &c., I have more than one on the same half sheet, but still keep up the arrangement as above. These leaves are then put loose in their proper places between the leaves of blue paper, which adds greatly to the beauty of the whole. It may be thought that these books are very unwieldy, but they are in fact no more so than Gerard's Herbal, or any

3 F

other volume of a similar size. This might be remedied by dividing the system into smaller portions, and having a corresponding number of books, which would perhaps be an improvement; in this case I would allow a leaf to each species throughout.—*Samuel King; 'Lane House, Luddenden, near Halifax, March* 13, 1843.

[The above is the only communication we have received in reply to the enquiry on the wrapper of the March Phytologist, relative to the best method of arranging a herbarium. We are obliged to Mr. King for his kind attention, and in our next number hope to give further information.—*Ed.*]

292. *Note on the supposed new British Cerastium.* In the number for this month (Phytol. 497) Mr. Edmonston has endeavoured to show that the Cerastium latifolium of Linnæus was not known to British botanists as an indigenous plant, until discovered by himself in Shetland; the plants of Wales and the Highlands, hitherto so named by the botanists of this country, being only a variety of C. alpinum. This idea is backed by a reference to the opinion of Mr. C. C. Babington, whose botanical acuteness, and particular study of the genus Cerastium, combine to render his opinion on the subject deserving of attention. After reading the paper of Mr. E., I examined living plants of the Cerastium alpinum and latifolium (of British authors), gathered on Ben Lawers in 1841, and now in my garden; also numerous specimens in my herbarium, from Wales and the Highlands, from Faroe, Norway, Switzerland, and Arctic America; and likewise the descriptions of them by various botanical authorities. The conclusion arrived at is, first, that the differential characters assigned to the two species (of Linnæus) by Mr. Edmonston are quite untenable; and secondly, that the Shetland plant is in all likelihood a mere form or variety of the same species as the C. latifolium (of British authors) found on many of the Highland mountains. I consider the characters assigned to C. alpinum by Mr. Edmonston to be untenable, because they would exclude not only many of the Highland plants commonly called C. latifolium, but also various specimens of undoubted C. alpinum preserved in my herbarium; while, on the contrary, his characters of C. latifolium (of Linnæus) apply to some of my specimens of C. alpinum, quite as well as they apply to my Swiss specimens of C. latifolium. To go no farther than the leaves (which indeed afford Mr. E. the strongest contrast — upon paper), I find this year's shoots of the Ben Lawers plants, both C. alpinum and C. latifolium, bearing leaves equally short, broad, and obtuse, as the leaves in Mr. Edmonston's figure of his Shetland plant. In some of my dried specimens of C. alpinum I observe the leaves are obtuse, while in others they are acute;

and in a specimen of C. latifolium (of British authors), gathered on Ben Lawers, there are lanceolate, ovate, and almost orbicular leaves from the same root. In Koch's Synopsis, the same terms are applied to the leaves of both species, namely, "elliptic or lanceolate." It is consequently evident that Mr. E. was describing only particular forms of these plants, when he set down the leaves of C. alpinum as "ovate, or ovate-lanceolate, acute," and those of C. alpinum as being "orbicular, obtuse." Each species produces both these forms of leaves. The other contrasted characters given by Mr. E. appear to be as little constant as those taken from the leaves.—*Hewett C. Watson; Thames Ditton, March* 28, 1843.

293. *Localities of Orchis hircina*, Scop. *and O. macra*, Lindl. For the information of your correspondent, Mr. E. Edwards (Phytol. 555), I send the following. Orchis hircina, *Scop.*, in consequence of the rapacity of collectors, is nearly if not entirely eradicated from the neighbourhood of Dartford, Kent. It was to be found to a certainty near Puddledock and Stanhill, in Wilmington parish, about twenty or thirty years ago, in the hedge-rows; also at Trulling Down, in the road to Greenstreet Green. Sir James Edw. Smith, Sir Wm. Hooker, Mr. Borrer, the late Professor Don, Mr. George Don, Mr. Joseph Smith, Mr. Anderson of Chelsea, and myself, have gathered it in these stations, and I dare say it is still to be met with at or near some one of the above-mentioned places. Orchis macra, *Lindl.*, I have gathered at Stonewood, near Bean; at the entrance to Lullingstone castle, and in a copse near the farm-house at Mapplescombe, in Kingsdown parish; and I think it is likely to be met with now at the latter place, if diligent search be made for it.— *Wm. Peete; Keston Heath, April* 6, 1843.

294. *On the proposed change in the name of Equisetum limosum.* To alter a name which is *now* generally adopted by botanists, solely because a different name was applied to the same species of plant a century ago, would surely be an adherence to the letter rather than to the spirit of that useful rule which says that priority must decide the name. The rule itself is highly convenient to prevent confusion in nomenclature, but surely, it is better to disregard the rule in any particular case, where an adherence to it would actually create confusion. Moreover, in the present instance, it seems doubtful whether the rule really sanctions a change. I deem it highly probable that the two Linnæan names, E. limosum and E. fluviatile, belong to two forms (unbranched and branched) of the one species which Smith and hosts of other botanists have known under the former name. If so, the

name of its variety (E. fluviatile) should not be substituted for the name of the species, (E. limosum). As to the plant now universally called E. fluviatile by British botanists, the same objection would not lie against a change for an older name than that erroneously applied to it by Smith and others. The species which is now known in Britain as E. fluviatile, must have been confounded with E. arvense by Hudson and others; for it is too frequent to have remained unknown. The E. fluviatile of Hudson must be the branched form of E. limosum; E fluviatile of Smith is probably the variety β. of Hudson's E. arvense.—*Hewett C. Watson; Thames Ditton, April* 10, 1843.

295. *Places of growth of Equisetum fluviatile of Smith.* There is a partial inaccuracy in the statement that Equisetum fluviatile "affects loose gravelly and sandy places unconnected with water," (Phytol. 533). It occurs occasionally in corn-fields and other places out of water, but is always (as far as my observation goes) short and stunted in such situations. The finest examples that I have met with were in the counties of Chester and Lancaster, growing on the red marl, by the sides of streams or in water with a deep muddy bottom. Indeed, it is a notion among the rustics of Cheshire, that horses get "bogged" by their endeavours to graze on this plant in the muddy pools of that county; and I have certainly seen a horse almost over head in mud in a small pond filled with the tall "horse-tails," which is the name given more particularly to the barren fronds of the present species. I may add also that I met with one locality for the same species in the Azores, and was ankle-deep in mud before I could reach a frond of it.—*Id.*

296. *The supposed locality of Geranium nodosum near Halifax.* The communication of Mr. S. Gibson (Phytol. 556) is interesting and satisfactory, as tending to establish the accuracy of localities for rare plants which were published in the 'New Botanist's Guide,' on the authority of specimens derived through the hands of Mr. Bowman;— but it seems that we must make an exception to this, in the case of Geranium nodosum, the locality of which was printed "Waterham, near Halifax." Mr. Gibson corrects this, by saying that the species was "G. pyrenaicum," and the locality "Washerlane, near Halifax." Doubtless I misread the label, which, even now, looks to me more like the name that I printed, than the one now given by Mr. Gibson; but every one must be aware of the difficulty of reading unfamiliar names unless very distinctly written. In regard to the species, however, I can only affirm that my specimen is certainly not G. pyrenaicum; but that it belongs to a section of the genus which includes our indigenous G. pratense and G. sylvaticum, as also that doubtful na-

tive, G. nodosum. The specimen is merely the top part of a stem, with the immature fruit after the fall of the petals; and though I cannot speak confidently with only this "fragmentary specimen" before me, I think it G. nodosum. But the name on Mr. Gibson's label is "G. pyrenaicum," so that there is either a mistake respecting the species, on the part of Mr. Gibson, or, it may be, an accidental substitution of a garden specimen of G. nodosum in place of a wild specimen of G. pyrenaicum. The question still remains, whether the *specimen* in my herbarium (that of G. nodosum, probably, but certainly not of G. pyrenaicum) was really gathered wild near Halifax? On receiving the specimen I wrote to Mr. Bowman for further information, but that gentleman was not able to say more than the label stated.—*Id.*

297. *Note on " Dail llosg y Tân."* In a late number of your Journal (Phytol. 521) enquiry was made as to the precise species of fern used by the Welsh peasantry, as a remedy for burns, under the above name, as alluded to in the memoir appended to the prize elegy (or " Marwnad,") to the memory of the late Lady Greenly. The mention of this enquiry to my friend "Tegid," the talented author of the poem, has procured for me, at his instance, from Lady Hall of Llanover, (another great promoter of Welsh literature), accredited specimens of the plant, which, as you will see from the enclosed frond, gathered over " Ffynnon Ofer," proves to be the Scolopendrium vulgare of botanists. Lady Hall remarks that it is in "some parts of *South* Wales" where this simple is known as " Dail llosgi Tân:" in fact, my enquiries on the subject in some parts of North Wales, availed nothing at all.—*W. L. Beynon; Torquay, April* 22, 1843.

ART. CXLI.—*Proceedings of Societies.*

LINNEAN SOCIETY.

January 17, 1843.—Edward Forster, Esq., V.P., in the chair.

Wm. Taylor, Esq., F.L.S., exhibited specimens of the oil, oil-cake, and seeds, of Camelina sativa.

Francis G. P. Neison, Esq., Wm. Maddocks Bust, M.D., and Wm. Osborn, Esq., were elected Fellows of the Society.

Read, a paper " On the Ovulum of Santalum," by W. Griffith, Esq.

February 7.—Edward Forster, Esq., V.P., in the chair.

The Rev. W. Hincks, F.L.S., exhibited a specimen of Neottia gemmipara, recently found by Dr. Wood, of Cork, very near the original locality named by Mr. Drummond. The specimen exhibited was in a much more advanced state than the one preserved in Sir J. E. Smith's herbarium, and figured in ' English Botany.'

Edward Forbes, Esq., Professor of Botany at King's College, London, was elected a Fellow of the Society.

Read, the conclusion of Mr. Hassall's paper " On the Fresh-water Confervæ."

February 21.—Edward Forster, Esq., V.P., in the chair.

Dr. Frederick Blundstone White, of Tetbury, Gloucestershire, and Edward Doubleday, Esq., were elected Fellows of the Society.

Read, "Observations on the Portraits of Linnæus," by the Rev. F. W. Hope, F.L.S. in illustration of which paper Mr. Hope exhibited an extensive collection of engraved portraits.

March 7.--The Lord Bishop of Norwich, President, in the chair.

J. O. Westwood, Esq., F.L.S., exhibited a wax impression of a medal of Linnæus, issued by the Sheffield Horticultural Society.

Mr. Westwood presented specimens of the aërial processes of the roots of Sonneratia acida, sent by Mr. Templeton, from Ceylon. They are described by Mr. Templeton as affording a wood of extremely light and close texture, admirably adapted for lining insect-boxes, on account of the facility with which it admits the finest pins, and the tenacity with which they are retained.

Thos. Corbyn Janson, Esq , and Wm. Hammond Solly, Esq., were elected Fellows of the Society.

Read, a continuation of Mr. Griffith's paper " On the Ovulum of Santalum, Loranthus, Viscum, &c."

March 21. -The Lord Bishop of Norwich, President, in the chair.

M. P. Edgworth, Esq., F.L.S., presented specimens of nineteen species of ferns from the Himalayas, new to the Society's collection.

Capt. Jones, R.N., presented specimens of Calicium hyperellum and Placodium canescens.

Mr. Kippist presented specimens of Cæsalpinia coriaria, used by the natives of Carthagena for tanning leather.

Mr. Janson exhibited flowering plants of the " hungry rice " of Sierra Leone (Paspalum exile, Phytol. 558), raised from seeds collected by Robert Clarke, Esq.

Mr. Arthur Henfrey was elected an Associate of the Society.

Read, a continuation of Mr. Griffith's paper " On the Ovulum of Santalum, &c."

April 4.—Edward Forster, Esq., V.P., in the chair.

M. Nicholas Lund presented a collection of dried plants, gathered by him during a tour in Finmark in 1841-2.

Hugh Cuming, Esq., presented various fruits and seeds collected in the Philippine Islands and Malacca.

J. Parkinson, Esq., presented a specimen of the Ambigo orange from Pernambuco.

Robt. Heward, Esq., presented specimens of Sphæria Robertsii, parasitical on the larvæ of a species of Hepialus from New Holland, collected by the late Allan Cunningham, F.L.S.

George Sutton, Esq., was elected a Fellow of the Society.

Read, a continuation of Mr. Griffith's paper " On the Ovulum of Santalum, &c."

April 18.—The Lord Bishop of Norwich, President, in the chair.

Edward Forster, Esq., presented a section of an unusually large stem of ivy.

Robt. Armstrong, M.D., Nathaniel Buckley, Esq., Charles Pope, M.D., and Thos. White, M.D., were elected Fellows of the Society.

Read, the conclusion of Mr. Griffith's paper " On the Ovulum of Santalum, &c."

March 9, 1843.—Dr. Neill in the chair. Dr. Seller, F.R.C.P., was elected a resident member. Numerous donations to the library and herbarium were laid on the table.

The following papers were read : —

1. Remarks on the mode of growth of the British fruticose Rubi, &c. By Mr. Edwin Lees, F.L.S.

2. Continuation of Remarks on the Diatomaceæ. By Mr. John Ralfs, M.R.C.S.L.

3. On Fumaria micrantha and F. calycina. By Mr. C. C. Babington, M.A., F.L.S.

4. On two new species of Jungermannia; and another new to Britain. By Thomas Taylor, M.D. Communicated by Mr. Wm. Gourlie, jun., Glasgow.

5. Notice of the new fossil plant, Lyginodendron Landsburgii, *Gourlie.* By Mr. Wm. Gourlie, jun.

Mr. James Macnab exhibited a magnificent cluster of the male catkins of a palm, from one of the South Sea islands, which Lady Harvey had obtained from the captain of a vessel, and kindly allowed to be shown to the Society. Its dimensions, when expanded, were about three feet by three and a half, and it somewhat resembled an ornamental grate-screen, formed of shavings.

This being the anniversary of the Society's public institution, the members and others present adjourned, at the close of business, to the Café Royal, where they sat down to an elegant supper; Dr. Neill, the president, in the chair, supported by Sir William Jardine, Dr. Greville, Mr. Ball of Dublin, Mr. Gourlie of Glasgow, &c.— Professor Graham, croupier. After supper, the usual loyal and appropriate toasts were drank, and the proceedings were further enlivened by occasional songs.

April 13.—Professor Graham in the chair. Numerous donations to the library and herbarium were laid on the table. The attention of the Society was chiefly directed to a donation by William Brown, Esq., R.N., consisting of a miscellaneous collection of plants and fruits from Canton river and Chusan—from the Cape and Prince's Island, including forty species of Ericeæ from Simond's Bay and Table Mountain.

The following papers were read : —

1. Two botanical visits to the Reeky Linn and Den of Airley, in April and June, 1842. By Mr. Wm. Gardiner, Dundee.

2. On the Diatomaceæ: No. VI. By Mr. Ralfs, Penzance.

April 21. Dr. W. H. Willshire, in the chair. Dr. Thomas Taylor, F.L.S., presented specimens of the following mosses.—*Trichostomum saxatile,* Taylor, MS. found near Dunkerron, Co. Kerry, Ireland, in 1841: and *Bryum recurvifolium,* Taylor, MS. found at Knockavolula, Co. Kerry, Ireland, 1842.

Mr. J. Reynolds, Treasurer, read the commencement of a paper, being "General Researches in the Physiology and Organogeny of Vegetables;" translated from a paper by M. Gaudichaud.—*G. E. D.*

April 19, 1843.—J. S. Bowerbank, Esq., in the chair.

Read, a paper by Arthur Hill Hassall, Esq., entitled, "Some further observations on the Decay of Fruit." The author refers to an opinion expressed by him in a former paper, that the well-known principle adopted by Liebig, that a body in the act of decomposition is capable of communicating the same to other bodies by a kind of induction, does not apply to the general form of decomposition occurring in fruit, principally from the circumstance of its attacking it in a highly vital condition, when it could not be supposed to be the subject of any spontaneous or chemical decomposition: and also to that of Dr. Lankester, expressed at a former meeting, that Liebig's views did not afford a sufficient explanation of every example of decay occurring in fruit, as he con_sidered that an apple, once removed from the tree, was no longer in a vital state, but that it immediately became a prey to a species of fermentation. He then proceeded to express his doubts as to the accuracy of the last supposition, inasmuch as it is pos_sible to preserve fruit, free from all visible deterioration, for many months after its re_moval from the tree. Still, admitting both these and Liebig's views to be correct, they do not, in his opinion, explain the reason why decay commences in a spot gradually extending itself over the surface of the fruit, and does not at once involve its entire substance and fabric. He therefore proceeded to show that these phenomena were to be referred to the operation of entophytal Fungi for a satisfactory solution, still admit_ting the existence of a second form of decomposition in fruit, this being comparatively of rare occurrence, and which appears to be the result of chemical affinities, in conse_quence of the fruit having ceased to exist, and to which the views of Liebig may in some cases, perhaps with propriety, be applied. He also stated that since the former meeting of the society he had repeated the experiment of inoculating fruit with the sporules of the Fungi, with the same success which attended the former trial, but that he had failed in inducing decay in sound fruit by the introduction of decayed matter destitute of Fungi in any state; still, however, he considered this might ensue in some cases in which the vitality of the fruit was either totally destroyed, or, at all events, much enfeebled; and even if an invariable consequence, still it would not in any way affect the statement made relative to the independent power possessed by Fungi in originating decay; and again, if these were proved not to do this, they would still be of as much importance in a practical point of view as ever; since, when inserted into fruit which is undergoing spontaneous decay, they produce marked and rapid effects, and speedily ensure its complete destruction. In conclusion, he stated that the apples employed in the experiments were of an exceedingly firm description, and that an equal number of each were inoculated with the sporules and with decayed matter.

Dr. Edward Jenner having again forwarded to the society some beautiful specimens of Fragilaria pectinalis and Diatoma flocculosum, and having had his attention directed to a report of the Proceedings of the Society, in which it was said that the specimens before sent were animalcules;—forwarded a paper in reply, wherein he states that the objects sent are considered by botanists to belong to Algæ. By Ehrenberg they are classed with his Infusoria and thought to be animalcules, as he supposes they increase by self-division; but this last fact is not sufficient to remove them from the vegetable kingdom, since many plants, such as the lily, crocus, &c., also increase their species by self-division. Three other genera, Achnanthes, Gomphonema and Cocconema, which are at present classed by botanists with plants, the author considers to be of a doubtful nature, thinking they may possibly be found to be Zoophytes. He also stated that the stomic cells mentioned by Ehrenberg, were the endochrome or colouring matter of the botanist, which, when ripened into sporidia, escapes through an opening in the frustule, being one of the methods by which the species are increased. The author also expressed his persuasion, that in the present imperfect state of our knowledge of these objects, great caution ought to be used in advancing any opinion respecting them.—*J. W.*

THE PHYTOLOGIST.

| No. XXV. | JUNE, MDCCCXLIII. | PRICE 1s. |

ART. CXLIII.— *A History of the British Equiseta.* By EDWARD NEWMAN. (Continued from p. 535).

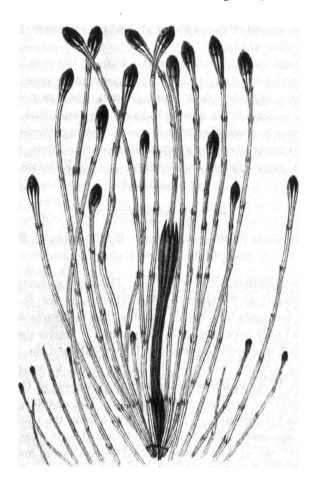

EQUISETUM PALUSTRE, VAR. POLYSTACHION.

THE variety of Equisetum palustre which appears to be universally known by the name of *polystachion*, is at once distinguished by its numerous catkins; these are usually and principally borne on the two

3 G

upper whorls of branches : the main stem generally terminating in a catkin of uniform size with the others, as represented in the figure, which is drawn from an exceedingly beautiful specimen, kindly lent me by Miss Griffiths.   At other times the stem bears a catkin of the normal size and form, while those on the branches are comparatively diminutive in size ; for specimens of the latter form I am indebted to several kind correspondents, particularly to Mr. Ashworth of Manchester.

Dillenius, in Ray's Synopsis,* gives a very faithful figure of this variety ; and several decided although less characteristic figures may be found in other works.

I may remark that the catkins in this form of the plant are usually small, and in the specimens which have come more especially under my notice, they are very black and compact, much more so than the single apical catkin of the normal form of the plant, and hence they much more nearly resemble those of the preceding species : in other respects this variety so nearly approaches the normal form, that a more minute description appears unnecessary.   EDWARD NEWMAN.

(To be continued).

ART. CXLIV.—*On Statice rariflora*.   By CHARLES C. BABINGTON, Esq., M.A., F.L.S., F.G.S., &c.

IN your last number (Phytol. 561) Mr. Heufrey has determined that the Statice found at Fareham by Mr. Notcutt, is the S. rariflora of Drejer ; but he states it to be his opinion that the Garlieston plant is different, and only a variety of S. Limonium.   Under these circumstances I think it as well to state that I have only had the opportunity of examining a minute scrap of the former plant, for which I am indebted to Mr. Watson, but that, through the kindness of Prof. Balfour, I possess excellent specimens of the latter.   The character of S. rariflora given in my Manual (P. 244), is drawn from the examination of the Scottish specimens, compared with two authentic examples of the Danish plant ; one of them gathered by Drejer himself, and presented to me by M. Sonder of Hamburgh, the other forming No. 2200 of Reich. Fl. Exsic., collected by Steenberg and authenticated by Drejer.   Both these specimens are small, and have leaves mostly resembling those of fig. 1, (Phytol. 561) ; whilst the Scottish plant usually has long leaves, similar to fig. 5.   There is, however, amongst those

* Syn. t. 5, f. 3.

sent me by my friend Balfour, one dwarf specimen, agreeing *in every respect* with those from Denmark, in one of which some of the leaves are considerably lengthened in proportion to their breadth, whilst the Scottish specimen referred to has a leaf (it only possesses two) even more ovate-spathulate than in Mr. Henfrey's fig. 1. It is therefore evident that the exact form of the leaves cannot be considered as a certain character. The calyx appears to me to be precisely similar in the Scottish and Danish plants, not differing materially from that of S. Limonium; for I fear that the want of denticulations, upon which I have laid some stress in my Manual, will not prove to be a constant character. A reference to the original descriptions, namely, that of Drejer in the Fl. Hafn. 121, or of Fries, who gives it as a probable species requiring examination, under the name of S. Limonium Bahusiensis, in his Mant. Prima, 10, will show that no great stress is laid by those distinguished botanists upon the form of the leaves or upon the calyx, but that the specific definition is founded upon the peculiar inflorescence, which is remarkably different in the two plants. In S. Limonium the stalk is simple in the lower part, scarcely ever dividing below the middle, the branches are very much divided and corymbose, and curved outwards into a horizontal or even deflexed position, the ultimate divisions are very short, with numerous closely-placed subimbricated flowers. This is the S. Limonium Scanica of Fries (Mant. Prima, 10), the S. Behen of Drejer (Fl. Hafn. 122), S. Limonium, Eng. Bot. 102. On the other hand, in S. rariflora the stalk is often divided far below the middle, but the branches are less compound, not at all corymbose, and rather curve inwards and upwards than outwards and downwards, the ultimate divisions are elongated and the flowers are at a considerable distance from each other.

Having, as I trust, shown that the Scottish plant is the true S. rariflora, it remains to be determined if the Fareham plant is or is not the same; and I must confess that the acute outer bracts and very narrow calyx-segments now lead me to suspect that it may prove different, in which case it will have to be identified with some continental species, or obtain a name as new. It is right to add, that I have reason to fear having been the misleader of both Mr. Watson and Mr. Henfrey, as (if I mistake not) the suggestion that Drejer's name was applicable to both the Scottish and English specimens originated with me; still I trust that pardon will be extended to me, in consideration of the fact that Mr. Watson intimated it to be his opinion that the plants were identical, and that although possessing good Scottish specimens, my example of the Fareham plant consists of five flowers and an inch of stalk. I of course therefore drew my ideas of the species from the Scottish plant.

CHARLES C. BABINGTON.

St. John's Coll. Cambridge,
May 6, 1843.

ART. CXLV. — *A Flora of the Neighbourhood of Sandringham, Norfolk.* By JAMES E. MOXON, Esq.

SANDRINGHAM, a hamlet of western Norfolk, situate about seven miles to the northward of Lynn Regis, and intermediate between Castle Rising and Snettisham, offers many inducements to the botanist, on account of the number and interest of the productions of its vicinity, mainly attributable to the variety of soils and situations. The geological features of this part of England are somewhat peculiar. Firstly, there is the chalk; which, after traversing various counties from Sussex, and being broken in upon by the extensive fens and marshes of Cambridgeshire, and of that part of Norfolk, denominated "marshland," again appears in the neighbourhood of Downham, and, occupying the most elevated portions of this part of the county, terminates at Hunstanton, its north-western extremity, situated at the entrance of the Wash, an extensive inlet of the German Ocean. Secondly, the silt, a marine deposit, occupying all the lower parts and valleys along the coast, and extending in some instances to a considerable distance inland. And thirdly, between this latter and the chalk, occurs the greensand formation, also a continuation of the Sussex beds, and terminating likewise at Hunstanton. Behind the chalk, a series of non-fossiliferous marls and clays extends in the direction of the interior of the county. In addition to this series of strata, the immediate neighbourhood of Sandringham offers a variety of localities. Heath, fen, marsh, woods, cultivated lands, meadows and hedge-rows, all nourishing their peculiar species, are spread around. Add to this the variety of soils; sand, clay, marl, loam, chalk and gravel, and likewise the sea-shore, shingly, sandy, and muddy; the salt marshes and ditches at Wolferton, Babingley &c.; and the river at Castle Rising;— all combine to render this a district abounding alike in rare and uncommon plants, as (considering its limits) in the number of species. Nor is this all; for lastly, the undulating character of the country in general, adds further to all these desiderata.

The size of the district examined, and to which the accompanying catalogue solely applies, is included within a radius of about three miles from Sandringham church. The following are its boundaries:—Northwards, Snettisham; East, the Fring and Harpley road, Anmer, Flitcham, Hillington, Congham; South, Roydon (including the fen), Wootton-heath and North Wootton; and to the West, the sea-coast as far north as Caen-hill wood, near Snettisham.

To avoid a more lengthened description, the nature of its general surface and the predominating vegetation, may be concisely defined as follows.

Divisions of Surface.	Area in Square Miles.	Character.	Predominant Vegetation.
1. Seashore (beyond the sea bank)	1¼	Stony, sandy, muddy.	Triticum repens, Salicornia herbacea, Atriplex portulacoides, Silene maritima, Glaucium luteum.
2. Salt marsh	5	Silt and clay	Various Gramineæ. Statice Armeria. In the ditches — Scirpus maritimus and Ruppia maritima.
3. Heath and commons, dry.	6	Sand....................	Calluna vulgaris, Erica cinerea, Ulex europæus, Agrostis vulgaris, Galium verum, G. saxatile, Senecio Jacobæa, Teucrium Scorodonia, Pteris Aquilina.
Ditto, marshy....	5	Peat, sand, clay	Erica Tetralix, Junci, Ranunculi, Myrica Gale, Aira cæsp., Carices.
Ditto, bog & fen	1	Sandy peat, with much Sphagnum.	Various Carices, Eriophorum angustifolium, Droseræ, Melica cærulea, Hypericum elodes, Anagallis tenella, Vaccinium Oxyco.
4. Woods, parks &c.	1½	Sand and clay.........	Festucæ, Phalaris, Agrostis vulgaris, Aira cæspitosa, Myrrhis temulenta, Heracleum, Scilla nut.
5. Meadows & pastures.	7¼	Sandy loams & marls on chalk: clays	Gramineæ (Bromus, Lolium, Poa, Phleum, Cynosurus &c.), Trifolia &c.
6. Cultivated lands, cornfields &c.	14¼	Ditto ....:	Chenopodium album, Stellaria media, Rumices, Polygona, Senecio vulgaris, Poa annua, &c.

41¼

The general accordance of these divisions with the position of the geological strata is also worthy of remark; the cultivated lands and pastures lying chiefly upon the chalk formation, the heaths and commons upon the greensand, and the marshes upon the silt.

In conclusion, the following tables will serve to illustrate the *botanical* character of the Flora and the general distribution of the species.

I.—*Number of Species.*

Exogens .... 64 Natural Orders, containing 384 species.
Gymnosperms... 1 1
Endogens 15 102
Acrogens (ferns 4 19
and allies only)——
84 506

II.—*Distribution of Species.*

	Exog.	Gymn.	Endo.	Acro.	Totals.
1. Trees and shrubs peculiar to dry ground	32	1	1	0—	34
... ... moist & watery places	21	0	0	0—	21
2. Plants peculiar to woods, hedges, bushy and shady					—— 55
places in general. Dry ground...................	42	0	6	5—	53
Moist and watery places.	10	0	9	9—	28
3. Plants peculiar to marine localities.					—— 81
Dry or sandy shore	4	0	3	0—	7
Muddy shore, salt marshes & salt-water ditches	16	0	5	0—	21
Salt water (submersed or floating plants)	0	0	2	0—	2
Generally distributed	1	0	0	0—	.1
					—— 31
4. Plants peculiar to other dry places	135	0	15	1—	151
5. Plants peculiar to other moist pl. marshes, fens, &c.	63	0	44	4—	111
6. Plants peculiar to water(submersed or floating plants.					
True aquatics).	6	0	2	0—	8
					——270
7. Plants more or less generally distributed	54	0	15	0—	69
					—— 69
	384	1	102	19	——506
Total of species....1. Dry ground............	213	1	25	6—	245
2. Moist ground &c. ...	110	0	58	13—	181
3. True water plants...	6	0	4	0—	10
4. Generally distributed	55	0	15	0—	70
					——506

III.—*Predominance of Natural Orders.*

Species.	Species.	Species.
	207	272
Compositæ 51		
Graminaceæ 43	Amentaceæ: Salicaceæ	13, Osmundaceæ 1 14
Leguminosæ 24	13, Corylaceæ 3, Be-	Alsinaceæ 13
Rosaceæ, 18. Pomeæ, 2.	tulaceæ 2, Myrica-	Polygonaceæ 12
Amygdaleæ, 2. San-	ceæ 1................... 19	Juncaceæ 12
guisorbeæ, 1......... 23	Cyperaceæ 18	Chenopodiaceæ 10
Scrophulariaceæ......... 23	Ranunculaceæ 14	Primulaceæ 10
Labiatæ 22	Cruciferæ 14	Stellatæ 8
Umbelliferæ 21	Filices: Polypodiaceæ	Boraginaceæ 8
207	272	359

Species. 359		Species. 437		Species. 480
Papaveraceæ 5, Fuma-		Valerianaceæ 3		Polygalaceæ 1, Ti-
riaceæ 2 7		Solanaceæ 3		liaceæ 1, Lythraceæ
Orchidaceæ............... 7		Alismaceæ 3		1, Celastraceæ 1,
Violaceæ 6		Typhaceæ 3		Portulaceæ 1, Oxa-
Onagraceæ 4 Circæeæ 1 5		Fluviales................. 3		lidaceæ 1, Crassula-
Hypericaceæ 5		Araliaceæ 2, Resedaceæ		ceæ 1, Ulmaceæ 1,
Silenaceæ 5		2, Aceraceæ 2, Li-		Callitrichaceæ 1,
Plantaginaceæ 5		naceæ 2, Malvaceæ		Vaccinaceæ 1, Cus-
Liliaceæ 5		2, Illecebraceæ 2,		cutaceæ 1, Aquifoli-
Euphorbiaceæ 4		Scleranthaceæ 2,		aceæ 1, Lobeliaceæ
Geraniaceæ 4		Convolvulaceæ 2,		1, Verbenaceæ 1,
Saxifragaceæ 4		Campanulaceæ 2,		Orobanchaceæ 1,Co-
Urticaceæ 4		Plumbaginaceæ 2,		niferæ 1, Amarylli-
Dipsaceæ 4		Lentibulaceæ2,Gen-		daceæ 1, Iridaceæ 1,
Equisetaceæ 4		tianaceæ 2, Oleaceæ		Butomaceæ 1, Di-
Droseraceæ............... 3		2, Juncaginaceæ 2, 28		oscoreaceæ 1, Ara-
Ericaceæ................... 3		Berberaceæ 1, Cornaceæ		ceæ 1, Pistiaceæ 1,
Caprifoliaceæ............ 3		1, Cucurbitaceæ 1,		Lycopodiaceæ 1 ... 26
437		480		Total...506

A Catalogue of the Flowering Plants growing in the neighbourhood of Sandringham.

Thalictrum majus. Dersingham heath, lo.

Anemone nemorosa. Woods and groves

Ranunculus Flammula. Moist gr. freqnt.

———— *Lingua.* Moist places, rare

———— *Ficaria.* Common.

———— *sceleratus.* Wolferton, ra. r.

———— *bulbosus, repens* and *acris.* Pastures, common.

———— *arvensis.* Cornfields, ra. rare.

———— *hederaceus.* Ditches &c. rare.

———— *aquatilis.* Ditto, common.

Caltha palustris. Plentiful.

——— *radicans.* Ingoldsthorpe common, 1842; a single plant.

Papaver Rhœas. Roadsides, flds. &c. abt.

——— *dubium.* Redbrink, Dersingham, 1840, rather sparingly.

——— *somniferum.* Roadside at Rising wood, abundant.

Glaucium luteum. Snettisham beach, pl.

Chelidonium majus. Occasionally.

Fumaria claviculata. Wolferton wood, lo.

——— *officinalis.* Not uncommon.

Hydrocotyle vulgaris. Moist or watery places on heaths &c. abundant.

Apium graveolens. Ditches in Wolferton salt marsh.

Sium angustifolium. Wade moor, Roydon fen.

——— *nodiflorum.* Roydon fen.

——— *repens.* Watery places, very freqnt.

Ægopodium Podagraria. Not uncommon.

Bunium flexuosum. Frequent.

Pimpinella Saxifraga. Roadsides &c. fr.

Œnanthe fistulosa. Watery places, freqt.

——— *pimpinelloides.* Snettisham; Ingoldsthorpe and Rising commons; frequent.

Æthusa Cynapium. Waste ground, not uncommon.

Angelica sylvestris. Woods and marshy places, common.

Pastinaca sativa. Frequent.

Heracleum Sphondylium. Common; abundant in woods.

Daucus Carota. Frequent.

Torilis Anthriscus. Common.

———— *nodosa.* Dersingham, not common.

Scandix Pecten-Veneris. Sandringham, ft.

Chærophyllum sylvestre. Frequent.

Myrrhis temulenta. Common; abundant in woods.

Conium maculatum. Sparingly.

Adoxa Moschatellina. Sandringham fir-woods, common.

Hedera Helix. Common.

Berberis vulgaris. Hedges &c. rare.

Epilobium hirsutum, parviflorum, montanum and *palustre.* Common.

Circæa Lutetiana. Wolferton wood, com.

Cornus sanguinea. Hedges, frequent.

Bryonia dioica. Hedges &c. common.

Nasturtium officinale. Watery places, co.

Barbarea vulgaris. Common.

Arabis Thaliana. Ditto.

———— *hirsuta.* Walls of Sandringham church-yard, roadsides &c. common.

Cardamine pratensis. Marshy meadows, very common.

Draba verna. Old walls, sandy ground, &c. very common.

Cochlearia officinalis. Wolferton & Snettisham, common.

Thlaspi Bursa-pastoris. Common.

Sisymbrium officinale. Common.

———————— *Sophia.* Babingley, Castle Rising &c., frequent.

Erysimum Alliaria. Frequent.

Senebiera Coronopus. Occasionally.

Brassica campestris. Frequent.

Sinapis arvensis. Common.

Reseda Luteola. Frequent.

———— *lutea.* Chalky fields and banks, common.

Viola odorata. Common.

———— *palustris.* Dersingham fen, frequent.

———— *canina.* Abundant.

———— *flavicornis.* Sandringham heath.

———— *tricolor* and *arvensis.* Frequent.

Drosera rotundifolia. Spongy bogs, abdt.

———— *longifolia.* Ditto, common.

———— *anglica.* Dersingham fen with both the preceding; Ingoldsthorpe common; Roydon fen: not uncommon.

Hypericum quadrangulum. Marshy places, common.

———— *perforatum* and *humifusum.* Common.

———— *pulchrum.* Frequent.

———— *elodes.* Dersingham and Babingley fens &c. abundant.

Acer Pseudo-platanus. Hedges, rare.

———— *campestre.* Frequent.

Polygala vulgaris. Common.

Linum catharticum. Cornfields near Sandringham chalk-pit, not uncommon.

Radiola Millegrana. Rising common, lo.

Malva sylvestris. Common.

———— *rotundifolia.* Frequent.

Tilia europæa. Hedges, occasionally.

Lythrum Salicaria. Watery places, freq.

Euphorbia Peplus. Common.

———— *Lathyris.* Waste ground at Sandringham, occasionally.

———— *helioscopia.* Fields and waste ground, frequent.

Mercurialis perennis. Wolferton wd. abt.

Euonymus europæus. Wolferton wood.

Montia fontana. Watery places, frequent.

Silene inflata. Frequent.

———— *maritima.* Snettisham beach, ple.

Lychnis Flos-cuculi. Watery places, com.

———— *dioica.* Hedges &c. frequent.

Agrostemma Githago. Cornfields, occas.

Sagina procumbens. Common.

———— *apetala.* With the preceding, oc.

Arenaria trinervis. Sparingly.

———— *serpyllifolia.* Frequent.

———— *rubra.* Sandy soil &c. abundt.

———— *marina.* Wolferton marsh, abt.

Cerastium vulgatum and *viscosum.* Comn.

———— *aquaticum.* Frequent.

Stellaria media. Very abundant.

———— *holostea.* Frequent.

———— *graminea.* Very common.

———— *uliginosa.* Dersingham heath.

Spergula arvensis. Wolferton, plentiful.

———— *nodosa.* Near Sandringham chalk-pit, common.

Geranium Robertianum and *molle.* Com.

———— *pusillum.* Frequent.

Erodium cicutarium. Sandy places, abt.

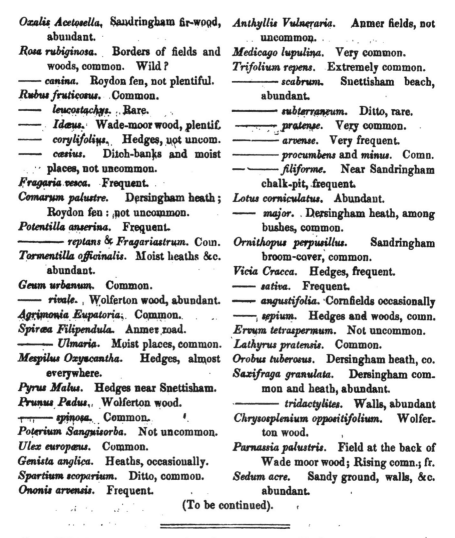

Oxalis Acetosella, Sandringham fir-wood, abundant.

Rosa rubiginosa. Borders of fields and woods, common. Wild?

—— *canina.* Roydon fen, not plentiful.

Rubus fruticosus. Common.

—— *leucostachys.* Rare.

—— *Idæus.* Wade-moor wood, plentif.

—— *corylifolius,* Hedges, not uncom.

—— *cæsius.* Ditch-banks and moist places, not uncommon.

Fragaria vesca. Frequent.

Comarum palustre. Dersingham heath; Roydon fen: not uncommon.

Potentilla anserina. Frequent.

—— *reptans* & *Fragariastrum.* Com.

Tormentilla officinalis. Moist heaths &c. abundant.

Geum urbanum. Common.

—— *rivale.* Wolferton wood, abundant.

Agrimonia Eupatoria. Common.

Spiræa Filipendula. Anmer road.

—— *Ulmaria.* Moist places, common.

Mespilus Oxyacantha. Hedges, almost everywhere.

Pyrus Malus. Hedges near Snettisham.

Prunus Padus. Wolferton wood.

—— *spinosa.* Common.

Poterium Sanguisorba. Not uncommon.

Ulex europæus. Common.

Genista anglica. Heaths, occasionally.

Spartium scoparium. Ditto, common.

Ononis arvensis. Frequent.

Anthyllis Vulneraria. Anmer fields, not uncommon.

Medicago lupulina. Very common.

Trifolium repens. Extremely common.

—— *scabrum.* Snettisham beach, abundant.

—— *subterraneum.* Ditto, rare.

—— *pratense.* Very common.

—— *arvense.* Very frequent.

—— *procumbens* and *minus.* Comn.

—— *filiforme.* Near Sandringham chalk-pit, frequent.

Lotus corniculatus. Abundant.

—— *major.* Dersingham heath, among bushes, common.

Ornithopus perpusillus. Sandringham broom-cover, common.

Vicia Cracca. Hedges, frequent.

—— *sativa.* Frequent.

—— *angustifolia.* Cornfields occasionally

—— *sepium.* Hedges and woods, comn.

Ervum tetraspermum. Not uncommon.

Lathyrus pratensis. Common.

Orobus tuberosus. Dersingham heath, co.

Saxifraga granulata. Dersingham common and heath, abundant.

—— *tridactylites.* Walls, abundant

Chrysosplenium oppositifolium. Wolferton wood.

Parnassia palustris. Field at the back of Wade moor wood; Rising comn.; fr.

Sedum acre. Sandy ground, walls, &c. abundant.

(To be continued).

ART. CXLVI. — *Journal of a short run into Badenoch, Strathspey, &c., from the 15th to the 21st of April, 1843.* By Mr. J. B. BRICHAN.

Grantown, April 17, 1843.

DEAR SIR,

I am at present on my return from a very short sojourn in Badenoch; and certainly the season of the year, and the weather when I left home, afford little or no foundation on which to construct any sort of article whatever for a botanical journal. It is quite possible, however, that the very *negation* of Botany may be made somewhat interesting to the lovers of Flora, especially if it be merely a temporary

negation, the effect of climate and season on the vegetable productions of a district in which, as the season advances, the hills and vales
are thickly clad with

" Bells and flow'rets of a thousand hues."

And such is the romantic Highland district in which I have just been.
The day before I entered it was bleak and wintry, and I was glad to
get myself housed for the night at Carr Bridge.   Next day was fresh
and mild, and the snow, which but the day before gave to everything
the appearance of winter, now only chequered mountain and plain,
and was fast yielding to the genial breath of what was really spring
weather.   Still, however, except where a small spot appeared in the
shape of a garden, no green thing but " the evergreen pine " cheered
the sight that was longing for the sweet gems of spring.   And perhaps
a more dismal spectacle could scarcely meet a botanist's eye, than the
one I witnessed between Carr Bridge and Aviemore.   The southern
slope of a tolerably high and rocky hill, was thinly sprinkled with
blasted firs, the remains of a forest consumed by fire, or dead from the
sterility of the soil.   It was some alleviation of the dreariness of such
a sight, that on each side of the road the native birch, yet scarcely
throwing out a catkin, scented the air with its peculiar odour.   As I
went on, with the magnificent group of the Cairngorms, partly clad in
snow, on the one hand, and bleak rocky hills on the other, the wide
wooded valley of Rothiemurchus spread before me, and waiting, as it
were, " in the hope of summer eves,"—I could not help thinking, and
being delighted with the thought, how many a beautiful flower the
summer's heat would call forth from the mountain cleft, and the sloping woodland, and the rich alluvial plain ; and with what ecstacy I—
for I *might* be there — would cull the precious gems, and breast the
mountain side in search of health, and beauty, and that peculiar pleasure which only botanists know.

As I entered Badenoch, the air was balmy as summer.   The heather—I prefer the word to the English *heath*—" exhaled its perfume,"
although far from its season of purple bells ; and the birch, the prevailing natural wood of the district, grew more thickly on every side
of me.   Here and there, a few alders and hazels were hanging out
their tassels, and a willow occasionally exhibited its golden flowers.
The beautiful and romantic Loch Alvie was slightly curled by the
breeze, but its vegetation as yet exhibited no signs of life. A few withered reeds were all that remained to show that a single green thing
had ever reared its head above its waters.   As I drove along the foot
of a wooded slope opposite Loch Insh, a solitary primrose cheered

my sight. It was the only wild thing in the shape of a flower that I had seen since I left Morayshire. Between Loch Insh and Kingussie there stretches a vast alluvial plain, in which are dug up the roots and trunks of fir trees, known in Scotland by the name of "moss fir." It is well known that the whole district was once covered with pine. Kingussie signifies "the head of the fir-wood," and Badenoch, "a clump or patch of trees." The plain must have formed, at one time, the bottom of a lake, of which Loch Insh is all that remains.

The weather, as I returned to-day, was equally delightful as before, and wood, and moor, and mountain, seemed "to listen for the rustle of their leaves." Vanessa Urticæ, our earliest butterfly, fluttered around me in considerable numbers. The district, in point of climate, seems not far behind the "How of Moray." Another week, or fortnight at most, would bring it up to the point which the lower and more temperate district had reached when I left it.

The Cairngorm group has frequently witnessed the enthusiastic rambler after "weeds" scaling its craggy sides, gathering its treasures, and laving his burning brow or parched throat in its cool and delicious fountains. I am not aware that the district of Badenoch has often been botanically perlustrated; but I am quite sure that its richly varied scenery — the extensive plain alluded to — its inferior ranges of hills — the rocky steeps of the "wild and majestic" Craig-dhu — the sweet sequestered Loch Uvie — and the beautiful Loch Laggan, which, though not so large or so well known, may vie with Lochness— would afford to the botanist who could devote a summer month to its examination, as rich a treat and as abundant a *harvest* as any other part of "the stern Scottish Highlands."

April 18. — Dalnashauch Inn, near Bridge of Avon. I have just passed over fourteen miles of excellent road between Grantown and this uncouthly named locality; and I must *bother* you a little more with almost the same negation of Botany which at present characterizes the district above described, and my description thereof. I am near the wildly situated Bridge of Avon (pronounced *Awn*), and hear the unceasing sound of that rapid mountain torrent which issues from Loch Avon, at the foot of Scotland's highest hill — Ben-mac-dui, and of which the honest old farmer, mentioned in Sir T. Lauder's 'Morayshire Floods,' said, alluding to a piece of ground of which the Avon had just deprived him, "I took it frae the Avon, and let the Avon hae its ain again." Certainly this has very little to do with Botany, and amounts to something very like a negation of it. But that I may not be altogether barren on the subject that is dear to all *phytological* hearts, let me just say that the beautiful drive I have last accomplished exhibits, as the prevailing natural production of Strathspey, "the gay green birk," at present of course quite purple — that there is a considerable degree of wild Highland grandeur in the immense slope between Castle Grant and the Spey, which is thickly clothed with Scotland's own pine—and that all along that winding, deep and rapid river, from Grantown to Ballindalloch, "the alders dank that fringe the pool" assert their peculiar right to the river's margin, and are at this moment clad with a profusion of brown or yellow catkins. The larch, which is planted in great abundance along the hilly slopes, is now throwing out its "tassels red" and sweet virgin green leaves. Its flowers are sometimes *white—quære*, why? They are here, as in Badenoch, though I have not mentioned them in my notes on that district, a week or a fortnight behind those in Moray's balmiest spot — Forres. I think I saw, about half way between this and Grantown, one solitary flower of Anemone nemorosa.

In the church-yard of Cromdale, three miles and a half from Grantown, lie the remains of M— C—, one of the sweetest flowers ever born in the Highlands, and transplanted to the lower part of the "Province." I crave this passing tribute to the memory of one, with whose surviving amiable sisters I am proud to say I am on intimate terms. In April, 1839, I accompanied her remains from Nairn to the interesting spot where they are now mouldering in silent dust. The weather, although it was later in the month than this, was keenly frosty, and the interests of the vegetable kingdom were as far from my thoughts as its beauties were from their summer perfection. At the moment I write, nothing is wanting to complete the interest of the landscape, but the green grass, the blooming heather, and the summer foliage of the forest. My yesterday's friend, Vanessa Urticæ, still flits across my path, enjoying Nature's hour of balm; and that big, humming, or, as we call him in Scotland, *bumming*, fellow, Bombus terrestris, wings his way as briskly as if summer itself were invigorating his powers of flight. The beauties of Strathspey are great, but they are "tame and domestic" in comparison with those of Badenoch. Let botanists visit Clova, if they will—but why not run a little farther north?

Manse of Kirkmichael. — April 18, ¼ to 11, P.M. Although half asleep, I cannot resist the temptation I feel to teaze you with the very important information that I left Dalnashauch to-day about 1 o'clock, and after crossing the lower extremity of Glenlivet, proceeded up the Avon through what is called Strathdown. About six miles of my road was so narrow that two vehicles could scarcely pass each other, with a steep slope above and below, the only protection against an overthrow being the natural birch-wood which skirted the way. The alder too was there. Strathdown is a narrow, wild, romantic valley, through which runs the beautifully clear Avon, and about the middle of which, on a bank of the river, some hundred feet above its bed, is situated the manse of Kirkmichael, from which I now write. The view from this spot is delightful, comprehending the summits of Cairngorm and Benavon—the climate not inferior to that of any part of my route above described. The ceaseless *sough* of the Avon here also greets my ear, and reminds me of my late residence on the banks of the Dee, which, in the colour and clearness of its waters it strongly resembles. In the dining room immediately below me is a flower-pot containing a few plants of mignonette in full flower. The seed was sown last spring — the young plants appeared in autumn — and the sweet-scented favourite of ladies and of bees has been in flower all the winter. Good night.

April 19. — I drove to-day to Tomintoul, six miles farther up the Avon than Kirkmichael, and within about twenty miles both of Ballater and Braemar, but could see nothing to neutralize the same monotonous negation of Botany I have already experienced, except a flower or two of Tussilago Farfara, which has been in blossom for the last month. Such, however, is the mildness of the season, that I have observed since yesterday a marked difference in the buds of the trees, now fast opening into green leaves.

April 20.—I am again at Dalnashauch. In my downward progress hitherward the only additional harbinger of summer that I observed was the bog-myrtle; all other things are *in statu quo.* I am about to proceed down the Spey to Rothes, to leave for the present — perhaps for ever — the interesting scenes through which I have passed, and which, though at all times worth seeing, are at this moment but opening into that state of mingled beauty, and sweetness, and grandeur, in which the devotee of phytological science finds his most appropriate and most delightful walk.

Forres. — April 21. The drive down the Spey from Inveravon to Craigellachie is partly through a bleak moorland tract, lying at the foot of Belrinnes, and partly through one lower and more cultivated, in which lie the church, manse and village of Aberlour, and in which, near the river-side, I saw, as evidences of the warmer temperature, patches of Anemone nemorosa on one side of the road, and low green meadows on the other. The iron bridge at Craigellachie is a fine specimen of art: at its northern extremity, where the road, at right angles to the line of the bridge, and parallel to the course of the river, is cut out of the face of a solid rock, grows Galium boreale, at this season scarcely visible. At this point I re-entered my own county—Moray. Two miles farther down lies the beautifully situated village of Rothes, with its old castle, romantic *burn* and fertile fields. On the inverted root of a tree, lying above a heap of stones, I observed a single head of Lamium purpureum, one of the earliest flowering and by no means least beautiful plants in the climate of Moray.

I left Rothes to-day—the weather still very fine. As I approached Elgin, the difference in the progress of the larch towards its summer foliage, began distinctly to appear; and Ulex europæus, now nearly in full bloom, showed the superior mildness of the climate of the "How of Moray." The gardens and shrubberies were half clad in green. After resting an hour or two, 1 left Elgin, passed the pine-clad Knock of Alves, the well-known station of the beautiful Linnæa, and was soon once more in Forres, where a week ago everything wore the appearance of winter, but where everything now is smiling into summer sweetness, and where, both in summer and winter, there blooms many a tender "bell and flow'ret," that may well vie with England's fairest.   I am, Dear Sir,

Your's faithfully,

J. B. BRICHAN.

To the Editor of 'The Phytologist.'

ART. CXLVII. — *Notice of 'A Visit to the Australian Colonies.* By JAMES BACKHOUSE.' London: Hamilton, Adams & Co. 1843.

(Continued from p. 577).

NEW SOUTH WALES.—The continent of New Holland, as far as visited by James Backhouse, seems less attractive to a botanist than the adjacent islands; still there is sufficient to invest it with an interest surpassing that possessed by most other countries with which we are on a similar footing of familiarity. There is nothing to compare with the gigantic forests which form so distinguishing a character of Tasmania, or with the wild luxuriance of Norfolk Island; yet the pines to which that island has given a name, may be seen, even at Sydney, towering here and there above the surrounding scenery, and proclaiming a climate different from our own; and there is a host of trees and plants altogether distinct from those of Tasmania.

On new-year's day, 1835, many beautiful native shrubs were in flower, including Lambertia formosa, Grevillea buxifolia and sericea, Epacris grandiflora, &c.: these grow in heathy soil, on the bushy ground covering the sandstone.

January 15.—"We walked to Elizabeth Bay, and met the Colonial Secretary* at his beautiful garden, which is formed on a rocky slope, on the margin of Port Jackson, of which it commands a fine view. Here are cultivated specimens of many of the interesting trees and shrubs of this colony, along with others from various parts of the world, intermixed with some growing in their native localities. Among the last, is a fine old rusty-leaved fig-tree, Ficus ferruginea, which is an evergreen, and has laurel-like leaves. A noble specimen of Acrostichum grande, a fern of very remarkable structure, from Moreton Bay, is attached to a log of wood, and secured by a chain to a limb of this fig-tree. The walks at this place are judiciously accommodated to the inequalities of the sinuous bay, and are continued round a point covered with native bush. Peaches are ripe in the open ground in abundance, and liberty to partake of them freely was kindly given by the open-hearted proprietor. Dendrobium speciosum and linguiforme, remarkable plants of the Orchis tribe, are wild here, upon the rocks; and D. tetragonum is naturalized on a branch of Avicenna tomentosa, covered with rock-oyster shells, and suspended in a tree near the shore. A fine patch of the elks-horn fern, Acrostichum alcicorne, retains its native station on a rocky point in the garden."—p. 239.

In a walk by the north shore on January 27, our traveller observed an old bushy fig-tree overhanging the water, some of its limbs almost covered with Acrostichum alcicorne and Dendrobium linguiforme; but we pass on to a description of Botany Bay, the account of which is less inviting than its far-famed name would lead us to anticipate, still not without instruction.

" Botany Bay, with its gay shrubs, might wear an imposing aspect to the first navigators of these seas, after a tedious voyage; but its shores are shallow, and not convenient for landing, and most of the land on the north side, is dreary sand and marsh, of little real value. The pieces that are worth anything, are of very limited extent, and are in few hands. One of the proprietors has established a woollen manufactory, which, from the price of labour in this country, is not likely to pay. He told us that the leaves of the wooden-pear, Xylomelum pyriforme, dye wool yellow, and that the branches of Leptospermum scoparium, answer the purposes of fustic-wood, and dye fawn-colour. A handsome species of grass-tree, Xanthorrhœa arborea, was in flower, in some of the sandy grounds; its root-stocks were surmounted by an elegant crest of rush-like leaves; from the centre of which the flower-stem arose to ten feet in height; somewhat less than the upper half of this, was densely covered with brown scales, giving it an appearance something like a bull-rush. From amongst these scales the small, white, star-like flowers emerged, as in the other species of this genus. The plants with large root-stocks had been destroyed, for fuel, for which purpose they are much valued. In this neighbourhood, as well as at Port Jackson, the sweet tea, Smi-

---

* Alexander MacLeay, Esq.

lax glyciphylla, abounds. It is a low, climbing plant, with narrow, heart-shaped leaves, having a taste something like Spanish liquorice. It was used instead of tea, by the early settlers, and formed the chief ingredient in their drink, on occasions of rejoicing."—p. 291.

The following notices of Zamia spiralis and three species of Loranthus are interesting.

July 18. — "In a bushy hollow we met with Zamia spiralis, a singular, palm-like plant, in fruit. The whole fruit has some resemblance to a pine-apple; but large nuts, in red coats, are fixed under the scales forming the outside. The Blacks place these nuts under stones, at the bottom of water, in order to extract some noxious principle from them; they are afterwards converted into food. In wet weather, an insipid, jelly-like gum, which is wholesome, and not unpalateable, exudes from the plant.

"20th. Three species of the genus Loranthus, which consists of plants allied to mistletoe, grow parasitically on trees in this neighbourhood. They have handsome blossoms, a little like honey-suckle, but with more green, than yellow or red in them. Two of them have external roots, adhering to the bark of the trees that support them, and incorporating themselves with it; but occasionally, one of these species happens to grow upon the other, and then it emits no external root! This is a striking instance of that power, sometimes exhibited by a plant, to adapt itself to circumstances, and which is called vegetable instinct."—p. 294.

The question of specific identity between the productions of countries so widely separated as Britain and her Australian possessions, is one of great and increasing interest. And when we find the plants and insects of a far distant land named unhesitatingly as identical with our own, we are apt to feel a desire for a more detailed explanation. We learn from the 'Narrative' before us, that "on the margins of the pools of the Bell River, there are reeds,—Arundo Phragmites, bull-rushes, — Typha latifolia, and some other aquatic plants, similar to those of England." Are we to infer that these are the aboriginal denizens of the spot, or have their seeds been accidentally introduced from the mother country? They occur in the valley of Wellington, formerly a penal settlement for educated prisoners, and still the residence of European missionaries.

On the afternoon of the 25th of September our traveller, still in Wellington Valley, ascended a hill about 500 feet high; from the summit is an extensive view over the adjoining country, which seems to be a continuation of open forest hills, many of them black and bare from fire.

" On the upper portion, there were she-oak, Casuarina quadrivalvis, and Grammitis rutæfolius, a small fern, both of which are common in V. D. Land, also a Cycas? a remarkable Eucalyptus, and Sterculia diversifolia. Upon the last, there was a remarkable Viscum, or mistletoe. Lower down the hill, the beautiful Acacia venusta formed a bush about six feet high; it bears heads of small, globular, golden blossoms."—p. 322.

(To be continued).

ART. CXLVIII.— *Varieties.*

298. *List of Jungermanniæ &c. found near Penzance.* I send for insertion in 'The Phytologist,' a list of Hepaticæ found in the neighbourhood of Penzance. Some of the Jungermanniæ may be considered amongst the scarcest and most interesting of our British species, and I believe we need not fear comparison with any district of the same extent in the kingdom. With about four exceptions, all the plants enumerated in the following list may be gathered within three miles of Penzance, and but one species is entirely barren.

Riccia glauca. Wheat-stubbles, Bologas &c., in fruit.

Anthoceros punctatus, a. and *β.* Trembath mills, plentifully in fruit.

Marchantia polymorpha. A single fertile plant.

Fegatella conica. Common, with male receptacles only.

Lunularia vulgaris. Very plentiful; with male receptacles at Trembath mills.

Jungermannia calycina, Taylor. Newlyn cliff, beautifully in fruit.

—— *epiphylla.* Common.

—— *Blasia.* Chyanháll moor, where it fruits plentifully every season.

—— *furcata.* Common in fruit.

—— *pinguis.* Chyanháll moor, in fruit.

—— —— *β.* Hayle sands, in fruit.

—— *multifida.* Common in fruit. I once gathered it with gemmæ of a yellowish green colour at the extremity of the fronds.

—— *Hibernica,* Hook. Hayle sands, in tolerable abundance, where it fruits very beautifully, with both male and female receptacles; the fruit-stalks generally attaining the height of two inches.

—— *Ralfsii,* Wilson, MS. Hayle sands, with anthers in the autumn and capsules in April. This is a beautiful and highly curious species.

—— *pusilla.* Chyanháll moor; in male and female fruit.

—— *Hookeri.* Chyanháll moor, very sparingly, but the specimens have generally male and female fruit.

Jungermannia asplenioides. Chyune, with anthers.

—— *punctata.* Gulval, with calyces; J. Ralfs, Esq.

—— *pumila.* Truro river, on shelving stones, in fruit, very scarce. I have also received it from the Rev. C. A. Johns, from near Helston.

—— *crenulata.* Chyanháll moor, common in fruit.

—— *emarginata.* Newlyn cliffs; upon moist rocks, in fruit.

—— *turbinata.* Near Hayle causeway, in fruit.

—— *excisa.* Dry hedges near Chyanháll moor, in fruit.

—— *ventricosa.* Newlyn cliffs, abundant in gemmæ; and at Carn Brea with calyces.

—— *bicuspidata.* Common in fruit.

—— *byssacea* Chyanháll moor; in fruit August and September.

—— *capitata?* Chyanháll moor, near the water's edge. I am not certain if this is not what I take for J. excisa, on dry hedges, somewhat altered by local circumstances.

—— *nemorosa.* Gulval, with fruit and gemmæ, J. Ralfs, Esq.

—— *undulata.* Bologas, on stones in rivers, moist rocks &c., and on the ground in Chyanháll moor, in fruit.

—— *resupinata.* Bologas, on hedges, abundant in fruit: a large state on a moist rock, Newlyn cliffs.

—— *albicans, complanata* and *scalaris.* Common, in fruit.

3 H

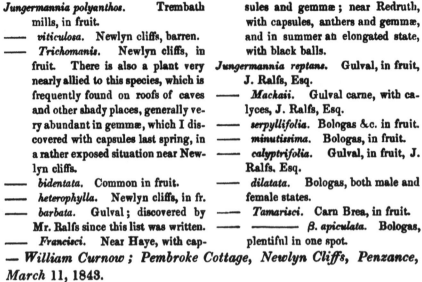

Jungermannia polyanthos. Trembath mills, in fruit.

—— *viticulosa.* Newlyn cliffs, barren.

—— *Trichomanis.* Newlyn cliffs, in fruit. There is also a plant very nearly allied to this species, which is frequently found on roofs of caves and other shady places, generally very abundant in gemmæ, which I discovered with capsules last spring, in a rather exposed situation near Newlyn cliffs.

—— *bidentata.* Common in fruit.

—— *heterophylla.* Newlyn cliffs, in fr.

—— *barbata.* Gulval; discovered by Mr. Ralfs since this list was written.

—— *Francisci.* Near Haye, with capsules and gemmæ; near Redruth, with capsules, anthers and gemmæ, and in summer an elongated state, with black balls.

Jungermannia reptans. Gulval, in fruit, J. Ralfs, Esq.

—— *Mackaii.* Gulval carne, with calyces, J. Ralfs, Esq.

—— *serpyllifolia.* Bologas &c. in fruit.

—— *minutissima.* Bologas, in fruit.

—— *calyptrifolia.* Gulval, in fruit, J. Ralfs. Esq.

—— *dilatata.* Bologas, both male and female states.

—— *Tamarisci.* Carn Brea, in fruit.

—— —— *β. apiculata.* Bologas, plentiful in one spot.

— *William Curnow; Pembroke Cottage, Newlyn Cliffs, Penzance, March 11, 1843.*

299. *Note on the absence of Jasione montana and Verbena officinalis from the vicinity of Edinburgh.* I was much surprised to see from the 2nd edition of the Edinburgh 'Catalogue of British Plants,' that neither Jasione montana nor Verbena officinalis have been found within thirty miles of that city. Not having been in Scotland, I cannot say whether this is correct, but to us Southrons it seems odd. The *absence* of plants, usually common, though not so striking to the eye, is, or ought to be, nearly as interesting as the *presence* of others usually rare.—*Geo. Sparkes; Bromley, Kent, April 18, 1843.*

300. *Note on British Plants occurring in foreign countries.* It would be extremely interesting if some botanist would publish a list of English plants, with the foreign countries in which they are found. The information might be taken from Don's book, as far as that book extends. Myrrhis odorata grows in *Spain*, although I have not met with it in England south of Derbyshire, nor do I find it in any of your southern local lists.—*Id.*

[The above note by Mr. Sparkes affords us an opportunity of mentioning Mr. Watson's admirable work on 'The Geographical Distribution of British Plants,' the first part of which, containing the Ranunculaceæ, appeared in the early part of the present year. Being printed for private distribution *only*, we have hitherto hesitated to give a detailed notice of this book; but our scruples having been removed by the perusal of a notice of a previous work on the same subject by Mr. Watson, in Loudon's 'Magazine of Natural History,' we hope, in an early number, to endeavour to do justice to its merits. About a dozen years ago, we received specimens of Myrrhis odorata, collected in a pasture field near Chipstead, Surrey, by a young lady, who stated that it was plentiful in the locality, although not cultivated in that neighbourhood.—*Ed.*]

externally with prominent transverse bars. The texture of the leaves and of the capsule is more dense and opaque, and the perichætial leaves and vaginula are considerably longer than in D. adiantoides. In a dried state also the two mosses have a different aspect, the foliage of D. taxifolium being less disposed to curl. I have never observed any such ambiguities as those which are pointed out by Mr. Sidebotham, (Phytol. 581).—*W. Wilson; Orford Mount, Warrington, May* 5, 1843.

304. *Dicranum osmundioides and bryoides.* The difference existing between these two species was long ago indicated by Meyrin, but his observations have been overlooked. The former of these two mosses has the calyptra mitriform, with the margin at the base inflexed. In D. bryoides the calyptra is constantly dimidiate, and the leaf has a colourless cartilaginous border ; characters which would be amply sufficient to distinguish the two species, even if their mutual resemblance were greater than it is. They differ also in the form of the capsule, and in the shape of the leaves. I may here remark that the blending of these two species, in ' Muscologia Britannica,' is solely attributable to their not having been compared together with due deliberation. Some species, in order to be properly understood, must be patiently studied in a living state, and at various stages of growth. If such a method be observed, the student will have no difficulty in persuading himself of the validity of those distinctions by which Polytrichum commune is separated from P. gracile, P. aloides from P. nanum, and Orthotrichum Rogeri from O. affine.— *Id.*

305. *Calla æthiopica,* L. One evening in April my attention was forcibly drawn to this plant (then in the parlour, not covered by glass), as it was copiously distilling water from the tips of its leaves. The evening gave no signs of an atmosphere overcharged with moisture, and at night the stars shone with brilliance. The roots of the plant had been well supplied with water in the forenoon. Mr. Quekett's admirable paper (Phytol. 218) recurred to mind, and I was induced to examine closely a phenomenon which presented itself to my notice for the first time, and in circumstances where I could not expect to see it. The exudation was observed on such leaves only as had the upper portion of the cylindrical points *faded* and *discoloured,* the line of demarcation between the faded and the living part indicating the spot where the fluid escaped; in every other respect the position of the outlet was altogether indeterminate, some leaves discharging their fluid at the base of the cylindrical point, others at a spot still lower down, and one at a considerable distance from the top, but close to the margin of the leaf, at a part which had become

discoloured. This led me to examine carefully the so-termed " marginal vein," which proved to be an *open continuous channel* of peculiar structure, extending from the base to the top of the leaf, altogether different from the vascular tissue constituting the proper veins, but no doubt in communication with them, and serving as the recipient and conductor of the fluid conveyed by them from the root. It is sufficiently wide to admit a bristle, which may be passed up it without the aid of a lens. On cutting a leaf across with scissors, the fluid escaped from the midrib and veins, but less plentifully than from the marginal channels. One leaf, which had recently expanded, and had its cylindrical point unfaded at the extremity, did not give out any fluid, nor was there any escape from the flower: with a microscope I could not detect any pore or valve, such as is stated to exist in Limnocharis Plumieri; hence I am inclined to infer that the exudation does not take place, even in a close humid atmosphere, until some previous natural decay of the cylindrical point of the leaf. Those who have the means of proving the contrary are requested to pursue the enquiry. It may be proper to add that an escape of fluid was observed at the back of two of the leaves, whose extremities were so curled as to admit of a lodgment of the water in that position, and that a quantity of the fluid, collected on glass and suffered to evaporate, left behind it only faint traces of mucilaginous matter. The opinion entertained by some, that spiral vessels convey air and not fluid, will derive little support from an examination of this plant. —*Id.*

306. *On the buds of Robinia Pseudacacia.* One fine evening last summer, during one of those welcome intervals of relaxation which are afforded by the visits of an esteemed friend, we happened to stroll under the shade of this tree, when a happy collision of ideas directed our joint attention to its well-known property of suddenly putting forth its foliage, late in the spring, without any previous warning. On cutting across one of the petioles, near its base, with a sharp penknife, I was struck with the appearance of an open cavity lined with stiff white pubescence; and as there was no trace of an external axillary bud, it instantly occurred to us that this tree resembled Platanus in its mode of vegetation. Longitudinal sections of this part were forthwith submitted to the microscope, when lo! instead of a solitary bud, no less than three were contained in the hollow base of each petiole, one placed immediately below the other, the lowest in a state of less perfect formation. It remained to enquire, at the proper season, whether or not two of the three buds were abortive, and I have recently directed my attention to this matter. The uppermost bud,

being the largest and most vigorous, is that which, in ordinary circumstances, develops itself into a branch ; but the one below it does certainly vegetate, and is visibly protruded from the cavity, so as to indicate a capability of immediately replacing the upper one, if that should be destroyed by frost or other accident. I wish some of your correspondents would examine other species, and also the Platanus. Professor Nuttall informs me that this mode of producing buds occurs in Psidium. May it not be a general property of tropical trees which do not form visible external winter-buds ?—*Id.*

307. *On the buds of Coniferæ.* Unlike dicotyledonous trees, which invariably produce a bud at the base of every leaf, either axillary or intra-petiolar, the Coniferæ have the greater part of their leaves destitute of buds, which are comparatively few, and scattered at distant intervals along the branches. On the axis of trees of this order, the branches exhibit a tendency to develope themselves in a peculiar manner, so that the tribe is characterised by the pyramidal or conical shape of the individuals belonging to it. No dicotyledonous tree, that I know of, produces its branches in whorls, such as we see in Pinus and Abies. Can any of your readers inform me of an exception to the rule in these genera; and whether any other genus is conformable to it ? In Larix this tendency does not appear to exist ; it is not very obvious in Taxus ; and even in full-grown trees of Pinus sylvestris it seems to be absent.—*Id.*

308. *Vinca major.* This appears to be one of the connecting links between the Apocynaceæ and the Asclepiadaceæ. No British author seems to have accurately described the pistillum, which has a very curious structure. The so-called stigma, described in Smith's ' English Flora,' although it bears considerable resemblance to a genuine stigma, cannot be considered to be such in reality, destitute as it is of stigmatic fluid, and moreover surrounded and fenced by a copious fringe of rough hairs forbidding the access of pollen. The style, with its appendages, is spindle-shaped, much and suddenly dilated in the thickened part, where there is a flattened edge nearly covered with a dense pubescence, under which is an orange-coloured reflexed membranous zone, covered with viscous fluid, and constituting, wholly or in part, the true stigma. The remainder of the style above the thickened part appears principally to serve the office of a pillar, to support a canopy formed of the crested extremities of the anthers, whose cells are lower in position than the stigma of Smith, but higher than the true stigma, the extent of which is somewhat doubtful. It may either comprehend the whole edge of the thickened part, which

has the pubescent zone imbued with the viscous fluid, abounding with active molecules, and easily diffusible in water; or it may consist only of the reflexed orange border underneath. The latter is rendered highly probable by the fact that pollen grains are constantly found attached to this part with their tubes fully developed, and some of them in positions scarcely accessible were it not for the curved form of the filaments, which give a favourable direction to the pollen as it falls from the anther. The central tissue, from this part to the base of the style, is composed of very slightly cohering oblong cells, of the same orange colour; and the whole may be compared to a funnel with a long narrow stalk, lying between two bundles of vascular tissue, which perforate the expanded part of the funnel on two opposite sides, and then pass up into the region above the true stigma, where they become greatly expanded and multiplied, so as to form a cylindrical congeries of spiral vessels, whose dark colour might lead to the supposition that they contained air, if it were not that there is no appearance of extricated bubbles on pressing portions of them when immersed in water. A portion of the placenta has the same orange tint, and is probably a continuation of the stigmatic tissue; but I sought in vain for pollen tubes in this region, and indeed they were not visible in the style, where they would have been so conspicuous had they been present. It may be that fecundation in this plant is of rare occurrence.—*Id.*

309. *On the glandular woody tissue of Coniferæ.* The structure of the circular glands being still a subject of debate among physiologists, permit me to send you a drawing of a longitudinal section of pine wood, perpendicular to the medullary rays, exhibiting sections of the glands. By way of explanation, let the contiguous walls of two cells, otherwise in perfect contact, be supposed to be separated here and there by small lens-shaped blisters, which are perforated in the middle, and the cavity filled up with a yellowish nucleus, having a slight depression in the centre of each convex face, immediately below the aperture of the membrane. The drawing represents the object magnified 300 diameters.—*Id.*

Section of Fir wood (Deal), Longitudinal and perpendicular to the medullary rays; 300 linear.

310. *On the spiral porous cells in the wood of the Yew tree.* The representation of these, after Kieser, in Lindley's Introduction, ed. 2, does not agree with my own observations. The spirally lined woody cells I believe to be perfectly distinct from the small oblique glands which are supposed to be situated between the spires, and I consider

that the glands belong entirely to the medullary rays, on which they are placed obliquely, and somewhat prominent beyond the membrane composing the ray. They are much smaller than the glands of the woody tissue, and of different shape.—*Id.*

311. *Note on the Locality of Pæonia Corallina.* I observed this plant growing in the rocky clefts of the Steep Holmes, in the Severn, in the summer of 1836, but it was then nearly destroyed by destructive visitors. When first added to the English Flora by the late Sir Francis B. Wright, in 1803, he observed it growing in great plenty on the island. A solitary plant was some time since observed growing in the centre of a large wood near Bath, Somerset, by Miss Lonsdale; but I am informed it has recently been dug up. — *T. B. Flower ;* 8, *Surrey St., Strand, May* 6, 1843.

312. *Note on a new British Lichen.* I wish to communicate to botanists by ' The Phytologist,' the discovery of a lichen new to the British Isles, namely, the Lecidea Wahlenbergii of Acharius, which I found on Ben Nevis, Inverness-shire, last July, upon the west side of the mountain, about three parts up, above Loch Nevis. This will be a very interesting addition to the next edition of the 'British Flora.'— *Fred. Bainbridge ;* 2, *Beulah Place, Harrogate, Yorkshire, May* 8, 1843.

313. *Note on Bryum androgynum,* Hedw. As the time of fruiting is not stated in the description of this moss in 'English Flora,' and it appears from the remark — " Fruit —— ? very rare," — to be of unfrequent occurrence in a fertile state ; I have thought that a notice of its discovery with capsules near London, would not be unacceptable to some of your readers. I found it in the early part of last April, with fruit not quite mature, growing on a shady bank on the south side of Abbey wood, near Erith, Kent, but in small quantity. Within a few feet was Tetraphis pellucida, also in fruit. — *W. Mitten;* 91, *Blackman St., Boro, May* 8, 1843.

314. *Note on Centranthus Calcitrapa.* In your September No. (Phytol. 309) appeared a critique by Mr. Babington, on some of the contents of the last Edinburgh Catalogue of phænogamous plants. Among other subjects, allusion is made to Centranthus Calcitrapa, in the following words:—"This has but slender claims to be considered a British plant, as it has only been found, in a naturalized state, at Eltham (?) in Kent." I do not presume to question the opinion of this eminent botanist, as to the slight claims possessed by this plant to rank among our native productions. The note of interrogation probably implies a doubt as to the exact locality in which the plant

has been found; I have therefore to inform your readers, that I met with it growing abundantly on the walls of Eltham church-yard, Kent, and on the walls of several gardens to the east of the church, about the middle of last June. After reading Mr. Babington's remarks on the subject, I went again to Eltham, in order to trace out, if possible, the source from whence this plant found its way to the above station. A gardener, of many years' standing at Eltham, had often seen it growing there, on walls, and had sometimes known it to spring up, self-sown, in gardens near the old palace; but was not aware that it had ever been cultivated in any gardens adjoining the church. Its inconspicuous size would not render it a very great favourite for the flower-border. He thought it likely that it might have been cultivated in a botanic garden, which long ago existed near the church; and that it might have escaped thence to the neighbouring walls. It is noticed in the 'Flora Metropolitana,' as being found on old walls at Eltham, on the authority of Mr. Wm. Pamplin. I accordingly wrote to Mr. Pamplin about it, and give the following extract from his prompt and obliging reply. " It is at least twenty years ago when I first gathered the Valeriana Calcitrapa at Eltham." " It was to Mr. Hewett C. Watson's Botanical Guide that I communicated the particulars above referred to, and I suppose it was from thence copied into Cooper's ' Flora Metropolitana.' In my communication to Mr. Watson's Guide, I ventured to suggest that this plant (so admirably adapted, with its feathery wings, for dispersion) might have escaped from Sherard's botanical garden at Eltham, originally, which I still think was most likely the fact." At any rate, whether indigenous or not, or whether likely or not to become extensively disseminated, it is a pretty and interesting little plant, and worthy of a place in the herbaria of our metropolitan botanists; who will, I have no doubt, if they extend their excursions to Eltham, find it very plentiful on many of the old walls near the church.— *Wm. Ilott; Bromley, Kent, May* 10, 1843.

315. *On the habitats of Equisetum fluviatile.* We have now before us two accounts of the habitats of Equisetum fluviatile; and as the two are so much at variance, and as it appears to me that the habits of that plant are not well understood, I hope I shall not be thought presumptuous in offering a few remarks on the subject. In the first place, Mr. Newman tells us that the plant " affects loose gravelly and sandy places, unconnected with water," (Phytol. 533). In the second place, Mr. Watson states, that so far as his observations go, when the plant grows in corn-fields and other places out of water, it is always

short and stunted; on the other hand, he says that the finest examples he has met with, were in the counties of Chester and Lancaster, "*growing on the red marl*," (ld. 588). Mr. Watson also relates some strange tales about the Cheshire horses getting "bogged," in their endeavours to graze on the plant; and that he (Mr. Watson) has seen a horse almost over head in mud, in a small pond (perhaps a mud-pond) filled with the tall horse-tails. I do not wish to contradict Mr. Watson, but merely to say that *water* or *mud* is not essential to the luxuriant growth of Equisetum fluviatile. At Broad-bank, four miles from Coln, in Lancashire, in a plot of ground which is appropriated to the growth of potatoes, we have the plant growing much higher than the fences. At Midge Hool, near Todmorden, we have it growing very fine, in a wood. And I think I shall be correct in saying that the plant is not common in the ponds in Lancashire, as I have been in the habit of visiting them for the last sixteen years, and never met with it in any of them. In the Manchester Flora, it is said to grow in "moist woods and hedge-banks, common." In the Yorkshire Flora there are four stations for E. fluviatile,—two of them in woods, and one of the others by a road-side. Francis observes that " the name *fluviatile* is not so applicable to this species as it would have been to some of the others, as it is rarely found on the banks of rivers or in ponds, nor do I remember ever having seen it growing in water." Withering, Smith and Hooker, say the plant grows on the banks of rivers, lakes, &c. Something might be said on the red marl, but as 'The Phytologist' is no medium for Geology, I will omit that altogether. How far Mr. Watson may be correct when he says that horses will go almost over head in *mud* to get at the plant, I know not; all that I can say is, that so far as my own observations go, horses will not eat the plant at all, if they can get anything else. I have never met with the fronds injured by horses, though growing in fields where they frequently feed. Lightfoot states, on the authority of Haller, that the plant was eaten by the Romans; he also tells us, on the authority of Linnæus, that horses refuse it. As Mr. Watson's is the first account we have, by British writers, of E. fluviatile growing in water, it would be well if he could give us a description of that part of the stem which grows *under* water. — *Samuel Gibson; Hebden Bridge, May* 10, 1843.

316. *Note on Geranium nodosum.* Mr. Watson still leaves undecided the question as to whether a Geranium which he has in his herbarium was gathered wild near Halifax, (Phytol. 588). In reply to that question I would say, that if the plant be one which *I* sent to

Mr. Bowman, it was gathered wild something less than two miles from Halifax, and is CERTAINLY *G. pyrenaicum*. The specimen I now enclose is the same as the one I sent to Mr. Bowman; and if that gentleman sent Mr. Watson some other, I know nothing of it. The specimen sent will serve to show how far I am correct in the name. After what has been said on this subject, it will be clearly seen that there is some mistake concerning this Geranium, but the question is, who is to bear the blame? Whoever may have made the mistake, I think Mr. Watson did wrong in publishing it, since he received the specimen under the name of *G. pyrenaicum*, and published it under that of *G. nodosum* in the first part of the New Guide, p. 278. In the second part, at p. 652, he says, " but I entertain scarcely any doubt as to the species;" and in 'The Phytologist,' p. 589, he says, " I THINK it G. nodosum." By this it will be seen that Mr. Watson does not know the plant, neither can he read the label, and therefore I think he ought not to have published it.—*Id*.

[It is evident that there is some mistake connected with the specimen of Geranium nodosum in Mr. Watson's possession, which it will now be difficult, if not impossible, to clear up. The specimen accompanying the above communication, is decidedly one of G. pyrenaicum; and it is labelled — " Geranium pyrenaicum. Washer-lane, near Halifax, 1830."—*Ed.*]

317. *Note on the Viviparous Grasses.* I am glad to see the subject of viviparous grasses taken up by Mr. Grindon (Phytol. 584), and will now give him the result of my observations on the Festuca vivipara of Smith. I have often been told that at a certain height upon our Yorkshire hills Festuca ovina is changed into the F. vivipara of Smith; and that Smith's F. vivipara is only a variety of F. ovina: this change, I was told, is effected by the damp atmosphere causing the seeds to germinate in the husk, before they fall to the ground. It had long puzzled me to know how it happened that the grass did not return to its natural form when growing at a lower elevation, as I have seen it cultivated in a garden for more than twenty years, and still retain its usual character—that of producing young plants instead of seeds. In June, 1842, I visited one of these hills (Fountains Fell) for the purpose of examining this grass; but on arriving at the place I did not find *all* the F. ovina changed into vivipara, but, on the contrary, there was an abundance of both, each retaining its usual character. Here I was under a second embarrassment, how to account for all this; however, I set to work, and got a number of specimens of F. vivipara and examined them, but not a single perfect flower was to be found: not one of those I examined had even the inner glume of the corolla,

and seeds were out of the question, as there were no organs of reproduction. I then collected a number of F. ovina, to try if I could find any of the flowers of that grass in a viviparous state ; and after I had examined them on the spot, I gathered specimens and brought them home, and examined them more minutely. This investigation convinced me that the two grasses are abundantly distinct, but how far they may be considered species, I shall leave to the better judgement of others. The two plants may be distinguished by the outer valve of the corolla of F. vivipara (when present) being more strongly ciliated on their edges, and, as Smith says, " keeled " on the back, not cylindrical as in F. ovina. The calyces are also strongly ciliated on their edges in F. vivipara, while, in F. ovina, they are, as Hooker describes them, subglabrous. I now enclose a specimen of each, and, by comparing them you will be able to see the difference.—*Id.*

318. *Note on Linaria Bauhinii.* It may be interesting to some of your readers to know that Linaria Bauhinii grows plentifully on the rocks of Coniston-water. It was discovered by John Roby, Esq., of Rochdale, in 1841.—*Id.*

319. *Note on Viviparous Grasses, &c.* I imagine that Mr. Grindon (Phytol. 584) takes the exception for the rule, in the matter of the viviparous grasses. That Festuca ovina, *e. vivipara*, produces heads of young plants, is certainly true, indeed, if such were not the case, I do not see why it should be introduced into lists as a *variety :* but, that the seeds of grasses do germinate in the husk in wet seasons, I think cannot be doubted, not only from the fact of the viviparous growth constantly taking place under such circumstances, in grasses and sedges also, but that it seems infinitely more probable than a change in the *nature* of the parts concerned. A circumstance came under my notice last year, which illustrates the effect of moisture on seeds while still in the seed-vessel. I collected some specimens of Drosera rotundifolia in flower, on Wimbledon common, and placed them in a Ward's case. The plants throve, and some time after I noticed on one of the heads, among the withered petals, a small green body. I watched this from day to day; two cotyledons were gradually protruded from the capsule, and at last a leaf, having the peculiar circinate vernation. The little plant then withered away, of course for want of nourishment, Nature not having intended Drosera for an aërial plant : whereas if it had been part of the original plant, there is no reason why it should not have gone on producing leaves, since it was not interfered with in any way. I may mention that it was in a small case, and kept very wet. It is, however, quite worth while to settle the point concerning *grasses*, by *facts ;*

and I shall be happy to communicate anything bearing on Mr. Grindon's question, which may come under my notice.—*Arthur Henfrey; Bot. Soc. London, May 12, 1843.*

320. *Note on a Locality of Equisetum fluviatile.* In my Reigate Flora I have recorded a single station for this plant, " on Reigate Hill, south side of the Wray lane, far from any water." This lane branches off from the Brighton road through Croydon and Reigate, at the turnpike-gate on Wray common, perhaps a mile and a half north-east of the latter town, and, after a gradual ascent in a north-westerly direction, joins the Brighton road through Sutton on the top of Reigate hill. Many years ago, the late Alderman Waithman, who then had a country house on Wray common, opened a pit close to Wray lane, on its south side, for the purpose of quarrying the upper green-sand, which was sent to London to be used as hearth-stone. After a short time, from various causes, the quarry was abandoned, but not until a high mound of loose sandy rubbish had accumulated at the entrance. About the year 1836, when I visited Reigate after a few years' absence, I found this mound covered with a most luxuriant crop of Equisetum fluviatile, looking like a fir-wood in miniature; the plants were two and three feet high. The locality is a very dry one; the nearest water is the large pond in Gatton park, and that is quite half a mile distant, and considerably below the spot in point of elevation. I do not remember having met with this plant in or near water, but Mr. Ilott has recorded a locality at Norwood, where it grows on a bank " and about a small pond close by," (Phytol. 295). — *Geo. Luxford ; 65, Ratcliff Highway, May 15, 1843.*

321. *On the arrangement of a Herbarium.* Excellent as my friend Mr. King may find his plan for the arrangement of a herbarium (Phytol. 585), prejudice in favour of a different one makes me think it open to one or two objections; and as it is highly desirable that a collection should be commenced upon a good system, inasmuch as changing it afterwards is attended with much inconvenience, allow me to suggest the following plan to your correspondent. Let every species have a separate folded sheet of white paper, and each genus a sheet of folded blue,* placing the former inside the latter, like drawings in a portfolio, all the creases or folds of the sheets being to the left hand. The name of the genus should be written on a slip of white paper, and pasted near the fold of the sheet, at equal distance from top and

* The exact size is immaterial, but it should not be smaller than foolscap nor larger than double folio post, which is ample, under skilful management, for any plant, to say nothing of the expense and unwieldiness of very large paper.

bottom. If *printed* labels are used, so much the better. By this means a genus can be found immediately, it being merely necessary to raise the sheets *seriatim* at their folds, with the fingers of the right hand, and glance over the labels. The names of the species should be written on the corresponding part of the white papers, which can be run through in the same way as the blue ones, when a particular plant is wanted; and thus we are never under the necessity of opening several papers when but one is required, nor liable to have the attention diverted from the object in view. The genera may be grouped together in sections, classes, or natural orders, according to the taste of the owner. Where British plants only are collected, the Linnæan system is unquestionably the better one to follow; but for the more extensive herbaria, which contain both indigenous and exotic plants, the natural arrangement* is decidedly preferable. The groups of genera should be tied together in fasciculi, a piece of stout blue pasteboard, an inch larger each way than the papers, being placed both at top and bottom, and *broad* white tape used for tying, to prevent any danger of cutting either paper or pasteboard. On the upper pasteboard should be fastened a slip of paper, containing the names of the orders, classes or genera comprised in the fasciculus. If there are *several* orders in the fasciculus, which it may be desirable to keep separate, pieces of thinner pasteboard may be used for that purpose. By giving a whole sheet to a species, any number of specimens may be successively introduced on *half sheets*, without intruding on space that can ill be spared, as must be the case in a *book* of limited size; for surely no modern botanist is content with a single specimen of any plant, and that probably only in flower. In a good collection, not only is the *fruit* necessary, but, in many cases, the *foliage* in its different stages or conditions. Of Tussilago Farfara, for instance, five or six specimens at least are necessary to illustrate the plant; and with regard to trees, a much larger number is requisite. I possess a series of Fagus sylvatica, comprising about twenty specimens, the first showing the expansion of the leaf-buds, and their beautiful rosy perules, and the last the ripe fruit, and yet am incomplete as regards many of the intermediate stages. In ferns, grasses and Carices, the necessity of numerous specimens is too obvious to be urged upon any one. Now the constant accumulation of specimens, independently of species, surely must be sadly at variance with such a plan of arrangement as Mr. King's; whereas by tying them in fasciculi, as above de-

* As given, for instance, in Hooker's edition of Smith's 'Introduction to Botany.'

scribed, all that is required is white paper in half sheets, and tape to meet the increase of bulk. The present plan is, moreover, so compact, and the fasciculi are so mathematically uniform in superficial measurement (that is, if a little dexterity and judgment are used in the mounting, so as to let the thick heads of thistles, for instance, lie in the corners of the papers), that they will lie upon each other like folio volumes. Another advantage over the book system is that a series of genera and species can be taken out and laid side by side for comparison, and be easily replaced when done with. If I rightly understand Mr. King's method, the latter does not admit of this; at all events, it cannot be practised with equal facility. Where a cabinet, purposely constructed for holding plants, is possessed, of course tape and pasteboards are not required, the papers lying loose on their respective shelves. In this case I would, nevertheless, still pursue the plan of having genera and species in separate blue and white papers, as above described.—*Leo. H. Grindon ; Manchester, May* 16, 1843.

322. *Mr. Babington's ' Manual of British Botany.'* We regret that we can do no more in the present number than announce to our readers the appearance of the above work, which we anticipate will ere long become the text-book and travelling companion of every British botanist. The generic and specific characters are of necessity condensed; but, judging from our hasty glance through the book, they appear to be clear and explicit, and are evidently drawn up with constant reference to the most recent information contained both in continental works of standard authority, and the British scientific journals. We must defer a regular notice until next month.—*Ed.*

ART. CXLIX.—*Proceedings of Societies.*

BOTANICAL SOCIETY OF LONDON.

May 19.—J. E. Gray, Esq., F.R.S., &c., President, in the chair. Donations to the library were announced from the American Academy of Sciences, Philadelphia, from the President, from Col. Jackson, Mr. Hogg, and Mr. E. Doubleday. British plants had been received from Mr. Edwin Lees and Miss Twining; and a collection of specimens from Western Australia, was presented by Mr. John Turner. Mr. William Andrews presented specimens of varieties of Saxifraga Geum, collected at the Great Blasquest Island, coast of Kerry, Ireland, one specimen of which had the nectaries thickly surrounding the ovary. Mr. A. Henfrey exhibited specimens of Leucojum æstivum, collected in Greenwich marshes. He also presented specimens of Dentaria bulbifera, collected at Harefield, Middlesex.

The following papers were read. " Notice of the discovery of two new species of British Fungi," by Dr. Philip B. Ayres :— Peziza corticalis, found on woodbine, between Stokenchurch, Oxfordshire, and Cadmore End; Hystericum rubrum, found on bean-stalks at Aston-Rowant and Tetsworth, Oxfordshire. Specimens were presented by Dr. Ayres. 2. "On the Groups into which the British Fruticose Rubi are divisible," by Mr. Edwin Lees, F.L.S., &c. This paper (which was illustrated by specimens and drawings) will be concluded at the next meeting, when a full report will be given.—*G. E. D.*

MICROSCOPICAL SOCIETY OF LONDON.

May 17.—J. S. Bowerbank, Esq., F.R.S. in the chair. Read, a paper from E. J. Quekett, Esq., "On the Nature of Vessels possessing longitudinal as well as spiral Fibres." Mr. Quekett stated that these vessels are not present in the majority of plants, and consequently have not been described until within the last few years. They have been found in plants belonging to very different orders, by various observers, and it is not improbable that future investigation may detect them in many other plants than those in which they have hitherto been observed. Thus they have been found by Schultz in Urania speciosa, by Mr. W. Wilson in Typha latifolia, in a plant of the gourd kind by Mr. Hassall, and by Mr. Quekett in Loasa and in Canna bicolor. In exogens these vessels do not appear to be either of the length or diameter they are in endogens, in which they seem to constitute the largest and longest of the vessels. In Loasa they do not exceed the $\frac{1}{250}$ of an inch in diameter, while in Urania and Canna they are nearly the $\frac{1}{100}$. In their structure they have a general resemblance to ordinary spiral vessels, having very frequently two or more fibres forming the same screw coiled in the interior of the vessel, as in compound spiral vessels. In addition to this spiral arrangement, there are longitudinal deposits of fibres whose number varies considerably, some vessels not having more than six or eight, while more than double that number may be detected in the larger vessels. Upon applying force, these longitudinal fibres are broken at the same point as the membrane, the broken edges of which project beyond the edge of the spiral ribband to which the vessels are reduced. Their terminations are very pointed, and they are applied to each other for some considerable space. At first Mr. Quekett supposed these vessels to be perfect of their kind, presenting higher marks of development than the true spiral vessels; subsequent observations however have enabled him to discover that a vessel presenting this longitudinal deposit of fibres, is only in a state preparatory to the complete development of a vessel exhibiting oval or quadrangular dots on its parietes: and he observes that although a tendency to produce longitudinal fibres may be seen in other annular and spiral vessels, it is only in the compound spiral that the ultimate conversion of a spiral vessel into one with dots regularly arranged in longitudinal fibres occurs. This is quite in accordance with the observations of Mohl and Schleiden, which prove that the spiral is the earliest type of every other form of vessel, whatever phases they may afterwards assume. He also described the various steps by which the spiral vessel becomes an annular one, and the reticulated vessel one with dots; and showed that the vessels forming the particular subject of this paper, are some of those in progress towards the dotted condition, as in some instances, in Canna bicolor, the various steps of this process can be witnessed in different parts of the same vessel. He concluded with some observations upon the dots on the vessels in woody exogens, showing that these last are of a more complex structure than those formed by the mere rounding of the meshes of the reticulated vessel.—*J. W.*

Lightning Source UK Ltd.
Milton Keynes UK
UKHW021330250219
337978UK00013B/1520/P